# Artificial Intelligence in Medical Image Processing and Segmentation

# Artificial Intelligence in Medical Image Processing and Segmentation

Editors

**Paolo Zaffino**
**Maria Francesca Spadea**

MDPI

Basel • Beijing • Wuhan • Barcelona • Belgrade • Novi Sad • Cluj • Manchester

*Editors*
Paolo Zaffino
Magna Graecia University
Catanzaro, Italy

Maria Francesca Spadea
Institute of Biomedical Engineering, Karlsruhe
Institute of Technology (KIT)
Karlsruhe, Germany

*Editorial Office*
MDPI
St. Alban-Anlage 66
4052 Basel, Switzerland

This is a reprint of articles from the Special Issue published online in the open access journal *Bioengineering* (ISSN 2306-5354) (available at: https://www.mdpi.com/journal/bioengineering/special_issues/artificial_intelligence_image_processing).

For citation purposes, cite each article independently as indicated on the article page online and as indicated below:

Lastname, A.A.; Lastname, B.B. Article Title. *Journal Name* **Year**, *Volume Number*, Page Range.

**ISBN 978-3-0365-8587-1 (Hbk)**
**ISBN 978-3-0365-8586-4 (PDF)**
**doi.org/10.3390/books978-3-0365-8586-4**

# Contents

bioengineering

MDPI

*Article*

# Recognizing Pediatric Tuberous Sclerosis Complex Based on Multi-Contrast MRI and Deep Weighted Fusion Network

**Dian Jiang [1,2,†], Jianxiang Liao [3,†], Cailei Zhao [4], Xia Zhao [3], Rongbo Lin [5], Jun Yang [1,2], Zhichen Li [1,2], Yihang Zhou [1,6], Yanjie Zhu [2,7], Dong Liang [1,2,7], Zhanqi Hu [3,*] and Haifeng Wang [2,7,*]**

1 Research Centre for Medical AI, Shenzhen Institutes of Advanced Technology, Chinese Academy of Sciences, Shenzhen 518000, China; dian.jiang@siat.ac.cn (D.J.); jun.yang@siat.ac.cn (J.Y.); zc.li@siat.ac.cn (Z.L.); yihang.zhou@outlook.com (Y.Z.); dong.liang@siat.ac.cn (D.L.)
2 University of Chinese Academy of Sciences, Beijing 100049, China; yj.zhu@siat.ac.cn
3 Department of Neurology, Shenzhen Children's Hospital, Shenzhen 518000, China; liaojianxiang@vip.sina.com (J.L.); 837623191@qq.com (X.Z.)
4 Department of Radiology, Shenzhen Children's Hospital, Shenzhen 518000, China; zhaocailei197866@163.com
5 Department of Emergency, Shenzhen Children's Hospital, Shenzhen 518000, China; 397126778@qq.com
6 Research Department, Hong Kong Sanatorium & Hospital, Hong Kong 999077, China
7 Paul C. Lauterbur Research Center for Biomedical Imaging, Shenzhen Institutes of Advanced Technology, Chinese Academy of Sciences, Shenzhen 518000, China
* Correspondence: huzhanqi1983@aliyun.com (Z.H.); hf.wang1@siat.ac.cn (H.W.)
† These authors contributed equally to this work.

**Abstract:** Multi-contrast magnetic resonance imaging (MRI) is wildly applied to identify tuberous sclerosis complex (TSC) children in a clinic. In this work, a deep convolutional neural network with multi-contrast MRI is proposed to diagnose pediatric TSC. Firstly, by combining T2W and FLAIR images, a new synthesis modality named $FLAIR_3$ was created to enhance the contrast between TSC lesions and normal brain tissues. After that, a deep weighted fusion network (DWF-net) using a late fusion strategy is proposed to diagnose TSC children. In experiments, a total of 680 children were enrolled, including 331 healthy children and 349 TSC children. The experimental results indicate that $FLAIR_3$ successfully enhances the visibility of TSC lesions and improves the classification performance. Additionally, the proposed DWF-net delivers a superior classification performance compared to previous methods, achieving an AUC of 0.998 and an accuracy of 0.985. The proposed method has the potential to be a reliable computer-aided diagnostic tool for assisting radiologists in diagnosing TSC children.

**Keywords:** tuberous sclerosis complex; children; convolutional neural network; multi-contrast MRI; rare neurodevelopmental disorder

**Citation:** Jiang, D.; Liao, J.; Zhao, C.; Zhao, X.; Lin, R.; Yang, J.; Li, Z.; Zhou, Y.; Zhu, Y.; Liang, D.; et al. Recognizing Pediatric Tuberous Sclerosis Complex Based on Multi-Contrast MRI and Deep Weighted Fusion Network. *Bioengineering* **2023**, *10*, 870. https://doi.org/10.3390/bioengineering10070870

Academic Editors: Paolo Zaffino, Maria Francesca Spadea and Kevin J. Otto

Received: 29 May 2023
Revised: 24 June 2023
Accepted: 12 July 2023
Published: 22 July 2023



## 1. Introduction

Tuberous sclerosis complex (TSC) is a rare neurodevelopmental disorder caused by mutations in the TSC1 and TSC2 genes [1,2]. It is characterized by angiofibromas of the face, epilepsy, an intellectual disability, and hamartomas in multiple organs including the heart, kidneys, brain, and lungs [3–5]. The majority of pediatric TSC patients experience their initial seizure in the first year of life [6–8], which has a severe impact on the lives of TSC children [9,10]. Therefore, it is urgent and valuable to develop valid and robust classification models for TSC children in a clinic.

Neurological symptoms are prevalent in nearly all children with TSC, and multi-contrast magnetic resonance imaging (MRI) is frequently employed for a clinical diagnosis [11]. To date, T2-weighted imaging (T2W) and fluid-attenuated inversion recovery (FLAIR) have been commonly utilized in a pediatric TSC diagnosis, allowing for the identification of lesions and facilitating high lesion-to-brain contrast visualization. But, the cerebrospinal fluid (CSF) signal is strong in T2W, which severely interferes with the visualization of periventricular TSC lesions. FLAIR imaging can suppress cerebrospinal fluid and sufficiently show the lesion–brain contrast clearly, and FLAIR also reduces the signal-to-noise ratio while pressing CSF [12]. Currently, it is not possible for a single MRI sequence to produce all the required tissue contrasts in a single contrast image due to the trade-offs that need to be made when choosing MRI pulse sequence parameters [13]. In recent studies, it has been demonstrated that a synthesized contrast that blends T2W and FLAIR imaging can augment the contrast of multiple sclerosis (MS) lesions, leading to an improved diagnostic efficacy [12,13]. However, to the best of our knowledge, there are not studies on applying a synthesis contrast combining T2W and FLAIR for diagnosing pediatric TSC so far.

Otherwise, deep learning has been studied as an advanced artificial intelligence technology that can automatically learn from medical image data and extract a large number of features [14]. Previously, deep learning models and multi-contrast MRIs have been successfully used for automatically detecting strokes [15] and classifying brain tissues [16]. Until now, convolutional neural networks (CNNs) have been applied to assist in tuber segmentation in TSC patients [17]. Sanchez et al. [18] used two types of contrast MRI, T2W and FLAIR, for the detection task of TSC tubers and achieved the receiver operating characteristic curve that can have an area under the curve (AUC) of 0.99. However, their approach employed a 2D network and solely relied on handpicked MRI slices with evident tubers as input to the network. This method failed to account for the spatial attributes of MRI and neglected the fact that not all TSC patients exhibit visible lesions. Additionally, their datasets were limited to merely 114 TSC patients and 114 controls. Alternatively, recent research suggests that 3D CNNs excel at capturing the spatial characteristics of MRI and effectively capitalize on the interplay between voxels. Consequently, they have been reported to yield superior results in predicting chronological age [19].

To further raise the performance of identifying TSC children in a clinic, a novel deep learning method, named the deep weighted fusion network (DWF-net), was proposed to effectively diagnose pediatric TSC lesions with multi-contrast MRIs. The proposed method has a synthesis contrast, named $FLAIR_3$, from the combination of T2W and FLAIR that can maximize the lesion–brain contrast of pediatric TSC lesions. Moreover, the proposed method has a 3D CNN strategy of the weighted late fusion model combined with multi-contrast MRI to automatically diagnose pediatric TSC. The experimental dataset has a total of 680 children, including 331 healthy and 349 TSC children. Experiments intuitively show that the new synthesis $FLAIR_3$ contrast and the weighted 3D CNN strategy can effectively improve the contrast saliency of pediatric TSC lesions, and the classification performance.

The proposed deep learning method is efficient in distinguishing TSC children from healthy children and presently achieves the best performance. The proposed method has great potential in helping clinical doctors diagnose TSC children and provides an effective research tool for pediatric doctors.

## 2. Methods

### 2.1. Optimal Combination of T2W and FLAIR

Cortical and subcortical nodules are the most common lesions in TSC children. The increased prominence of lesions is crucial for clinical doctors to diagnose pediatric TSC [20]. The T2W signal is related to water content, and most of the lesions have stronger T2W signals than surrounding normal tissues, often exhibiting a bright state. Therefore, the location and size of the pediatric TSC lesions can be seen from the T2W sequence. However, the outline of the lesion is relatively vague in the T2W sequence, and it is difficult to clearly outline the outline of the lesion. Moreover, there was a strong cerebrospinal fluid (CSF) signal interference in T2W. FLAIR, also known as water-suppression imaging, suppresses (darkens) CSF hyperintensity in T2W, thereby making lesions adjacent to CSF clear (brightened). Compared with the T2W sequence, the FLAIR sequence can better represent the surroundings of the lesion and clearly show the lesion area. FLAIR is a T2W scan that selectively suppresses CSF by reversing pulses. However, CSF signal suppression comes at the expense of reducing the signal-to-noise ratio [12]. $FLAIR_2$ and $FLAIR_3$ have been proposed to combine T2W and FLAIR to improve lesion visualization in MS disease [12,13]. Inspired by [12,13], we propose to optimize the combination of T2W and FLAIR as a new modality named $FLAIR_3$ in pediatric TSC disease as follows [13]:

$$FLAIR_3 = FLAIR^{\alpha} \times T2W^{\beta} \quad s.t.\ \alpha + \beta = 3 \tag{1}$$

where the optimized $\alpha$ is 1.55 and $\beta$ is 1.45 based on the signal equations of FLAIR and T2W [13], which can optimally balance the lesion contrast between FLAIR and T2W.

### 2.2. Late Fusion Strategies

Some recent studies [21] have shown that the late fusion model could grasp the data distribution effectively and finally achieve the best classification performance. Inspired by [22,23], a weighted late fusion strategy was used to combine multi-contrast MRI for classification tasks in pediatric TSC patients. First, T2W, FLAIR, and $FLAIR_3$ were fed into a feature extractor. We propose a deep weighted network (DWF net) that takes the scores of the T2W, FLAIR, and $FLAIR_3$ models as input, and outputs the final classification with a simple and efficient weighted average integration method, as follows:

$$S_{DWF} = W_1 \times S_{T2W} + W_2 \times S_{FLAIR} + W_3 \times S_{FLAIR3} \quad s.t.\ \textstyle\sum_{i=1}^{3} W_i = 1 \tag{2}$$

where $S_{T2W}$, $S_{FLAIR}$, and $S_{FLAIR3}$ represent the classification scores of T2W, *FLAIR*, and $FLAIR_3$ models, respectively. $S_{DWF}$ denotes the final output prediction scores of the proposed DWF-net. $W_1$, $W_2$, and $W_3$ are the weights of the prediction scores of the three multi-contrast MRIs.

To explore the optimal fusion between multi-contrast MRI and to enhance the AUC of the proposed DWF-net, the experiments were performed for values of $W_1$ between 0 and 1, and $W_2$ from 0.1 to $1-W_1$ with a step of 0.1; $W_3$ is $1-W_1-W_2$. The weight-searching algorithm is shown in Algorithm 1.

**Algorithm 1** The weight searching algorithm for fusion

**Input:** The prediction scores $S_{T2W}$, $S_{FLAIR}$, and $S_{FLAIR3}$ of three input images and corresponding ground truth $y$ on testing set.
**Output:** The weight ($W_1$, $W_2$, and $W_3$) with best AUC on testing set.

Initialize $AUC_{best} \leftarrow 0$.
**for** $i$: =0 to 10 **do**
  **for** $j$: =0 to 10–$i$ **do**
    $k \leftarrow 10-i-j$
    $S_{temp} = (i \times S_{T2W} + j \times S_{FLAIR} + k \times S_{FLAIR3}) \times 0.1$
    $AUC_{temp}$ = Compare ($S_{temp}$, $y$)
    **if** $AUC_{temp} > AUC_{best}$ **then**
      $AUC_{best} \leftarrow AUC_{temp}$
      $W_1 \leftarrow i \times 0.1$
      $W_2 \leftarrow j \times 0.1$
      $W_3 \leftarrow k \times 0.1$
    **end for**
  **end for**
**end for**
**Return** $W_1$, $W_2$, and $W_3$

### 2.3. Network Architectures

The proposed DWF-net method for pediatric TSC patients was implemented using two different 3D CNN architectures. The following sections describe two different 3D CNN models.

ResNet was proposed in 2015 and has been widely applied in detection, segmentation, recognition, and other fields [24]. In addition, ResNet has demonstrated a stable and excellent classification performance in image classification among different variants of various 3D CNNs [24]. Therefore, the first 3D CNN model we consider is 3D-ResNet, which uses a shortcut connection to make a reference for the input of each layer and learns to form a residual function. The residual function is easier to optimize, making the number of network layers much deeper, and can easily obtain a higher accuracy from deeper depths.

For the second 3D CNN model, we utilized the 3D-EfficientNet architecture [25] as our feature extractor. This classification network is known for its efficiency in improving accuracy and reducing the training time and network parameters. The EfficientNet was designed using a neural architecture search and employs the mobile inverted bottleneck convolution (MBConv) module as its core structure. This module, similar to depth-wise separable convolution, minimizes parameters significantly. In addition, the attention idea of the squeeze-and-excitation network (SENet) is also introduced [26] in EfficientNet. The attention mechanism of SENet allows the model to focus more on channel features that are most informative, while suppressing those unimportant channel features, thereby improving the model performance.

As shown in Figure 1a, for the pediatric TSC identification tasks with one single MRI modality, the 3D-ResNet34 and 3D-EfficientNet were used as a feature extractor. When DWF-net was used, two or three modalities were applied as inputs, as shown in Figure 1b. Table 1 displays the 10 models that were trained in this study, each with distinct architectures and inputs.

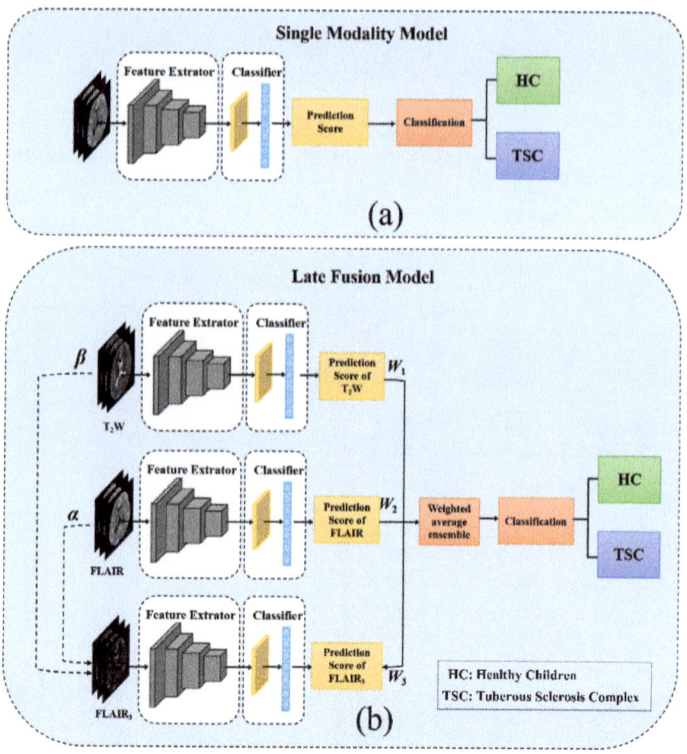


**Figure 1.** Overall network structure, (**a**) single modality model pipeline, (**b**) schematic of the proposed DWF-net pipeline. The two dotted lines represent the optimal combination of T2W and FLAIR to generate $FLAIR_3$.

Table 1. Detailed information on ten network structures.

| Model Name | Input Modality | Method |
|---|---|---|
| Eff_FLAIR | FLAIR only | 3D-EfficientNet |
| Eff_T2W | T2W only | 3D-EfficientNet |
| Eff_FLAIR$_3$ | FLAIR$_3$ only | 3D-EfficientNet |
| Eff_FLAIR_T2W | FLAIR + T2W | DWF_net |
| Eff_DWF_net | FLAIR + T2W + FLAIR$_3$ | DWF_net |
| Res_FLAIR | FLAIR only | 3D-ResNet34 |
| Res_T2 | T2W only | 3D-ResNet34 |
| Res_FLAIR$_3$ | FLAIR$_3$ only | 3D-ResNet34 |
| Res_FLAIR_T2W | FLAIR + T2W | DWF_net |
| Res_DWF_net | FLAIR + T2W + FLAIR$_3$ | DWF_net |

## 3. Materials and Experiments

### 3.1. Dataset

In this study, all pediatric volunteers were from Shenzhen Children's Hospital. The study was approved by the Ethics Committee of Shenzhen Children's Hospital (No.2019005). Written informed consent was obtained from all pediatric volunteers and/or their parents. In total, 349 TSC children and 331 healthy children (HC) were included in this study. Inclusion criteria for pediatric TSC patients were (1) aged 0–20 years, (2) no other neurological disorders, and (3) clinically diagnosed with TSC. (4) T2W and FLAIR images are complete and clear. Inclusion criteria for healthy children were (1) aged 0–20 years, (2) without any neurological disorder, (3) clinically defined normal or non-specific findings during routine clinical care. (4) T2W and FLAIR images are complete and clear. Figure 2 shows the exclusion and inclusion criteria of our study.

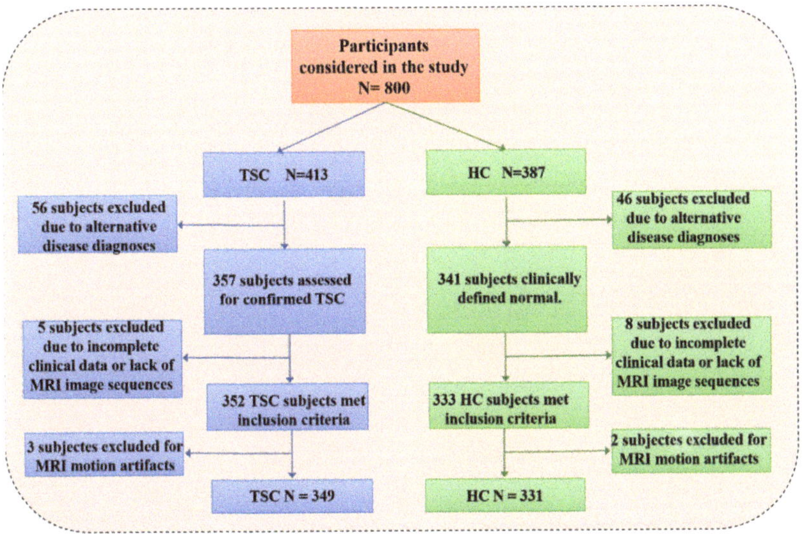


Figure 2. Study exclusion and inclusion criteria of the pediatric dataset.

The data were randomly split into train-validation-test sets in a 7:1:2 ratio. To ensure that every group had the same class proportion, stratified random sampling was employed. Training, validation, and testing datasets had no overlap of patients.

### 3.2. Data Processing

Firstly, a FMRIB Linear Image Registration Tool (FLIRT) of FSL (http://fsl.fmrib.ox.ac.uk (accessed on 1 January 2021.)) was used to register T2W into the FLAIR space, and mutual information was used as the cost function. In neuroimaging studies, the lesions are usually located in the brain tissue, and the skull part is an irrelevant site. When brain MRI images are used for classification network research, the brain tissue of the region of interest is often the input. HD-bet is an algorithm for extracting brain tissue [27], which can remove irrelevant images such as of the neck and eyeball. Therefore, in the second step, the deep learning tool HD-bet is used to strip the skull in MRI. Subsequently, all 3D MRI images were resized to 128 × 128 × 128, and the image intensity was normalized to the range of 0 to 1 using the min–max normalization formula:

$$x_{Normalized} = \frac{x - Min(x)}{Max(x) - Min(x)} \quad (3)$$

where *Max*(*x*) and *Min*(*x*) represent the highest and lowest values of the brain-extracted MRI images, respectively, and $x_{Normalized}$ refers to the normalized MRI images. Finally, T2W and FLAIR were combined and transformed into $FLAIR_3$. The flowchart illustrating the data preprocessing can be found in Figure 3.

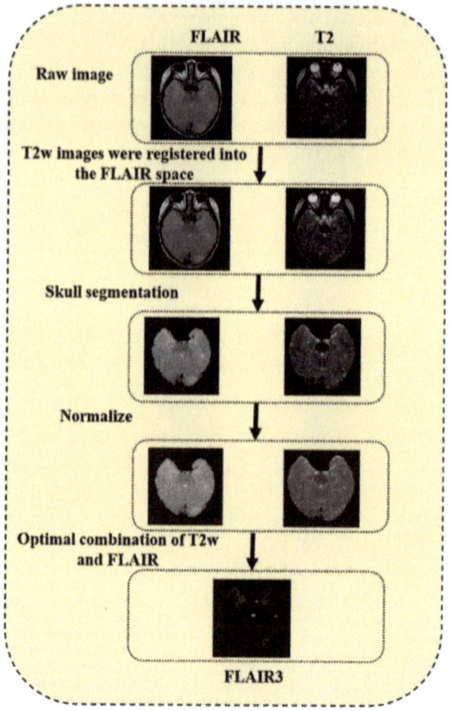


**Figure 3.** Flowchart of the data preprocessing.

### 3.3. Baseline and Effectiveness of Skull Stripping

In this study, we compared 10 different proposed 3D CNN models with a 2D-InceptionV3 model [18] (baseline model) to evaluate the effectiveness of the proposed deep learning methods. The 2D-InceptionV3 model was exclusively trained on our FLAIR data, with the maximum transverse slice of the FLAIR chosen as the input. Furthermore, we conducted a series of experiments on FLAIR images and T2W images with and without skull-stripping preprocessing to assess the effectiveness of the skull-stripping methodology.

### 3.4. Comparison of Normalization Methods

Typically, normalization methods often have a significant impact on the performance of deep learning models. The min–max normalization and Z-score normalization are most used in medical image normalization. While the min–max normalization approach is appropriate for most kinds of data and can effortlessly maintain the initial data distribution structure, it is not ideal for handling sparse data and is prone to being affected by outliers. The Z-score normalization method employs the mean and standard deviation of the original data to normalize it. The following formula illustrates this:

$$x_{Normalized} = \frac{x - Mean(x)}{std(x)} \tag{4}$$

When Mean(x)= 0, std(x) = 1, that is, the mean is 0 and the standard deviation is 1, meaning that the processed data conform to the standard normal distribution. This Z-score method is suitable for most types of data, but it is a centralized method, which will change the distribution structure of the original data, and it is also not suitable for the processing of sparse data. To explore the effectiveness of the normalization operation, we conducted three sets of experiments on both T2W and FLAIR images when using the same network, which are without the normalization method, the Z-score normalization, and the min–max normalization, respectively.

### 3.5. Model Training and Evaluation

For our experiments, we used the same partitioning for the training set, validation set, and test set across all models. Each model was trained using a learning rate of 0.0001, SGD optimization, a batch size of 4, and 50 epochs, with the binary cross-entropy loss function. To implement the training, validation, and testing process, we used Python version 3.8.10 and PyTorch version 1.9.0 environments.

For each cohort, we calculated the area under the curve (AUC) of the receiver operating characteristic (ROC), accuracy (ACC), sensitivity (SEN), and specificity (SPE) to evaluate the classification performance of all models. These metrics rely on the true positive (TP), which counts the total number of correct positive classifications, and the true negative (TN), which represents the total number of accurate negative classifications. The false positive (FP) accounts for the total number of positive classifications that are incorrect, while the false negative (FN) represents the total number of negative classifications that are incorrect. We obtained the ACC, SEN, and SPE through the following formulas:

Accuracy (ACC): The percentage of the whole sample that is correctly classified:

$$\mathrm{ACC} = \frac{\mathrm{TP} + \mathrm{TN}}{\mathrm{TP} + \mathrm{TN} + \mathrm{FP} + \mathrm{FN}} \tag{5}$$

Sensitivity (SEN): The percentage of the total sample that is true and correctly classified:

$$\mathrm{SEN} = \frac{\mathrm{TP}}{\mathrm{TP} + \mathrm{FN}} \tag{6}$$

Specificity (SPE): The percentage of the total sample that is negative and correctly classified:

$$\mathrm{SPE} = \frac{\mathrm{TN}}{\mathrm{TN} + \mathrm{FP}} \tag{7}$$

### 3.6. Statistical Analysis

For this research, categorical variables were presented using the frequency and percentage, while continuous variables were expressed as the mean ± standard deviation. Continuous variables were analyzed using the F-test, while categorical variables underwent a chi-square analysis. Statistical significance was defined as $p < 0.05$. All statistical analyses were performed using the scikit learn, scipy, and stats libraries in Python 3.8.10.

## 4. Results

### 4.1. Clinical Characteristics of Patients

All of the 680 child subjects' primary clinical features are listed in Table 2. Among the 349 TSC patients, 188 (53.9%) were identified as male, averaging 45.5 months in age. Moreover, among the 331 HC, 183 (55.3%) were identified as male, averaging 733 months in age. There was a significant difference in the average age between the HC group and the TSC group, with a *p*-value less than 0.05. There was no significant difference in gender.

**Table 2.** The main clinical characteristics of all 680 child subjects.

| | TSC | HC | *p*-Value |
|---|---|---|---|
| Number | 349 | 331 | - |
| Male, number (%) | 188 (53.9%) | 183 (55.3%) | 0.711 |
| Age at imaging, mean ± SD (months) | 45.5 ± 46.6 | 73.3 ± 49.2 | <0.001 |

### 4.2. Visualization Results of $FLAIR_3$

Figure 4 shows FLAIR, T2W, and $FLAIR_3$ images of a TSC child and a healthy child. On three MRI images of the TSC child, it can be observed that the contrast between the lesions and brain tissue on FLAIR is not clear enough, there is a severe interference of cerebrospinal fluid on T2W, and the contrast and clarity of the lesions on the newly generated $FLAIR_3$ image are significantly improved (TSC lesion as shown by the red arrow). In addition, $FLAIR_3$ inhibits cerebrospinal fluid and can clearly locate the TSC lesion.

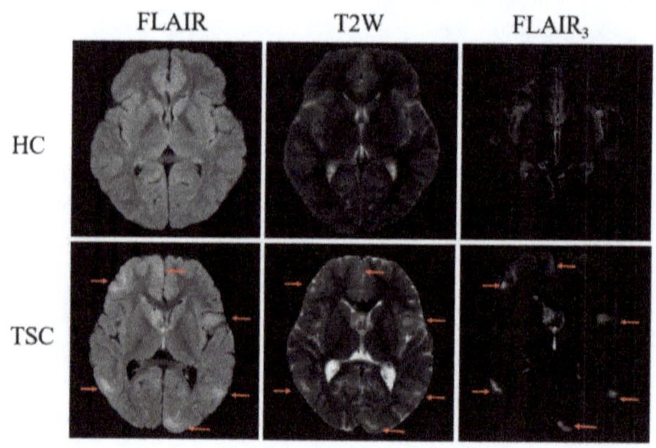


**Figure 4.** Representative MRI from a TSC child and a healthy child, including T2W, FLAIR, and the proposed $FLAIR_3$ (the red arrow highlights the TSC lesion).

### 4.3. Performance of the Models

The performance of DWF-net varies with the weight of $W_1$, $W_2$, and $W_3$ as shown in Figure 5. The feature extractor in Figure 5a is 3D-EfficientNet, and the best AUC performance of 3D-EfficientNet is 0.989 ($W_1 = 0.0$, $W_2 = 0.3$, $W_3 = 0.7$). Among the models evaluated, Res_DWF_net (with weight parameters $W_1 = 0.2$, $W_2 = 0.3$, $W_3 = 0.5$), which employs 3D-ResNet as a feature extractor and a late fusion strategy as depicted in Figure 5b, achieves the highest performance. This model has an accuracy of 0.985 and an AUC of 0.998, outperforming other models.

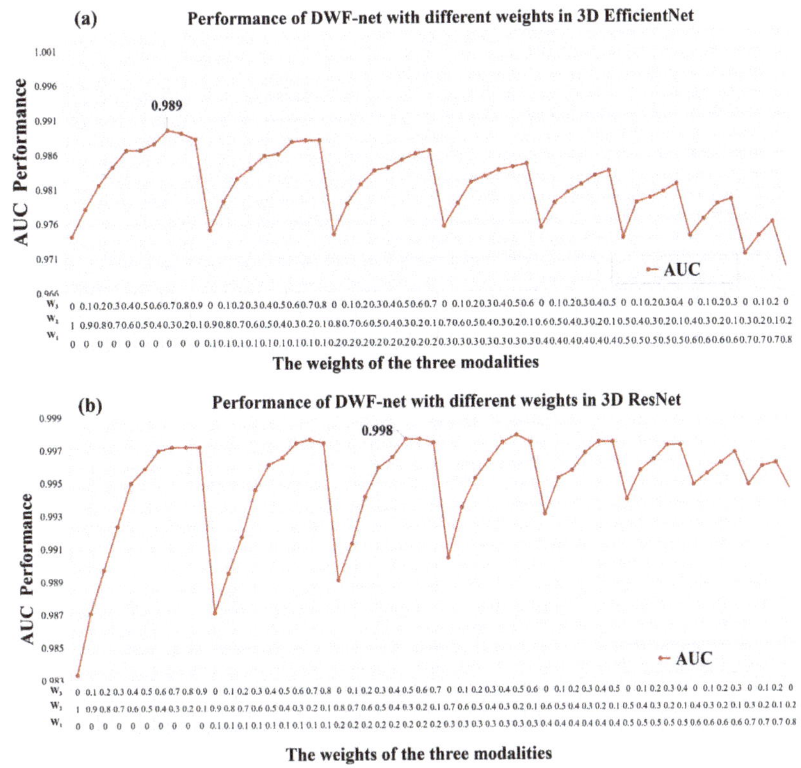


**Figure 5.** The performance of DWF-net with different weights. The feature extractor in (**a**) is 3D-EfficientNet, and the feature extractor in (**b**) is 3D-ResNet. The horizontal axis represents the weight of $W_1$, $W_2$, and $W_3$, and the vertical axis represents the performance of AUC.

The results for all the compared models in the testing dataset are presented in Table 3. When using 3D-EfficientNet, $FLAIR_3$ achieves an AUC performance of 0.987 and the AUC of Eff_FLAIR_T2W is 0.974, and the AUC of $FLAIR_3$ is higher than Eff_FLAIR_T2W. $FLAIR_3$ achieves an AUC performance of 0.997 when using 3D-ResNet as the feature extraction network. When the feature extraction network is 3D ResNet, the AUC of Res_FLAIR_T2W is 0.994, and the AUC of $FLAIR_3$ is higher than Res_FLAIR_T2W.

**Table 3.** Detailed performance of different models in pediatric testing datasets.

| Input Modality | Model Name | AUC | ACC | SEN | SPE |
|---|---|---|---|---|---|
| FLAIR + T2W | InceptionV3 [18] | 0.933 | 0.851 | 0.812 | 0.893 |
| FLAIR only | Eff_FLAIR | 0.974 | 0.911 | 0.869 | 0.954 |
| T2W only | Eff_T2W | 0.971 | 0.919 | 0.869 | 0.970 |
| $FLAIR_3$ | Eff_$FLAIR_3$ | 0.987 | 0.926 | 0.884 | 0.970 |
| FLAIR + T2W | Eff_FLAIR_T2W | 0.974 | 0.933 | 0.928 | 0.939 |
| FLAIR + T2W + $FLAIR_3$ ($W_1 = 0.0$, $W_2 = 0.3$, $W_3 = 0.7$) | Eff_DWF_net | **0.989** | **0.963** | **0.942** | **0.985** |
| FLAIR only | Res_FLAIR | 0.994 | 0.970 | **0.986** | 0.955 |
| T2W only | Res_T2W | 0.983 | 0.956 | 0.913 | 0.999 |
| $FLAIR_3$ | Res_$FLAIR_3$ | 0.997 | 0.978 | 0.957 | 0.999 |
| FLAIR + T2W | Res_FLAIR_T2W | 0.994 | 0.970 | 0.942 | 0.999 |
| FLAIR + T2W + $FLAIR_3$ ($W_1 = 0.2$, $W_2 = 0.3$, $W_3 = 0.5$) | Res_DWF_net | **0.998** | **0.985** | 0.971 | **0.999** |

When using the same single-modal MRI as inputs, 3D-ResNet outperforms 3D-EfficientNet. Additionally, the AUC performance of the $FLAIR_3$ model outperforms the

T2W-only model and FLAIR-only model. The baseline network (InceptionV3) achieves an AUC performance of 0.952, and the performance of our all-3D network exceeds the AUC performance of the baseline network of InceptionV3.

ROC curves for all models of the testing cohort are shown in Figure 6a–c, and Figure 6d shows the classification performance for all models of the testing cohort.

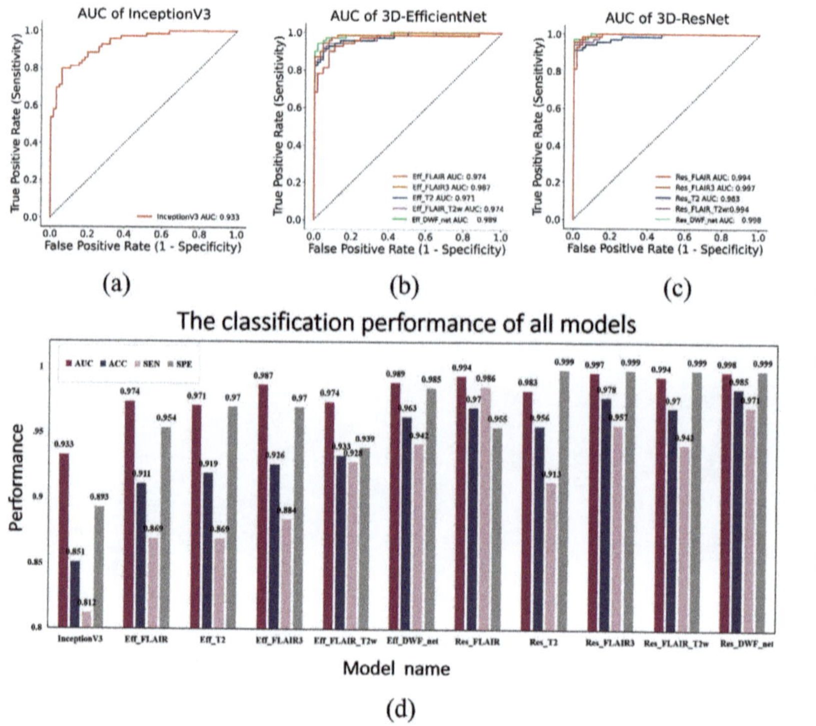


**Figure 6.** (**a**–**c**) represent the ROC curves for all models of the testing cohort. (**d**) represents the classification performance for all models of the testing cohort. The horizontal axis shows the model name, while the vertical axis represents the performance regarding AUC, ACC, SEN, and SPE.

## *4.4. Results of Skull Stripping*

The classification performance of FLAIR and T2W images, with or without skull dissection, is presented in Table 4. The table demonstrates that if the network structure and input modality remain constant and the skull dissection preprocessing is not carried out, the classification performance of 3D ResNet and 3D EfficientNet will show a decline.

**Table 4.** The results of with/without skull stripping in T2W and FLAIR.

| Modality | Model Name | Preprocessing | AUC | ACC | SEN | SPE |
|---|---|---|---|---|---|---|
| FLAIR only | 3D-EfficientNet | Without skull stripping | 0.898 | 0.829 | 0.754 | 0.909 |
| | | Skull stripping | **0.974** | **0.911** | **0.869** | **0.954** |
| | 3D-ResNet | Without skull stripping | 0.959 | 0.881 | 0.855 | 0.909 |
| | | Skull stripping | **0.994** | **0.970** | **0.986** | **0.955** |
| T2W only | 3D-EfficientNet | Without skull stripping | 0.968 | 0.916 | **0.881** | 0.951 |
| | | Skull stripping | **0.971** | **0.919** | 0.869 | **0.970** |
| | 3D-ResNet | Without skull stripping | 0.914 | 0.829 | 0.797 | 0.863 |
| | | Skull stripping | **0.983** | **0.956** | **0.913** | **0.999** |

### 4.5. Comparison of Normalization Methods

Table 5 and Figure 7 depict the classification performance of three normalization methods, including without normalization, Z-score normalization, and min–max normalization on FLAIR images and T2W images. The horizontal axis represents the different normalization techniques, while the vertical axis represents their corresponding performance. In instances where the input modality and network structure remain constant, it is worth noting that the without-normalization method has the poorest AUC performance. Furthermore, the AUC performance of the min–max normalization technique is better than the Z-score normalization technique.

**Table 5.** The classification performance of with/without skull stripping in FLAIR images and T2W images.

| Modality | Model Name | Preprocessing | AUC | ACC | SEN | SPE |
|---|---|---|---|---|---|---|
| FLAIR only | 3D-EfficientNet | Without normalization | 0.951 | 0.899 | 0.863 | 0.936 |
| | | Z-score | 0.965 | 0.867 | 0.754 | **0.984** |
| | | Min–max | **0.974** | **0.911** | **0.869** | 0.954 |
| | 3D-ResNet | Without normalization | 0.985 | 0.933 | 0.971 | 0.893 |
| | | Z-score | 0.914 | 0.867 | 0.797 | 0.933 |
| | | Min–max | **0.994** | **0.970** | **0.986** | **0.955** |
| T2W only | 3D-EfficientNet | Without normalization | 0.950 | 0.911 | 0.884 | 0.939 |
| | | Z-score | 0.967 | **0.933** | **0.898** | 0.969 |
| | | Min–max | **0.971** | 0.919 | 0.869 | **0.970** |
| | 3D-ResNet | Without normalization | 0.974 | 0.918 | **0.927** | 0.909 |
| | | Z-score | 0.982 | 0.918 | 0.884 | 0.954 |
| | | Min–max | **0.983** | **0.956** | 0.913 | **0.999** |

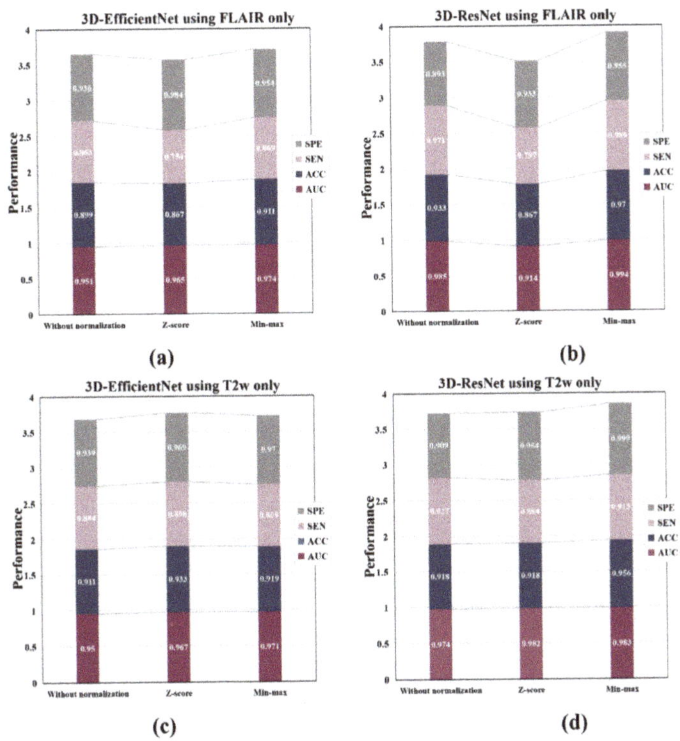


**Figure 7.** The classification performance of the without-normalization method, the Z-score normalization, and the min–max normalization in FLAIR images and T2W images. (**a**) 3D-EfficientNet as a network feature extractor, FLAIR as the network input. (**b**) 3D-ResNet as a network feature extractor, FLAIR as the network input. (**c**) 3D-EfficientNet as a network feature extractor, T2W as the network input. (**d**) 3D-ResNet as a network feature extractor, T2W as the network input.

## 5. Discussion

The main objective of the proposed approach is to identify TSC children at an early stage using a 3D CNN model in conjunction with multi-contrast MRI in an automated manner. Initially, the approach incorporates $FLAIR_3$ as a novel modality for diagnosing pediatric TSC lesions and optimizes the T2W and FLAIR combination to enhance the lesion–brain contrast in a clinic. The findings indicate that $FLAIR_3$ has the ability to enhance the prominence of TSC lesions, while also enhancing classification accuracy and providing a more intuitive understanding of our deep learning model. Otherwise, the proposed method used two networks as feature extractors; one is 3D-EfficientNet, which is a parameter-efficient deep convolutional neural network framework, and the other classification network is 3D-ResNet, which is a classical residual network. Previously, the $FLAIR_3$ modality was only used in MS disease [13], but the proposed methods generalized it to pediatric TSC disease and demonstrated that $FLAIR_3$ was able to better visualize TSC lesions. Furthermore, a multi-modal fusion network for multi-contrast MRI data was proposed, which can feed $FLAIR_3$ as a new modality into the proposed DWF-net network, finally achieving a state-of-the-art classification performance in identifying children with pediatric TSC. And the dataset has no PET and EEG as input, and only has just the structural MRI that can be easily and wildly collected at any hospital, which helpfully maximizes the potential applicability of the proposed approach in clinical practice. In summary, the proposed method also has innovations in the following aspects: 1) the use of a weighted fusion algorithm to maximize the fusion multi-contrast MRI and optimize weights to improve performance; 2) firstly proposes to use a $FLAIR_3$ image to position and visualize the lesions in a clinical diagnosis of TSC. 3) The utilization of $FLAIR_3$ as the complementary imaging input to maximize the information extracted from the structure MRI.

In comparison to the 2D CNN model InceptionV3 discussed in [18], the proposed 3D CNN models exhibit an enhanced classification performance. Some previous studies are also consistent with our conclusion that 3D networks perform better than 2D networks [19,28]. We believe that the performance improvement of the 3D network is mainly due to the full use of the spatial features of MRI voxels, which can extract more information. In this study, the proposed late fusion method can improve the classification performance compared to a single modality using a 3D CNN approach, implying that combining multiple contrasting MRI can exploit complementary visual information between multiple sequences. This result is consistent with a recent study by Han Peng et al. [29], which demonstrated that combining models from diverse modalities with complementary information leads to a superior performance. The success of the ensemble strategy is not only attributed to the number of large models but also to independent information gathered from different modalities. Additionally, recent research has revealed that the late fusion method outperforms the early fusion technique [30,31]. In addition, Jonsson et al. used a majority voting strategy to form the final predictions and achieved performance gains with multimodal inputs [22]. In our experimental results, our findings indicate that when utilizing the same MRI modality as network inputs, all models with 3D-ResNet feature extractors outperform the 3D-EfficientNet model. One possible explanation is that 3D-ResNet has more network parameters than 3D-Effectient, and the network structure is more complicated. Therefore, 3D-ResNet can extract more high-level image feature information than 3D-EfficientNet.

Surprisingly, our experiments have successfully demonstrated the effectiveness of $FLAIR_3$ in a pediatric TSC diagnosis, and the AUC performance of the $FLAIR_3$-only model outperforms the T2W-only model and FLAIR-only model when using the same network. We found that the use of 3D-EfficientNet results in a better AUC score for the Eff_$FLAIR_3$ model compared to the Eff_FLAIR_T2W model and that the Res_$FLAIR_3$ model outperforms the Res_FLAIR_T2W model when using the feature extraction network 3D ResNet. This could imply that $FLAIR_3$ can provide more information. When the late fusion strategy is used, the weight $W_3$ of $FLAIR_3$ is the largest. A reasonable note is that $FLAIR_3$ can enhance the lesion-to-brain contrast and the TSC lesion is clearer in $FLAIR_3$ than in T2W and FLAIR, so $FLAIR_3$ can offer more low-dimensional visual lesion information for deep learning during the feature extraction stage. Such low-dimensional visual information may be very helpful for our deep learning algorithms, which could increase the interpretability of our deep learning algorithms [32].

Moreover, skull stripping plays a crucial role in computational neuro-imaging by being a vital preprocessing step that has a direct impact on subsequent analyses [33–35]. In this study, we found that both the 3D-ResNet and 3D-EfficientNet models perform better when utilizing MRI with skull stripping applied as the input. This may be due to the fact that the pixel value of the skull is significantly higher than that of the brain tissue [30,36], which allows for more information to be extracted during the feature selection phase. However, it is important to note that such information may be irrelevant for our deep learning methods and may even reduce their performance [37].

Furthermore, image normalization is critical to develop powerful deep learning methods [38,39]. In this study, the experiments included normalization, no normalization, min–max normalization, and Z-score normalization. All of the results showed that the AUC performance without the normalization method is the worst; the AUC performance of the min–max normalization is better than the Z-score normalization when the input modality and network structure are the same. Therefore, we suggest that in future similar studies, the min–max normalization method can be used as a primary choice to normalize the MRI images.

Otherwise, many experts considered that tubers are stable in size and appearance after birth and that the proportion to the whole brain will not obviously change with age [40]. The myelination process in a clinic has three stages, namely before 7–8 months of age, 7–8 months to 2 years of age, and after 2 years of age. So, the TSC situation of MRI after 2 years of age should be the same as before, but myelination after 2 years of age may not have affected our MRI images [41]. But these are statistical results, and there are some different situations for different TSC patients. In a clinic, MRI should be scanned several times under the age of 2 to reflect dynamic changes in epileptic lesions. Here, we did not exclude children under 2 years of age for being close to real clinical situations. The deep learning method we proposed can be promoted in a clinic and only needs to collect FLAIR and T2W images of a patient. Our method is simple and effective in a clinic and can be used as a computer-aided tool to help doctors diagnose TSC patients. In the future, further situations of TSC patients should be evaluated.

## 6. Conclusions

In summary, a novel deep learning method of the weighted late fusion model was proposed to effectively diagnose pediatric TSC lesions with multi-contrast and synthesis-contrast $FLAIR_3$ MRI. The collected dataset of pediatric TSC disease has a total of 680 children, including 331 healthy and 349 TSC children. The current testing results illustrated that the proposed approach can attain a state-of-the-art AUC of 0.998 and accuracy of 0.985. As such, this method can act as a robust foundation for future studies regarding pediatric TSC patients.

## 7. Patents

The work reported in this manuscript has resulted in a patent.

**Author Contributions:** Data curation, C.Z., X.Z., R.L., Y.Z. (Yihang Zhou) and Z.H.; Formal analysis, J.L., J.Y., Z.L., Y.Z. (Yihang Zhou) and D.L.; Investigation, Z.H., H.W. and D.L.; Methodology, D.J., J.L., D.L., Z.H. and H.W.; Resources, D.J. and R.L.; Software, J.Y., Y.Z. (Yanjie Zhu) and H.W.; Validation, D.J., C.Z., X.Z. and H.W.; Writing—original draft, D.J. and Z.L.; Writing—review and editing, Z.H., H.W. and D.L. All authors have read and agreed to the published version of the manuscript.

**Funding:** This study received support from various sources, including the Sanming Project of Medicine in Shenzhen (SZSM201812005), the Guangdong High-level Hospital Construction Fund (Shenzhen Children's Hospital, ynkt2021-zz11), the Pearl River Talent Recruitment Program of Guangdong Province (2019QN01Y986), the Key Laboratory for Magnetic Resonance and Multimodality Imaging of Guangdong Province (2020B1212060051), the National Natural Science Foundation of China (62271474, 6161871373, 81729003, and 81901736), the Strategic Priority Research Program of Chinese Academy of Sciences (XDB25000000 and XDC07040000), the Science and Technology Plan Program of Guangzhou (202007030002), the Key Field R&D Program of Guangdong Province (2018B030335001), and the Shenzhen Science and Technology Program (JCYJ20210324115810030, KQTD20180413181834876, and JCYJ20220530160005012).

**Institutional Review Board Statement:** This study was approved by the Ethics Committee of Shenzhen Children's Hospital (No.2019005).

**Informed Consent Statement:** Informed consent was obtained from all subjects involved in this study. Written informed consent was obtained from the patient(s) to publish this paper.

**Data Availability Statement:** All of our data is from Shenzhen Children's Hospital and the data are unavailable due to privacy or ethical restrictions.

**Acknowledgments:** We thank the Shenzhen Institute of Advanced Technology, Chinese Academy of Sciences for providing experimental equipment.

**Conflicts of Interest:** The authors declare no conflict of interest. The funders had no role in the design of the study; in the collection, analyses, or interpretation of data; in the writing of the manuscript; or in the decision to publish the results.

## References

1. Chu-Shore, C.J.; Major, P.; Camposano, S.; Muzykewicz, D.; Thiele, E.A. The natural history of epilepsy in tuberous sclerosis complex. *Epilepsia* **2009**, *51*, 1236–1241. [CrossRef] [PubMed]
2. Liu, A.J.; Lusk, J.B.; Ervin, J.; Burke, J.; O'Brien, R.; Wang, S.H.J. Tuberous sclerosis complex is a novel, amyloid-independent tauopathy associated with elevated phosphorylated 3R/4R tau aggregation. *Acta Neuropathol. Commun.* **2022**, *10*, 27. [CrossRef] [PubMed]
3. Henske, E.P.; Jóźwiak, S.; Kingswood, J.C.; Sampson, J.R.; Thiele, E.A. Tuberous sclerosis complex. *Nat. Rev. Dis. Primers* **2016**, *2*, 16035. [CrossRef] [PubMed]
4. Sato, A.; Tominaga, K.; Iwatani, Y.; Kato, Y.; Wataya-Kaneda, M.; Makita, K.; Nemoto, K.; Taniike, M.; Kagitani-Shimono, K. Abnormal White Matter Microstructure in the Limbic System Is Associated with Tuberous Sclerosis Complex-Associated Neuropsychiatric Disorders. *Front. Neurol.* **2022**, *13*, 782479. [CrossRef] [PubMed]
5. Liu, L.; Yu, C.; Yan, G. Identification of a novel heterozygous TSC2 splicing variant in a patient with Tuberous sclerosis complex: A case report. *Medicine* **2022**, *101*, e28666. [CrossRef]
6. Miszewska, D.; Sugalska, M.; Jóźwiak, S. Risk Factors Associated with Refractory Epilepsy in Patients with Tuberous Sclerosis Complex: A Systematic Review. *J. Clin. Med.* **2021**, *10*, 5495. [CrossRef]
7. Okanishi, T.; Akiyama, T.; Tanaka, S.-I.; Mayo, E.; Mitsutake, A.; Boelman, C.; Go, C.; Snead, O.C.; Drake, J.; Rutka, J.; et al. Interictal high frequency oscillations correlating with seizure outcome in patients with widespread epileptic networks in tuberous sclerosis complex. *Epilepsia* **2014**, *55*, 1602–1610. [CrossRef]
8. Zhang, K.; Hu, W.-H.; Zhang, C.; Meng, F.-G.; Chen, N.; Zhang, J.-G. Predictors of seizure freedom after surgical management of tuberous sclerosis complex: A systematic review and meta-analysis. *Epilepsy Res.* **2013**, *105*, 377–383. [CrossRef]
9. Yang, J.; Zhao, C.; Su, S.; Liang, D.; Hu, Z.; Wang, H.; Liao, J. Machine Learning in Epilepsy Drug Treatment Outcome Prediction Using Multi-modality Data in Children with Tuberous Sclerosis Complex. In *Proceedings of the 2020 6th International Conference on Big Data and Information Analytics (BigDIA), Shenzhen, China, 4–6 December 2020*; IEEE: Manhattan, NY, USA, 2020. [CrossRef]
10. De Ridder, J.; Verhelle, B.; Vervisch, J.; Lemmens, K.; Kotulska, K.; Moavero, R.; Curatolo, P.; Weschke, B.; Riney, K.; Feucht, M.; et al. Early epileptiform EEG activity in infants with tuberous sclerosis complex predicts epilepsy and neurodevelopmental outcomes. *Epilepsia* **2021**, *62*, 1208–1219. [CrossRef]
11. Russo, C.; Nastro, A.; Cicala, D.; De Liso, M.; Covelli, E.M.; Cinalli, G. Neuroimaging in tuberous sclerosis complex. *Childs Nerv. Syst.* **2020**, *36*, 2497–2509. [CrossRef]
12. Wiggermann, V.; Hernandez-Torres, E.; Traboulsee, A.; Li DK, B.; Rauscher, A. FLAIR$^2$: A Combination of FLAIR and T2 for Improved MS Lesion Detection. *Am. J. Neuroradiol.* **2016**, *37*, 259–265. [CrossRef]
13. Gabr, R.E.; Hasan, K.M.; Haque, M.E.; Nelson, F.M.; Wolinsky, J.S.; Narayana, P.A. Optimal combination of FLAIR and T2-weighted MRI for improved lesion contrast in multiple sclerosis. *J. Magn. Reson. Imaging* **2016**, *44*, 1293–1300. [CrossRef] [PubMed]
14. Lyu, Q.; Shan, H.; Steber, C.; Helis, C.; Whitlow, C.; Chan, M.; Wang, G. Multi-Contrast Super-Resolution MRI Through a Progressive Network. *IEEE Trans. Med. Imaging* **2020**, *39*, 2738–2749. [CrossRef] [PubMed]
15. Cetinoglu, Y.K.; Koska, I.O.; Uluc, M.E.; Gelal, M.F. Detection and vascular territorial classification of stroke on diffusion-weighted MRI by deep learning. *Eur. J. Radiol.* **2021**, *145*, 110050. [CrossRef] [PubMed]

16. Srikrishna, M.; Pereira, J.B.; Heckemann, R.A.; Volpe, G.; van Westen, D.; Zettergren, A.; Kern, S.; Wahlund, L.-O.; Westman, E.; Skoog, I.; et al. Deep learning from MRI-derived labels enables automatic brain tissue classification on human brain CT. *Neuroimage* **2021**, *244*, 118606. [CrossRef]
17. Park, D.K.; Kim, W.; Thornburg, O.S.; McBrian, D.K.; McKhann, G.M.; Feldstein, N.A.; Maddocks, A.B.; Gonzalez, E.; Shen, M.Y.; Akman, C.; et al. Convolutional neural network-aided tuber segmentation in tuberous sclerosis complex patients correlates with electroencephalogram. *Epilepsia* **2022**, *63*, 1530–1541. [CrossRef]
18. Sánchez Fernández, I.; Yang, E.; Calvachi, P.; Amengual-Gual, M.; Wu, J.Y.; Krueger, D.; Northrup, H.; Bebin, M.E.; Sahin, M.; Yu, K.-H.; et al. Deep learning in rare disease. Detection of tubers in tuberous sclerosis complex. *PLoS ONE* **2020**, *15*, e0232376. [CrossRef]
19. Cole, J.H.; Poudel, R.P.; Tsagkrasoulis, D.; Caan, M.W.; Steves, C.; Spector, T.D.; Montana, G. Predicting brain age with deep learning from raw imaging data results in a reliable and heritable biomarker. *Neuroimage* **2017**, *163*, 115–124. [CrossRef]
20. Moffa, A.P.; Grilli, G.; Perfetto, F.; Specchiulli, L.P.; Vinci, R.; Macarini, L.; Zizzo, L. Neuroimaging features of tuberous sclerosis complex and Chiari type I malformation: A rare association. *J. Pediatr. Neurosci.* **2018**, *13*, 224–228. [CrossRef]
21. Liang, G.; Xing, X.; Liu, L.; Zhang, Y.; Ying, Q.; Lin, A.L.; Jacobs, N. Alzheimer's Disease Classification Using 2D Convolutional Neural Networks. In *Proceedings of the 2021 43rd Annual International Conference of the IEEE Engineering in Medicine & Biology Society (EMBC), Guadalajara, Mexico, 1–5 November 2021*; IEEE: Manhattan, NY, USA, 2021; pp. 3008–3012.
22. Jonsson, B.A.; Bjornsdottir, G.; Thorgeirsson, T.E.; Ellingsen, L.M.; Walters, G.B.; Gudbjartsson, D.F.; Stefansson, H.; Ulfarsson, M.O. Brain age prediction using deep learning uncovers associated sequence variants. *Nat. Commun.* **2019**, *10*, 1–10. [CrossRef]
23. Eweje, F.R.; Bao, B.; Wu, J.; Dalal, D.; Liao, W.H.; He, Y.; Luo, Y.; Lu, S.; Zhang, P.; Peng, X.; et al. Deep Learning for Classification of Bone Lesions on Routine MRI. *EBioMedicine* **2021**, *68*, 103402. [CrossRef]
24. He, K.; Zhang, X.; Ren, S.; Sun, J. Deep Residual Learning for Image Recognition. In Proceedings of the IEEE Conference on Computer Vision and Pattern Recognition, Las Vegas, NV, USA, 27–30 June 2016; pp. 770–778.
25. Tan, M.; Le, Q. Efficientnet: Rethinking model scaling for convolutional neural networks. In Proceedings of the International Conference on Machine Learning, Long Beach, CA, USA, 9–15 June 2019; pp. 6105–6114.
26. Hu, J.; Shen, L.; Sun, G. Squeeze-and-Excitation Networks. In Proceedings of the IEEE Conference on Computer Vision and Pattern Recognition, Salt Lake City, UT, USA, 18–23 June 2018; pp. 7132–7141.
27. Isensee, F.; Schell, M.; Pflueger, I.; Brugnara, G.; Bonekamp, D.; Neuberger, U.; Wick, A.; Schlemmer, H.-P.; Heiland, S.; Wick, W.; et al. Automated brain extraction of multisequence MRI using artificial neural networks. *Hum. Brain Mapp.* **2019**, *40*, 4952–4964. [CrossRef] [PubMed]
28. Kim, H.; Lee, Y.; Kim, Y.H.; Lim, Y.M.; Lee, J.S.; Woo, J.; Jang, S.K.; Oh, Y.J.; Kim, H.W.; Lee, E.J.; et al. Deep Learning-Based Method to Differentiate Neuromyelitis Optica Spectrum Disorder from Multiple Sclerosis. *Front. Neurol.* **2020**, *11*, 599042. [CrossRef] [PubMed]
29. Peng, H.; Gong, W.; Beckmann, C.F.; Vedaldi, A.; Smith, S.M. Accurate brain age prediction with lightweight deep neural networks. *Med. Image Anal.* **2020**, *68*, 101871. [CrossRef] [PubMed]
30. De Luna, A.; Marcia, R.F. Data-Limited Deep Learning Methods for Mild Cognitive Impairment Classification in Alzheimer's Disease Patients. In Proceedings of the 2021 43rd Annual International Conference of the IEEE Engineering in Medicine & Biology Society (EMBC), Guadalajara, Mexico, 1–5 November 2021; pp. 2641–2646.
31. Jian, J.; Li, Y.A.; Xia, W.; He, Z.; Zhang, R.; Li, H.; Zhao, X.; Zhao, S.; Zhang, J.; Cai, S.; et al. MRI-Based Multiple Instance Convolutional Neural Network for Increased Accuracy in the Differentiation of Borderline and Malignant Epithelial Ovarian Tumors. *J. Magn. Reason. Imaging* **2021**, *56*, 173–181. [CrossRef]
32. Banerjee, S.; Dong, M.; Lee, M.-H.; O'Hara, N.; Juhasz, C.; Asano, E.; Jeong, J.-W. Deep Relational Reasoning for the Prediction of Language Impairment and Postoperative Seizure Outcome Using Preoperative DWI Connectome Data of Children with Focal Epilepsy. *IEEE Trans. Med. Imaging* **2020**, *40*, 793–804. [CrossRef]
33. Thakur, S.P.; Doshi, J.; Pati, S.; Ha, S.M.; Sako, C.; Talbar, S.; Kulkarni, U.; Davatzikos, C.; Erus, G.; Bakas, S. Skull-Stripping of Glioblastoma MRI Scans Using 3D Deep Learning. In *Brainlesion: Glioma, Multiple Sclerosis, Stroke and Traumatic Brain Injuries*; Springer: Cham, Switzerland, 2020; Volume 11992, pp. 57–68.
34. Fischmeister, F.P.; Höllinger, I.; Klinger, N.; Geissler, A.; Wurnig, M.C.; Matt, E.; Rath, J.; Robinson, S.D.; Trattnig, S.; Beisteiner, R. The benefits of skull stripping in the normalization of clinical fMRI data. *NeuroImage Clin.* **2013**, *3*, 369–380. [CrossRef]
35. Kleesiek, J.; Urban, G.; Hubert, A.; Schwarz, D.; Maier-Hein, K.; Bendszus, M.; Biller, A. Deep MRI brain extraction: A 3D convolutional neural network for skull stripping. *Neuroimage* **2016**, *129*, 460–469. [CrossRef]
36. Fatima, A.; Shahid, A.R.; Raza, B.; Madni, T.M.; Janjua, U.I. State-of-the-Art Traditional to the Machine- and Deep-Learning-Based Skull Stripping Techniques, Models, and Algorithms. *J. Digit. Imaging* **2020**, *33*, 1443–1464. [CrossRef]
37. Jiang, D.; Hu, Z.; Zhao, C.; Zhao, X.; Yang, J.; Zhu, Y.; Liang, D.; Wang, H. Identification of Children's Tuberous Sclerosis Complex with Multiple-contrast MRI and 3D Convolutional Network. In Proceedings of the 2022 44th Annual International Conference of the IEEE Engineering in Medicine & Biology Society (EMBC), Glasgow, UK, 11–15 July 2022; pp. 2924–2927.
38. Zheng, Y.; Jiang, Z.; Zhang, H.; Xie, F.; Hu, D.; Sun, S.; Shi, J.; Xue, C. Stain Standardization Capsule for Application-Driven Histopathological Image Normalization. *IEEE J. Biomed. Health Inform.* **2021**, *25*, 337–347. [CrossRef]

39. Isaksson, L.J.; Raimondi, S.; Botta, F.; Pepa, M.; Gugliandolo, S.G.; De Angelis, S.P.; Marvaso, G.; Petralia, G.; DE Cobelli, O.; Gandini, S.; et al. Effects of MRI image normalization techniques in prostate cancer radiomics. *Phys. Medica* **2020**, *71*, 7–13. [CrossRef] [PubMed]
40. Curatolo, P. Intractable epilepsy in tuberous sclerosis: Is the tuber removal not enough? *Dev. Med. Child Neurol.* **2010**, *52*, 987. [CrossRef]
41. Davis, P.E.; Filip-Dhima, R.; Sideridis, G.; Peters, J.M.; Au, K.S.; Northrup, H.; Bebin, E.M.; Wu, J.Y.; Krueger, D.; Sahin, M.; et al. Presentation and Diagnosis of Tuberous Sclerosis Complex in Infants. *Pediatrics* **2017**, *140*, e20164040. [CrossRef] [PubMed]

bioengineering

MDPI

*Article*

# Mask-Transformer-Based Networks for Teeth Segmentation in Panoramic Radiographs

**Mehreen Kanwal [1,†], Muhammad Mutti Ur Rehman [2,†], Muhammad Umar Farooq [3] and Dong-Kyu Chae [3,*]**

1 DeepChain AI&IT Technologies, Islamabad 45570, Pakistan; mehreen@deepchain.pk
2 Department of Computer and Software Engineering, National University of Science and Technology, Islamabad 43701, Pakistan; mmutti.ce41ceme@student.nust.edu.pk
3 Department of Computer Science, Hanyang University, Seoul 04763, Republic of Korea; umarfarooq@hanyang.ac.kr
* Correspondence: dongkyu@hanyang.ac.kr
† These authors contributed equally to this work.

**Abstract:** Teeth segmentation plays a pivotal role in dentistry by facilitating accurate diagnoses and aiding the development of effective treatment plans. While traditional methods have primarily focused on teeth segmentation, they often fail to consider the broader oral tissue context. This paper proposes a panoptic-segmentation-based method that combines the results of instance segmentation with semantic segmentation of the background. Particularly, we introduce a novel architecture for instance teeth segmentation that leverages a dual-path transformer-based network, integrated with a panoptic quality (PQ) loss function. The model directly predicts masks and their corresponding classes, with the PQ loss function streamlining the training process. Our proposed architecture features a dual-path transformer block that facilitates bi-directional communication between the pixel path CNN and the memory path. It also contains a stacked decoder block that aggregates multi-scale features across different decoding resolutions. The transformer block integrates pixel-to-memory feedback attention, pixel-to-pixel self-attention, and memory-to-pixel and memory-to-memory self-attention mechanisms. The output heads process features to predict mask classes, while the final mask is obtained by multiplying memory path and pixel path features. When applied to the UFBA-UESC Dental Image dataset, our model exhibits a substantial improvement in segmentation performance, surpassing existing state-of-the-art techniques in terms of performance and robustness. Our research signifies an essential step forward in teeth segmentation and contributes to a deeper understanding of oral structures.

**Keywords:** teeth segmentation; panoramic radiographs; mask-transformer-based networks; panoptic segmentation

**Citation:** Kanwal, M.; Ur Rehman, M.M.; Farooq, M.U.; Chae, D.-K. Mask-Transformer-Based Networks for Teeth Segmentation in Panoramic Radiographs. *Bioengineering* **2023**, *10*, 843. https://doi.org/10.3390/bioengineering10070843

Academic Editors: Paolo Zaffino, Maria Francesca Spadea and Antonio Brunetti

Received: 3 May 2023
Revised: 10 July 2023
Accepted: 12 July 2023
Published: 17 July 2023



## 1. Introduction

Teeth segmentation is pivotal in the clinical diagnosis of oral diseases, offering essential precision in surgical planning through the accurate delineation of teeth boundaries [1,2]. In orthodontics, real-time information regarding teeth movement and root depths is crucial for immediate assessment of a patient's dental alignment and for accelerating the orthodontic treatment cycle [3]. The prerequisite for achieving this is the precise segmentation of teeth in dental panoramic X-ray images [4], which has additional applications in forensic identification [5], age estimation, and the analysis of hidden dental structures, including benign or malignant masses [6]. Dentistry extensively utilizes radiographic images for diagnosis, given their comprehensive visualization of the internal structure of the mouth [7]. Extra-oral radiographs, encompassing panoramic and cephalometric images, capture the complete dentition and surrounding areas, providing critical insights into a patient's teeth, as demonstrated in Figure 1. However, manual and semi-automated segmentation approaches for teeth and tissues in these radiographs often prove time consuming, tedious, and subjective, with their efficacy heavily reliant on the dentist's expertise. Additionally, segmentation in low-quality image settings presents even greater challenges. Given these circumstances, the development of an automatic, accurate, and efficient teeth segmentation method is paramount.

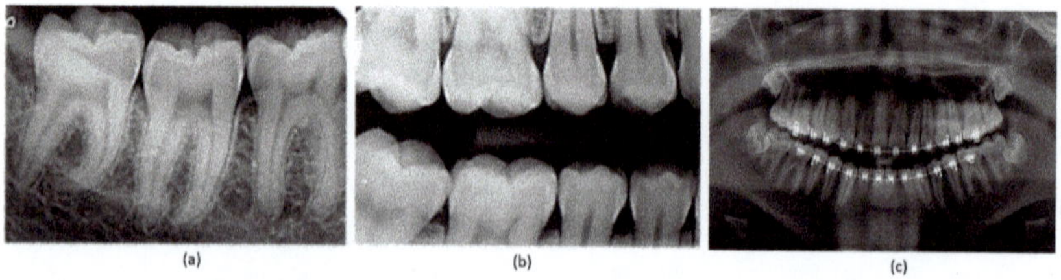


**Figure 1.** Types of X-ray images: (**a**) periapical X-ray; (**b**) bitewing X-ray; (**c**) panoramic X-ray.

Traditionally, teeth segmentation has been approached through semantic and instance segmentation techniques [8,9]. While semantic segmentation classifies each pixel into predefined classes without distinguishing between object instances, instance segmentation offers a more comprehensive understanding by segmenting objects and distinguishing each tooth object instance. Both category and instance labels are crucial in this context, which has become a focal point in dental research. However, both proposal-based and proposal-free instance segmentation approaches have their limitations. They often struggle with differentiating object instances within the same category, particularly when objects overlap, and preserving pixel-wise location information, which often results in coarse mask boundaries.

Numerous attempts have been made to develop a highly accurate automatic teeth segmentation algorithm [10,11]. However, teeth segmentation remains challenging due to fuzzy boundaries caused by low contrast and noisy dental panoramic X-ray images. The diversity of teeth conditions across different patients and the presence of dental instruments, such as metal racks and dental implants, pose significant obstacles to achieving accurate teeth segmentation. Recognizing these challenges, this research introduces a novel approach based on panoptic segmentation [12]. Panoptic segmentation unifies the typically disjoint tasks of semantic segmentation (identifying and classifying objects in an image) and instance segmentation (segmenting individual instances of each object), offering a more holistic and precise tooth and oral tissue segmentation strategy [13,14]. Several studies have shown the effectiveness of panoptic segmentation for optimizing the performance of deep-learning-based models [15–18].

We propose a panoptic-segmentation-based approach for instance teeth segmentation and surrounding tissue semantic segmentation. Panoptic segmentation, a unified framework for semantic and instance segmentation, yields better Dice scores for teeth segmentation by providing an improved context understanding, better discrimination of close or touching instances, and consistent pixel-level labeling. This approach reduces false positives and negatives by correctly segmenting teeth instances and accurately labeling non-teeth regions, enhancing the overlap between prediction and ground truth, which the Dice score measures. Our model employs a mask transformer to predict non-overlapping masks and their corresponding semantic segmentation labels directly. The panoptic quality (PQ) style loss is utilized to optimize the output masks and classes. More specifically, we design the similarity metric between consecutive teeth-labeled masks as the product of their masks and class similarity, inspired by the PQ definition. Moreover, the innovative strategies proposed by groundbreaking works that use attention mechanisms, such as [19,20], motivated us to incorporate attention modules into our proposed network.

We introduce a novel architecture to effectively train and infer using the mask transformer. Unlike traditional architectures [21,22] where the transformer is placed on top of a convolutional neural network (CNN) [23], we adopt a dual-path framework that effectively merges CNNs with transformers [24–27]. This allows CNN layers to read and write into global memory by incorporating memory-to-pixel attention (M2P), memory path self-attention (M2M), pixel–path axial self-attention (P2P), and pixel-to-memory attention (P2M). As a result, the transformer can be inserted at any position in the CNN to enable communication with the global memory at any layer. The proposed architecture also employs a stacked hourglass-style decoder [28,29] to aggregate multi-scale features and produce a high-resolution output, which is then multiplied with the global memory feature to predict the mask. The proposed framework significantly improves segmentation performance and demonstrates the potential to be employed for teeth numbering. Rigorously evaluated on the publicly available UFBA-UESC dental image dataset, our experimental results demonstrate that the proposed model significantly outperforms existing state-of-the-art techniques in terms of segmentation performance and robustness.

This paper is organized as follows: Section 2 provides the background and related work. Section 3 offers a detailed description of the network and dataset. Section 3.4 is dedicated to the experimental setup, and then Section 4 presents the results and discussion. Finally, Section 5 concludes the paper and provides the future directions.

## 2. Related Work

There have been numerous attempts by researchers to develop teeth segmentation techniques that can be applied to various types of radiographic images, such as panoramic, periodical, and bitewing imaging. Silva et al. [30] presented a comparison of various segmentation techniques applied in dental imaging, categorizing solutions into five groups and evaluating them based on accuracy, specificity, precision, recall, and F1-score. However, all these techniques struggled to fully segment the teeth due to the presence of the bone structure inside the buccal cavity.

Classic image processing techniques have been utilized to address these challenges. For instance, to counteract the problem of low contrast, Lin et al. [31,32] first enhanced the image to distinguish between teeth and gums before applying edge extraction methods for segmentation. In a similar vein, Chandran et al. [33] improved the quality of dental images by applying CLAHE, followed by the Otsu threshold method for teeth segmentation. Level set methods have been utilized by studies [34,35] to enhance the root contrast, thus improving segmentation. Horizontal and vertical integral projection methods have also been deployed, although their performance was not satisfactory [36,37].

Recently, deep learning (DL)-based techniques have garnered attention across various industrial applications due to their impressive performance [38–40]. These applications span object classification [41], segmentation [42–44], counting [45], medical image enhancement [46,47], and object detection [48]. Specifically, in tasks such as object detection and segmentation, DL-based methods have revolutionized the field [49]. As a result, several DL-based techniques have been employed to enhance teeth segmentation in dental panoramic X-ray images. While some studies have focused solely on the semantic segmentation of teeth, limiting the level of detail for further processing steps in most automatic dental analyses [30,50,51], others have identified teeth alongside segmentation, providing more information for automatic analysis. However, these instance segmentation techniques, which typically consist of two stages, ROI/fuzzy boundary detection and teeth segmentation, increase the complexity and are more prone to errors due to their cascading nature. The errors from the first stage can propagate to the second, limiting the performance of these methods. Additionally, the information obtained from instance segmentation may not be sufficient for a comprehensive teeth analysis, as apart from intra-teeth segmentation, it is crucial to accurately segment the teeth from other oral tissues.

For instance, Jader et al. [11] employed the mask-region-based convolutional neural network (Mask-R-CNN) for instance segmentation. Their method, evaluated on a diverse set of 1500 images, achieved an accuracy of 98%, an F1-score of 88%, a precision of 94%, a recall of 84%, and a specificity of 99% over 1224 unseen images, considerably outperforming 10 unsupervised methods. However, the method was limited to teeth detection and did not account for other issues such as dentures and areas with missing teeth. Similarly,

Zhang et al. [52] utilized deep-learning-based methods to detect and classify teeth, merging the Faster R-CNN and region-based fully convolutional networks (R-FCN) to identify common patient issues such as tooth loss, decay, and fillings. Similarly, Koch et al. [50] employed the U-Net architecture in conjunction with an FCN for semantic segmentation of dental panoramic radiographs and explored ways to improve segmentation performance, such as network ensembling, test-time augmentation, bootstrapping of low-quality annotations, and data symmetry exploitation. In their study, Lee et al. [53] utilized data augmentation techniques such as rotation, flipping, Gaussian blur, and shear transformation to generate 1024 training samples from 30 radiographs. They implemented a fully deep learning method using the Mask R-CNN model through a fine-tuning process to detect and localize tooth structures, achieving an F1 score of 0.875 and a mean IoU of 0.877. Muresan et al. [54] proposed a novel approach for automatic teeth detection and dental problem classification using panoramic X-Ray images. They utilized a CNN model trained on their collected data and employed image pre-processing techniques to refine segmentation, resulting in an F1 score of 0.93.

Building upon previous efforts, Zhao et al. [55] introduced a dual-stage scheme, TSAS-Net, to address specific issues like fuzzy tooth boundaries resulting from poor contrast and intensity distribution in dental panoramic X-rays. The method, tested on a dataset of 1500 radiographs, achieved an impressive accuracy of 96.94%, a Dice score of 92.72%, and a recall of 93.77%. Kong et al. [56] have made a substantial contribution to the scientific community by introducing a publicly available dataset that includes 2602 panoramic dental X-ray images. Each image in the dataset is paired with expertly annotated segmentation masks, thereby significantly enriching this resource. Harnessing the power of this dataset, they engineered a proficient encoder–decoder network named EED-Net. This network is specifically designed for the autonomous segmentation of the maxillofacial region, demonstrating their innovative application of the dataset. Arora et al. [57] recently introduced a multimodal encoder-based architecture, designed to extract a variety of features from panoramic radiographs. These extracted features were subsequently processed through a deconvolutional block to generate the final segmentation mask. By achieving precision and recall rates of 95.01% and 94.06%, respectively, this approach outperformed other leading methods.

In a different approach, Almalki et al. [58] utilized self-supervised learning methods, such as SimMIM and UM-MAE, to boost model efficiency in comprehending a limited number of available dental radiographs. Their SimMIM method yielded the highest performance, achieving 90.4% and 88.9% in detecting teeth and dental restorations and instance segmentation, respectively. This outperformed the random initialization baseline by an average precision increase of 13.4 and 12.8. However, the method's requirement for extensive parameter fine-tuning creates challenges in achieving optimal results. Recently, Hou et al. [59] proposed the Teeth U-Net model. This model combines a Squeeze-Excitation Module in both the encoder and decoder, supplemented by a dense skip connection, in an attempt to bridge the semantic gap. The model also includes a Multi-scale Aggregation attention Block (MAB) in the bottleneck layer to effectively extract teeth shape features and adaptively fuse multi-scale features. To incorporate dental feature information from a broader field of view, they devised a Dilated Hybrid self-Attentive Block (DHAB) at the bottleneck layer. This block is designed to suppress irrelevant background region information without increasing the network parameters. Although the study showcased competitive performance on a private dataset, it has not yet been evaluated on publicly available datasets.

Table 1 summarizes the strides made by the aforementioned studies towards accurately segmenting teeth in panoramic radiographs.

**Table 1.** Summary of previously published methods for teeth segmentation in panoramic radiographs.

| Authors, Year | Technique | Contribution/Advantages | Limitations |
|---|---|---|---|
| Jader et al. [11], 2018 | Instance segmentation for panoramic X-ray images | Introduced a new instance segmentation technique for teeth segmentation with promising results. | Struggles with overlapping and adjacent teeth. |
| Zhang et al. [52], 2018 | Label tree with cascade network structure for teeth recognition | Improved teeth recognition using a novel label tree and cascade network structure. | Inefficient with teeth suffering from severe pathologies. |
| Koch et al. [50], 2019 | U-Nets for dental panoramic radiographs segmentation | Developed an accurate tooth segmentation technique based on U-Nets. Demonstrated improved performance. | Difficulty in segmenting teeth with complex structures or deformities. |
| Lee et al. [53], 2020 | Deep convolutional neural network for tooth segmentation automation | Employed a deep convolutional neural network for automated tooth segmentation. Enhanced both efficiency and accuracy. | Limitations when dealing with noisy or poor-quality images. |
| Muresan et al. [54], 2020 | Deep learning and image processing techniques for teeth detection and dental problem classification | Introduced a novel approach using deep learning and image processing techniques for teeth detection and dental problem classification. | Struggles with dental problems underrepresented in the training data. |
| Zhao et al. [55], 2020 | TSASNet: Two-Stage Attention Segmentation Network for tooth segmentation | Developed TSASNet, a Two-Stage Attention Segmentation Network for tooth segmentation, showing enhanced results. | Inefficient with teeth of unusual shapes or sizes. |
| Kong et al. [56], 2020 | Efficient encoder–decoder network for automated maxillofacial segmentation | Proposed an automated segmentation method for maxillofacial regions in dental X-ray images. Showed improved efficiency and accuracy. | Difficulty with radiographs containing artifacts or of poor quality. |
| Shubhangi et al. [60], 2022 | CNNs combined with classical image processing methods | Performed teeth segmentation and numbering using a histogram-based plurality vote process. | Computationally expensive, posing challenges for real-time applications. |
| Arora et al. [57], 2023 | Multimodal encoder-based architecture | Achieved superior teeth segmentation performance. | Limited to semantic segmentation. |
| Datta et al. [61], 2023 | Combination of neutrosophic logic and a fuzzy c-means algorithm | Demonstrated competitive performance. | Relies on conventional image processing techniques, which might lack robustness. |
| Almalki et al. [58], 2023 | Self-supervised learning methods (i.e., SimMIM and UM-MAE) for dental panoramic radiographs | SimMIM, a masking-based method, outperformed UM-MAE and supervised and random initialization methods for teeth and dental restoration detection and instance segmentation. | Parameter fine-tuning, including mask ratio and pre-training epochs, substantially influence segmentation performance. |
| Hou et al. [59], 2023 | UNet with dense skip connection and attention units | Used dense skip connections and attention units to handle the irregular shape of teeth. Introduced Multi-scale Aggregation Attention Block (MAB) and Dilated Hybrid self-Attentive Block (DHAB) at the bottleneck layer. | Lacks performance analysis on public datasets, making a fair comparison challenging. |

## 3. Materials and Methods

### 3.1. Datasets

Silva et al. [30] released the UFBA-UESC Dental Images Dataset, which initially contained 1500 panoramic images along with semantic segmentation of teeth. Jader et al. [11] later introduced instance segmentation, leading to the creation of the UFBA-UESC Dental Images Deep dataset. This new dataset comprises a total of 276 images designated for training and validation. Further development by Silva et al. [7] involved the addition of tooth number information, resulting in a cumulative dataset of 543 images, inclusive of those from the UFBA-UESC Deep dataset. Named the DNS (Detection, Numbering, and Segmentation) Panoramic Images, this dataset comes equipped with binary masks and annotations in the COCO format. Detailed information about the UFBA-UESC Dental Images Dataset's characteristics is depicted in Table 2.

**Table 2.** UFBA-UESC Dental Images Dataset characteristics. Note that ✓and – represent the presence and absence of category, respectively.

| Category | Restoration | Appliance | Teeth Numbers | Image Numbers |
|---|---|---|---|---|
| 1 | ✓ | ✓ | 32 | 73 |
| 2 | ✓ | – | 32 | 220 |
| 3 | – | ✓ | 32 | 45 |
| 4 | – | – | 32 | 140 |
| 5 | – | – | 18 | 120 |
| 6 | – | – | 37 | 170 |
| 7 | ✓ | ✓ | 27 | 115 |
| 8 | ✓ | – | 29 | 457 |
| 9 | – | ✓ | 28 | 45 |
| 10 | – | – | 28 | 115 |
| Total | – | – | – | 1500 |

For our study, we adjusted the annotations of the DNS Panoramic Images dataset for panoptic segmentation. We achieved this by merging the provided semantic and instance labels and converting them into TFRecords for our experiment. This dataset served for both training and validation, with 500 images set aside for the training set and 43 images allocated for validation. Testing images were sourced from the original UFBA-UESC Dental Images dataset.

Our research utilized the UFBA-UESC Dental Images Deep dataset [7]. This dataset is accessible through a reasonable request made to the corresponding author (https://github.com/IvisionLab/dns-panoramic-images-v2 (accessed on 2 May 2023)). Table 3 provides comprehensive details regarding the dataset, such as the presence of thirty-two teeth, restorations, and appliances, as well as the total number of images used for numbering, instance segmentation, and SS. We excluded images from categories 5 and 6 due to the presence of implants and deciduous teeth.

**Table 3.** Dataset characteristics used in this work. Note that ✓and – represent the presence and absence of the corresponding category, respectively.

| Category | 32 Teeth | Restoration | Appliance | Number and Instance Segmentation | Segmentation |
|---|---|---|---|---|---|
| 1 | ✓ | ✓ | ✓ | 23 | 57 |
| 2 | ✓ | ✓ | – | 174 | 80 |
| 3 | ✓ | – | ✓ | 42 | 11 |
| 4 | ✓ | – | – | 92 | 68 |
| 7 | – | ✓ | ✓ | 36 | 87 |
| 8 | – | ✓ | – | 128 | 355 |
| 9 | – | – | ✓ | 14 | 33 |
| 10 | – | – | – | 34 | 87 |
| Total | – | – | – | 543 | 778 |

### 3.2. Network Architecture

The proposed model employs a network architecture comprised of three primary components: a Transformer block, a stacked decoder, and output heads. This end-to-end instance segmentation model predicts masks and their corresponding classes directly. In this study, we utilize Mask Transformer-Based Networks (M-TransNet) integrated with PQ Loss [62]. These networks function as instance segmentation models inspired by panoptic segmentation. The M-TransNet directly predicts class-labeled masks for panoptic segmentation, with PQ-style loss employed to train the model. This section also introduces the dual-path transformer architecture and the auxiliary losses that significantly facilitate the model's training. A complete network diagram is displayed in Figure 2.

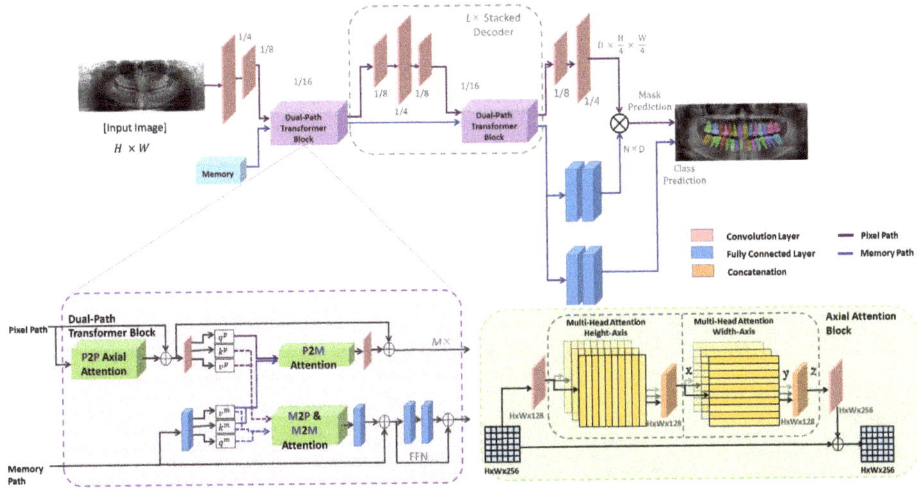


**Figure 2.** The structure of the proposed framework. An image and global memory are input into a dual-path transformer, which directly generates a collection of masks and classes (excluding residual connections). A dual-path transformer block is designed with all four types of attention (M2P, M2M, P2M, and P2P) between the two paths. On the right bottom side, the structure of the axial-attention block is illustrated. The axial attention mechanism decomposes the 2D attention into two 1D attentions; one applied along the height axis of the image, and the other applied along the width axis. By doing so, it significantly reduces the complexity from quadratic to linear, which makes it more computationally efficient.

#### 3.2.1. Architecture Formulation

The overarching goal of panoptic segmentation is to segment every object in an image $I \in \mathbb{R}^{H \times W \times 3}$ and assign a class label to each mask. The ground truth for a panoptic segmentation model can be expressed as:

$$\{y_i\}_{i=1}^{K} = \{(m_i, c_i)\}_{i=1}^{K} \tag{1}$$

where $K$ represents the total number of non-overlapping ground truth masks $m_i \in 0,1^{H \times W}$ and $c_i$ denotes the class label for each $m_i$. The output from our proposed network should precisely mirror the ground truth, thereby predicting the mask of each object alongside the class labels.

$$\{\hat{y}_i\}_{i=1}^{N} = \{(\hat{m}_i, \hat{p}_i(c))\}_{i=1}^{N} \tag{2}$$

where $N$ remains constant and is greater than $K$, with $\hat{p}_i(c)$ representing the probability of mask $m_i$ being associated with class $c$. The network is optimized to assign an empty class to masks where $N$ exceeds $K$. The class label for each mask can be predicted by taking the argmax of class probabilities:

$$\hat{c}_i = \arg\max_c(\hat{p}_i(c)) \tag{3}$$

Similarly, the mask-ID can be assigned to each pixel by applying argmax again:

$$\hat{z}h, w = \arg\max_i(\hat{m}_i, h, w)\ \forall h \in 1, 2, \ldots, H, \quad \forall w \in 1, 2, \ldots, W \tag{4}$$

Each argmax is filtered using a confidence threshold. Masks or pixels with a low confidence score are removed.

#### 3.2.2. Transformer Block

The dual-path transformer module comprises two paths: a CNN path and a memory path. The CNN path processes the input image and extracts features, while the memory path stores information about the objects and their relationships within the scene. The two paths communicate through a set of attention mechanisms, which allows the model to selectively attend to different parts of the input and memory.

The CNN path within the dual-path transformer module is a standard convolutional neural network that processes the input image and extracts features. The features are passed through a series of convolutional layers, followed by a set of axial-attention blocks that implement pixel-to-pixel (P2P) self-attention. The output of the CNN path is a feature map encoding information about the input image.

The memory path in the dual-path transformer module is a memory-augmented transformer that stores information about the objects and their relationships within the scene. The memory is initialized with a set of learned object queries, which are used to attend to the input feature map and extract object features. These object features are then stored in the memory, along with their corresponding object queries. The memory is updated at each time step using a set of memory update operations, which enable the model to reason about the relationships between different objects in the scene.

The two paths in the dual-path transformer module communicate through a set of attention mechanisms. These mechanisms enable the model to selectively attend to different parts of the input and memory, allowing the model to reason about the relationships between different parts of the image and memory.

By using a dual-path transformer module, the architecture effectively combines the strengths of both CNNs and transformers for panoptic segmentation. The CNN path extracts rich visual features from the input image, while the memory path reasons about the relationships between different objects in the scene. The attention mechanisms facilitate communication between the two paths, allowing the model to selectively attend to the most relevant information for the task at hand.

#### 3.2.3. Attention Mechanisms

The attention module in the network is a key component of the memory-augmented transformer. It allows the model to selectively focus on different parts of the input image and memory, based on their relevance to the task at hand. Specifically, the attention module computes a set of attention weights for each position in the input feature map or memory, based on its similarity to other positions. These weights are then used to compute a weighted sum of the feature map or memory, which is passed through a feedforward network to produce the final output.

The dual-path transformer block employs four types of attention to facilitate communication between the CNN path and the memory path:

- Memory-to-pixel (M2P) attention: This type allows the model to attend to the memory from the pixel path. It computes attention weights for each position in the input feature map, based on its similarity to the memory.
- Memory-to-memory (M2M) self-attention: This type allows the model to attend to the memory from the memory path. It computes attention weights for each position in the memory, based on its similarity to other positions in the memory.

- Pixel-to-memory (P2M) feedback attention: This type allows the model to attend to the memory from the pixel path, but also allows the memory to attend back to the pixel path. It computes attention weights for each position in the memory, based on its similarity to the input feature map.
- Pixel-to-pixel (P2P) self-attention: This type allows the model to attend to the input feature map from the pixel path. It computes attention weights for each position in the input feature map, based on its similarity to other positions in the input feature map. In the network, P2P self-attention is implemented as axial-attention blocks, which are more efficient than global 2D attention on high-resolution feature maps.

3.2.4. Decoder Block and Output Heads

The decoder block is stacked $L$ times, iterating through output strides (4, 8, and 16 [63,64]) multiple times at each decoding resolution. It merges features by performing bilinear resizing, simple summation, and applying either convolutional blocks or transformer blocks before moving to the next resolution. While it shares similarities with feature pyramid networks [65,66] designed for pyramidal anchor predictions [67], the purpose of our decoder block is solely to aggregate multi-scale features without directly using intermediate pyramidal features for prediction.

The output heads are designed to make predictions from the processed features. Following the stacked decoder, two fully connected layers (2FC) and a softmax function predict mask classes using the memory feature of length $N$. For mask prediction, the decoder block is followed by 2FC to obtain a memory path mask feature ($f$). The decoder output at stride 4 passes through two convolution layers (2Conv) to generate the normalized pixel path feature ($g$). The predicted mask is then obtained from the multiplication of $f$ and $g$, where $f \in \mathbb{R}^{N \times D}$ and $g \in \mathbb{R}^{D \times \frac{H}{4} \times \frac{W}{4}}$.

3.2.5. Combining Outputs for Panoptic Segmentation

The network directly predicts class-labeled masks using a mask transformer, which outputs a set of instance masks and a semantic mask. The instance masks represent the pixels belonging to each object instance in the scene, while the semantic mask represents the pixels belonging to each semantic class.

To obtain the final panoptic segmentation, the instance masks and the semantic mask are combined using a post-processing step. Specifically, the instance masks are first grouped into object instances using a clustering algorithm, such as mean-shift or DBSCAN. The resulting object instances are then assigned a unique instance ID, used to distinguish them from other object instances in the radiographs.

Next, the semantic mask is merged with the instance masks to obtain the final panoptic segmentation of teeth. This is achieved by assigning each pixel in the semantic mask to the object instance to which it belongs, based on the instance ID of the corresponding pixel in the instance masks.

*3.3. Loss Function*

For training, we used a main loss function and auxiliary losses. Panoptic segmentation comprises two main tasks: segmentation and recognition. Therefore, an optimal loss function should check the quality of both. Our main loss function is a product of recognition quality (RQ) and segmentation quality (SQ). The loss function basically maximises a similarity metric over matched masks. One-to-one bipartite matching between the predicted and ground truth masks is performed first, followed by the computation of the similarity metric that can be given as:

$$\mathrm{sim}(y_i, \hat{y}_j) = \hat{p}_j(c_i) \times \mathrm{Dice}(m_i, \hat{m}_j) \tag{5}$$

where $\mathrm{sim}(\cdot, \cdot)$ is the mask similarity metric between class-labelled ground truth mask $y_i = (m_i, c_i)$ and predicted mask $\hat{y}_j = (\hat{m}_j, \hat{p}_j(c))$. The similarity metric ranges between 0 and 1. The value will be 0 when the class is wrong or the masks do not overlap, while it will be 1 when both the classes and masks match precisely. For mask matching, each predicted mask is matched with the ground truth until maximum total similarity is achieved using one-to-one bipartite matching, which is given as:

$$\hat{\sigma} = \underset{\sigma \in \mathfrak{S}_N}{\arg\max} \sum_{i=1}^{K} \mathrm{sim}\left(y_i, \hat{y}_{\sigma(i)}\right) \tag{6}$$

where $\{\hat{y}_i\}_{i=1}^{N}$ and $\{y_i\}_{i=1}^{K}$ are the prediction and ground truth sets, respectively, and $\sigma \in \mathfrak{S}_N$ is the permutation of $N$ elements that best assigns the predictions to obtain maximum similarity. Considering the similarity metric and the mask-matching process, the loss function can be given as:

$$\begin{aligned} \mathcal{L}_{\mathrm{PQ}}^{\mathrm{pos}} = & \sum_{i=1}^{K} \underbrace{\hat{p}_{\hat{\sigma}(i)}(c_i)}_{\text{weight}} \cdot \underbrace{\left[-\mathrm{Dice}\left(m_i, \hat{m}_{\hat{\sigma}(i)}\right)\right]}_{\text{Dice loss}} \\ & + \sum_{i=1}^{K} \underbrace{\mathrm{Dice}\left(m_i, \hat{m}_{\hat{\sigma}(i)}\right)}_{\text{weight}} \cdot \underbrace{\left[-\log \hat{p}_{\hat{\sigma}(i)}(c_i)\right]}_{\text{Cross-entropy loss}} \end{aligned} \tag{7}$$

Intuitively, we optimize the dice loss weighed by class correctness and the cross-entropy loss weighted by mask correctness as we want both class and mask to be correct at the same time. Apart from $\mathcal{L}_{\mathrm{PQ}}^{\mathrm{pos}}$ for positive masks, we define a cross-entropy term $\mathcal{L}_{\mathrm{PQ}}^{\mathrm{neg}}$ for negative (unmatched) masks:

$$\mathcal{L}_{\mathrm{PQ}}^{\mathrm{neg}} = \sum_{i=K+1}^{N} \left[-\log \hat{p}_{\hat{\sigma}(i)}(\varnothing)\right] \tag{8}$$

This term trains the model to predict $\varnothing$ for negative masks. We balance the two terms by $\alpha$ as a common practice to weight positive and negative samples:

$$\mathcal{L}_{\mathrm{PQ}} = \alpha \mathcal{L}_{\mathrm{PQ}}^{\mathrm{pos}} + (1-\alpha)\mathcal{L}_{\mathrm{PQ}}^{\mathrm{neg}} \tag{9}$$

where $\mathcal{L}_{\mathrm{PQ}}$ denotes our final PQ-style loss. In addition to the PQ-style loss, we also use three other losses: (1) Instance discrimination, used while learning feature maps. This loss helps cluster decoder features into instances. (2) Mask ID cross entropy, helps classify each pixel into $N$ masks. (3) Semantic segmentation loss, helps in separating the final mask features.

### 3.4. *Experimental Setup*

#### 3.4.1. Training

All experiments were conducted using the UFBA-UESC dataset. The proposed network was implemented with the Tensorflow framework. Training was performed on an NVIDIA RTX Titan GPU for 500 epochs.

#### 3.4.2. Evaluation Parameters

The following evaluation metrics were used to compare our results with state-of-the-art segmentation models, where the *F*1 *score* was mainly used as a reference since it can give a better estimation of overall performance.

$$Accuracy = \frac{TP + TN}{TP + FN + TN + FP} \tag{10}$$

$$Specificity = \frac{TN}{TN + FP} \tag{11}$$

$$Precision = \frac{TP}{TP + FP} \tag{12}$$

$$Recall = \frac{TP}{TP + FN} \tag{13}$$

$$F1\ Score = \frac{2 \times Precision \times Recall}{Precision + Recall} \tag{14}$$

## 4. Results

We evaluate the performance of our proposed network on the UFBA-UESC Dental Images dataset. Our analysis includes both quantitative and qualitative assessments, comparing our results to those of other state-of-the-art techniques. This section provides a comprehensive discussion of our evaluation results. Figure 3 presents a visual comparison of instance segmentation results produced by various networks (i.e., PANet, HTC, Mask R-CNN, ResNet, and our approach) alongside the ground truth.

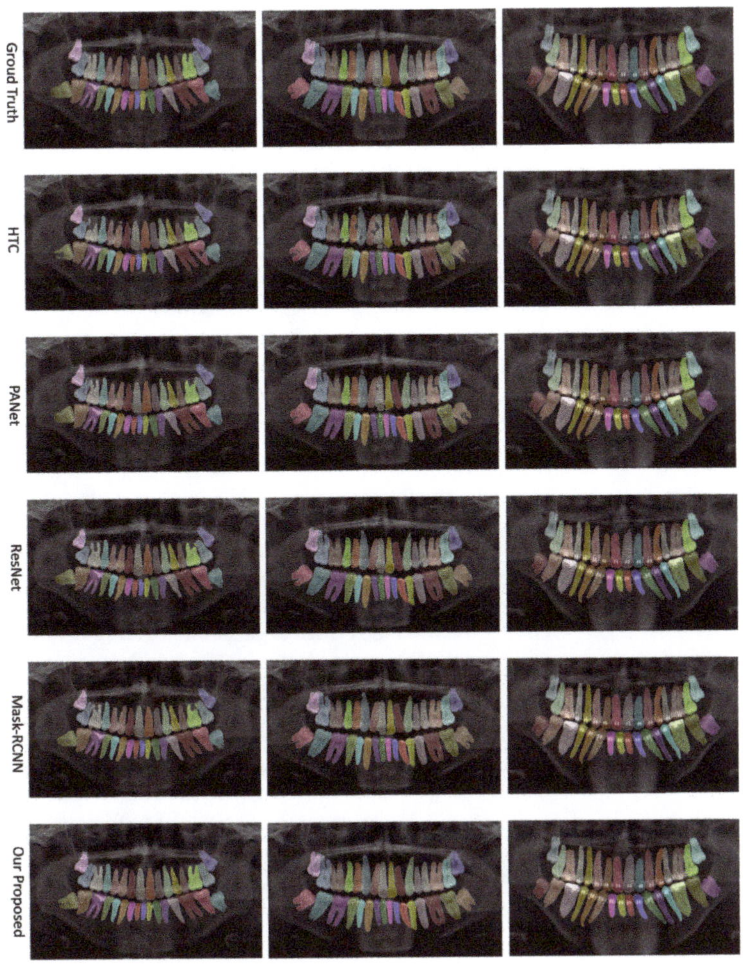


**Figure 3.** Comparison of teeth instance segmentation results for various networks—PANet, HTC, Mask R-CNN, ResNet, and our proposed approach—alongside the ground truth.

### 4.1. Ablation Study

We also performed an ablation study to understand the contribution of different components of our network better. This study focused on a subset of the dataset and examined changes in the *F1-score, Precision,* and *Recall* as we removed different components. We have summarized the results in Table 4.

**Table 4.** Ablation study results.

| Component Removed | Accuracy | F1-Score | Precision | Recall |
|---|---|---|---|---|
| None (Full model) | 97.25 | 93.47 | 95.13 | 93.92 |
| Transformer Block | 95.68 | 91.34 | 92.81 | 90.53 |
| Stacked Decoder | 95.04 | 90.12 | 91.57 | 88.84 |
| Output Heads | 94.12 | 88.90 | 90.36 | 87.66 |
| Pixel-to-Memory | 95.32 | 90.77 | 92.20 | 89.48 |
| Memory-to-Pixel | 95.56 | 91.22 | 92.62 | 89.97 |

The ablation study provides valuable insights into the performance impact of each network component. For instance, the transformer block greatly enhances the performance by enabling efficient bi-directional communication between the pixel path CNN and memory path. Similarly, the stacked decoder, which plays a critical role in aggregating multi-scale features, helps to improve the accuracy of the segmentation output. The output heads are responsible for predicting mask classes and have a direct impact on the network's performance. The pixel-to-memory (P2M) feedback attention, a component of the transformer block, allows for the selective aggregation of information from memory, enabling the model to capture context-aware features, thus leading to improved teeth segmentation. Both the memory-to-pixel (M2P) and memory-to-memory (M2M) self-attention mechanisms demonstrated their significance by capturing long-range dependencies within the memory path and providing global context information.

### 4.2. Qualitative Analysis

To further substantiate our comparison, we visualized the results from our proposed model. Figure 3 displays the instance segmentation results of various networks compared to the ground truth. Our method demonstrates closer alignment with the ground truth, indicating better performance in teeth instance segmentation tasks compared to the other methods. Notably, our proposed network maintains a consistent performance across all teeth, unlike the other networks. The synergistic benefits of the two tasks, SS and affinity pyramid, primarily drive the improvement in instance segmentation performance. Figure 4 depicts the results of panoptic segmentation with the background class (semantic segmentation) and tooth classes (instance segmentation). Figure 5 presents the precision–recall curve, which is the average of precision and recall for all classes. Panoptic segmentation improves the Dice score by also considering the surrounding tissues of teeth; thus, the loss also takes into account the background segmentation to yield better results.

### 4.3. Comparison with State-of-the-Art Models

Next, we compared our model with state-of-the-art approaches in the context of instance segmentation and SS. Table 5 demonstrates that our proposed framework outperforms all previously proposed methods. Mask R-CNN [30] and the TSAS-Net [55] have both been utilized for teeth segmentation, while PANet [7] has achieved state-of-the-art results. However, our approach surpasses these existing methods by capturing hidden patterns more effectively and providing more accurate segmentation of human teeth, even in challenging scenarios like overlapping teeth masks.

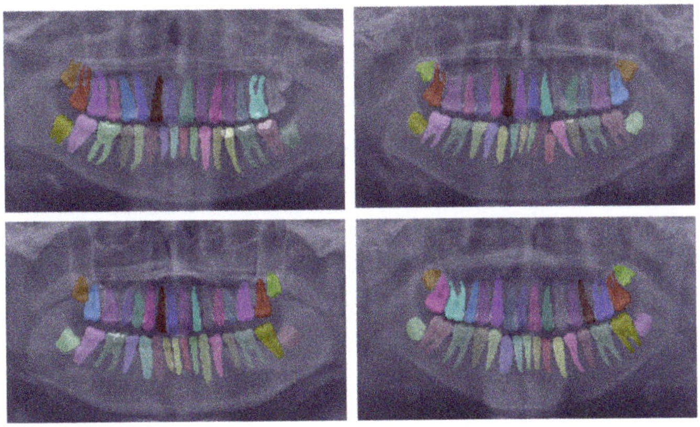

**Figure 4.** Showcasing the best panoptic segmentation results that encompass both the semantic segmentation of the background class and the instance segmentation of the teeth classes.

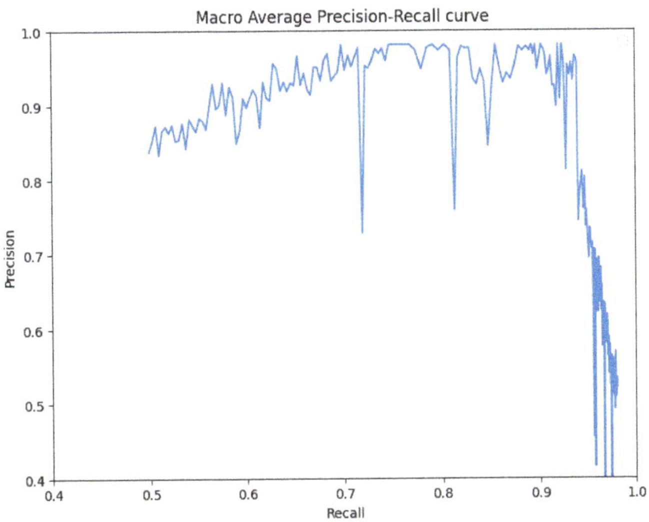


**Figure 5.** Precision–recall curve.

**Table 5.** Comparison with state-of-the-art methods, the best results are indicated in bold.

| Method | Accuracy | Specificity | Precision | Recall | F1-Score | mAvP | AvP50 | AvP75 |
|---|---|---|---|---|---|---|---|---|
| Mask R-CNN [30] | 92.08 | 96.12 | 83.73 | 76.19 | 79.44 | 66.4 ± 0.7 | 96.9 ± 0.2 | 85.1 ± 1.0 |
| TSAS-Net [55] | 96.94 | 97.81 | 94.97 | 93.77 | 92.72 | 70.9 ± 0.1 | 97.7 ± 0.1 | 89.7 ± 0.5 |
| PANet [7] | 96.7 | 98.7 | 94.4 | 89.1 | 91.6 | 71.3 ± 0.3 | 97.5 ± 0.3 | 88.0 ± 0.2 |
| HTC | 96 | 98.5 | 93.7 | 85.9 | 89.6 | 63.7 ± 1.4 | 97.0 ± 0.0 | 82.2 ± 2.0 |
| UNet | 96.04 | 97.68 | 89.89 | 90.18 | 89.33 | 67.0 ± 0.5 | 96.3 ± 0.2 | 87.7±0.9 |
| Ours | **97.25** | **97.65** | **95.13** | **93.92** | **93.47** | **71.5 ± 0.2** | **98.1 ± 0.4** | **89.2 ± 0.1** |

We further evaluated the performance of our proposed method in comparison to previously published studies related to teeth segmentation in panoramic radiographs. Table 6 summarizes the results, which underscore the remarkable performance of our proposed scheme. Given the impressive performance of our framework, as substantiated by our experimental results, we assert that our proposal has established a new state of the art in teeth segmentation.

**Table 6.** Comparison with previously published studies, the best results are indicated in bold.

| Method | Accuracy | Specificity | Precision | Recall | F1-Score |
|---|---|---|---|---|---|
| Wirtz et al. [51] | – | – | 79 | 82.7 | 80.3 |
| Lee et al. [53] | – | – | 85.8 | 89.3 | 87.5 |
| Arora et al. [57] | 96.06 | **99.92** | 95.01 | 93.06 | 91.6 |
| Fatima et al. [68] | – | – | 86 | 87 | 84 |
| Karaoglu et al. [69] | – | – | 93.33 | 93.33 | 93.16 |
| Proposed Method | **97.25** | 97.65 | **95.13** | **93.92** | **93.47** |

### 4.4. Limitations

Our proposed method seeks to achieve instance segmentation of teeth in panoramic radiographs by leveraging an end-to-end model specifically designed for panoptic segmentation. This innovative approach unifies semantic and instance segmentation tasks, introducing a dual-path architecture that adds a global memory path to the conventional CNN path. This unique setup facilitates direct communication across all CNN layers. The architecture explicitly crafted for panoptic segmentation leverages novel objectives, providing equal treatment to both semantic regions and instance objects. As a result, the proposed scheme significantly enhances the instance segmentation performance of teeth in panoramic radiographs. Despite these notable advancements, the proposed approach does introduce certain challenges. One key limitation lies in its additional computational complexity, which may impede real-time clinical applications. Furthermore, our evaluation of the proposed method relies solely on a single dataset. This limited scope constrains a comprehensive assessment of the scheme's generalization capabilities, restricting its potential for a more universally applicable evaluation.

## 5. Conclusions and Future Directions

We have applied a panoptic segmentation strategy to conduct instance segmentation of teeth in panoramic radiographs. Our approach uniquely intertwines the instance segmentation of teeth with the semantic segmentation of the background, enhancing intra-teeth classification and enabling our architecture to accurately distinguish teeth from oral tissue. Our method incorporates an end-to-end deep learning model, which leverages a mask transformer to predict class-labelled masks directly. This is accomplished via a dual-path architecture that introduces an additional global memory path alongside the CNN path, thus enabling direct communication with any CNN layer. We trained our model utilizing a panoptic-quality-inspired loss through bipartite matching. As a result, our proposed framework attains a significantly improved segmentation performance, which also proves beneficial for teeth numbering. The proposed method underwent rigorous evaluation on the publicly accessible UFBA-UESC Dental Image dataset. The experimental results validate that our proposed model outstrips existing state-of-the-art techniques in terms of segmentation performance and robustness.

Looking ahead, our future work aims to further enhance the dual-path-based mask transformer architecture. A key focus will be enabling the numbering of teeth in panoramic radiographs, a crucial factor for accurate tooth identification that significantly aids in diagnosis, treatment planning, and effective communication among dental professionals.

**Author Contributions:** Conceptualization, M.K., M.M.U.R. and M.U.F.; methodology, M.K., M.M.U.R. and M.U.F.; validation, M.K., M.M.U.R. and M.U.F.; formal analysis, M.K., M.M.U.R. and M.U.F.; investigation, M.K., M.M.U.R. and M.U.F.; data curation, M.U.F.; writing—original draft preparation, M.K., M.M.U.R., M.U.F. and D.-K.C.; writing—review and editing, D.-K.C.; visualization, M.M.U.R. and M.U.F.; supervision, D.-K.C.; project administration, D.-K.C. All authors have read and agreed to the published version of the manuscript.

**Funding:** This work was partly supported by (1) the Institute of Information and Communications Technology Planning and Evaluation (IITP) grant funded by the Korean government (MSIT) (no. 2020-0-01373, Artificial Intelligence Graduate School Program (Hanyang University)) and (2) the Bio and Medical Technology Development Program of the National Research Foundation (NRF) funded by the Korean government (MSIT) (no. NRF-2021M3E5D2A01021156).

**Institutional Review Board Statement:** Not applicable.

**Informed Consent Statement:** Not applicable.

**Data Availability Statement:** The data used in this study are openly available in [7] via https://github.com/IvisionLab/dns-panoramic-images-v2 (accessed on 2 May 2023).

**Conflicts of Interest:** The authors declare no conflicts of interest.

## References

1. Nomir, O.; Abdel-Mottaleb, M. Computer-aided diagnostic tool for early detection of periodontal diseases using digital panoramic dental images. *Proc. SPIE Int. Soc. Opt. Eng.* **2007**, *6511*, 65111I.
2. Huang, T.L.; Huang, T.H.; Hsieh, Y.H.; Lee, C.W. Tooth segmentation on dental meshes using morphologic skeleton. *Comput. Methods Programs Biomed.* **2013**, *109*, 69–78.
3. Van Dessel, J.; Nicolielo, L.P.; Huang, Y.; Coudyzer, W.; Salmon, B.; Lambrichts, I.; Maes, F.; Jacobs, R. Automated segmentation of teeth and interproximal contact points from cone beam computed tomography images. *Dento Maxillo Facial Radiol.* **2015**, *44*, 20140315.
4. Al, A.; Ijaz, U.; Song, Y.J.; Lee, S.; Park, S.; Lee, K.W.; Seo, W.B.; Park, K.W.; Han, J.W.; Lee, H. Deep learning for segmentation of 49 regions in 2D and 3D panoramic dental X-ray images. *Dento Maxillo Facial Radiol.* **2018**, *47*, 20170389.
5. Chen, Y.; Mapar, M.; Mohamed, W.A.; Cohen, L.; Jacobs, R.; Huang, T.H.; RamachandraRao, S. Dental biometrics: Human identification using dental radiographs. *Proc. IEEE* **2017**, *105*, 387–398.
6. Khocht, A.; Janal, M.; Turner, B.; Rams, T.E.; Haffajee, A.D. Assessment of periodontal bone level revisited: A controlled study on the diagnostic accuracy of clinical evaluation methods and intra-oral radiography. *J. Clin. Periodontol.* **2008**, *35*, 776–784.
7. Silva, B.; Pinheiro, L.; Oliveira, L.; Pithon, M. A study on tooth segmentation and numbering using end-to-end deep neural networks. In Proceedings of the 2020 33rd SIBGRAPI Conference on Graphics, Patterns and Images (SIBGRAPI), Porto de Galinhas, Brazil, 7–10 November 2020 ; IEEE: Piscataway, NJ, USA, 2020; pp. 164–171.
8. Yang, M.; Yu, K.; Zhang, C.; Li, Z.; Yang, K. Denseaspp for semantic segmentation in street scenes. In Proceedings of the IEEE Conference on Computer Vision and Pattern Recognition, Salt Lake City, UT, USA, 18–23 June 2018; pp. 3684–3692.
9. Xu, X.; Chiu, M.T.; Huang, T.S.; Shi, H. Deep affinity net: Instance segmentation via affinity. *arXiv* **2020**, arXiv:2003.06849.
10. Singh, N.K.; Raza, K. Progress in deep learning-based dental and maxillofacial image analysis: A systematic review. *Expert Syst. Appl.* **2022**, *199*, 116968. [CrossRef]
11. Jader, G.; Fontineli, J.; Ruiz, M.; Abdalla, K.; Pithon, M.; Oliveira, L. Deep instance segmentation of teeth in panoramic X-ray images. In Proceedings of the 2018 31st SIBGRAPI Conference on Graphics, Patterns and Images (SIBGRAPI), Parana, Brazil, 29 October–1 November 2018; IEEE: Piscataway, NJ, USA, 2018; pp. 400–407.
12. Kirillov, A.; He, K.; Girshick, R.; Rother, C.; Dollár, P. Panoptic segmentation. In Proceedings of the IEEE/CVF Conference on Computer Vision and Pattern Recognition, Long Beach, CA, USA, 15–20 June 2019; pp. 9404–9413.
13. Li, X.; Chen, D. A survey on deep learning-based panoptic segmentation. *Digit. Signal Process.* **2022**, *120*, 103283. [CrossRef]
14. Chuang, Y.; Zhang, S.; Zhao, X. Deep learning-based panoptic segmentation: Recent advances and perspectives. *IET IMage Process.* **2023**. [CrossRef]
15. Li, Z.; Wang, W.; Xie, E.; Yu, Z.; Anandkumar, A.; Alvarez, J.M.; Luo, P.; Lu, T. Panoptic segformer: Delving deeper into panoptic segmentation with transformers. In Proceedings of the IEEE/CVF Conference on Computer Vision and Pattern Recognition, New Orleans, LA, USA, 18–24 June 2022; pp. 1280–1289.
16. Xiong, Y.; Liao, R.; Zhao, H.; Hu, R.; Bai, M.; Yumer, E.; Urtasun, R. Upsnet: A unified panoptic segmentation network. In Proceedings of the IEEE/CVF Conference on Computer Vision and Pattern Recognition, Long Beach, CA, USA, 15–20 June 2019; pp. 8818–8826.
17. Grandio, J.; Riveiro, B.; Lamas, D.; Arias, P. Multimodal deep learning for point cloud panoptic segmentation of railway environments. *Autom. Constr.* **2023**, *150*, 104854. [CrossRef]
18. Cheng, B.; Collins, M.D.; Zhu, Y.; Liu, T.; Huang, T.S.; Adam, H.; Chen, L.C. Panoptic-deeplab: A simple, strong, and fast baseline for bottom-up panoptic segmentation. In Proceedings of the IEEE/CVF Conference on Computer Vision and Pattern Recognition, Seattle, WA, USA, 13–19 June 2020; pp. 12475–12485.
19. Tang, C.; Liu, X.; Zheng, X.; Li, W.; Xiong, J.; Wang, L.; Zomaya, A.Y.; Longo, A. DeFusionNET: Defocus blur detection via recurrently fusing and refining discriminative multi-scale deep features. *IEEE Trans. Pattern Anal. Mach. Intell.* **2020**, *44*, 955–968. [CrossRef] [PubMed]
20. Tang, C.; Liu, X.; An, S.; Wang, P. BR2 Net: Defocus Blur Detection Via a Bidirectional Channel Attention Residual Refining Network. *IEEE Trans. Multimed.* **2020**, *23*, 624–635. [CrossRef]

21. Tian, Z.; Shen, C.; Chen, H. Conditional convolutions for instance segmentation. In Proceedings of the European Conference on Computer Vision, Glasgow, UK, 23–28 August 2020; Springer: Cham, Switzerland, 2020; pp. 282–298.
22. Carion, N.; Massa, F.; Synnaeve, G.; Usunier, N.; Kirillov, A.; Zagoruyko, S. End-to-end object detection with transformers. In Proceedings of the European Conference on Computer Vision, Glasgow, UK, 23–28 August 2020; Springer: Cham, Switzerland, 2020; pp. 213–229.
23. LeCun, Y.; Bottou, L.; Bengio, Y.; Haffner, P. Gradient-based learning applied to document recognition. *Proc. IEEE* **1998**, *86*, 2278–2324. [CrossRef]
24. Dey, M.S.; Chaudhuri, U.; Banerjee, B.; Bhattacharya, A. Dual-path morph-UNet for road and building segmentation from satellite images. *IEEE Geosci. Remote Sens. Lett.* **2021**, *19*, 1–5. [CrossRef]
25. Cheng, Y.; Wei, F.; Bao, J.; Chen, D.; Zhang, W. Adpl: Adaptive dual path learning for domain adaptation of semantic segmentation. *IEEE Trans. Pattern Anal. Mach. Intell.* **2023**, *45*, 9339–9356. [CrossRef]
26. Chen, Y.; Li, J.; Xiao, H.; Jin, X.; Yan, S.; Feng, J. Dual path networks. *Adv. Neural Inf. Process. Syst.* **2017**, *30*, 4470–4478.
27. Wang, Y.; Chen, C.; Ding, M.; Li, J. Real-time dense semantic labeling with dual-Path framework for high-resolution remote sensing image. *Remote Sens.* **2019**, *11*, 3020. [CrossRef]
28. Ronneberger, O.; Fischer, P.; Brox, T. U-net: Convolutional networks for biomedical image segmentation. In Proceedings of the International Conference on Medical Image Computing and Computer-Assisted Intervention, Munich, Germany, 5–9 October 2015; Springer: Cham, Switzerland, 2015; pp. 234–241.
29. Newell, A.; Yang, K.; Deng, J. Stacked hourglass networks for human pose estimation. In Proceedings of the European Conference on Computer Vision, Amsterdam, The Netherlands, 11–14 October 2016; Springer: Cham, Switzerland, 2016; pp. 483–499.
30. Silva, G.; Oliveira, L.; Pithon, M. Automatic segmenting teeth in X-ray images: Trends, a novel data set, benchmarking and future perspectives. *Expert Syst. Appl.* **2018**, *107*, 15–31. [CrossRef]
31. Lin, P.L.; Lai, Y.H.; Huang, P.W. An effective classification and numbering system for dental bitewing radiographs using teeth region and contour information. *Pattern Recognit.* **2010**, *43*, 1380–1392. [CrossRef]
32. Lin, P.L.; Lai, Y.H.; Huang, P.W. Dental biometrics: Human identification based on teeth and dental works in bitewing radiographs. *Pattern Recognit.* **2012**, *45*, 934–946. [CrossRef]
33. Chandran, V.; Nizar, G.S.; Simon, P. Segmentation of dental radiograph images. In Proceedings of the Third International Conference on Advanced Informatics for Computing Research, Shimla, India, 15–16 June 2019; pp. 1–5.
34. Shin, S.; Kim, Y. A Study on Automatic Tooth Root Segmentation For Dental CT Images. *J. Soc. e-Bus. Stud.* **2014**, *19*, 45–60. [CrossRef]
35. Gan, Y.; Xia, Z.; Xiong, J.; Zhao, Q.; Hu, Y.; Zhang, J. Toward accurate tooth segmentation from computed tomography images using a hybrid level set model. *Med. Phys.* **2015**, *42*, 14–27. [CrossRef] [PubMed]
36. Nomir, O.; Abdel-Mottaleb, M. Fusion of matching algorithms for human identification using dental X-ray radiographs. *IEEE Trans. Inf. Forensics Secur.* **2008**, *3*, 223–233. [CrossRef]
37. Wanat, R.; Frejlichowski, D. A problem of automatic segmentation of digital dental panoramic X-ray images for forensic human identification. In Proceedings of the CESCG 2011: The 15th Central European Seminar on Computer Graphics, Vinicné, Slovakia, 2–4 May 2011; pp. 1–8.
38. Ullah, Z.; Usman, M.; Latif, S.; Khan, A.; Gwak, J. SSMD-UNet: Semi-supervised multi-task decoders network for diabetic retinopathy segmentation. *Sci. Rep.* **2023**, *13*, 9087. [CrossRef]
39. Ullah, Z.; Usman, M.; Gwak, J. MTSS-AAE: Multi-task semi-supervised adversarial autoencoding for COVID-19 detection based on chest X-ray images. *Expert Syst. Appl.* **2023**, *216*, 119475. [CrossRef]
40. Usman, M.; Rehman, A.; Shahid, A.; Latif, S.; Byon, S.S.; Kim, S.H.; Khan, T.M.; Shin, Y.G. MESAHA-Net: Multi-Encoders based Self-Adaptive Hard Attention Network with Maximum Intensity Projections for Lung Nodule Segmentation in CT Scan. *arXiv* **2023**, arXiv:2304.01576.
41. Hossain, M.S.; Al-Hammadi, M.; Muhammad, G. Automatic fruit classification using deep learning for industrial applications. *IEEE Trans. Ind. Inform.* **2018**, *15*, 1027–1034. [CrossRef]
42. Usman, M.; Lee, B.D.; Byon, S.S.; Kim, S.H.; Lee, B.i.; Shin, Y.G. Volumetric lung nodule segmentation using adaptive roi with multi-view residual learning. *Sci. Rep.* **2020**, *10*, 12839. [CrossRef]
43. Rehman, A.; Usman, M.; Shahid, A.; Latif, S.; Qadir, J. Selective Deeply Supervised Multi-Scale Attention Network for Brain Tumor Segmentation. *Sensors* **2023**, *23*, 2346. [CrossRef]
44. Usman, M.; Shin, Y.G. DEHA-Net: A Dual-Encoder-Based Hard Attention Network with an Adaptive ROI Mechanism for Lung Nodule Segmentation. *Sensors* **2023**, *23*, 1989. [CrossRef] [PubMed]
45. Shi, Z.; Zhang, L.; Sun, Y.; Ye, Y. Multiscale multitask deep NetVLAD for crowd counting. *IEEE Trans. Ind. Inform.* **2018**, *14*, 4953–4962. [CrossRef]
46. Usman, M.; Latif, S.; Asim, M.; Lee, B.D.; Qadir, J. Retrospective motion correction in multishot MRI using generative adversarial network. *Sci. Rep.* **2020**, *10*, 4786. [CrossRef] [PubMed]
47. Latif, S.; Asim, M.; Usman, M.; Qadir, J.; Rana, R. Automating motion correction in multishot MRI using generative adversarial networks. *arXiv* **2018**, arXiv:1811.09750.
48. Usman, M.; Rehman, A.; Shahid, A.; Latif, S.; Byon, S.S.; Lee, B.D.; Kim, S.H.; Shin, Y.G.; et al. MEDS-Net: Self-Distilled Multi-Encoders Network with Bi-Direction Maximum Intensity projections for Lung Nodule Detection. *arXiv* **2022**, arXiv:2211.00003.

49. Latif, S.; Usman, M.; Manzoor, S.; Iqbal, W.; Qadir, J.; Tyson, G.; Castro, I.; Razi, A.; Boulos, M.N.K.; Weller, A.; et al. Leveraging data science to combat COVID-19: A comprehensive review. *IEEE Trans. Artif. Intell.* **2020**, *1*, 85–103. [CrossRef]
50. Koch, T.L.; Perslev, M.; Igel, C.; Brandt, S.S. Accurate segmentation of dental panoramic radiographs with U-Nets. In Proceedings of the 2019 IEEE 16th International Symposium on Biomedical Imaging (ISBI 2019), Venice, Italy, 8–11 April 2019; IEEE: Piscataway, NJ, USA, 2019; pp. 15–19.
51. Wirtz, A.; Mirashi, S.G.; Wesarg, S. Automatic teeth segmentation in panoramic X-ray images using a coupled shape model in combination with a neural network. In Proceedings of the International Conference on Medical Image Computing and Computer-Assisted Intervention, Granada, Spain, 16–20 September 2018; Springer: Cham, Switzerland, 2018; pp. 712–719.
52. Zhang, K.; Wu, J.; Chen, H.; Lyu, P. An effective teeth recognition method using label tree with cascade network structure. *Comput. Med. Imaging Graph.* **2018**, *68*, 61–70. [CrossRef]
53. Lee, J.H.; Han, S.S.; Kim, Y.H.; Lee, C.; Kim, I. Application of a fully deep convolutional neural network to the automation of tooth segmentation on panoramic radiographs. *Oral Surg. Oral Med. Oral Pathol. Oral Radiol.* **2020**, *129*, 635–642. [CrossRef]
54. Muresan, M.P.; Barbura, A.R.; Nedevschi, S. Teeth Detection and Dental Problem Classification in Panoramic X-Ray Images using Deep Learning and Image Processing Techniques. In Proceedings of the 2020 IEEE 16th International Conference on Intelligent Computer Communication and Processing (ICCP), Cluj-Napoca, Romania, 3–5 September 2020; IEEE: Piscataway, NJ, USA, 2020; pp. 457–463.
55. Zhao, Y.; Li, P.; Gao, C.; Liu, Y.; Chen, Q.; Yang, F.; Meng, D. TSASNet: Tooth segmentation on dental panoramic X-ray images by Two-Stage Attention Segmentation Network. *Knowl.-Based Syst.* **2020**, *206*, 106338. [CrossRef]
56. Kong, Z.; Xiong, F.; Zhang, C.; Fu, Z.; Zhang, M.; Weng, J.; Fan, M. Automated Maxillofacial Segmentation in Panoramic Dental X-Ray Images Using an Efficient Encoder-Decoder Network. *IEEE Access* **2020**, *8*, 207822–207833. [CrossRef]
57. Arora, S.; Tripathy, S.K.; Gupta, R.; Srivastava, R. Exploiting multimodal CNN architecture for automated teeth segmentation on dental panoramic X-ray images. *Proc. Inst. Mech. Eng. Part H J. Eng. Med.* **2023**, *237*, 395–405. [CrossRef] [PubMed]
58. Almalki, A.; Latecki, L.J. Self-Supervised Learning with Masked Image Modeling for Teeth Numbering, Detection of Dental Restorations, and Instance Segmentation in Dental Panoramic Radiographs. In Proceedings of the IEEE/CVF Winter Conference on Applications of Computer Vision, Waikoloa, HI, USA, 2–7 January 2023; pp. 5594–5603.
59. Hou, S.; Zhou, T.; Liu, Y.; Dang, P.; Lu, H.; Shi, H. Teeth U-Net: A segmentation model of dental panoramic X-ray images for context semantics and contrast enhancement. *Comput. Biol. Med.* **2023**, *152*, 106296. [CrossRef]
60. Shubhangi, D.; Gadgay, B.; Fatima, S.; Waheed, M. Deep Learning and Image Processing Techniques applied in Panoramic X-Ray Images for Teeth Detection and Dental Problem Classification. In Proceedings of the 2022 International Conference on Emerging Trends in Engineering and Medical Sciences (ICETEMS), Nagpur, India, 18–19 November 2022; IEEE: Piscataway, NJ, USA, 2022; pp. 64–68.
61. Datta, S.; Chaki, N.; Modak, B. A novel technique for dental radiographic image segmentation based on neutrosophic logic. *Decis. Anal. J.* **2023**, *7*, 100223. [CrossRef]
62. Wang, H.; Zhu, Y.; Adam, H.; Yuille, A.L.; Chen, L. MaX-DeepLab: End-to-End Panoptic Segmentation with Mask Transformers. *arXiv* **2020**, arXiv:2012.00759.
63. Liu, C.; Chen, L.C.; Schroff, F.; Adam, H.; Hua, W.; Yuille, A.L.; Fei-Fei, L. Auto-deeplab: Hierarchical neural architecture search for semantic image segmentation. In Proceedings of the IEEE/CVF Conference on Computer Vision and Pattern Recognition, Long Beach, CA, USA, 15–20 June 2019; pp. 82–92.
64. Chen, L.C.; Zhu, Y.; Papandreou, G.; Schroff, F.; Adam, H. Encoder-decoder with atrous separable convolution for semantic image segmentation. In Proceedings of the European Conference on Computer Vision (ECCV), Munich, Germany, 8–14 September 2018; pp. 801–818.
65. Lin, T.Y.; Dollár, P.; Girshick, R.; He, K.; Hariharan, B.; Belongie, S. Feature pyramid networks for object detection. In Proceedings of the IEEE Conference on Computer Vision and Pattern Recognition, Honolulu, HI, USA , 21–26 July 2017; pp. 2117–2125.
66. Liu, S.; Qi, L.; Qin, H.; Shi, J.; Jia, J. Path aggregation network for instance segmentation. In Proceedings of the IEEE Conference on Computer Vision and Pattern Recognition, Salt Lake City, UT, USA, 18–23 June 2018; pp. 8759–8768.
67. Liu, W.; Anguelov, D.; Erhan, D.; Szegedy, C.; Reed, S.; Fu, C.Y.; Berg, A.C. Ssd: Single shot multibox detector. In Proceedings of the European Conference on Computer Vision, Amsterdam, The Netherlands, 11–14 October 2016; Springer: Cham, Switzerland, 2016; pp. 21–37.
68. Fatima, A.; Shafi, I.; Afzal, H.; Mahmood, K.; Díez, I.d.l.T.; Lipari, V.; Ballester, J.B.; Ashraf, I. Deep Learning-Based Multiclass Instance Segmentation for Dental Lesion Detection. *Healthcare* **2023**, *11*, 347. [CrossRef]
69. Karaoglu, A.; Ozcan, C.; Pekince, A.; Yasa, Y. Numbering teeth in panoramic images: A novel method based on deep learning and heuristic algorithm. *Eng. Sci. Technol. Int. J.* **2023**, *37*, 101316. [CrossRef]

*bioengineering*

Article

# Multi-Stage Classification of Retinal OCT Using Multi-Scale Ensemble Deep Architecture

**Oluwatunmise Akinniyi [1], Md Mahmudur Rahman [1], Harpal Singh Sandhu [2], Ayman El-Baz [2] and Fahmi Khalifa [3,4,*]**

1 Department of Computer Science, School of Computer, Mathematical and Natural Sciences, Morgan State University, Baltimore, MD 21251, USA; olaki58@morgan.edu (O.A.); md.rahman@morgan.edu (M.M.R.)
2 Bioengineering Department, University of Louisville, Louisville, KY 20292, USA; harpal.sandhu@louisville.edu (H.S.S.); ayman.elbaz@louisville.edu (A.E.-B.)
3 Electronics and Communications Engineering Department, Mansoura University, Mansoura 35516, Egypt
4 Electrical and Computer Engineering Department, Morgan State University, Baltimore MD 21251, USA
* Correspondence: fahmi.khalifa@morgan.edu

**Abstract:** Accurate noninvasive diagnosis of retinal disorders is required for appropriate treatment or precision medicine. This work proposes a multi-stage classification network built on a multi-scale (pyramidal) feature ensemble architecture for retinal image classification using optical coherence tomography (OCT) images. First, a scale-adaptive neural network is developed to produce multi-scale inputs for feature extraction and ensemble learning. The larger input sizes yield more global information, while the smaller input sizes focus on local details. Then, a feature-rich pyramidal architecture is designed to extract multi-scale features as inputs using DenseNet as the backbone. The advantage of the hierarchical structure is that it allows the system to extract multi-scale, information-rich features for the accurate classification of retinal disorders. Evaluation on two public OCT datasets containing normal and abnormal retinas (e.g., diabetic macular edema (DME), choroidal neovascularization (CNV), age-related macular degeneration (AMD), and Drusen) and comparison against recent networks demonstrates the advantages of the proposed architecture's ability to produce feature-rich classification with average accuracy of 97.78%, 96.83%, and 94.26% for the first (binary) stage, second (three-class) stage, and all-at-once (four-class) classification, respectively, using cross-validation experiments using the first dataset. In the second dataset, our system showed an overall accuracy, sensitivity, and specificity of 99.69%, 99.71%, and 99.87%, respectively. Overall, the tangible advantages of the proposed network for enhanced feature learning might be used in various medical image classification tasks where scale-invariant features are crucial for precise diagnosis.

**Keywords:** ensemble learning; OCT; pyramidal network; feature fusion; scale-adaptive

**Citation:** Akinniyi, O.; Rahman, M.M.; Sandhu, H.S.; El-Baz, A.; Khalifa, F. Multi-Stage Classification of Retinal OCT Using Multi-Scale Ensemble Deep Architecture. *Bioengineering* **2023**, *10*, 823. https://doi.org/10.3390/bioengineering10070823

Academic Editors: Paolo Zaffino and Maria Francesca Spadea

Received: 30 May 2023
Revised: 1 July 2023
Accepted: 7 July 2023
Published: 10 July 2023



## 1. Introduction

Specialized non-invasive imaging techniques are extensively utilized in clinical research to detect/diagnose retinal diseases that may lead to vision loss. In practice, different image types are exploited for that purpose, including optical coherence tomography (OCT), fundus photography, OCT angiography (OCTA), etc. The OCT-based imaging technique in particular is widely exploited in clinical practice due to its ability to produce high-resolution cross-sectional images of the retina, which greatly help in the assessment of several retinal diseases [1,2]. However, due to the complexity and variability of the image features, accurate classification of OCT images is challenging. Developing an accurate diagnostic system for diseases is clinically essential for personalized medicine [3]. Furthermore, retinal disease diagnosis is a critical target since it is almost entirely subjective and the appropriate treatment path to effectively manage retina diseases relies on the accuracy of the diagnosis.

Retinal image diagnosis has shown an increased interest recently from various research groups. A large volume of research work has shown promising results in improving the accuracy and efficiency of OCT-based image analysis [4]. The accuracy of OCT image classification has shown considerable promise when using machine learning (ML) techniques. Particularly, the use of deep learning (DL) can optimize solutions to several complex classification problems [5]. DL-based techniques have the potential to perform efficient classification as well as segmentation of various structures (e.g., drusen) and grading of OCT images [6–9].

In recent years, several ML/DL research papers have been published on retinal image classification for various diseases, e.g., age-related macular degeneration (AMD), diabetic retinopathy (DR), diabetic macular edema (DME), and choroidal neovascularization (CNV). A few papers have proposed ensemble methods to improve the overall accuracy of retinal image classification tasks for macular diseases (e.g., AMD, CNV, DR, DME, etc.) by combining multiple DL models. For example, multi-step techniques for DR diagnosis using OCT were proposed by Elgafi et al. [10]. The system sequentially segments the retinal layers, extracts 3D retinal features, and uses a multilayer perceptron (MLP) for classification using the extracted features. In a leave-one-subject-out evaluation, their system achieved an accuracy of 96.81%. A similar approach with the addition of a feature selection step using the Firefly algorithm was proposed in Reference [11] by Özdaş et al. Multiple binary classifications were conducted using two public datasets and achieved a mean accuracy of 0.957 and 0.954, respectively. A multi-scale convolutional mixture of expert (MCME) ensemble models was proposed in Reference [12] by Rasti et al. to separate the normal retina from DME and dry AMD. The authors also introduced a new cost function for discriminative and fast learning. The system has been evaluated on a total of 193 subjects and demonstrated a precision rate and area under the curve (AUC) of 98.86% and 0.9985, respectively. Ai et al. [13] proposed a fusion network (FN)-based disease detection algorithm for retinal OCT images. They utilized InceptionV3, Inception-ResNet, and Xception DL algorithms as base classifiers, each accompanied by an attention mechanism. Multiple prediction–fusion strategies were employed to output the final prediction results. Comparison to other algorithms showed improved accuracy in the classification of the diseases. A shallow network of only five layers was introduced by Ara et al. in Reference [14] for OCT-B scan classification. The authors investigated the effects of image augmentation as well as deeper networks on final classification. The approach reduced computational time by 16.5% based on the model size, and data augmentation yielded improved accuracy.

A study by Tvenning et al. [15] utilized a DL-based method for AMD identification on OCT scans. The neural architecture, so-called OptiNet, integrates classical DL networks and different parallel layer-wise modules created from filter features. The systems have been evaluated on 600 AMD cases and documented the ability of the deep network to detect alterations in retinal scan regions that correspond to the retinal nerve fiber and choroid layers, which can be linked to AMD. Another CNN-based approach for macular disease classification was proposed by Mishra et al. [16]. the authors introduced a deformation-aware attention-based module to encode crucial morphological variations of retinal layers. The proposed module was integrated into a transfer-learning(TL)-based deep network. The main advantage of the proposed approach is that it is void of preprocessing steps, and the results showed superior performance over competing methods. Another attention-based architecture was proposed by Huang et al. in Reference [17]. Due to the ability of their global attention block (GAB) to focus on lesion locations in the OCTs, the authors proposed a lightweight classification network model. Evaluation on the public UCSD dataset has demonstrated superior classification compared to commonly used attention mechanisms. S.-Paima et al. [18] developed a two-stage multi-scale method for classifying AMD-related pathologies using different backbone models. Hierarchical features were extracted from the input images. This end-to-end model employed a single convolutional neural network (CNN) model to extract different-sized features which were then fused for classification. Two sets of datasets were used: 12,649 images from NCH and 108,312 images from UCSD [19]. Using pre-trained ImageNet weights, the model accuracy was 92.0% $\pm$ 1.6%, which was boosted 93.4% $\pm$ 1.4% in stage two by fine-tuning the model.

A multi-scale deep feature fusion (MDFF) approach was introduced by Das et al. [20]. The model leveraged the fusion of features from multiple scales—thereby capturing the inter-scale variations in images in order to introduce discriminative and complementary features—and employed transfer learning to reduce training parameters. TL, however, reduces dependence and has poor adaptation to the differences among different datasets. Similarly, Li et al. [21] used a deep TL-based method to fine-tune the pre-trained VGG-16 in order to classify 109,312 images and thereby obtained a prediction accuracy of 98.6%. The validation dataset was also used as the testing dataset, so the reported performance could be biased, and training the model on inadequate amounts of data makes it susceptible to overfitting.

Wang et al. tested and evaluated five neural network structures for OCT diagnosis [22] (DenseNet121, ResNet50, DPN92, ResNext101, CliqueNet), and VGG16, VGG19, inception-V3 neural networks, and support vector machine (SVM) methods were added in order to improve experimental comparisons. The network was fine-tuned using features extracted from the OCT dataset, and evaluation was carried out using two public datasets of 3231 and 5084 images, respectively. The dataset used for this experiment consists of eyeball images, not just retina images from OCT; thus, the pre-processing required for the screening of images and the size of the block is time-consuming, and training takes much longer.

Smitha et al. [23] introduced a GAN-based system for retinal disorder diagnosis in which the discriminator classifies the image into normal or abnormal categories. Their method employed denoising enhancement of the retinal layers as a pre-processing step. Two datasets were used for evaluation. Overall accuracy was 83.12% on a small dataset (3980 images: DME, dry AMD, and NORMAL) with low training parameters and 92.42% on a larger dataset (83,605 images: CNV, DME, NORMAL, and Drusen) with larger training parameters. The shortcomings of this method are that segmentation output greatly depends on the quality of the ground-truth images and that image denoising has a high probability of overfitting and thus does not enhance the generalization ability of the classifier. Tsuji et al. [24] constructed a network that utilized the capsule network to improve classification accuracy. Their architrave was built on six convolutional layers (CL) and one primary capsule network. Additionally, four CLs were added to the capsule network architecture of two CLs and one fully connected (FC) layer. Their method achieved an accuracy of 99.6%. The network requires a fixed-input image of 512 × 512. Resizing utilized linear interpolation, which causes some undesirable softening of details and can still produce somewhat jagged images.

In order to detect and grade the severity of DR, Reddy et al. [25] introduced a hybrid deep architecture that utilized a modified grey wolf optimizer with variable weights and attention modules to extract disease-specific features. The hybrid system aided in the joint DR–DME classification on the publicly available IDRiD dataset and achieved detection accuracy rates of 96.0%, 93.2%, and 92.23% for DR, DME, and joint DR-DME, respectively. Upadhyay et al. designed a cohesive CNN approach. The shallow-network (five-layered) layers were cohesively linked to allow for a smooth flow of image features, and batch normalization was instilled along with every activity layer. The approach obtained an accuracy of 97.19% for retinal disease detection for four-class classification [26]. A hybrid fully dense fusion CNN (FD-CNN) architecture was developed by Kayadibi et al. [27] to detect retinal diseases. They first employed a dual hybrid speckle reduction filter to diminish OCTs speckle noise followed by the FD-CNN to extract features. The classification was performed by deep SVM (D-SVM) and deep K-nearest neighbor (D-KNN) classifiers. The hybrid FD-CNN showed significant performance improvement compared to the single performance of CNN.

In summary, the existing literature proposes various techniques, and it is important to note that the results of these papers vary depending on the specific task, dataset, and the DL technique used. Most of the existing literature used larger datasets while using pre-trained models, and some methods employed direct fusion for multi-scale predictions. Furthermore, features related to the higher-order reflectivity of the OCT images were not utilized in conjunction with deeper features, and cascaded classification was not investigated. This paper proposes a multi-stage classification of OCT image features that integrates discriminatory features through a multi-resolution feature pyramid with a scale adaptation module. The proposed cascaded multi-stage classification system is divided into two main steps (Figure 1). First, a scale adaptation network module is used to obtain various image scales for ensemble learning. Second, a transfer learning approach is utilized to extract features from OCT images using a pyramidal structure that allows for the extraction of differently scaled features from the same image dataset. Finally, the extracted features from three different scales of input images are fused to produce a single feature for classification. This fused feature has a rich concentration of local and global features at different levels. Using the one-vs.-rest (OVR) classifier, a binary classification of normal vs. abnormal (CNV, DME, or Drusen) is trained at the first stage, and the abnormal outputs are further passed through the same classification pipeline using different classifier algorithms to differentiate the classes in the second stage.

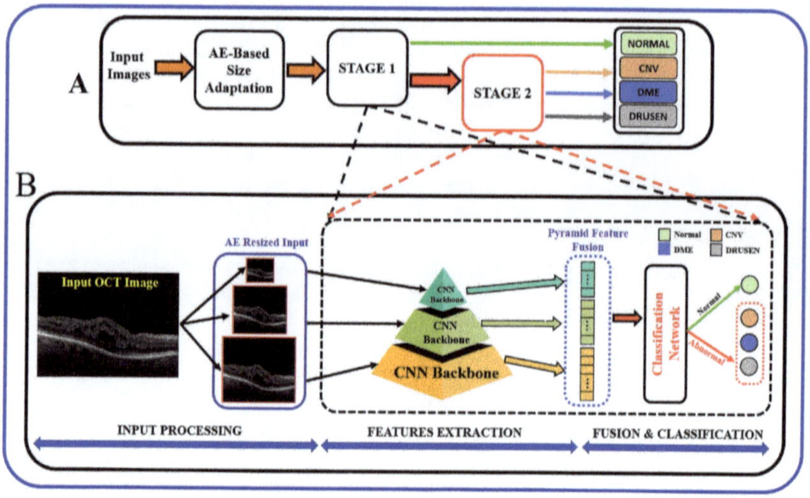


**Figure 1.** Schematic of the proposed multi-stage (**A**) and multi-resolution deep architecture model (**B**) for retinal disorders diagnosis using OCT scans.

The main contributions of this work are as follows: (i) we designed a multi-scale, pyramidal, feature-rich input, as compared to single-scale, through the ensemble/fusion of multi-resolution features for classification; (ii) in order to extract prominent features from the input image, we adopted a scale-adaptive network architecture for generating the multi-scale input images instead of using image resizing; (iii) we utilized a transfer learning technique to extract the features in order to facilitate intermediate feature learning; (iv) we used a two-stage classification approach for a global (binary: normal vs. abnormal) and multi-disease classification overall pipeline fusing both lower- and higher-scale features; (v) we improved classification accuracy for both binary and multi-class scenarios using cross-validation despite the great overlap among the extracted features from the OCT images.

This manuscript is partitioned into four sections. An introduction to OCT and its role in retinal disease diagnosis in modern CAD systems is given in Section 1. This is followed by a relevant review of the recent literature work on this topic as well as the paper's contributions. The materials and methods used along with specifics on the structure of the developed pyramidal architecture are fully detailed in Section 2. The dataset used, the employed performance criteria, the experimental design, the network parameters, the results, and a discussion are given in Section 3. At last, Section 4 provides work conclusions and limitations and future work suggestions.

## 2. Materials and Methods

In order to obtain better predictive performance, we developed a two-stage framework that includes pyramidal feature extraction, multiresolution feature ensemble, and classification. The input to the designed system is retinal OCT images obtained from two publicly available datasets. The proposed architecture provides both global (normal vs. abnormal) and stratified abnormal classifications. The proposed network architecture is schematized in Figure 1 with details described below.

OCT images that are collected from different imaging systems have different sizes, and using TL for the pre-trained network requires downscaling of the input images to fit the employed pre-trained model's input. Unfortunately, downscaling exhibits the loss of important information from images. In order to account for this, we developed an autoencoder (AE)-based resizing module that accepts OCT images of any size and resizes them for use with pre-trained backbones when applying transfer learning. AE networks are considered unsupervised methods (no labels) that learn a latent-space (compressed) representation of the training data. The main advantage of AE neural architecture is its ability to filter out the noise and irrelevant information while reconstructing its output with minimal information losses. In our design, the AE module aims to resize the input images for use as an input in a pre-trained feature extraction ensemble architecture.

The AE module is used to generate three different image scales for the proposed pyramidal feature extraction and ensemble learning (i.e., $224 \times 224 \times 3$, $112 \times 112 \times 3$ and $56 \times 56 \times 3$). The module architecture is shown in Figure 2. The encoding path consists of consecutive convolution and pooling layers, which produce the feature map $F_{AE}$ of size $224 \times 224 \times 3$. $F_{AE}$ is then processed through CL, transposed convolutional, and reshaped to $224 \times 224 \times 12$. Original and processed $F_{AE}$s are integrated using the concatenation layer to produce both high and low-resolution images. The former is generated from $F_{AE}$ and is fed to the pyramidal feature extraction network. The latter is required for the module training phase in order to ensure that the reconstruction error between the module's output and the original input image is minimal, i.e., the network learns important features from the inputs and discards redundancy and noise.

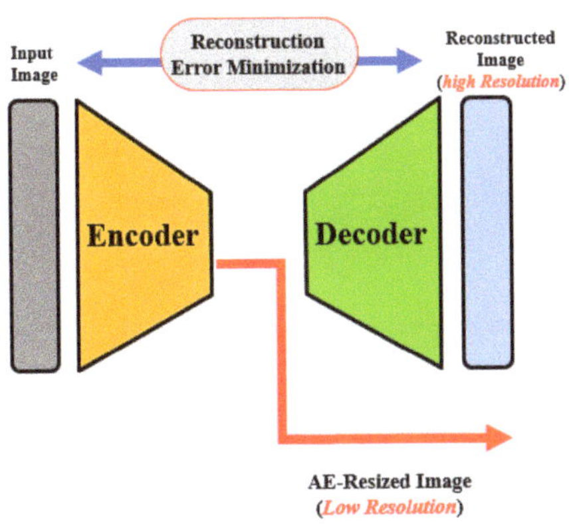


**Figure 2.** Illustration of the autoencoder-based size adaptation network.

For AE module training, a custom loss that combines two pseudo-Huber loss functions and a log-cosh loss function for high resolution and low resolution, respectively, is used.

Pseudo-Huber loss is more robust against outliers. Its behaviors for small and large errors resemble squared and absolute losses, respectively, and are defined mathematically as [28]:

$$_{P}Huber(x) = \delta^2 \left( \sqrt{1 + \left(\frac{x}{\delta}\right)^2} - 1 \right) \quad (1)$$

Here, $x$ is the difference between the actual and predicted values and $\delta$ is a tunable hyper-parameter. On the other hand, the log-cosh loss function $logcosh(x) = log(cosh(x))$ is similar to Huber loss, but it is double differentiable everywhere [29]. Again, $x$ is the difference between the actual and predicted values.

Following the AE-based resizing, the feature extraction step is performed for both global or binary (normal vs. abnormal) as well as for multiclass (CNV vs. DME vs. Drusen) classification of OCT images. At this stage, extraction of discriminating features from the retinal images is performed using pyramidal DL-based architecture. In order to achieve feature-rich classification as compared to single-level networks, a pyramidal DL system is proposed to extract various information to help with multi-class classification tasks; see Figure 1A. Namely, retinal images are resized using the AE module at three different scales (224 × 224, 112 × 112 and 56 × 56). Then, each of the pyramidal CNNs constructs a hierarchical representation of the input images that is then used to build a feature vector which in turn is eventually fused as a feature for the classification task. Although encoders in a wide variety of famous DL networks create a pyramidal feature that can be fused [18], the performance depends on fusion techniques. Thus, we chose to fuse the features of several networks in order to improve the semantic representation of the proposed model.

The proposed architecture, Figure 1, can be seen as a multiresolution feature ensemble in which each CNN path utilizes transfer learning. Transfer learning is a great way to obtain significant results in a classification problem with low data volume. We adopted the pre-trained DenseNet201 model [30] in this work as the backbone of our pyramidal network. DenseNet has performed brilliantly on a variety of datasets and applications where direct connections from all previous layers to all following layers are established; Figure 3. This not only provides ease of training by facilitating feature reuse by different layers and improving connectivity but also increases the variance in later-layer inputs and thus enhances performance [31].

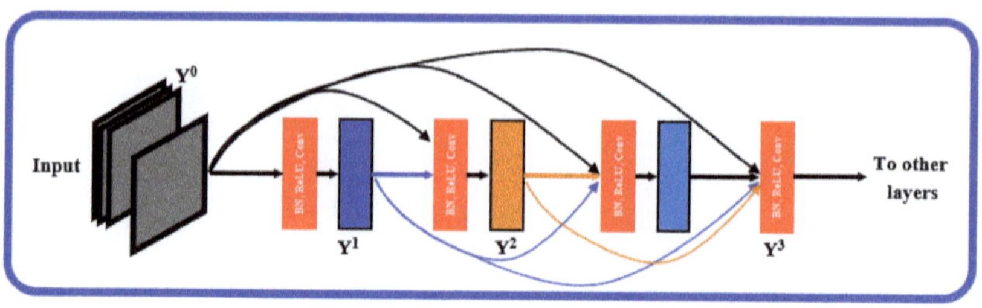


**Figure 3.** layered dense block representing direct connections between layers.

Dense blocks are formed in the network design for downsampling purposes and are separated by layers known as transition layers. The latter help the network to learn intermediate features and consists of batch normalization (BN), 1 × 1 convolution layers, and finally, a 2 × 2 average pooling layer. The BN stabilizes and speeds up the training process. A given feature map at layer $l$ can be described mathematically as $Y^l = R_1([Y^0, Y^1, \ldots\ldots, Y^{l-1}])$ where: $R_1$: is a non-linear transformation comprised of BN, a nonlinearity, and a convolution of 3 × 3. $\left[Y^0, Y^1, \ldots\ldots, Y^{l-1}\right]$ refers to the feature map concatenation corresponding to layers 0 through $(l-1)$ that are incorporated in a single layer.

Another hyperparameter, $k$, specifies the growth rate, or the rate at which the layer's size in individual blocks of the network grows. It can be visualized as a regulator controlling the flow of information in successive layers to reach state-of-the-art results. For instance, when $k = 11$, a filter size of 11 is used at each layer in an individual block. Generally, DenseNet performs well when smaller $k$ are used, as the architecture considers feature maps as the network's global state. As a result, each subsequent layer has access to all previous layers' feature maps. Each layer adds $k$ feature maps to the global state, with the total number of input feature maps at the $l$-th layer $(\text{FM})^l$ is defined as $(\text{FM})' = k^0 + k(l-1)$, where the channels in the input layer are determined by $k^0$.

In order to enhance computational efficiency, a $1 \times 1$ convolution layer is added before each $3 \times 3$ convolution layer (see Figure 4) to reduce the number of input feature maps, which is often greater than the number of $k$ output feature maps [32]. The global pooling layer pools the input features' overall spatial locations at the end of each DenseNet path. The resulting vectors are then used to obtain the feature representations of the training and testing images and are fused for classification.

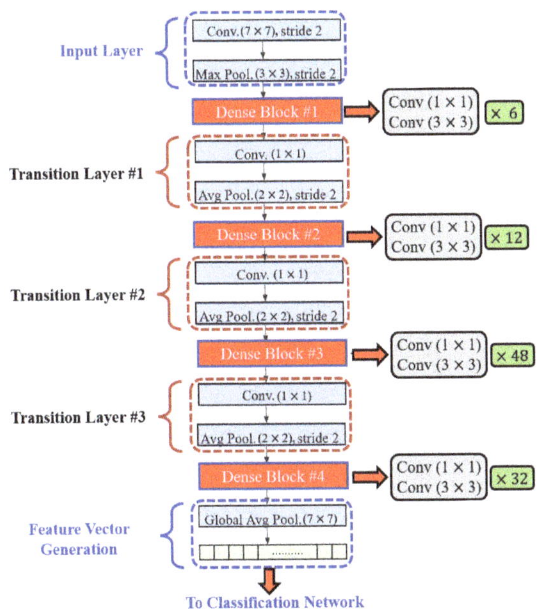


**Figure 4.** Layered Architecture of DenseNet201.

Finally, once all feature vectors for all three CNNs are constructed, they are fused (concatenated) to form predictor variables in a classification network. Features are extracted from pyramidal CNNs at the last layer just before the fully-connected layer. Since we used a pre-trained model, the number of features is typically fixed and is not affected by the input image size or other factors during inference. The size of the feature vectors for the three scales was 1920 individually (5760 after fusion). For classification, we used different classifiers in the first stage (binary) to classify the dataset into normal and abnormal as well as in the second stage (multiclass) to further differentiate the abnormal into three different classes. Namely, we used multilayer perceptron (MLP), logistic regression (LR), SVM, decision tree (DT), random forest (RF), and Naïve Bayes (NB) [33,34]. LR is a predictive analysis classifier that uses the Sigmoid function to predict input features and corresponding weight into a probabilistic output. SVM finds a hyperplane in N-dimensional space (N is a number of features) that distinctly classifies the data points of classes using the maximum margin. Although commonly used in data mining to reach a goal, DT is a supervised learning tree-structured classifier that predicts the value of a target variable by learning simple decision rules inferred from the data features. Similarly to DT, RF builds decision trees from various samples and takes the average to improve the predictive accuracy of that dataset. Finally, NB is a probabilistic ML classifier built on the Bayes theorem that predicts the probability of belonging to the "A" class given that "B" has occurred. The features are independent of each other, bringing about the name Naïve.

## 3. Experimental Results and Discussion

Evaluation to assess the proposed system is performed using various experiments on a UCSD dataset, and both binary and multi-class classification stages have been conducted. The first classification stage (binary) classifies the image as either a normal or abnormal retina, and the second (or the multi-class) stage stratifies the input image as either DME, CNV, or Drusen. The pyramidal CNNs were trained on publicly available datasets [19]. The dataset contains OCT images (Spectralis OCT, Heidelberg Engineering, Germany) from retrospective cohorts of adult patients provided by the Shiley Eye Institute of the University of California San Diego, the California Retinal Research Foundation, Medical Center Ophthalmology Associates, the Shanghai First People's Hospital, and Beijing Tongren Eye Center [19]. About 108K OCTs in total for four classes (CNV: 37,206, DME: 11,349, Drusen: 8617, normal: 51,140) and the testing set containing 1000 retinal OCT images (250 from each class) are available from Reference [35]. We used Jupyter Notebook to implement the software on a Dell Precision 3650 Tower ×64-based workstation with an Intel Core(TM) eight-core CPU running at 2.50 GHz, 64 GB RAM, and with NVIDIA RTX A5000 GPU.

The multilayer perceptron (MLP) pyramidal networks were trained over 50 epochs with a batch size of 128. Additionally, a 5-fold cross-validation strategy was utilized as an unbiased estimator to assess the performance of our ensemble model against other methods. The use of cross-validation partially reduces problems of overfitting or selection bias and also provides insights on how deep architecture will generalize to an independent dataset. Both training and testing data were mixed and cross-validation was employed on the total dataset. All of the dense layers for both the first and second stages used the rectified linear unit (ReLU) as their activation function. Binary cross-entropy for the first stage and sparse categorical cross-entropy for the second stage were utilized as the loss function. An Adam optimizer was employed with a learning rate starting at 0.001, and this was reduced automatically during the training phase in order to improve results whenever the loss metric had stopped improving on both stages. Total network parameters of 1,665,197 out of 1,665,397 parameters were used for training in the first stage and 3,041,711 out of 3,041,911 for the second stage.

We first investigated the first stage for the global (i.e., binary) classification of the retinal images as normal or abnormal. This step mimics human perception of separate groups. Evaluation of the proposed pipeline performance is conducted using known classification metrics, such as accuracy, sensitivity, specificity, and AUC of the receiver operating curve (ROC). Those metrics are defined in terms of experiments' outcomes of true positive (TP), true negative (TN), false positive (FP), and false negative (FN) as follows:

$$\text{Acc} = \frac{\text{TN} + \text{TP}}{\text{TN+TP+FN+FP}}, \text{Sen} = \frac{\text{TP}}{\text{TP+FN}}, \text{and Spc} = \frac{\text{TN}}{\text{TN+FP}} \tag{2}$$

Different ML classifiers were further employed for both stages, and our overall MLP model accuracy performance for both stages is demonstrated in Table 1 for the 5 folds. For the ML classifiers, default parameters were used for the classification. SVM (kernel = 'rbf' and decision function = 'OVR'), DT (criterion = 'gini', splitter = 'best', none for others), RF (criterion = 'gini', estimator = '100'), NB (priors = 'none', smoothing = '1e-9') but for LR (solver = 'liblinear').

As can readily be seen, MLP performed best (97.79% accuracy in the first stage and 96.83% in the second stage) among the other classifiers. This is mainly due to its capability to learn complex nonlinear patterns by amplifying relevant aspects of input data and suppressing irrelevant information [36]. Additionally, confusion matrices were used as an alternative quantitative evaluation. Figure 5 shows our network's confusion matrix for different classifiers in the first stage using 5-fold cross-validation. Network evaluation and monitoring benefit from confusion matrices. From the obtained confusion matrix, other indices such as precision, f1 score, and recall can be derived. For the assessment evaluation of classification models, both the confusion matrix and related metrics are typically employed together.

**Table 1.** Performance of different classifiers for the proposed cascaded classifications all well as for all-at-once (four classes) classification using 5-fold cross validation on the UCSD dataset. LR: logistic regression; SVM: support vector machine; DT: decision tree, RF: random forest; NB: Naïve Bayes, and MLP: multilayer perceptron.

| | First Stage (Binary) | | | | Second Stage (3-Classes) | | | 4-Classes | | |
|---|---|---|---|---|---|---|---|---|---|---|
| **Classifiers** | Acc% | Sen% | Spc% | AUC% | Acc% | Sen% | Spc% | Acc% | Sen% | Spc% |
| MLP | 97.79 | 95.55 | 99.72 | 99.86 | 96.83 | 97.75 | 98.87 | 94.26 | 96.29 | 98.74 |
| LR | 89.23 | 87.00 | 95.77 | 97.47 | 89.34 | 88.69 | 93.99 | 85.95 | 86.08 | 94.91 |
| SVM | 90.33 | 85.80 | 96.29 | 97.98 | 89.47 | 89.68 | 94.56 | 86.53 | 85.72 | 94.79 |
| DT | 80.14 | 69.72 | 89.32 | 78.80 | 69.92 | 67.15 | 81.67 | 65.22 | 65.55 | 85.15 |
| RF | 85.40 | 92.53 | 90.20 | 97.04 | 84.62 | 84.57 | 91.61 | 81.10 | 80.11 | 92.82 |
| NB | 73.71 | 54.04 | 94.70 | 87.90 | 67.46 | 67.46 | 81.00 | 63.82 | 63.75 | 84.35 |

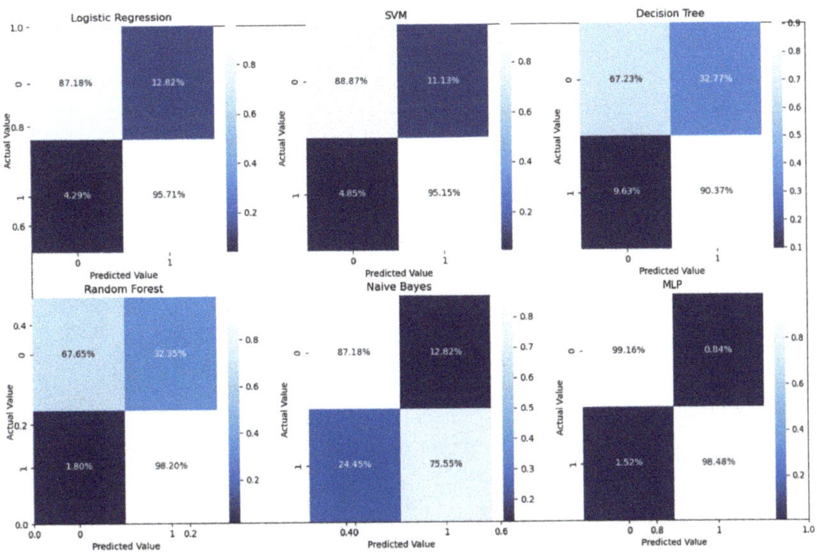


**Figure 5.** Confusion matrices for the first stage using 5-fold cross validation on the UCSD dataset.

Binary classification is an initial step in any treatment procedure by retina specialists. However, personalized medicine would require the determination of the disease and, more appropriately, its grade. Thus, the second set of experiments investigated multi-class classification (DME vs. CNV vs. Drusen). The results for different classifiers are summarized in the middle part of Table 1, and the second stage confusion matrices are depicted in Figure 6. Moreover, in order to demonstrate the efficacy of the pipeline to separate the four classes, we performed an additional experiment using cross-validation on the UCSD dataset. The model accuracy using the evaluation metrics is given in the right part of Table 1, and the confusion matrices are given in Figure 7. Besides accuracy metrics, the system's accuracy and robustness are confirmed using the receiver operating characteristics (ROC) curves in Figure 8. The figure depicts the ROCs for the proposed cascaded classification network for the first stage (Figure 8a), the second stage (Figure 8b), and all-at-once classification (Figure 8c).

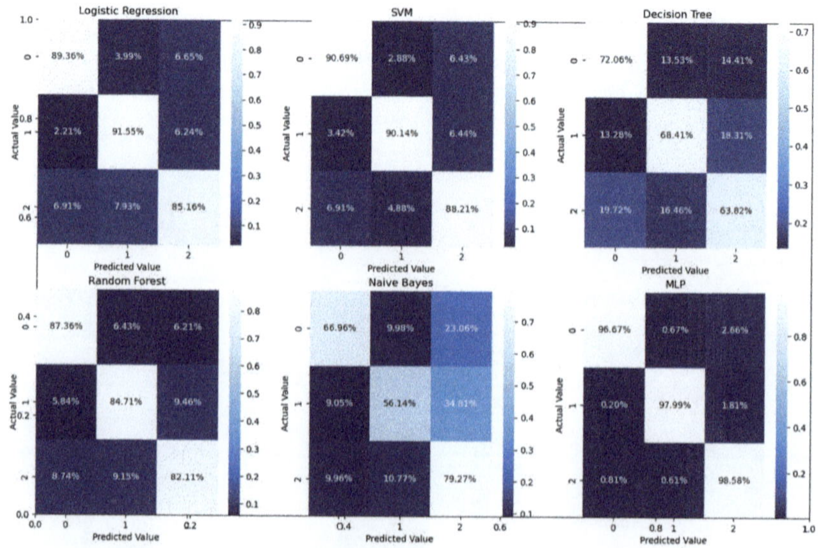


**Figure 6.** Confusion matrices for different classifiers for the second stage (i.e., three classes using 5-fold cross-validation on UCSD data set.

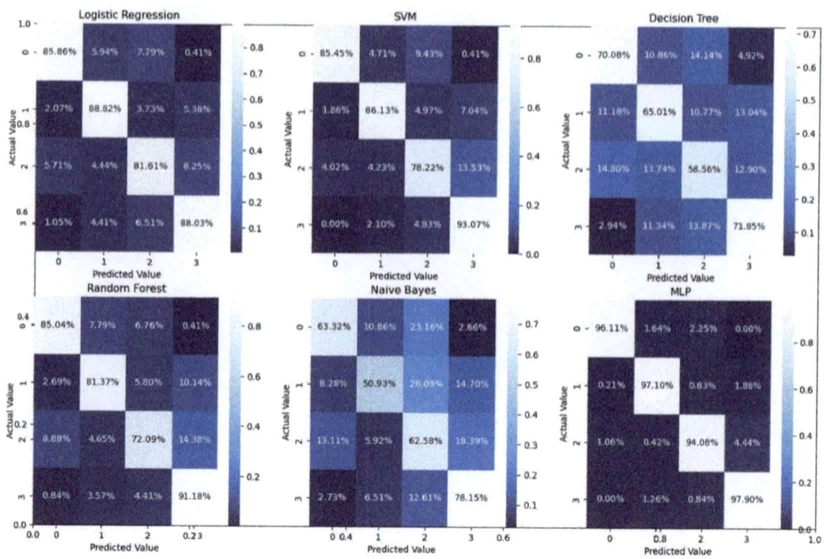


**Figure 7.** Confusion matrices for the four classes using 5-fold cross-validation on the UCSD dataset.

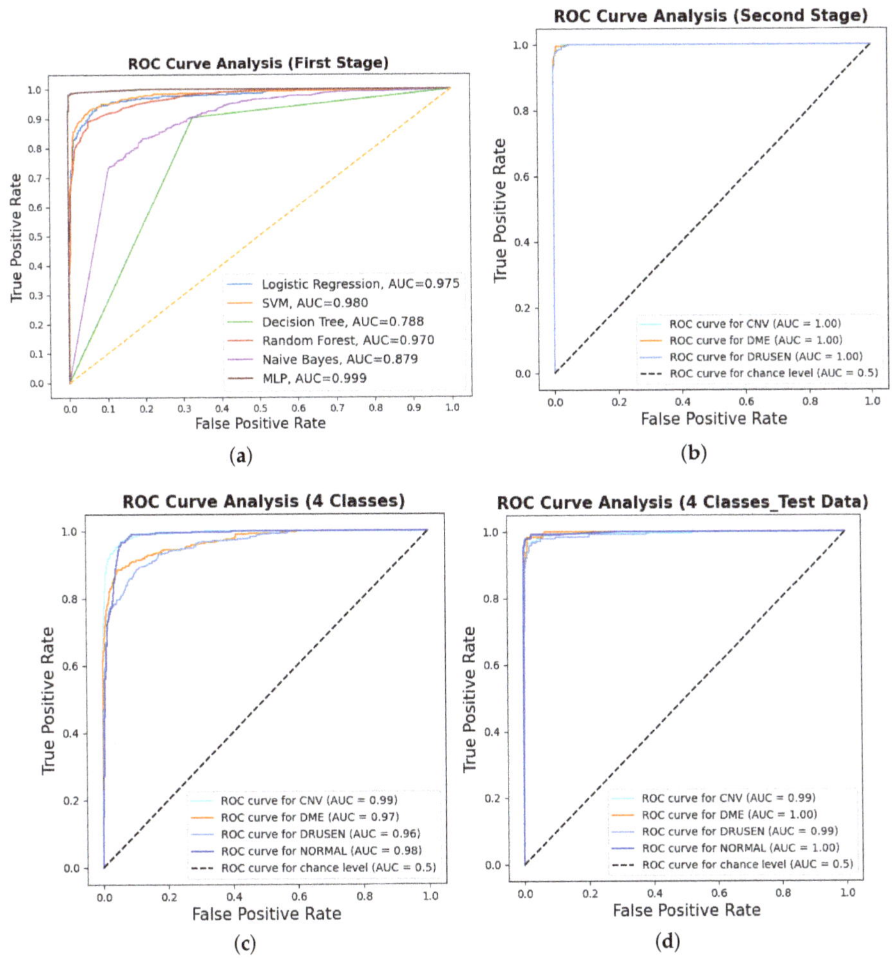


**Figure 8.** The receiver operating characteristic (ROC) curves for the proposed cascaded framework using cross-validation on the UCSD dataset: (**a**) binary classification using different classifiers; (**b**) second-stage classification OVR using the MLP. Furthermore, the figure shows the ROCs for all-at-once four-class classification using the MLP for (**c**) cross-validation and (**d**) test dataset only.

According to Table 1 and the confusion matrices in Figures 5–7, binary classification demonstrated the highest accuracy compared with the second stage and all-at-once classification. This is an important aspect of the presented cascaded classification structure that aligns with clinical diagnostics and emulates the process of a physician's diagnosis. Specifically, the system is designed to initially classify patients into broad groups with a high level of confidence, such as distinguishing between normal and abnormal cases or identifying AMD versus DME. Once patients have been stratified and critical cases have been identified, physicians can then conduct a more comprehensive evaluation using other available clinical signs and biomarkers. This allows for a refined differential diagnosis, moving beyond OCT-based signs alone and towards an accurate and specific diagnosis. Although there is the recent advantage of multi-scale DL-based fusion workflows in many applications, including retinal applications, separating a large number of classes (sub types or grades) at once is a challenging task. This explains the slight reduction in accuracy when the system separates all four classes at once. This, however, can be enhanced in practice by integrating other available clinical signs/biomarkers/images for challenging and complicated retinal diseases, including other diseases.

Our ultimate goal was to design and evaluate a versatile system that can be extended to detect various retinal diseases. In order to explore the benefits of TL, we conducted an additional experiment in which we evaluated several well-known ImageNet-based pre-trained feature extractor architectures as replacements for DenseNet201. The architectures we tested included VGG16, VGG19, Xception, and InceptionV3. The features extracted from these architectures were then fused and used for classification. The results of this experiment are presented in Table 2. The accuracy of the different backbones showed slight variations, with the VGG architectures performing particularly well. These findings demonstrate the potential of our cascaded architecture to leverage various pre-trained models, which can be further improved through fine-tuning. Consequently, our system can be extended to detect other retinal diseases not covered by the datasets used in this study.

**Table 2.** Performance of different feature extractors for the proposed cascaded classifications all well as for all-at-once (four classes) classification using 5-fold cross-validation on the UCSD dataset and multilayer perceptron.

| | First Stage (Binary) | | | Second Stage (3-Classes) | | | 4-Classes | | |
|---|---|---|---|---|---|---|---|---|---|
| **Classifiers** | **Acc%** | **Sen%** | **Spc%** | **Acc%** | **Sen%** | **Spc%** | **Acc%** | **Sen%** | **Spc%** |
| Xception | 95.91 | 98.96 | 95.64 | 91.96 | 96.86 | 98.43 | 93.15 | 97.97 | 99.32 |
| InceptionV3 | 95.34 | 89.16 | 99.50 | 92.21 | 95.64 | 97.80 | 91.76 | 94.24 | 98.01 |
| VGG19 | 95.94 | 97.57 | 99.31 | 93.88 | 95.40 | 97.67 | 93.39 | 97.01 | 99.67 |
| VGG16 | 97.26 | 98.55 | 99.99 | 93.92 | 96.15 | 99.58 | 94.65 | 99.16 | 96.72 |
| DenseNet201 | 97.79 | 95.55 | 99.72 | 96.83 | 97.75 | 98.87 | 94.26 | 96.29 | 98.74 |

All of the above experiments employed cross-validation for the cascaded as well as all-at-once classifications for the four categories in the UCSD dataset. In addition to that, we have further conducted an additional experiment for four-class classification using the train/test data split of the UCSD dataset. The overall accuracies, confusion matrices, and ROCs for the examined classifiers for the four-class classification on the test dataset are given in Table 3, Figures 8d and 9. The results are consistent with the results in Table 1 with a slight accuracy increase of 2%.

**Table 3.** Four-class classification performance using the UCSD test dataset only. LR: logistic regression; SVM: support vector machine; DT: decision tree, RF: random forest; NB: Naïve Bayes; MLP: multilayer perceptron.

| | Metric | | |
|---|---|---|---|
| **Classifier** | **Acc%** | **Sen%** | **Spc%** |
| MLP | 96.17 | 96.17 | 98.69 |
| RF | 95.45 | 94.83 | 98.22 |
| LR | 93.28 | 93.29 | 97.66 |
| SVM | 91.73 | 95.97 | 98.63 |
| DT | 75.92 | 78.51 | 91.64 |
| NB | 79.54 | 79.55 | 92.23 |

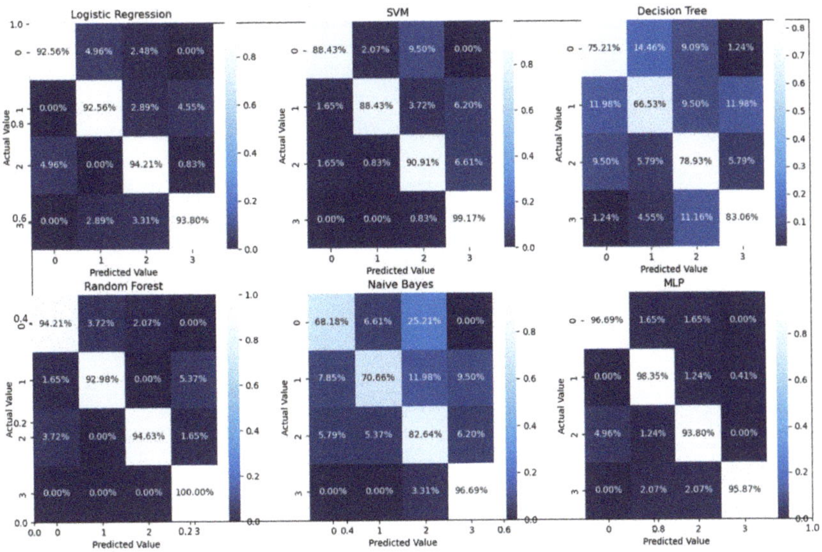

**Figure 9.** Confusion matrices for different classifiers for the four classes using UCSD test data only.

Moreover, the advantage of our system for retinal diseases'/disorders' diagnosis has been compared with standard and recent literature methods. All of the compared networks were tested on the available images in order to compare their abilities for both the multi-class and binary stages. For the first-stage classification, our network performance was compared with traditional methods pre-well-trained on the Imagenet dataset [37] mainly to show the effect of the ensemble learning and scale adaptation network on the overall performance. The comparison included the DenseNet121 by Huang et al. [30], the ResNet101 by Szegedy et al. [38], and the method by Haggag et al. [39], which was designed for retinal image analysis. Since the UCSD dataset does not have ground truth for the retinal layers to compute other local and global feature images, we only used the grayscale images in Reference [39]. For the pre-trained network, the top layer was removed and replaced by a fully connected layer with a dropout of 40% and a final node of the sigmoid activation function for classification. A summary of the performance metrics is given in Table 4. Statistical significance tests were performed using a paired Student's $t$-test to assess the accuracy of the proposed method in comparison to the other methods. The results indicated that our method is statistically significantly better than the compared methods ($p$-value $< 10^{-4}$). Further, an ablation experiment was conducted to verify the effect of the scale adaptation module on the classification performance. For the first- and second-stage classification, our network showed and average accuracy of 95.76% and 94.93%. The overall enhancement ($\sim$2%) was promising, and future work should be conducted to explore other module improvements.

For the four-class comparison, our architecture was compared with well-known CNN models and multiple well-known classification frameworks that reported accuracy on the UCSD dataset. The comparative accuracy is demonstrated in Table 5, and the confusion matrices for the different classifiers are shown in Figure 7. As can readily be seen in Tables 1 and 5, the proposed pipeline showed improved performance compared to its counter and off-the-shelf networks. This is also confirmed using Student's $t$-test, ($p$-values $< 10^{-4}$) similar to the binary classification.

To verify our system performance on other datasets in addition to the UCSD dataset, we tested our approach on the Duke dataset [40], which contains a total of 3231 OCT images for three classes: normal (1407), AMD (723), and DME (1101) patients. The dataset does not have any training and testing splits, so we followed the same approach as was used by Kayadibi et al. in [27], where the train–test split was 90% and 10%, respectively. The proposed pyramidal cascaded architecture results compared with other methods tested on the same dataset are given in Table 6. The results document the better performance of our architecture. These results are encouraging, and we ultimately plan to expand our system in future work to be able to be even more specific, such that we identify not purely signs (e.g., macular edema or CNV), but could actually distinguish between different causes of cystoid macular edema (CME) based on OCT features, such as retinal vein occlusion, diabetic macular, or uveitic macular edema.

**Table 4.** Comparisons with other related work for binary classification on the UCSD data set.

| Method | Acc% | Sen% | Spc% |
|---|---|---|---|
| Haggag et al. [39] | 90.1 | 87.7 | 92.61 |
| Huang et al. [30] | 92.30 | 89.01 | 94.61 |
| Szegedy et al. [38] | 89.12 | 82.3 | 85.18 |
| Proposed | 97.79 | 95.55 | 99.72 |

**Table 5.** Comparisons with other related work for four-class classification using 5-fold cross-validation.

| Applied Method | Acc% | Sen% | Spc% |
|---|---|---|---|
| Fang et al. (JVCIR) [41] | 87.3 | 84.7 | 95.8 |
| Fang et al. [42] | 90.1 | 86.8 | 96.6 |
| S.-Paima et al. [18] | 93.9 | 93.4 | 98.0 |
| Proposed | 94.3 | 96.3 | 98.7 |

**Table 6.** Overall accuracy in comparison with other works tested on the Duke data set.

| Applied Method | Acc% | Sen% | Spc% |
|---|---|---|---|
| Thomas et al. [43] | 96.66 | — | — |
| Amaladevi and Jacob [44] | 96.20 | 96.20 | 99.89 |
| Kayadibi and Güraksın [27] | 97.50 | 97.64 | 98.91 |
| Proposed | 99.69 | 99.71 | 99.87 |

## 4. Conclusions

We have developed a multi-level, multi-resolution feature ensemble architecture for the classification of retinal disorders. The proposed pipeline mimics the human perception of global diagnosis followed by stratification of the suspected cases. The scale-adaptation networks help to produce multi-scale inputs while retaining valuable information when downscaling. Additionally, the pyramidal layout helps extract various information to help with the binary and multi-class classification stages of the three retinal disorders. In summation, the proposed architecture not only provides global diagnosis but also automatically distinguishes between different retinal diseases, thus allowing for earlier treatment of the patient's condition. Despite promising results, some limitations of this work should be addressed in future work. First, the proposed system should be evaluated on more challenging retinal datasets with different diseases for rigorous evaluation. Second, we used only pre-trained CNNs for feature extraction, and thus, more evaluation using visual transformers should be investigated.

Future research venues will explore integrating the architecture into more-complex retinal disorders' pipelines to include, for example, sub-grades of disease (such as dry and wet AMD) for accurate and precision medicine. Further, integration of explainable AI modules (e.g., Grad-CAM, LIME, etc.) to gain further insights into the reasoning behind the systems' output will be explored. Finally, a weighted fusion of the multi-scale features will be thoroughly investigated as well as the study of additional higher-order features using spatial models.

**Author Contributions:** Conceptualization, F.K. and A.E.-B.; methodology, M.M.R., A.E.-B. and F.K.; software, O.A. and F.K.; validation, H.S.S., A.E.-B. and M.M.R.; formal analysis, A.E.-B., H.S.S. and F.K.; investigation, H.S.S., A.E.-B. and F.K.; resources, M.M.R. and F.K.; data curation, O.A. and F.K.; writing—original draft preparation, review and editing, O.A. and F.K.; visualization, O.A. and F.K.; supervision, M.M.R. and F.K.; project administration, F.K.; funding acquisition, M.M.R., A.E.-B. and F.K. All authors have read and agreed to the published version of the manuscript.

**Funding:** This work is supported by the Center for Equitable Artificial Intelligence and Machine Learning Systems (CEAMLS) at Morgan State University under Project #11202202 and by the "CyBR-MSI: IRR" sub-award grant from the American Society for Engineering Education (ASEE) funded by NSF (Award ID # 2139136).

**Institutional Review Board Statement:** Not applicable.

**Informed Consent Statement:** This work used publicly available data.

**Data Availability Statement:** The datasets are publicly available at https://data.mendeley.com/datasets/rscbjbr9sj (accessed on 1 February 2023) and https://people.duke.edu/~sf59/Srinivasan_BOE_2014_dataset.htm (accessed on 1 Febuary 2023).

**Conflicts of Interest:** The authors declare no conflict of interest.

## References


1. Park, J.; Lee, K.P.; Kim, H.; Park, S.; Wijesinghe, R.E.; Lee, J.; Han, S.; Lee, S.; Kim, P.; Cho, D.W.; et al. Biocompatibility evaluation of bioprinted decellularized collagen sheet implanted in vivo cornea using swept-source optical coherence tomography. *J. Biophotonics* **2019**, *12*, e201900098. [CrossRef] [PubMed]
2. Wijesinghe, R.E.; Park, K.; Kim, P.; Oh, J.; Kim, S.W.; Kim, K.; Kim, B.M.; Jeon, M.; Kim, J. Optically deviated focusing method based high-speed SD-OCT for in vivo retinal clinical applications. *Opt. Rev.* **2016**, *23*, 307–315. [CrossRef]
3. MacEachern, S.J.; Forkert, N.D. Machine learning for precision medicine. *Genome* **2021**, *64*, 416–425. [CrossRef] [PubMed]
4. Karn, P.K.; Abdulla, W.H. On Machine Learning in Clinical Interpretation of Retinal Diseases Using OCT Images. *Bioengineering* **2023**, *10*, 407. [CrossRef]
5. Khan, M.; Silva, B.N.; Han, K. Efficiently processing big data in real-time employing deep learning algorithms. In *Deep Learning and Neural Networks: Concepts, Methodologies, Tools, and Applications*; IGI Global: Hershey, PA, USA, 2020; pp. 1344–1357.
6. Haggag, S.; Khalifa, F.; Abdeltawab, H.; Elnakib, A.; Ghazal, M.; Mohamed, M.A.; Sandhu, H.S.; Alghamdi, N.S.; El-Baz, A. An Automated CAD System for Accurate Grading of Uveitis Using Optical Coherence Tomography Images. *Sensors* **2021**, *21*, 5457. [CrossRef]
7. ElTanboly, A.; Ismail, M.; Shalaby, A.; Switala, A.; El-Baz, A.; Schaal, S.; Gimel'farb, G.; El-Azab, M. A computer-aided diagnostic system for detecting diabetic retinopathy in optical coherence tomography images. *Med. Phys.* **2017**, *44*, 914–923. [CrossRef]
8. Murugeswari, S.; Sukanesh, R. Investigations of severity level measurements for diabetic macular oedema using machine learning algorithms. *Ir. J. Med Sci.* **2017**, *186*, 929–938. [CrossRef]
9. Miri, M.S.; Abràmoff, M.D.; Kwon, Y.H.; Sonka, M.; Garvin, M.K. A machine-learning graph-based approach for 3D segmentation of Bruch's membrane opening from glaucomatous SD-OCT volumes. *Med. Image Anal.* **2017**, *39*, 206–217. [CrossRef]
10. Elgafi, M.; Sharafeldeen, A.; Elnakib, A.; Elgarayhi, A.; Alghamdi, N.S.; Sallah, M.; El-Baz, A. Detection of Diabetic Retinopathy Using Extracted 3D Features from OCT Images. *Sensors* **2022**, *22*, 7833. [CrossRef]
11. Özdaş, M.B.; Uysal, F.; Hardalaç, F. Classification of Retinal Diseases in Optical Coherence Tomography Images Using Artificial Intelligence and Firefly Algorithm. *Diagnostics* **2023**, *13*, 433. [CrossRef]
12. Rasti, R.; Rabbani, H.; Mehridehnavi, A.; Hajizadeh, F. Macular OCT classification using a multi-scale convolutional neural network ensemble. *IEEE Trans. Med. Imaging* **2017**, *37*, 1024–1034. [CrossRef]
13. Ai, Z.; Huang, X.; Feng, J.; Wang, H.; Tao, Y.; Zeng, F.; Lu, Y. FN-OCT: Disease detection algorithm for retinal optical coherence tomography based on a fusion network. *Front. Neuroinform.* **2022**, *16*, 876927. [CrossRef]
14. Ara, R.K.; Matiolański, A.; Dziech, A.; Baran, R.; Domin, P.; Wieczorkiewicz, A. Fast and efficient method for optical coherence tomography images classification using deep learning approach. *Sensors* **2022**, *22*, 4675. [CrossRef]
15. Tvenning, A.O.; Hanssen, S.R.; Austeng, D.; Morken, T.S. Deep learning identify retinal nerve fibre and choroid layers as markers of age-related macular degeneration in the classification of macular spectral-domain optical coherence tomography volumes. *Acta Ophthalmol.* **2022**, *100*, 937–945. [CrossRef]
16. Mishra, S.S.; Mandal, B.; Puhan, N.B. MacularNet: Towards fully automated attention-based deep CNN for macular disease classification. *SN Comput. Sci.* **2022**, *3*, 142. [CrossRef]

17. Huang, X.; Ai, Z.; Wang, H.; She, C.; Feng, J.; Wei, Q.; Hao, B.; Tao, Y.; Lu, Y.; Zeng, F. GABNet: Global attention block for retinal OCT disease classification. *Front. Neurosci.* **2023**, *17*, 1143422. [CrossRef]
18. Sotoudeh-Paima, S.; Jodeiri, A.; Hajizadeh, F.; Soltanian-Zadeh, H. Multi-scale convolutional neural network for automated AMD classification using retinal OCT images. *Comput. Biol. Med.* **2022**, *144*, 105368. [CrossRef]
19. Kermany, D.S.; Goldbaum, M.; Cai, W.; Valentim, C.C.; Liang, H.; Baxter, S.L.; McKeown, A.; Yang, G.; Wu, X.; Yan, F.; et al. Identifying medical diagnoses and treatable diseases by image-based deep learning. *Cell* **2018**, *172*, 1122–1131. [CrossRef]
20. Das, V.; Dandapat, S.; Bora, P.K. Multi-scale deep feature fusion for automated classification of macular pathologies from OCT images. *Biomed. Signal Process. Control.* **2019**, *54*, 101605. [CrossRef]
21. Li, F.; Chen, H.; Liu, Z.; Zhang, X.; Jiang, M.; Wu, Z.; Zhou, K. Deep learning-based automated detection of retinal diseases using optical coherence tomography images. *Biomed. Opt. Express* **2019**, *10* 12, 6204–6226. [CrossRef]
22. Wang, D.; Wang, L. On OCT image classification via deep learning. *IEEE Photonics J.* **2019**, *11*, 3900714. [CrossRef]
23. Smitha, A.; Jidesh, P. Detection of retinal disorders from OCT images using generative adversarial networks. *Multimed. Tools Appl.* **2022**, *81*, 29609–29631. [CrossRef]
24. Tsuji, T.; Hirose, Y.; Fujimori, K.; Hirose, T.; Oyama, A.; Saikawa, Y.; Mimura, T.; Shiraishi, K.; Kobayashi, T.; Mizota, A.; et al. Classification of optical coherence tomography images using a capsule network. *BMC Ophthalmol.* **2020**, *20*, 114. [CrossRef] [PubMed]
25. Reddy, V.P.C.; Gurrala, K.K. Joint DR-DME classification using deep learning-CNN based modified grey-wolf optimizer with variable weights. *Biomed. Signal Process. Control.* **2022**, *73*, 103439. [CrossRef]
26. Upadhyay, P.K.; Rastogi, S.; Kumar, K.V. Coherent convolution neural network based retinal disease detection using optical coherence tomographic images. *J. King Saud Univ. Comput. Inf. Sci.* **2022**, *34*, 9688–9695. [CrossRef]
27. Kayadibi, İ.; Güraksın, G.E. An Explainable Fully Dense Fusion Neural Network with Deep Support Vector Machine for Retinal Disease Determination. *Int. J. Comput. Intell. Syst.* **2023**, *16*, 28. [CrossRef]
28. Charbonnier, P.; Blanc-Féraud, L.; Aubert, G.; Barlaud, M. Deterministic edge-preserving regularization in computed imaging. *IEEE Trans. Image Process.* **1997**, *6*, 298–311. [CrossRef]
29. Saleh, R.A.; Saleh, A.K.M.E. Statistical properties of the log-cosh loss function used in machine learning. *arXiv* **2022**, arXiv:2208.04564.
30. Huang, G.; Liu, Z.; Van Der Maaten, L.; Weinberger, K.Q. Densely connected convolutional networks. In Proceedings of the IEEE Conference on Computer Vision and Pattern Recognition, Honolulu, HI, USA, 21–26 July 2017; pp. 4700–4708.
31. Jaiswal, A.; Gianchandani, N.; Singh, D.; Kumar, V.; Kaur, M. Classification of the COVID-19 infected patients using DenseNet201 based deep transfer learning. *J. Biomol. Struct. Dyn.* **2021**, *39*, 5682–5689. [CrossRef]
32. Lodhi, B.; Kang, J. Multipath-DenseNet: A Supervised ensemble architecture of densely connected convolutional networks. *Inf. Sci.* **2019**, *482*, 63–72. [CrossRef]
33. Bonaccorso, G. *Machine Learning Algorithms*; Packt Publishing Ltd.: Birmingham, UK, 2017.
34. Mohammed, M.; Khan, M.B.; Bashier, E.B.M. *Machine Learning: Algorithms and Applications*; CRC Press: Boca Raton, FL, USA, 2016.
35. Large Dataset of Labeled Optical Coherence Tomography (OCT) and Chest X-Ray Images. 2018. Available online: https://data.mendeley.com/datasets/rscbjbr9sj./ (accessed on 1 February 2023).
36. LeCun, Y.; Bengio, Y.; Hinton, G. Deep learning. *Nature* **2015**, *521*, 436–444. [CrossRef] [PubMed]
37. Deng, J.; Dong, W.; Socher, R.; Li, L.J.; Li, K.; Fei-Fei, L. Imagenet: A large-scale hierarchical image database. In Proceedings of the 2009 IEEE Conference on Computer Vision And Pattern Recognition, Miami, FL, USA, 20–25 June 2009; pp. 248–255.
38. Szegedy, C.; Ioffe, S.; Vanhoucke, V.; Alemi, A.A. Inception-v4, inception-resnet and the impact of residual connections on learning. In Proceedings of the Thirty-First AAAI Conference on Artificial Intelligence, San Francisco, CA, USA, 4–9 February 2017.
39. Haggag, S.; Elnakib, A.; Sharafeldeen, A.; Elsharkawy, M.; Khalifa, F.; Farag, R.K.; Mohamed, M.A.; Sandhu, H.S.; Mansoor, W.; Sewelam, A.; et al. A Computer-Aided Diagnostic System for Diabetic Retinopathy Based on Local and Global Extracted Features. *Appl. Sci.* **2022**, *12*, 8326. [CrossRef]
40. Srinivasan, P.P.; Kim, L.A.; Mettu, P.S.; Cousins, S.W.; Comer, G.M.; Izatt, J.A.; Farsiu, S. Fully automated detection of diabetic macular edema and dry age-related macular degeneration from optical coherence tomography images. *Biomed. Opt. Express* **2014**, *5*, 3568–3577. [CrossRef] [PubMed]
41. Fang, L.; Jin, Y.; Huang, L.; Guo, S.; Zhao, G.; Chen, X. Iterative fusion convolutional neural networks for classification of optical coherence tomography images. *J. Vis. Commun. Image Represent.* **2019**, *59*, 327–333. [CrossRef]
42. Fang, L.; Wang, C.; Li, S.; Rabbani, H.; Chen, X.; Liu, Z. Attention to lesion: Lesion-aware convolutional neural network for retinal optical coherence tomography image classification. *IEEE Trans. Med. Imaging* **2019**, *38*, 1959–1970. [CrossRef]
43. Thomas, A.; Harikrishnan, P.; Krishna, A.K.; Palanisamy, P.; Gopi, V.P. A novel multiscale convolutional neural network based age-related macular degeneration detection using OCT images. *Biomed. Signal Process. Control.* **2021**, *67*, 102538. [CrossRef]
44. Amaladevi, S.; Jacob, G. Classification of Retinal Pathologies using Convolutional Neural Network. *Int. J. Adv. Trends Comput. Sci. Eng.* **2020**, *9*, 3865–3869.

bioengineering

MDPI

Article

# Automated Prediction of Osteoarthritis Level in Human Osteochondral Tissue Using Histopathological Images

**Ateka Khader and Hiam Alquran**

Department of Biomedical Systems and Informatics Engineering, Hijjawi Faculty for Engineering Technology, Yarmouk University, Irbid 21163, Jordan
* Correspondence: heyam.q@yu.edu.jo

**Abstract:** Osteoarthritis (OA) is the most common arthritis and the leading cause of lower extremity disability in older adults. Understanding OA progression is important in the development of patient-specific therapeutic techniques at the early stage of OA rather than at the end stage. Histopathology scoring systems are usually used to evaluate OA progress and the mechanisms involved in the development of OA. This study aims to classify the histopathological images of cartilage specimens automatically, using artificial intelligence algorithms. Hematoxylin and eosin (HE)- and safranin O and fast green (SafO)-stained images of human cartilage specimens were divided into early, mild, moderate, and severe OA. Five pre-trained convolutional networks (DarkNet-19, MobileNet, ResNet-101, NasNet) were utilized to extract the twenty features from the last fully connected layers for both scenarios of SafO and HE. Principal component analysis (PCA) and ant lion optimization (ALO) were utilized to obtain the best-weighted features. The support vector machine classifier was trained and tested based on the selected descriptors to achieve the highest accuracies of 98.04% and 97.03% in HE and SafO, respectively. Using the ALO algorithm, the F1 scores were 0.97, 0.991, 1, and 1 for the HE images and 1, 0.991, 0.97, and 1 for the SafO images for the early, mild, moderate, and severe classes, respectively. This algorithm may be a useful tool for researchers to evaluate the histopathological images of OA without the need for experts in histopathology scoring systems or the need to train new experts. Incorporating automated deep features could help to improve the characterization and understanding of OA progression and development.

**Keywords:** osteoarthritis; histopathological; hematoxylin eosin; safranin O fast green; DarkNet-19; MobileNet; NasNet; ResNet-101; ShuffleNet; PCA; ALO

**Citation:** Khader, A.; Alquran, H. Automated Prediction of Osteoarthritis Level in Human Osteochondral Tissue Using Histopathological Images. *Bioengineering* **2023**, *10*, 764. https://doi.org/10.3390/bioengineering10070764

Academic Editors: Paolo Zaffino and Maria Francesca Spadea

Received: 24 May 2023
Revised: 21 June 2023
Accepted: 23 June 2023
Published: 25 June 2023



## 1. Introduction

Osteoarthritis (OA) is the leading cause of pain and disability in working-age adults and the elderly [1,2]. OA is not a process of mechanical wear and tear as previously thought; instead, it is a whole-organ disease that is driven by the disruption of the balance of cartilage homeostasis, inflammatory mediators, genetic factors, and innate immunity [3–5]. Joint destruction in the knee can be severe in OA patients and can lead to total knee replacement (TKR). A better understanding of the pattern and initiation of OA in the knee could help in the understanding of OA progression and influence the selection of therapies.

The histopathology of cartilage is usually used to evaluate the in situ state of the cartilage tissue. Microscopic histopathological grading of osteochondral tissue is usually used to evaluate OA development ex vivo. The most common OA grading systems are the Osteoarthritis Research Society International (OARSI) [6] and Histological-Histochemical Grading System (HHGS) scoring systems [7]. Although the HHGS score system is the most often used for the histological scoring of osteoarthritic cartilage, it is usually used to evaluate the more severe OA specimens [8]. OARSI is the best choice for mild or earlier phases of OA and for investigating the progression of OA. In general, a sensitive grading system that is able to detect early OA and its progression could be of great interest for drug development and OA research [9]. Moreover, the identification of early OA and the progression of OA is important in the development of early interferences and therapeutic techniques that could prevent the progression of OA [10].

Manual histopathological scoring systems can be time-consuming and need pathologists with years of experience and/or the training of new scorers [11]. Automatic OA evaluation and assessment based on histopathological image classification are very limited. Manual scoring systems are widely used for evaluation of the OA histopathological images. Machine learning and deep learning have aided massive data analyses, pattern identification, decision-making, and the production of accurate predictions [12]. Machine learning and deep learning were used for the histopathological grading of different tissues, using magnetic resonance imaging (MRI) [13,14], optical microscopy [15], and ultrasound [16].

The prediction and classification of the OA progression of the osteochondral tissue using machine learning and deep learning have been proposed in the literature; these methods were based on magnetic resonance imaging (MRI) [17,18] and radiography [19]. A deep convolutional neural network (CNN) was used to automatically diagnose hip OA using 420 hip X-ray images [20]. The results showed that the CNN model had 95% sensitivity and 92.8% accuracy as compared to the conventional manual assessment by physicians. In another study, deep learning was used for the automatic segmentation and subregional assessment of MRI images of articular cartilage and compared to manual segmentation [21]. Tiulpin et al. studied the use of deep learning and leveraged an ensemble of residual networks with 50 layers to predict OARSI and Kellgren–Lawrence (KL) grades of OA from knee radiographs [22]. The detection of the presence of OA using their model yielded an average precision of 0.98 and an area under the ROC curve (AUC) of 0.98.

However, few studies have looked at automation in the grading of histopathological samples. Rytky et al. used regularized linear and logistic regression models for the histopathological grading of osteochondral specimens imaged with contrast-enhanced microcomputed tomography (microCT) [23]. The models were trained against the manually graded histopathological samples to predict the grades of degeneration for the articular cartilage of the surface, deep, and calcified cartilage zone. They found that the model could detect the degeneration in the surface zone with an average precision of 0.89 (AUC of 0.92) while the detection of degeneration in the deep zone was the lowest, with an average precision of 0.46 (AUC of 0.62) [23]. Power et al. used supervised deep learning to automate the grading system for the histological images of engineering cartilage tissue [24]. Safranin O and fast green (SafO) was used for staining the engineered tissue; then, two experts graded the images. Transfer learning using a pre-trained DenseNet model was used to automate the scoring of the histological images; the scoring resulted in errors comparable to inter-user errors [24].

In this study, we aim to automate the classification of histopathological grading into early, mild, moderate, and severe OA using machine learning and deep learning techniques. The histological images of the osteochondral specimens were obtained from Venkata et al. [25]. The current methods could be improved with the development of methods for the analysis and grading of osteochondral histological samples, particularly as most researchers use manual grading for the histological samples. The developed methods could be used not only for the OA histological samples harvested after total knee arthroplasty but also for tissue engineering models of articular cartilage.

## 2. Materials and Methods

The method proposed in this paper is shown in Figure 1; then, each block is explained in the following sections.

As is clear in Figure 1, the histopathological images passed through various stages: from deep learning structures, the extraction of feature maps, and the employing of PCA to the weighting optimization algorithm. The evaluation criteria are calculated in each stage. The corresponding sections clarify the novelty of the proposed approach.

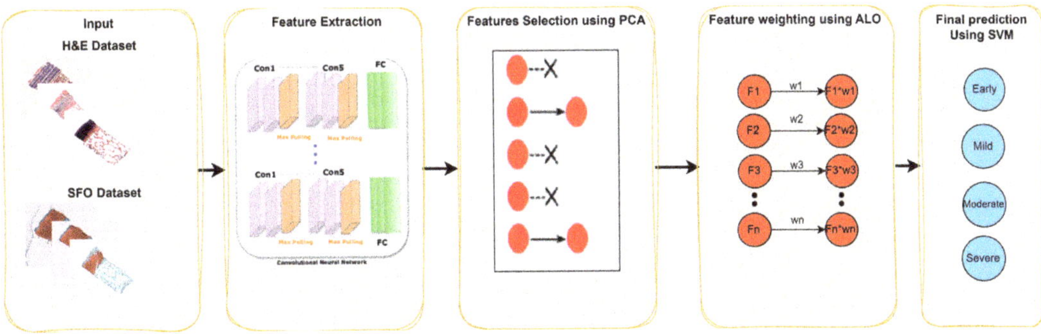


**Figure 1.** The proposed method for distinguishing the severity levels for both hematoxylin and eosin (HE) and safranin O and fast green (SafO) histopathological images.

*2.1. Database*

The osteochondral images were obtained from the database of Venkata et al. [25] (Available: https://doi.org/10.18735/77ye-yh24 (accessed on 2 February 2023)). Briefly, the samples were harvested from 90 patients undergoing total knee arthroplasty. Two osteochondral specimens (4 × 4 × 8 mm) were obtained, one from the medial (CM) and one from the lateral (CL), from the lateral femoral condyle. The specimens were stained with hematoxylin and eosin (H&E) or safranin O and fast green (SafO). SafO staining is usually used for staining glycosaminoglycans [26] and hematoxylin and eosin (H&E) staining is usually used for staining nuclei and extracellular proteins [27]. The samples were previously graded according to the OARSI grading system by three scorers 3 times (separated by at least 3 months) [25]. According to the average grades of the scorers, we divided the images of HE and SafO into early, mild, moderate, and severe OA, as shown in Figures 2 and 3. In the OARSI scoring system, the score for early is less than 3.4, for mild it is 2.4–8.6, for moderate it is 8.6–15.4, and for severe it is 15.4–24 [28].

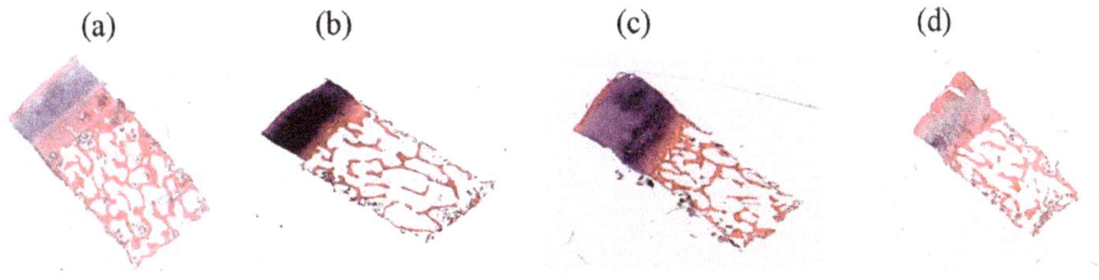


**Figure 2.** Representative HE-stained images of cartilage specimens, indicating (**a**) early, (**b**) mild, (**c**) moderate, and (**d**) severe OA.

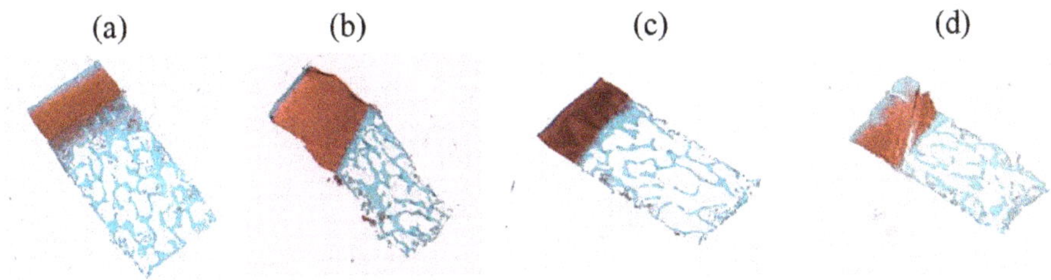


**Figure 3.** Representative SafO-stained images of cartilage specimens, indicating (**a**) early, (**b**) mild, (**c**) moderate, and (**d**) severe OA.

### 2.2. Deep Learning Features

Deep learning features represent the graphical descriptors for each class. They are inherent to the categories themselves. In this paper, several pre-trained deep learning models are employed to differentiate various levels of OA in two types of stained histological images (HE and SafO). The utilization of pre-trained convolutional neural networks (CNNs) to discriminate between two kinds of histological images does not provide accurate results. Therefore, the proposed method combines deep learning, machine learning, and optimization techniques to achieve high accuracy in predicting OA levels. The proposed method depends mainly on extracting the most representative features from the last fully connected model in each CNN. The deep learning structures were trained on the ImageNet database to classify 1000 classes. The transfer learning technique that was utilized to maintain the established structures is compatible with the desired problem statement, which focused on anticipating four levels of histological OA images. The transfer learning was made applicable by augmenting the input size of the image to make it suitable for the input layer of each one. Moreover, removing the last fully connected layer reduced it to four levels. The deep descriptors for each model were extracted from the last fully connected layer. Each one supplied four representative attributes for four levels for both types of stained images (HE and SafO) [29,30]. The utilized networks were ResNet-101, MobileNet, ShuffleNet, NasNet, and DarkNet-19. The idea behind using various structures is based on the ability of each one to extract features in a different manner and to learn in various ways, either in deep or in multiscale resolution. This leads to the obtaining of more representative features that can accurately represent the histopathological OA images.

#### 2.2.1. DarkNet-19

The DarkNet-19 is a type of CNN that consists of 19 convolutional layers, followed by a max-pooling layer and then two fully connected layers. DarkNet architecture is similar to that of VGGNet but with fewer parameters. It is applied to computer vision tasks such as object detection, image classification, and segmentation. Moreover, it was introduced as a part of YOLO (You Only Look Once), which is designed for tracking real-time objects [31].

#### 2.2.2. NasNet

NasNet stands for neural search architecture networks. This CNN is a well-known predefined convolutional neural network, which is trained over the ImageNet dataset with over 1000 classes from nature. The NasNet internal structure consists of a multi-series of cells. There are two types of cells: normal and reduction cells. The normal cells are responsible for extracting the graphical descriptors and producing the feature maps via convolutional filters. On the other hand, the reduction cell is in charge of reducing the size of the feature map's width and height by a factor of 2. NasNet is ended by a SoftMax layer that allows obtaining the probability of classification task [31].

#### 2.2.3. ResNet-101

Residual neural networks are convolutional neural networks pre-trained over the ImageNet database; there are various versions based on the number of convolutional layers (Res-18,50, and 101). This kind of CNN is distinguished by its residual block property, which overcomes the vanishing gradient that appears due to deep learning. The skip connections lead to the bypassing of some of the neural layers and the feeding of the output of one layer as the input to the next level, which provides a different path for the gradient in backpropagation. That is the architecture of the residual block. ResNets consist of the stacking of such blocks. By transfer learning, the input image must be augmented to be compatible with ResNet input size $224 \times 224 \times 3$, and the last fully connected layer must be replaced by another one that is suitable for the intended classification task [31,32].

#### 2.2.4. ShuffleNet

ShuffleNet is one of the most well-known pre-trained CNNs; it is appropriate for mobile applications. ShuffleNet executes two types of convolution to achieve a high level of accuracy. They are the point-wise convolution and the channel convolution; they lead to reduced time computation and make the results more accurate. The ShuffleNet structure consists of the stacking of shuffle netblocks; each one includes a point-wise convolutional layer and a depth-wise layer. The resultant output is passed to the ReLU layer for mapping purposes. The transfer learning is performed by augmenting the input data to be $224 \times 224 \times 3$ and replacing the last fully connected layer to make it compatible with the number of intended classes [31].

#### 2.2.5. MobileNet

MobileNet is a pre-trained CNN designed for mobile and embedded devices. It is organized based on one depth-wise separable convolution that yields a reduction in the number of required parameters to maintain a good performance. The idea behind the depth-wise separable convolution is to split the convolution operation into two separate operations: a depth-wise convolution and a pointwise convolution. In a depth-wise convolution, each channel of the input is convolved with a separate filter, resulting in a set of feature maps. Then, a pointwise convolution is devoted to combining the attribute maps into the output by utilizing a $1 \times 1$ filter to convolve across all channels.

The MobileNet architecture consists of a series of convolutional layers, followed by global average pooling and a fully connected layer. The depth-wise separable convolution is performed in all these layers to obtain an efficient performance. The MobileNet structure may be adjusted by modifying the number of layers, filter sizes, and other hyperparameters [31,33].

### *2.3. Features Engineering*

The features were extracted from each of the previously mentioned CNNs, four features for each CNN; the total number of extracted features from each type of stained image (HE or SafO) was 20 features. The extracted features underwent further processing techniques: through reduction by choosing the most significant or by weighting them using one of the most common optimization methods, which is known as the ant lion optimization technique.

#### 2.3.1. Principal Component Analysis

Principal component analysis (PCA) is well-known in data pre-processing and machine learning and is considered to be a feature selection algorithm. PCA transforms a high-dimensional dataset into a lower-dimensional space by identifying the principal components which explain the maximum variance in the datasets. PCA reduces the dimension of that dataset by preserving the most important information and discarding the redundant data task [29–32].

The principal components define the direction of the maximum variance in the extracted features. The following steps describe the process required to perform the PCA algorithm.

1. Standardization: this step is performed by standardizing each column feature that makes the mean for each feature zero, and the variance is unity.
2. Covariance matrix: this step is performed by constructing the covariance matrix, which is a square matrix that reflects the variance between each pair of features; its diagonal represents the variance for each feature and the off-diagonal represents the covariance between each pair of features.
3. Computation of the principal components: this step is performed by computing the eigenvector, which explains the direction of maxim variance, and the eigenvalue that quantifies the amount of maximum variance.

4. Selection of the principal components: the principal components are selected based on 95% of the majority variance of the features.
5. Mapping between the selected principal components and the features: this is performed by projecting the standardized features onto the best principal components.

2.3.2. Feature Weighting Using ALO

Feature weighting represents the features that are more important than others when optimizing the classification problem; it reveals the role of each feature in the classification pattern by distinguishing by weight. The linear weight is proposed for the feature space to obtain a specific weight for the features; then, the new feature represents the original feature multiplied by its weight, as shown in the following equation:

$$New_{Feature} = Weight \times Old\_Feature \tag{1}$$

Ant lion optimization (ALO) is a metaheuristic optimization algorithm that is used for tuning the parameters to achieve high accuracy. In this paper, we explored feature weights and the optimal value of k in the k-nearest neighbors (k-NN) algorithm; simultaneously, we used the accuracy of k-NN as a fitness function. The difference between PCA and ALO is that the former reveals the significant features and discards the less influential features. All the selected attributes have the same weight, which leads to an equal impact on the classification results. On the other hand, in this paper, the cascading of these two optimization techniques was the key to improving and obtaining the highest accuracies. The selected features were passed to the ALO algorithm to achieve an optimized weight for each one that was significant.

The ALO algorithm can be updated to search for a combination of feature weights and k values that optimize the performance of the k-NN model. The approach is performed using the accuracy of k-NN as a fitness function [34].

The steps of ALO are as follows:

1. Initialize the population of ant lions randomly.
2. Evaluate the accuracy of each ant lion in the population based on both weight and k-value.
3. Define the king ant lion based on the highest accuracy.
4. Move the ant lions towards the king ant lion using a certain formula that simulates the hunting behavior of the ant lions.
5. Calculate the accuracy for the new position.
6. Repeat steps 3–5 until the stopping criterion is met.
7. The results are the optimized weights.

### 2.4. Support Vector Machine

Support vector machines (SVMs) are popular supervised machine learning algorithms used in medical diagnosis. SVM is superior for both linear and non-linear separable data. SVM is used in the medical diagnosis field for discriminating between various classes, such as cancer, diabetics heart arrhythmia, cervical cancer, brain tumors, liver cancer, corneal ulcer, etc.

It is based on finding the optimal margin region for different classes and mapping the features to higher dimensional space using kernels to make the data separable in higher dimensional space. The kernel choice function has a significant impact on the performance of the classifier, in addition to the choosing of the relevant features. SVM is a powerful tool for medical diagnosis, and it is applied for different applications due to its reliability and high performance [35,36]. In this paper, we employed deep learning, feature engineering, and an SVM machine learning classifier to predict OA levels in human osteochondral tissue using histopathological images. The novel combination between them leads to build a reasonable system that can infer significant deep features and can weight them to obtain a reliable scoring diagnosis.

## 3. Results

The two types of stained images were passed to five pre-trained CNN models. The classification procedure was performed in four scenarios. First, deep learning classification was used to classify the four levels of OA. Second, deep learning features were extracted for each CNN and a support vector machine classifier was used to distinguish between the four levels for each type of stained image. Third, feature engineering techniques were applied to evaluate the most significant features from five CNNs using PCA. The last scenario reveals the importance of the feature weighting method by applying the ALO algorithm to give weight to each selected feature. The following subsections are devoted to discussing the obtained results in each scenario. The evaluation criteria that were used in this paper are those in [37].

$$accuracy = \frac{TP + TN}{TP + TN + FP + FN} \quad (2)$$

$$Recall = \frac{TP}{TP + FN} \quad (3)$$

$$Precision = \frac{TP}{TP + FP} \quad (4)$$

$$Specificity = \frac{TN}{TN + FP} \quad (5)$$

$$F1 - score = \frac{2 \times Precision \times Recall}{Precision + Recall} \quad (6)$$

### 3.1. Pre-Trained Model Classification

Table 1 represents the accuracy for both the HE and the SafO images using DarkNet-19, MobileNet, NasNet, ResNet-101, and ShuffleNet. As is clear from Table 1, the accuracy of utilizing deep learning for HE images does not exceed 70.6% using NasNet. Moreover, the sensitivity and precision are too low, which leads to the F1 score being too low. Therefore, the deep convolution networks could not distinguish between various types of severity levels. For the SafO images, the accuracy ranged between 73.3% and 80% for the different CNN classifiers, among which DarkNet-19 had the highest accuracy. The obtained results were not promising; therefore, a hybrid model is recommended to extract the deep features and then pass them to a machine learning classifier to outperform the classification results.

**Table 1.** The accuracy using different CNN structures for HE and SafO images.

| Images \ CNN | DarkNet-19 | MobileNet | NasNet | ResNet-101 | ShuffleNet |
|---|---|---|---|---|---|
| HE | 69.6% | 61.8% | 70.6% | 69.6% | 64.7% |
| SafO | 80.2% | 77.2% | 73.3% | 76.2% | 74.3% |

### 3.2. Deep Features with SVM

Four features were extracted from the last fully connected layer for each CNN. The deep features were passed to the SVM classifier. Tables 2 and 3 show the performance of the classification for the HE images; the performance was enhanced except in the case of DarkNet-19. The enhancement comes from employing deep learning features and machine learning classifiers. The reason behind the worst performance of DarkNet-19 was the failure of DarkNet to extract the representative features for the four classes. The improved accuracy was 96% for the ShuffleNet features with the 3rd polynomial SVM classifier. The recall was the highest for the MobileNet features for the early class level. Moreover, the precision was also the best in MobileNet. The highest precision that was obtained was 100% for the severe class in MobileNet, NasNet, and ShuffleNet. On top of that, Figure 4 illustrates the receiver operating curve for each classification procedure. Each figure represents the relation between the true positive and the false positive rates. As the area under the curve (AUC) increases, the classifier has a high performance in distinguishing the particular classes. All the suggested CNNs had the AUC in all the classes, except DarkNet, which failed to extract the representative features for each class.

**Table 2.** The accuracy using different CNN structures with SVM classifier for HE and SafO images.

| Images \ CNN | DarkNet-19 | MobileNet | NasNet | ResNet-101 | ShuffleNet |
|---|---|---|---|---|---|
| HE | 60.8% | 99% | 95.1% | 94.1% | 96.1% |
| SafO | 95% | 98% | 95% | 94.1% | 95% |

**Table 3.** The precision and sensitivity using different CNN features with SVM classifier for HE and SafO images.

| | | HE Images | | SafO Images | |
|---|---|---|---|---|---|
| Class | | Sensitivity | Precision | Sensitivity | Precision |
| Early | DarkNet-19 | 11.8% | 16.7% | 100% | 85.7% |
| | MobileNet | 100% | 100% | 94.1% | 94.1% |
| | NasNet | 89.9% | 94.1% | 88.2% | 88.2% |
| | ResNet-101 | 82.4% | 100% | 94.4% | 94.4% |
| | ShuffleNet | 94.1% | 94.1% | 94.1% | 88.9% |
| Mild | DarkNet-19 | 85.7% | 69.2% | 93.5% | 98.3% |
| | MobileNet | 98.4% | 100% | 98.4% | 98.4% |
| | NasNet | 95.2% | 96.8% | 96.8% | 95.3% |
| | ResNet-101 | 98.4% | 95.4% | 98.4% | 92.4% |
| | ShuffleNet | 100% | 95.4% | 96.8% | 96.8% |
| Moderate | DarkNet-19 | 29.4% | 55.6% | 94.1% | 94.1% |
| | MobileNet | 100% | 94.4% | 100% | 100% |
| | NasNet | 100% | 98.5% | 100% | 100% |
| | ResNet-101 | 88.2% | 88.2% | 76.5% | 100% |
| | ShuffleNet | 83.3% | 100% | 94.4% | 100% |
| Severe | DarkNet-19 | 20% | 16.7% | 100% | 100% |
| | MobileNet | 100% | 100% | 100% | 100% |
| | NasNet | 100% | 100% | 75% | 100% |
| | ResNet-101 | 100% | 83.3% | 100% | 100% |
| | ShuffleNet | 100% | 100% | 100% | 100% |

The same procedure was applied for the SafO images; the performances of each classifier with SVM are shown in Tables 2 and 3. The performance of the DarkNet was much better than in the HE cases. The accuracy for all the CNN features with SVM ranged from 94.1% to 98% for ResNet-101 and MobileNet, respectively. The worst sensitivity was obtained for the ResNet-101 features for the moderate class. Nevertheless, the recall was almost high in all the classes for each network descriptor. The lowest positive predictive value for all the classes was greater than 85%. This indicates the ability of the extracted features to help in differentiating between various levels of severity.

Moreover, for more analysis and clarification, the ROC curve (Figure 5) explains the impact of applying a hybrid process between deep learning and machine learning. The improvement of the AUC for each class, early, mild, moderate, and severe, reflects the ability of the proposed procedure to determine the kind of severity level for osteochondral tissue using SafO-stained images of human cartilage specimens, which imply cartilage structure, cell glycosaminoglycan content, and tide-mark integrity for the four types of severity levels, as we mentioned before: early, mild, moderate, and severe OA. To improve the performance of the proposed procedure using feature engineering techniques, the simplest method is to combine all the features from all the CNNs and then pass them to the kernel SVM to improve the results. The huge dimensions of using twenty features may lead to an increase in the computation time cost, which leads to the use of the principal component analysis (PCA). PCA is one of the most familiar methods for feature reduction that indicate up to 95% variance of the features. The proposed approach is to mix the benefits from all the CNNs and then find the significant features. The next section describes the results for PCA.

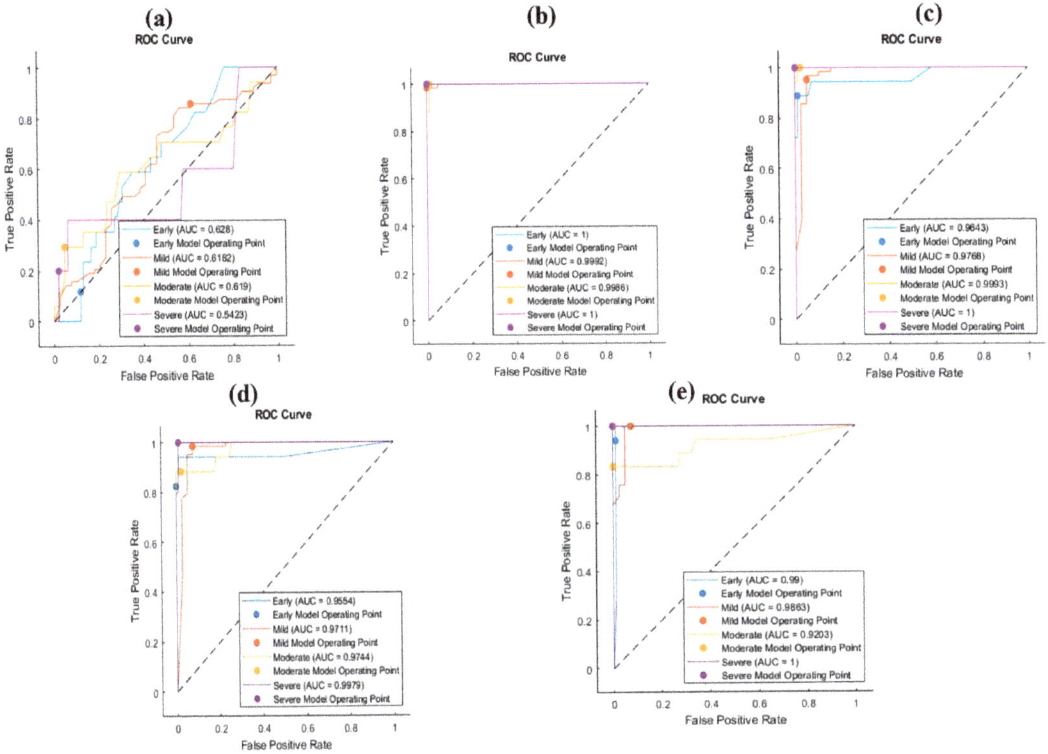


**Figure 4.** ROC curves of HE images for (**a**) deep DarkNet-19 features with SVM, (**b**) deep DarkNet-19 features with SVM, (**c**) deep DarkNet-19 features with SVM (**d**) deep DarkNet-19 features with SVM, and (**e**) deep DarkNet-19 features with SVM.

### *3.3. Principal Component Analysis (PCA)*

All the features from the previous CNNs were fused and utilized to classify the images; then, PCA was devoted to the prediction of the most significant features. The twenty features from five CNNs were further processed under PCA to find the most significant subset features. Then, the most significant features passed to the SVM. The best obtained ten features for the HE images were:

1. Four features from MobileNet.
2. Three features from ShuffleNet.
3. Two features from NasNet.
4. One feature from ResNet-101.

The most significant features did not involve any features from the DarkNet which was expected since the accuracy was low for the DarkNet. Figures 6 and 7 show the confusion matrix of the PCA of all the features from all the convolution neural networks and the corresponding ROC curve for the HE and SafO images, respectively. Figure 6 describes the resultant confusion matrix and its corresponding ROC curve for the HE images. The accuracy was 98% for all the classes. On the other hand, the sensitivity for all the categories was 100%, except for the moderate level, which was 89%. However, the precision was 100% for the early and moderate levels, whereas it was 98.4% and 83.3% for the mild and severe levels, respectively. The AUC was 1 for the early and severe classes. On the other hand, the AUC was 0.995 for the mild class and 0.981 for the severe class. The obtained features using MobileNet performed better than those using the ten features. Therefore, after applying PCA for all the fused features, the most significant were the MobileNet features. They improved the previous results obtained using MobileNet features only.

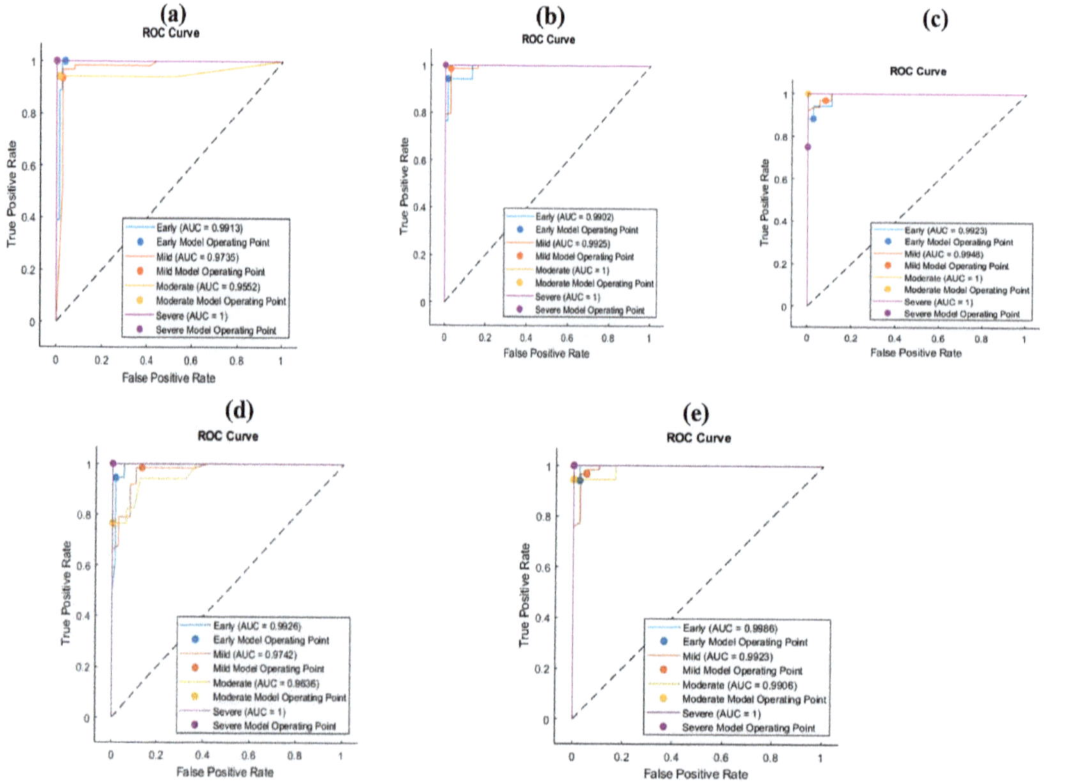


**Figure 5.** ROC curves of SaFO images for (**a**) deep DarkNet-19 features with SVM, (**b**) deep DarkNet-19 features with SVM, (**c**) deep DarkNet-19 features with SVM (**d**) deep DarkNet-19 features with SVM, and (**e**) deep DarkNet-19 features with SVM.

The same procedure was applied to the fused features that were extracted from the SafO-stained images. The most significant features with 95% variance were ordered as follows:

1. Three features from MobileNet.
2. Three features from ShuffleNet.
3. Two features from NasNet
4. Two features from DarkNet

The ordering of the significant features satisfied the obtained results that employed features from each CNN individually. The highest accuracy appeared in MobileNet, then ShuffleNet. The worst accuracy was obtained using the ResNet-101 features. Therefore, they were not counted as significant features. Figure 7 describes the obtained results for the SafO-stained images using the most significant ten features.

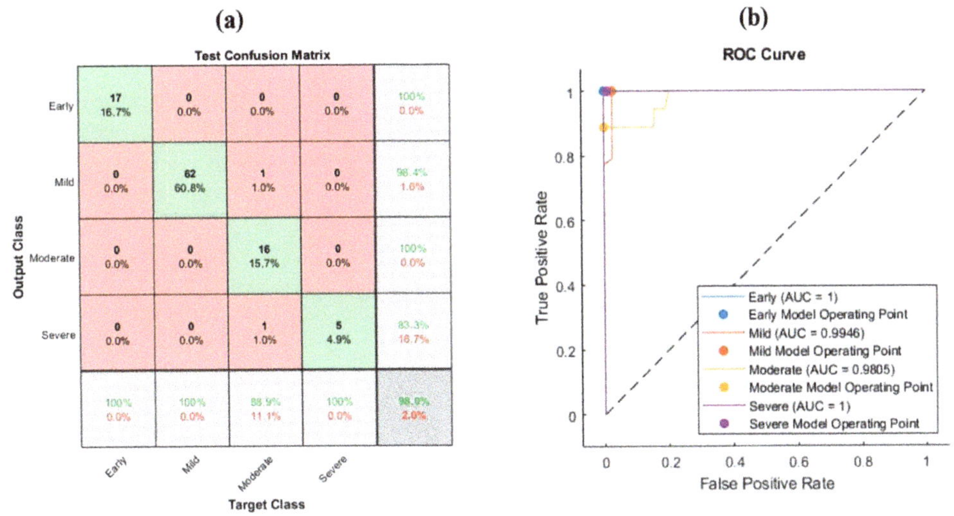


**Figure 6.** Feature fusion for HE images with PCA: (**a**) confusion matrix and (**b**) ROC curve.

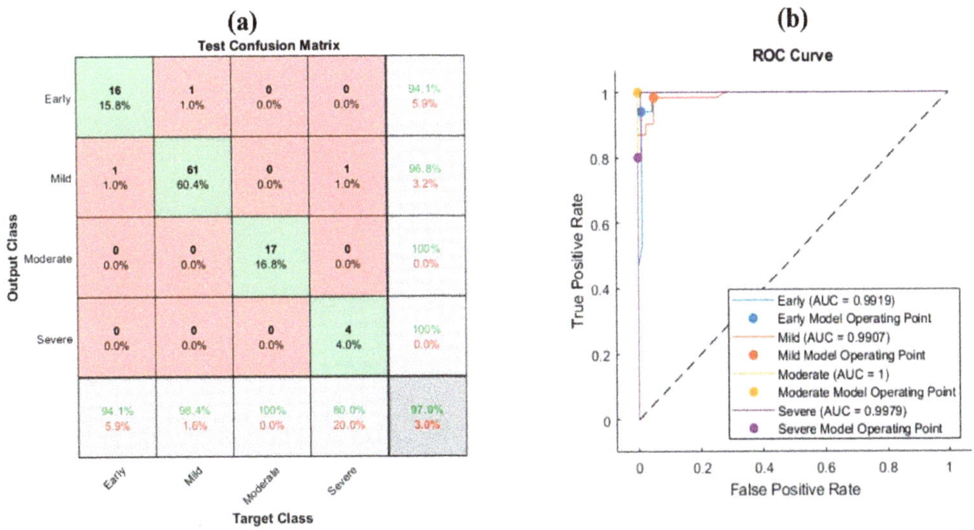


**Figure 7.** Feature fusion for SafO images with PCA: (**a**) confusion matrix and (**b**) ROC curve.

The obtained accuracy was 97%. The highest recall was in the moderate category, whereas the lowest sensitivity was in the severe class. On top of that, the best precision was maintained in the moderate and severe classes. The lowest positive predictive value was in the early class. The area under the curve for all the classes was almost 1.

### 3.4. Ant Lion Optimization (ALO)

The ant lion optimization method combines the weights for each feature alongside the objective function, which is the loss of the convergence. The iterative procedure is performed to achieve the plateau of loss. This leads to the best weights for the features. The range of weights for each feature is [0–1]. The algorithm was applied to both kinds of images for all the extracted deep features. Figure 8 shows the convergence loss function versus the number of iterations for the HE images. As is clear from the figure, the maximum iteration is 100, and the convergence is constant after 60 iterations. The corresponding equation shows the optimized weight for each feature.

$$y = 0.522642 \times F1 + 0.503514 \times F2 + 0.093848 \times F3 + 0.482934 \times F4 + 0.11463 \times F5 + 0.167205 \times F6 + 0.750722 \times F7 + 0.770949 \times F8 + 0.159337 \times F9 + 0.364798 \times F10 \quad (7)$$

where $y$ represents the label of the image, and F1–F10 are the ten most significant features.

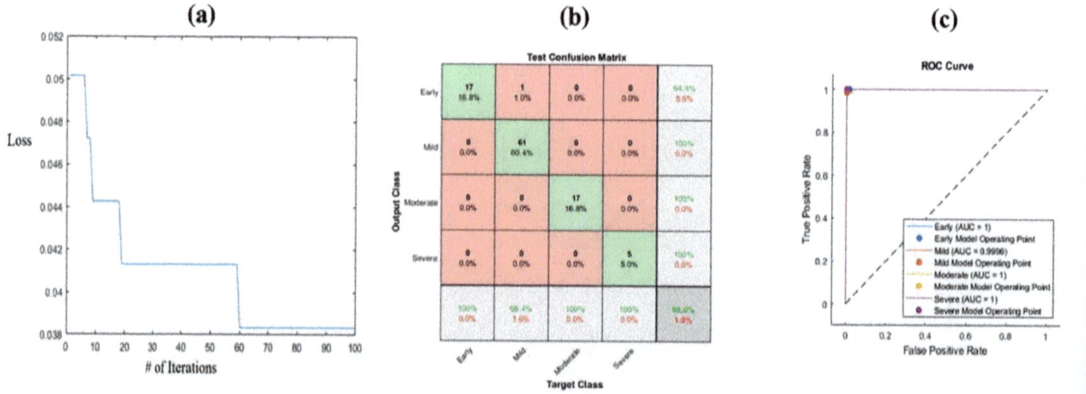


**Figure 8.** ALO algorithm for HE images: (**a**) convergence of the algorithm, (**b**) confusion matrix, and (**c**) ROC curve. Where # represents the number.

The confusion matrix of the obtained results is described in Figure 8b. The weighting features enhanced the accuracy to 99%. The sensitivity and precision were almost 100% for all the classes, except that the recall was 98.8% for the mild level and 94.4% for the early class. The ROC curve is illustrated in Figure 8c. The area under the curve was 1 for all the classes. The F1 score values were 0.97, 0.991, 1, and 1 for the early, mild, moderate, and severe classes, respectively (Table 4). The specificity values were 98.8%, 100%, 100%, and 100% for the early mild, moderate, and severe classes, respectively. As is clear from Table 4 and Figure 8, ALO has a higher performance than PCA in all the classes.

**Table 4.** The performance of feature engineering on HE-stained images.

| Class | Feature Engineering | Sensitivity | Precision | Specificity | F1 Score |
|---|---|---|---|---|---|
| Early | PCA | 100% | 100% | 100% | 1 |
| | ALO | 100% | 98.4% | 98.8% | 0.97 |
| Mild | PCA | 100% | 98.4% | 97.5% | 0.991 |
| | ALO | 100% | 98.4% | 97.5% | 0.991 |
| Moderate | PCA | 88.9% | 100% | 100% | 0.941 |
| | ALO | 100% | 100% | 100% | 1 |
| Severe | PCA | 100% | 83.3% | 99% | 0.909 |
| | ALO | 100% | 100% | 100% | 1 |

The same procedure was applied for the SafO images; Figure 9a shows the number of iterations for the ALO algorithm versus the loss function. After 80 iterations, the loss function was constant, and the optimized weighted features were maintained. The optimized weights were:

$$y = 0.216401 \times F1 + 0.898295 \times F2 + 0.92736 \times F3 + 0.110877 \times F4 + 0.416086 \times F5 + 0.749176 \times F6 + 0.386958 \times F7 + 0.67024 \times F8 + 0.030166 \times F9 + 0.584659 \times F10 \quad (8)$$

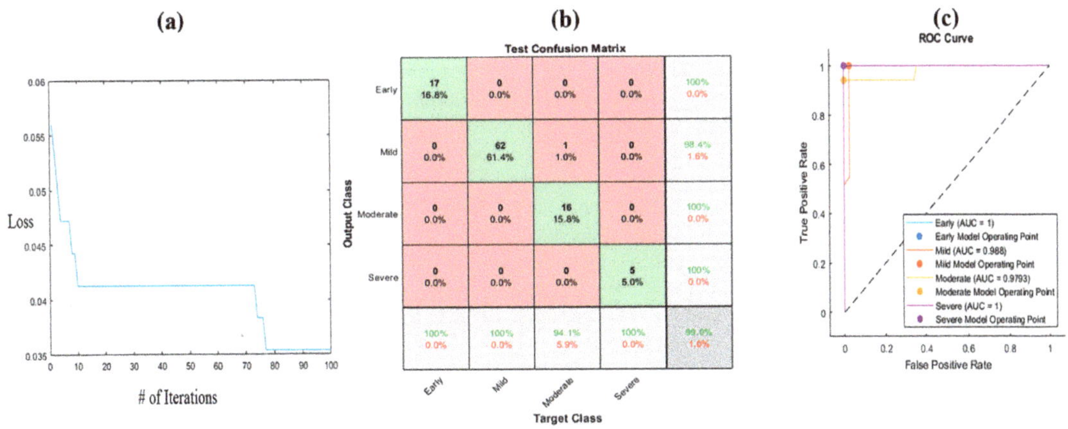


**Figure 9.** ALO algorithm for SafO images: (**a**) convergence of the algorithm, (**b**) confusion matrix, and (**c**) ROC curve. Where # represents the number.

The achieved accuracy in the SafO images was the same as in the HE images (99%). The highest sensitivity was 100% in the early, mild, and severe categories. However, the highest precision was in the early, moderate, and severe levels. Figure 9c describes the AUC for the weighted features and the SVM classifier. The AUC was 1 in both the early and the severe classes, while the AUC was 0.979 in the moderate class and 0.988 in the mild class. The specificity was computed for all the levels, as follows: 100%, 97.4%, 100%, and 100% for the early, mild, moderate, and severe classes, respectively (Table 5). Furthermore, the F1 score values were 1, 0.971, 1, and 0.889 for the early, mild, moderate, and severe categories, respectively, using the PCA classifier, while the F1 score values were 1, 0.991, 0.97, and 1 for the early, mild, moderate, and severe categories, respectively, using the ALO classifier. As with the HE images, the ALO classifier performed better compared with PCA for the SafO images.

**Table 5.** The impact of feature engineering on SafO images.

| Class | Feature Engineering | Sensitivity | Precision | Specificity | F1 Score |
|---|---|---|---|---|---|
| Early | PCA | 94.1% | 94.1% | 98.8% | 1 |
| | ALO | 100% | 100% | 100% | 1 |
| Mild | PCA | 98.4% | 96.8% | 94.8% | 0.971 |
| | ALO | 100% | 98.4% | 97.4% | 0.991 |
| Moderate | PCA | 100% | 100% | 100% | 1 |
| | ALO | 94.1% | 100% | 100% | 0.97 |
| Severe | PCA | 80% | 100% | 100% | 0.889 |
| | ALO | 100% | 100% | 100% | 1 |

## 4. Discussion

In this study, we showed that machine learning and deep learning can be used to automatically classify the osteochondral histopathological images into early, mild, moderate, and severe OA. The manual histopathological scoring systems are time-consuming and need a trained scorer to grade the images. This study used five CNN models, including ResNet-101, MobileNet, ShuffleNet, NasNet, and DarkNet-19, to extract features from HE and SafO histopathological images of different levels of OA. As deep learning was insufficient to classify the OA images, we employed the deep features with a machine learning classifier to enhance the classification results, and we then optimized these features using various engineering methods, such as PCA and ALO. Although the deep learning method was first used in this manuscript to predict the severity of OA, the histopathological OA images were very complex due to the many changes that happen in both the cartilage and the subchondral bone during OA progression, such as the network of collagen fibers, the subchondral bone structure, the proliferation of chondrocytes, the size of cartilage change, and the proteoglycans loss, which results in surface cracking [38]. All of these make it very difficult for deep learning procedures alone to classify histopathological OA images. So, in this study, combinations of multiple algorithms were used with machine learning classifiers and various engineering methods, such as PCA and ALO. Combinations of different feature engineering approaches have been utilized in different studies due to the complexity of the images, the tissue, the type of images, and the sizes [39–42].

The results showed that the F1 score values were 0.97, 0.991, 1, and 1 for the early, mild, moderate, and severe classes, respectively, for the HE-stained images using the ALO classifier. For the SafO images, the F1 score values were 1, 0.991, 0.97, and 1 for the early, mild, moderate, and severe categories, respectively, using the ALO classifier. This study had a limitation in the dataset in that there was a very small number of images for the severe class. Only 14 images were available for the HE staining and another 14 images for the SafO staining for the severe class. So, we focused on reporting the F1 score since the data were imbalanced [43].

Few studies have utilized artificial intelligence to score or classify osteochondral or cartilage histopathological images. In another study, a machine learning technique was used to automatically grade 3D histopathological images of osteochondral samples to predict the degeneration of surface, deep, and calcified cartilage zones [23]. The samples were imaged using defect contrast-enhanced microCT. Transfer learning using a pre-trained ResNet-34 encoder was used. The model was able to predict the degeneration in the surface zone (AUC of 0.92 and AP of 0.89), followed by the calcified zone (AUC of 0.71 and AP of 0.65) and the deep zone (AUC of 0.62 and AP of 0.46) [23]. In another study, a deep learning technique was used to automate the grading of the histological images of engineered cartilage, in which the grading was classified into four categories [24]. Transfer learning using a pre-trained DenseNet model was used for feature extraction to automatically score the histological images of engineered cartilage. It was found that the RMSEs for the model prediction were in a similar range as the inter-user of 0.71 [24]. In our study, using the ALO algorithm for HE images, the specificity values were 98.8%, 100%, 100%, and 100% for early mild, moderate, and severe classes, respectively, and the AUC was 1 for all the classes. Using the ALO algorithm for the SafO images, the specificity values were 100%, 97.4%, 100%, and 100% for the early, mild, moderate, and severe classes, respectively, and the AUC values were 1, 0.988, 0.979, and 1 for the early, mild, moderate, and severe classes, respectively.

Machine and deep learning have recently been used to investigate OA development and progression using MRI or X-ray images [44–47]. Ashinsky et al. used machine learning to investigate the development of OA using the MRI images of 68 patients. A hierarchy of algorithms representing morphology (WND-CHRM) was used to classify the development of OA with 75% accuracy [17]. In another study, the T2 relaxation time of the MRI images of the 4384 subjects with and without OA was analyzed using DenseNet and random forests to distinguish OA [45]. The DenseNet training model attained a sensitivity equal to 74.53% and a specificity equal to 76.13%, which was comparable to the random forest model with a sensitivity of 67.01% and a specificity of 71.79%. Tolpadi et al. used a DenseNet CNN to predict the total knee replacement (TKR) from the MRI images and the clinical and demographic information of patients with OA and patients without OA [48]. Their model was able to predict the TKR with the AUCs of $0.834 \pm 0.036$ and $0.943 \pm 0.057$ for patients with OA and without OA, respectively.

In OA, the integrity of collagen and glycosaminoglycan, which give the cartilage the mechanical properties, is compromised [49]. The articular cartilage has a complex structure without blood vessels or nerves, making it difficult to repair or to treat the cartilage defect. So, the progression of OA has been investigated by many researchers using a manual grading system [25,50,51]. Saarakkala et al. studied the collagen and proteoglycan changes during OA progression using the OARSI histopathology grading system [52]. Then, a composition-based finite element (FE) model was employed to study the tissue function. Mantripragada et al. investigated the scoring of polarized light microscopy (PLM) images as a potential method to understand early OA as compared with the standard histopathological methods [50]. They found that adding a PLM scoring system helped in the characterization of early and mild OA. OA progression and development have also been studied in many animal models of human OA [53–55]. A whole joint microCT image scoring and histologic scoring systems of a Hartley guinea pig, which is considered a model of human OA, were investigated to determine the changes in articular cartilage and bone [55]. The grading was conducted by two experts using the OARSI guidelines. So, automating the grading system of histopathological methods could help in understanding OA progression and development.

## 5. Conclusions

The proposed methods revealed the ability of the integration between deep learning, machine learning, and feature engineering in scoring the severity levels of OA. The deep learning models help the researcher in the classification and extraction of the representative features of each category. The feature engineering method enhanced the performance of the classification results, which focused on obtaining the most important attribute in addition to giving them a specific weight. The best results obtained in this study were obtained by using PCA followed by ALO then SVM classifiers. To the best of our knowledge, this is the first study that handles the combination between PCA and ALO to obtain the best classification. Moreover, this is the first study that discusses the employment of artificial intelligence in OA microscopic histopathological images. In this study, we were able to build an artificial intelligence model that could distinguish the different stages of the OA from the osteochondral histopathological images without the need of human experts, which could be of great interest to the researchers and scientific community. Furthermore, the model could be modified for the evaluation of tissue engineering cartilage formation instead of using the manual grading system.

**Author Contributions:** Conceptualization, A.K. and H.A.; methodology, A.K. and H.A.; software, A.K. and H.A.; validation, A.K.; formal analysis, A.K. and H.A.; investigation, A.K.; resources, A.K.; data curation, A.K.; writing—original draft preparation, A.K. and H.A.; writing—review and editing, A.K.; visualization, A.K. and H.A.; supervision, A.K.; project administration, A.K. and H.A. All authors have read and agreed to the published version of the manuscript.

**Funding:** This research received no external funding.

**Institutional Review Board Statement:** Not applicable.

**Informed Consent Statement:** Not applicable.

**Data Availability Statement:** The data presented in this study are available upon request from the corresponding author.

**Acknowledgments:** The authors thank Yarmouk University for its support.

**Conflicts of Interest:** The authors declare no conflict of interest.

## References

1. Hubertsson, J.; Petersson, I.F.; Thorstensson, C.A.; Englund, M. Risk of sick leave and disability pension in working-age women and men with knee osteoarthritis. *Ann. Rheum. Dis.* **2013**, *72*, 401–405. [CrossRef] [PubMed]
2. Lawrence, R.C.; Felson, D.T.; Helmick, C.G.; Arnold, L.M.; Choi, H.; Deyo, R.A.; Gabriel, S.; Hirsch, R.; Hochberg, M.C.; Hunder, G.G.; et al. Estimates of the prevalence of arthritis and other rheumatic conditions in the United States. Part II. *Arthritis Rheumatol.* **2008**, *58*, 26–35. [CrossRef]
3. Loeser, R.F.; Goldring, S.R.; Scanzello, C.R.; Goldring, M.B. Osteoarthritis: A disease of the joint as an organ. *Arthritis Rheumatol.* **2012**, *64*, 1697–1707. [CrossRef] [PubMed]

4. Jiang, Y. Osteoarthritis year in review 2021: Biology. *Osteoarthr. Cartil.* **2022**, *30*, 207–215. [CrossRef]
5. Fujii, Y.; Liu, L.; Yagasaki, L.; Inotsume, M.; Chiba, T.; Asahara, H. Cartilage homeostasis and osteoarthritis. *Int. J. Mol. Sci.* **2022**, *23*, 6316. [CrossRef] [PubMed]
6. Pritzker, K.P.H.; Gay, S.; Jimenez, S.A.; Ostergaard, K.; Pelletier, J.P.; Revell, P.A.; Salter, D.; van den Berg, W.B. Osteoarthritis cartilage histopathology: Grading and staging. *Osteoarthr. Cartil.* **2006**, *14*, 13–29. [CrossRef]
7. Rutgers, M.; van Pelt, M.J.; Dhert, W.J.; Creemers, L.B.; Saris, D.B. Evaluation of histological scoring systems for tissue-engineered, repaired and osteoarthritic cartilage. *Osteoarthr. Cartil.* **2010**, *18*, 12–23. [CrossRef]
8. Custers, R.J.; Creemers, L.B.; Verbout, A.J.; van Rijen, M.H.; Dhert, W.J.; Saris, D.B. Reliability, reproducibility and variability of the traditional Histologic/Histochemical Grading System vs. the new OARSI Osteoarthritis Cartilage Histopathology Assessment System. *Osteoarthr. Cartil.* **2007**, *15*, 1241–1248. [CrossRef]
9. Pollard, T.C.; Gwilym, S.E.; Carr, A.J. The assessment of early osteoarthritis. *J. Bone Jt. Surg. Br.* **2008**, *90*, 411–421. [CrossRef]
10. Favero, M.; Ramonda, R.; Goldring, M.B.; Goldring, S.R.; Punzi, L. Early knee osteoarthritis. *RMD Open* **2015**, *1*, e000062. [CrossRef]
11. Pauli, C.; Whiteside, R.; Heras, F.L.; Nesic, D.; Koziol, J.; Grogan, S.P.; Matyas, J.; Pritzker, K.P.; D'Lima, D.D.; Lotz, M.K. Comparison of cartilage histopathology assessment systems on human knee joints at all stages of osteoarthritis development. *Osteoarthr. Cartil.* **2012**, *20*, 476–485. [CrossRef] [PubMed]
12. Serey, J.; Alfaro, M.; Fuertes, G.; Vargas, M.; Durán, C.; Ternero, R.; Rivera, R.; Sabattin, J. Pattern Recognition and Deep Learning Technologies, Enablers of Industry 4.0, and Their Role in Engineering Research. *Symmetry* **2023**, *15*, 535. [CrossRef]
13. Dou, T.; Zhou, W. 2D and 3D convolutional neural network fusion for predicting the histological grade of hepatocellular carcinoma. In Proceedings of the 2018 24th International Conference on Pattern Recognition (ICPR), Beijing, China, 20–24 August 2018; pp. 3832–3837.
14. Zhang, L.; Ren, Z. Comparison of CT and MRI images for the prediction of soft-tissue sarcoma grading and lung metastasis via a convolutional neural networks model. *Clin. Radiol.* **2020**, *75*, 64–69. [CrossRef]
15. Lee, J.H.; Shih, Y.T.; Wei, M.L.; Sun, C.K.; Chiang, B.L. Classification of established atopic dermatitis in children with the in vivo imaging methods. *J. Biophotonics* **2019**, *12*, e201800148. [CrossRef] [PubMed]
16. Azizi, S.; Bayat, S.; Yan, P.; Tahmasebi, A.; Nir, G.; Kwak, J.T.; Xu, S.; Wilson, S.; Iczkowski, K.A.; Lucia, M.S.; et al. Detection and grading of prostate cancer using temporal enhanced ultrasound: Combining deep neural networks and tissue mimicking simulations. *Int. J. Comput. Assist. Radiol. Surg.* **2017**, *12*, 1293–1305. [CrossRef]
17. Ashinsky, B.G.; Bouhrara, M.; Coletta, C.E.; Lehallier, B.; Urish, K.L.; Lin, P.C.; Goldberg, I.G.; Spencer, R.G. Predicting early symptomatic osteoarthritis in the human knee using machine learning classification of magnetic resonance images from the osteoarthritis initiative. *J. Orthop. Res.* **2017**, *35*, 2243–2250. [CrossRef] [PubMed]
18. Du, Y.; Almajalid, R.; Shan, J.; Zhang, M. A Novel Method to Predict Knee Osteoarthritis Progression on MRI Using Machine Learning Methods. *IEEE Trans. NanoBiosci.* **2018**, *17*, 228–236. [CrossRef]
19. Leung, K.; Zhang, B.; Tan, J.; Shen, Y.; Geras, K.J.; Babb, J.S.; Cho, K.; Chang, G.; Deniz, C.M. Prediction of Total Knee Replacement and Diagnosis of Osteoarthritis by Using Deep Learning on Knee Radiographs: Data from the Osteoarthritis Initiative. *Radiology* **2020**, *296*, 584–593. [CrossRef]
20. Xue, Y.; Zhang, R.; Deng, Y.; Chen, K.; Jiang, T. A preliminary examination of the diagnostic value of deep learning in hip osteoarthritis. *PLoS ONE* **2017**, *12*, e0178992. [CrossRef] [PubMed]
21. Panfilov, E.; Tiulpin, A.; Nieminen, M.T.; Saarakkala, S.; Casula, V. Deep learning-based segmentation of knee MRI for fully automatic subregional morphological assessment of cartilage tissues: Data from the Osteoarthritis Initiative. *J. Orthop. Res.* **2022**, *40*, 1113–1124. [CrossRef]
22. Tiulpin, A.; Saarakkala, S. Automatic Grading of Individual Knee Osteoarthritis Features in Plain Radiographs Using Deep Convolutional Neural Networks. *Diagnostics* **2020**, *10*, 932. [CrossRef] [PubMed]
23. Rytky, S.J.O.; Tiulpin, A.; Frondelius, T.; Finnilä, M.A.J.; Karhula, S.S.; Leino, J.; Pritzker, K.P.H.; Valkealahti, M.; Lehenkari, P.; Joukainen, A.; et al. Automating three-dimensional osteoarthritis histopathological grading of human osteochondral tissue using machine learning on contrast-enhanced micro-computed tomography. *Osteoarthr. Cartil.* **2020**, *28*, 1133–1144. [CrossRef] [PubMed]
24. Power, L.; Acevedo, L.; Yamashita, R.; Rubin, D.; Martin, I.; Barbero, A. Deep learning enables the automation of grading histological tissue engineered cartilage images for quality control standardization. *Osteoarthr. Cartil.* **2021**, *29*, 433–443. [CrossRef] [PubMed]
25. Mantripragada, V.P.; Piuzzi, N.S.; Muschler, G.F.; Erdemir, A.; Midura, R.J. A comprehensive dataset of histopathology images, grades and patient demographics for human Osteoarthritis Cartilage. *Data Brief* **2021**, *37*, 107129. [CrossRef]
26. Schmitz, N.; Laverty, S.; Kraus, V.B.; Aigner, T. Basic methods in histopathology of joint tissues. *Osteoarthr. Cartil.* **2010**, *18*, S113–S116. [CrossRef]
27. Cooper, J.A.; Mintz, B.R.; Palumbo, S.L.; Li, W.J. 4—Assays for determining cell differentiation in biomaterials. In *Characterization of Biomaterials*; Jaffe, M., Hammond, W., Tolias, P., Arinzeh, T., Eds.; Woodhead Publishing: Sawston, UK, 2013; pp. 101–137.
28. Mantripragada, V.P.; Piuzzi, N.S.; Zachos, T.; Obuchowski, N.A.; Muschler, G.F.; Midura, R.J. High occurrence of osteoarthritic histopathological features unaccounted for by traditional scoring systems in lateral femoral condyles from total knee arthroplasty patients with varus alignment. *Acta Orthop.* **2018**, *89*, 197–203. [CrossRef]

29. Alquran, H.; Al-Issa, Y.; Alsalatie, M.; Mustafa, W.A.; Qasmieh, I.A.; Zyout, A. Intelligent Diagnosis and Classification of Keratitis. *Diagnostics* **2022**, *12*, 1344. [CrossRef] [PubMed]
30. Alquran, H.; Mustafa, W.-A.; Qasmieh, I.-A.; Yacob, Y.-M.; Alsalatie, M.; Al-Issa, Y.; Alqudah, A.-M. Cervical Cancer Classification Using Combined Machine Learning and Deep Learning Approach. *Comput. Mater. Contin.* **2022**, *72*, 5117–5134. [CrossRef]
31. Tawalbeh, S.; Alquran, H.; Alsalatie, M. Deep Feature Engineering in Colposcopy Image Recognition: A Comparative Study. *Bioengineering* **2023**, *10*, 105. [CrossRef]
32. Alquran, H.; Alsalatie, M.; Mustafa, W.A.; Abdi, R.A.; Ismail, A.R. Cervical Net: A Novel Cervical Cancer Classification Using Feature Fusion. *Bioengineering* **2022**, *9*, 578. [CrossRef]
33. Srinivasu, P.N.; SivaSai, J.G.; Ijaz, M.F.; Bhoi, A.K.; Kim, W.; Kang, J.J. Classification of Skin Disease Using Deep Learning Neural Networks with MobileNet V2 and LSTM. *Sensors* **2021**, *21*, 2852. [CrossRef]
34. Singh, D.; Singh, B. Hybridization of feature selection and feature weighting for high dimensional data. *Appl. Intell.* **2019**, *49*, 1580–1596. [CrossRef]
35. Alquran, H.; Qasmieh, I.A.; Alqudah, A.M.; Alhammouri, S.; Alawneh, E.; Abughazaleh, A.; Hasayen, F. The melanoma skin cancer detection and classification using support vector machine. In Proceedings of the 2017 IEEE Jordan Conference on Applied Electrical Engineering and Computing Technologies (AEECT), Aqaba, Jordan, 11–13 October 2017; pp. 1–5.
36. Yacob, Y.M.; Alquran, H.; Mustafa, W.A.; Alsalatie, M.; Sakim, H.A.M.; Lola, M.S.H. pylori Related Atrophic Gastritis Detection Using Enhanced Convolution Neural Network (CNN) Learner. *Diagnostics* **2023**, *13*, 336. [CrossRef] [PubMed]
37. Alawneh, K.; Alquran, H.; Alsalatie, M.; Mustafa, W.A.; Al-Issa, Y.; Alqudah, A.; Badarneh, A. LiverNet: Diagnosis of Liver Tumors in Human CT Images. *Appl. Sci.* **2022**, *12*, 5501. [CrossRef]
38. Sulzbacher, I. Osteoarthritis: Histology and pathogenesis. *Wien. Med. Wochenschr.* **2013**, *163*, 212–219. [CrossRef] [PubMed]
39. Yang, J.; Shi, R.; Wei, D.; Liu, Z.; Zhao, L.; Ke, B.; Pfister, H.; Ni, B. MedMNIST v2—A large-scale lightweight benchmark for 2D and 3D biomedical image classification. *Sci. Data* **2023**, *10*, 41. [CrossRef] [PubMed]
40. Chen, C.; Qi, S.; Zhou, K.; Lu, T.; Ning, H.; Xiao, R. Pairwise attention-enhanced adversarial model for automatic bone segmentation in CT images. *Phys. Med. Biol.* **2023**, *68*, 035019. [CrossRef] [PubMed]
41. Chen, C.; Liu, B.; Zhou, K.; He, W.; Yan, F.; Wang, Z.; Xiao, R. CSR-Net: Cross-Scale Residual Network for multi-objective scaphoid fracture segmentation. *Comput. Biol. Med.* **2022**, *137*, 104776. [CrossRef]
42. Das, A.; Mohapatra, S.K.; Mohanty, M.N. Design of deep ensemble classifier with fuzzy decision method for biomedical image classification. *Appl. Soft Comput.* **2022**, *115*, 108178. [CrossRef]
43. Al-Badarneh, I.; Habib, M.; Aljarah, I.; Faris, H. Neuro-evolutionary models for imbalanced classification problems. *J. King Saud Univ. Comput. Inf. Sci.* **2022**, *34*, 2787–2797. [CrossRef]
44. Jamshidi, A.; Leclercq, M.; Labbe, A.; Pelletier, J.-P.; Abram, F.; Droit, A.; Martel-Pelletier, J. Identification of the most important features of knee osteoarthritis structural progressors using machine learning methods. *Ther. Adv. Musculoskelet. Dis.* **2020**, *12*, 1759720X20933468. [CrossRef]
45. Pedoia, V.; Lee, J.; Norman, B.; Link, T.M.; Majumdar, S. Diagnosing osteoarthritis from T2 maps using deep learning: An analysis of the entire Osteoarthritis Initiative baseline cohort. *Osteoarthr. Cartil.* **2019**, *27*, 1002–1010. [CrossRef]
46. Hayashi, D.; Roemer, F.W.; Guermazi, A. Magnetic resonance imaging assessment of knee osteoarthritis: Current and developing new concepts and techniques. *Clin. Exp. Rheumatol.* **2019**, *37*, 88–95.
47. Martel-Pelletier, J.; Paiement, P.; Pelletier, J.-P. Magnetic resonance imaging assessments for knee segmentation and their use in combination with machine/deep learning as predictors of early osteoarthritis diagnosis and prognosis. *Ther. Adv. Musculoskelet. Dis.* **2023**, *15*, 1759720X231165560. [CrossRef]
48. Tolpadi, A.A.; Lee, J.J.; Pedoia, V.; Majumdar, S. Deep learning predicts total knee replacement from magnetic resonance images. *Sci. Rep.* **2020**, *10*, 6371. [CrossRef]
49. Sophia Fox, A.J.; Bedi, A.; Rodeo, S.A. The Basic Science of Articular Cartilage: Structure, Composition, and Function. *Sports Health* **2009**, *1*, 461–468. [CrossRef]
50. Mantripragada, V.P.; Gao, W.; Piuzzi, N.S.; Hoemann, C.D.; Muschler, G.F.; Midura, R.J. Comparative Assessment of Primary Osteoarthritis Progression Using Conventional Histopathology, Polarized Light Microscopy, and Immunohistochemistry. *Cartilage* **2021**, *13*, 1494s–1510s. [CrossRef]
51. Mantripragada, V.P.; Piuzzi, N.S.; Zachos, T.; Obuchowski, N.A.; Muschler, G.F.; Midura, R.J. Histopathological assessment of primary osteoarthritic knees in large patient cohort reveal the possibility of several potential patterns of osteoarthritis initiation. *Curr. Res. Transl. Med.* **2017**, *65*, 133–139. [CrossRef]
52. Saarakkala, S.; Julkunen, P.; Kiviranta, P.; Mäkitalo, J.; Jurvelin, J.S.; Korhonen, R.K. Depth-wise progression of osteoarthritis in human articular cartilage: Investigation of composition, structure and biomechanics. *Osteoarthr. Cartil.* **2010**, *18*, 73–81. [CrossRef] [PubMed]
53. Nagira, K.; Ikuta, Y.; Shinohara, M.; Sanada, Y.; Omoto, T.; Kanaya, H.; Nakasa, T.; Ishikawa, M.; Adachi, N.; Miyaki, S.; et al. Histological scoring system for subchondral bone changes in murine models of joint aging and osteoarthritis. *Sci. Rep.* **2020**, *10*, 10077. [CrossRef] [PubMed]

54. Namhong, S.; Wongdee, K.; Suntornsaratoon, P.; Teerapornpuntakit, J.; Hemstapat, R.; Charoenphandhu, N. Knee osteoarthritis in young growing rats is associated with widespread osteopenia and impaired bone mineralization. *Sci. Rep.* **2020**, *10*, 15079. [CrossRef] [PubMed]
55. Radakovich, L.B.; Marolf, A.J.; Shannon, J.P.; Pannone, S.C.; Sherk, V.D.; Santangelo, K.S. Development of a microcomputed tomography scoring system to characterize disease progression in the Hartley guinea pig model of spontaneous osteoarthritis. *Connect. Tissue Res.* **2018**, *59*, 523–533. [CrossRef] [PubMed]

*bioengineering*

*Article*

# Comparison of Artificial Intelligence-Based Applications for Mandible Segmentation: From Established Platforms to In-House-Developed Software

**Robert R. Ileșan [1,*,†], Michel Beyer [1,2,†], Christoph Kunz [1] and Florian M. Thieringer [1,2]**

1 Department of Oral and Cranio-Maxillofacial Surgery, University Hospital Basel, 4031 Basel, Switzerland; michel.beyer@usb.ch (M.B.); christoph.kunz@usb.ch (C.K.); florian.thieringer@usb.ch (F.M.T.)
2 Medical Additive Manufacturing Research Group (Swiss MAM), Department of Biomedical Engineering, University of Basel, 4123 Allschwil, Switzerland
* Correspondence: robert.ilesan@unibas.ch
† These authors contributed equally to this work.

**Abstract:** Medical image segmentation, whether semi-automatically or manually, is labor-intensive, subjective, and needs specialized personnel. The fully automated segmentation process recently gained importance due to its better design and understanding of CNNs. Considering this, we decided to develop our in-house segmentation software and compare it to the systems of established companies, an inexperienced user, and an expert as ground truth. The companies included in the study have a cloud-based option that performs accurately in clinical routine (dice similarity coefficient of 0.912 to 0.949) with an average segmentation time ranging from 3′54″ to 85′54″. Our in-house model achieved an accuracy of 94.24% compared to the best-performing software and had the shortest mean segmentation time of 2′03″. During the study, developing in-house segmentation software gave us a glimpse into the strenuous work that companies face when offering clinically relevant solutions. All the problems encountered were discussed with the companies and solved, so both parties benefited from this experience. In doing so, we demonstrated that fully automated segmentation needs further research and collaboration between academics and the private sector to achieve full acceptance in clinical routines.

**Keywords:** artificial intelligence; mandible; segmentation; 3D virtual reconstruction; CBCT; CT; Convolutional Neural Networks; comparison; in-house; software; patch size; Cranio-Maxillofacial surgery; DICOM

**Citation:** Ileșan, R.R.; Beyer, M.; Kunz, C.; Thieringer, F.M. Comparison of Artificial Intelligence-Based Applications for Mandible Segmentation: From Established Platforms to In-House-Developed Software. *Bioengineering* **2023**, *10*, 604. https://doi.org/10.3390/bioengineering10050604

Academic Editors: Paolo Zaffino and Maria Francesca Spadea

Received: 25 April 2023
Accepted: 16 May 2023
Published: 17 May 2023



## 1. Introduction

The segmentation of anatomical structures is a process that virtually reconstructs the region of interest from medical images in three dimensions. It helps the physician prepare for surgical interventions and virtual surgical planning (VSP), visualize and interact with the patient's anatomy (through three-dimensional (3D) printing or augmented and virtual reality (AR/VR)), and improve the medical outcome [1–6]. Until recently, the segmentation process was either manual, where the anatomical structure was labeled slice by slice, or semi-automatic, where the software identifies the region of interest and excludes other anatomical structures based on the selected threshold, marked points, or other user inputs [7–10]. Both segmentation types are subjective, time-intensive, and require specialized personnel. Artificial intelligence (AI)-based technologies are gradually being integrated into the clinical routine, and some companies already offer fully automated cloud-based solutions [11,12]. The most common techniques used for automatic segmentation are Statistical Shape Analysis [13] and Convolutional Neuronal Networks (CNNs) [14]. The last-mentioned technique has proven itself to be especially helpful for automatic segmentation [15–17]; for biomedical image segmentation, the U-Net architecture exhibits state-of-the-art performance [18]. In some cases, both techniques are combined to further improve segmentation accuracy [19]. Especially in the Cranio-Maxillofacial (CMF) field, due to the complex anatomy of the face, AI-based segmentation solutions could be advantageous and lead to fully automated virtual surgical planning workflows.

*Related Work*

Previously conducted research has shown promising results for fully automated segmentation using different Convolutional Neural Network (CNN) architectures. Verhelst P.J. et al. [12] proposed a system for mandible segmentation in which two different 3D U-Net CNNs were trained in two phases with 160 cone-beam computed tomography (CBCT) images of the skull from orthognathic surgery patients. The automatically generated mandibles were compared to user-refined AI segmentations and semi-automatic ones, obtaining dice similarity coefficients of 0.946 and 0.944, respectively.

In a different approach, Lo Giudice A. et al. [20] proposed a fully convolutional deep encoder–decoder network that was trained on the MICCAI Head and Neck 2015 dataset and fine-tuned on 20 additional CBCT images. The segmentations were cut so that only the mandibular bone was considered for the assessment. The achieved dice similarity coefficient in comparison to the manual segmentations was 0.972. Apart from the mandibles, other anatomical structures of the skull were also automatically segmented with CNNs. One paper, which was published by Li Q. et al. [21], proposed a method that used a deep Convolutional Neural Network to segment and identify teeth from CBCT images. Another publication, from Kwak G.H. et al. [22], presented an automatic inferior alveolar canal detection system with different U-Net variants (3D SegNet, 2D U-Net, and 3D U-Net), where the three-dimensional U-Net performed best.

Deep learning technologies have improved in terms of performance and accuracy in recent years due to the growing accessibility of new technologies and global digitalization. This has encouraged the development of automatic diagnosis software in dentistry, as shown by Ezhov M. et al. [16], who evaluated a deep learning-based system to determine its real-time performance on CBCT images for five different applications (segmentation of jaw and teeth, tooth localization, numeration, periodontitis module, caries localization, and periapical lesion localization). The same researchers developed an AI-based evaluation tool for the pharyngeal airway in obstructive sleep apnea patients [17].

Other researchers, such as Yang W.F. et al. [11], used Mimics Viewer (Materialise) to segment the skull bones automatically. Compared to the ground truth, the segmented maxilla and mandible achieved dice similarity coefficient scores of 0.924 and 0.949, respectively. Although strenuous, Magnetic Resonance Imaging (MRI) segmentation of soft tissue has gained importance for VSP, as shown by Musatian S.A. et al. [23], who presented solutions for orbit and brain tumor segmentation based on CNNs. One software that is used in this study for semi-automatic segmentation is Brainlab IPlan.

Considering the gains of the last decade's affordable computing power and a better understanding of AI programming, we decided to develop an automatic segmentation software and assess its performance in the clinical routine. The main research question was to determine how close non-professional medical personnel in the field of CMF/AI for automated segmentation applications could achieve the level of established companies (including the leading players and known start-ups). For that, we set up a research protocol that included the development of in-house segmentation software, followed by comparing an expert and an inexperienced user with a good anatomical understanding of the selected companies.

We use brand names that are/can be protected but are not marked with ®.

## 2. Materials and Methods

Our research protocol consists of setting up a fully automatic in-house segmentation software and comparing it with segmentation applications developed by established companies and manual segmentations performed by an inexperienced user with good anatomical understanding (surgeon with less than 50 segmentations) with regard to the ground truth performed by an expert (researcher with over 500 segmentations). We selected 210 head and neck DICOM (Digital Imaging and Communications in Medicine) files, where the mandibles were manually segmented. The comparison was made with twenty selected and anonymized DICOMs (ten computed tomography (CT) and ten cone-beam computed tomography (CBCT) images, with and without artifacts), where the expert provided the ground truth. For the analysis, we used standard surface- and volume-based metrics. For all segmentation steps, the time was measured (segmentation duration and postprocessing time: filling, smoothing, and exporting). The CNN development timeline is shown in Figure 1.

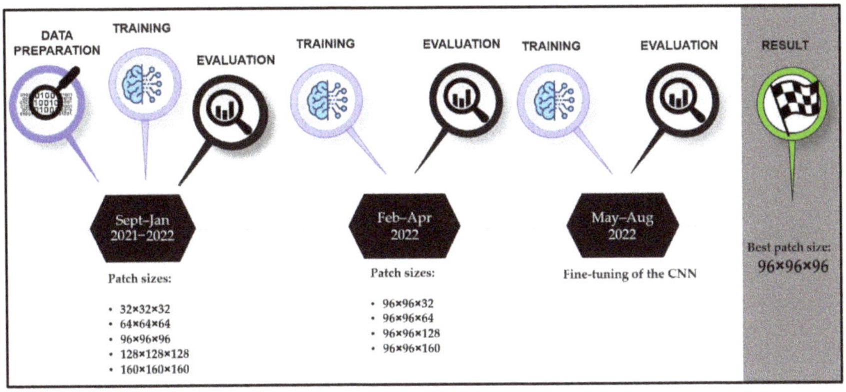


**Figure 1.** Timeline of the CNN development.

### 2.1. Statistical Analysis

The accuracy of the mandible segmentations was measured using the dice similarity coefficient (DSC), average surface distance (ASD), Hausdorff distance (HD), relative volume difference (RVD), volumetric overlap error (VOE), false positive rate (FPR), and false negative rate (FNR). The formulas for the calculation of these metrics are shown in Table 1.

**Table 1.** List of the metrics used in this study and their formula.

| Metric | Formula | Legend |
|---|---|---|
| Dice similarity coefficient (DSC) | $DSC = \frac{2\|A \cap B\|}{\|A\|+\|B\|} = \frac{2TP}{2TP+FP+FN}$ | The dice similarity coefficient measures the similarity between two sets of data. |
| Average surface distance (ASD) | $ASD = \frac{1}{n_A+n_B}\left(\sum_{i=1}^{n_A}\min_{b\in B}\|\|a_i-b\|\|_2+\sum_{j=1}^{n_B}\min_{a\in A}\|\|b_j-a\|\|_2\right)$ | The average surface distance is the average of all the distances between the surfaces of the ground truth and the volume. |
| Hausdorff distance (HD) | $d_H = \max\left\{\sup_{x\in X} d(x,Y), \sup_{y\in Y} d(X,y)\right\}$ | The Hausdorff distance is the maximum distance between the ground truth and the volume. |
| Relative volume difference (RVD) | $RVD = \frac{\|B\|-\|A\|}{\|A\|}$ | The relative volume difference measures the absolute size difference of the ground truth to the volume as a fraction of the ground truth. |
| Volumetric overlap error (VOE) | $VOE = 1 - \frac{DSC}{2-DSC}$ | The volumetric overlap error is the corresponding error metric of the dice similarity coefficient. |
| False positive rate (FPR) | $FPR = \frac{FP}{FP+TN}$ | The false positive rate is the probability that a positive result is given when the true value is negative. |
| False negative rate (FNR) | $FNR = \frac{FN}{FN+TP}$ | The false negative rate or miss rate is the probability that the analysis misses a true positive. |

### 2.2. CNN Development

#### 2.2.1. Training and Validation Data

For the training and validation of the Convolutional Neural Network (CNN), we relied on open-source data containing 504 DICOMs (Fluorodeoxyglucose-Positron Emission Tomography (FDG-PET) and CT images) of 298 patients that were diagnosed with cancer in the head and neck area. The databank is offered by the McGill University, Montreal, Canada, and the data acquisition took place between April 2006 and November 2014 [24]. A total of 160 DICOM files were selected to obtain heterogeneity regarding gender distribution, resolution, artifacts, and dentition, as shown in Table 2. The number of slices varies between 90 and 348, with an average of 170.5. The pixel spacing in the X and Y directions varies from 0.88 × 0.88 mm to 1.37 × 1.37 mm, whereas the slice thickness varies from 1.5 mm to 3.27 mm. The extended list is shown in Annex S1. The DICOM files were distributed among two datasets: the training dataset with 120 samples (60 with artifacts and 60 without artifacts) and the validation dataset with 40 samples (20 with artifacts and 20 without artifacts). Exclusion criteria were images of patients with brackets and osteosynthesis materials (screws and plates).

**Table 2.** List of characteristics of the images used for the training of the Convolutional Neural Network.

| Nr. Studies | With Artifacts | Without Artifacts—With Teeth | Without Artifacts—Without Teeth (Edentulous) |
|---|---|---|---|
| Female | 33 | 12 | 19 |
| Male | 47 | 28 | 21 |
| Male and Female | 80 | 40 | 40 |

#### 2.2.2. Test Data

For the test dataset, 10 CT and 10 CBCT images from the University Hospital of Basel were selected. Both subgroups contained five DICOM files with metallic artifacts and five without. The number of slices ranges from 169 to 489, with a mean value of 378. The pixel spacing in X and Y directions ranges from 0.25 × 0.25 mm to 0.59 × 0.59 mm, with a mean value of 0.35 × 0.35 mm, and the slice thickness varies from 0.25 mm to 3.0 mm, with a mean value of 0.71 mm. None of the CT images have an isotropic voxel spacing (voxel spacing and slice thickness have the same value), whereas 9 out of 10 CBCTs have isotropic spacing. These images are representative of the ones used in the clinical routine; therefore, they differ greatly in aspects such as image dimension, voxel spacing, layer thickness, noise, etc. The same exclusion criteria were applied for the test dataset as for the training dataset. All images were anonymized.

#### 2.2.3. Segmentation

The DICOMs for the training and validation were imported into Mimics Innovation Suite (Version 24.0, Materialise NV, Leuven, Belgium), whereas the test samples were imported later into Mimics Innovation Suite Version 25.0. A semi-automatic segmentation workflow was applied using the Threshold, Split Mask, Region Grow, Edit Mask, Multiple Slice Edit, Smart Fill, and Smooth Mask tools. The teeth were included in the segmentation, and the masks were filled (i.e., they do not contain any voids). The mandible and the inferior nerve canal were labeled as a single mask and exported as a Standard Tessellation Language (STL) file.

#### 2.2.4. Model Architecture

For the automatic segmentation of the mandible, the Medical Image Segmentation with Convolutional Neural Networks (MIScnn) Python library, Version 1.2.1 to 1.4.0 [25], was used. As architecture, a 3D U-Net, a Convolution Neural Network, was selected (Figure 2), which was developed for biomedical image segmentation [26]. The number of filters in the first layer (N filters) was set to 32, the number of layers of the U-Net structure (depth) was set to 4 as an activation function, the sigmoid function was used, and batch normalization was activated. The dice cross-entropy function was chosen as a loss function, which is a sum of the soft Dice Similarity Coefficient and the Cross-Entropy [27]. As normalization, the Z-score function was applied, and the image was resampled using a voxel spacing of 1.62 × 1.62 × 3.22 mm. The clipping subfunction was implemented to clip pixel values in a range between 50 and 3071 of the Hounsfield scale. The learning rate was set to 0.0001 at the beginning of the training, but through the Keras Callback function, it was reduced to 0.00001 once no further improvement was observed, with a patience of 10 epochs. Scaling, rotation, elastic deformation, mirroring, brightness, contrast changes, and Gaussian noise were used for data augmentation (a method to increase the number of training samples by slightly modifying/newly creating DICOMs from existing data to avoid overfitting and to improve the performance of the CNN). The models were trained for 1000 epochs with a NVIDIA RTX 3080 GPU (12 GB of VRAM), 64 GB of RAM, and an i9-11950H processor. The training time was about 100 h per model.

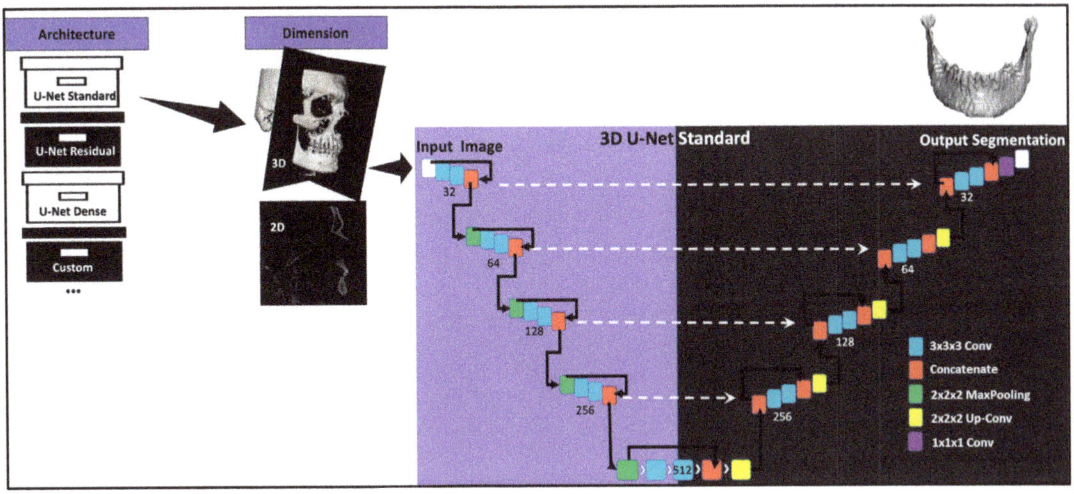


**Figure 2.** Architecture of the Convolutional Neural Network.

The CNN was trained in a two-phase approach. Firstly, the model was trained using five different cubical patch sizes (32 × 32 × 32, 64 × 64 × 64, 96 × 96 × 96, 128 × 128 × 128, and 160 × 160 × 160). In the second phase, the height of the best-performing input volume (96 × 96 × 96) was modified along the Z axis. Five further models with patch sizes of 96 × 96 × 32, 96 × 96 × 64, 96 × 96 × 128, and 96 × 96 × 160 were trained. The results are displayed in Table 3. The model trained with the 96 × 96 × 96 patch size (Figure 3) was the best-performing and, hence, was further improved by training it with 50 additional CT images from the University Hospital, Basel, and its performance was tested on the test dataset.

**Table 3.** The patch sizes with which the CNNs were trained; the reached dice similarity coefficient (DSC) and its standard deviation (SD); and the epoch when it was reached.

| Patch Size | Max. DSC | SD | Epoch |
|---|---|---|---|
| 32×32 × 32 | 0.222 | 0.073 | 545 |
| 64 × 64 × 64 | 0.838 | 0.110 | 840 |
| 96 × 96 × 32 | 0.857 | 0.067 | 635 |
| 96 × 96 × 64 | 0.902 | 0.060 | 1000 |
| **96 × 96 × 96** | **0.916** | **0.033** | **975** |
| 96 × 96 × 128 | 0.878 | 0.087 | 995 |
| 96 × 96 × 160 | 0.852 | 0.147 | 810 |

**Table 3.** *Cont.*

| Patch Size | Max. DSC | SD | Epoch |
|---|---|---|---|
| 128 × 128 × 128 | 0.907 | 0.038 | 915 |
| 160 × 160 × 160 | 0.860 | 0.077 | 725 |

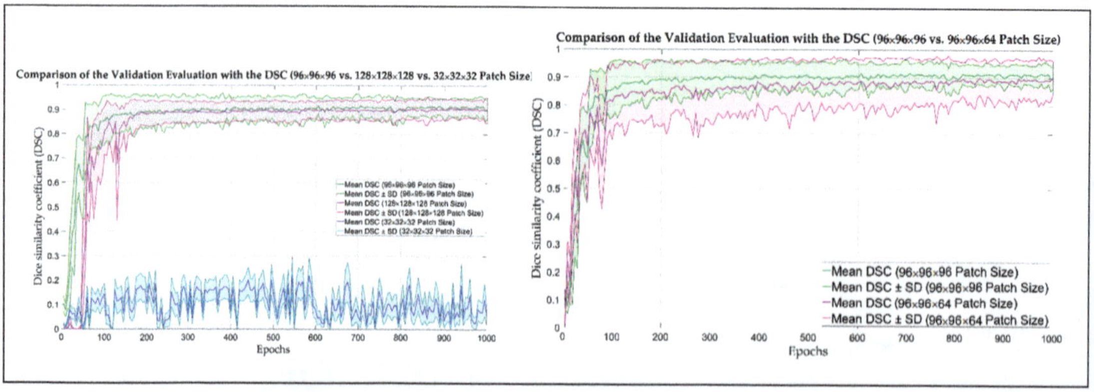


**Figure 3.** Graph of the evolution of the dice similarity coefficient (DCS) and its standard deviation (SD) of the validation samples for different patch size.

## 2.3. *Software Comparison*

### 2.3.1. Relu

Relu (Figure 4) is an established start-up that offers fully automated cloud-based segmentation for CBCT and CT images for applications in the Cranio-Maxillofacial field. The segmented anatomical structures are the toothless mandible, the mandibular teeth (each tooth individually), the inferior alveolar canal, the toothless maxillary complex, the maxillary teeth (each tooth individually), the maxillary sinuses, the pharynx, and the soft tissue. The bone segmentations include cortical and cancellous structures. Relu is ISO 13485 compliant and has a CE mark pending.

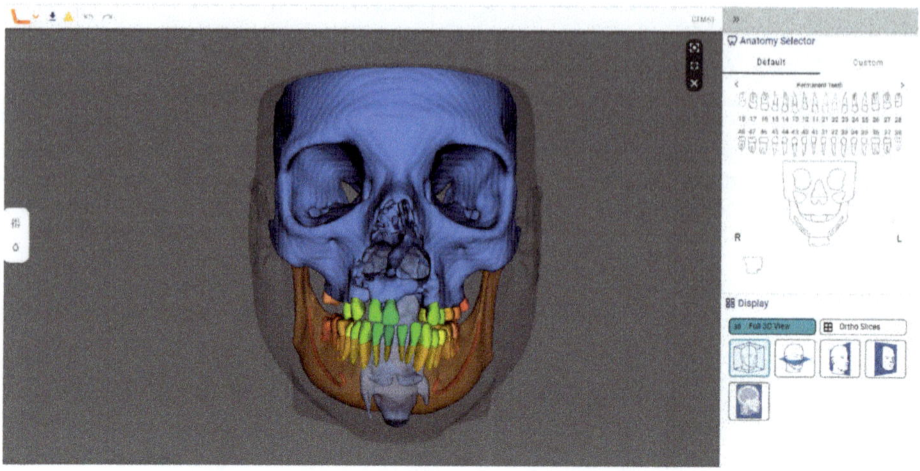


**Figure 4.** Relu's user interface (CT w/A 1 displayed).

For the segmentation of the mandible, the anonymized DICOM files of the test dataset were uploaded onto the cloud system (the company names it web application) and the segmentations were requested, but only for the mandible, mandibular teeth, and the inferior nerve canal, since these are the analyzed structures. After the segmentation was completed, these structures were combined directly in the cloud and downloaded as one STL file. This was then imported into Mimics (Version 25.0) and transformed into a mask, which was then manually filled with the "Smart Fill" tool. Afterward, the part was transformed into an object using the "Calculate Part tool", smoothed for 4 iterations with the "Smooth" tool at a factor of 0.4, and finally exported as an STL file.

With Relu, we encountered problems in 3 of the 20 test DICOMs during the segmentation process regarding voxel spacing, image orientation, and cropping. All transmitted problems were solved by the support team.

#### 2.3.2. Materialise Mimics Viewer

The Materialise Viewer (Figure 5) is a cloud-based platform for online visualization and segmentation of DICOM files. Fully automatic segmentation can be requested for CMF CBCT, heart CT, shoulder CT, hip CT, knee CT, knee MRI, and all bones CT. The Mimics Automatic Algorithms are part of the FDA 510(k) of Mimics Medical and standalone CE-marked medical devices.

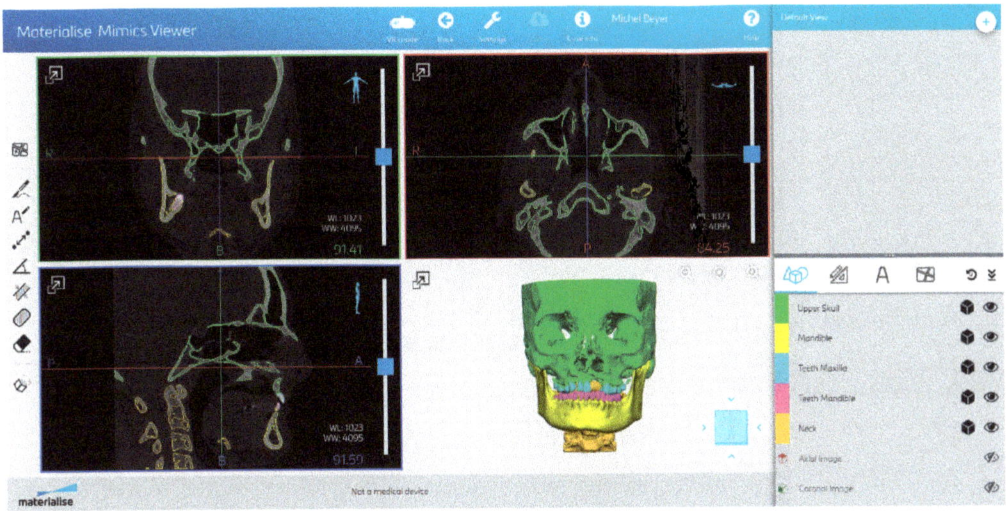


**Figure 5.** Materialise Viewer's user interface (CT w/A 1 displayed).

For the segmentation of the mandible, the CMF CBCT segmentation algorithm was used, which was designed to segment both CBCT and CT. The anonymized DICOM files of the test dataset were inserted into a Mimics file, which was then uploaded onto Mimics Viewer and the segmentation was requested. The output of the fully automatic segmentation was a Mimics file containing five segmented parts, which are called "Upper skull", "Mandible", "Teeth Maxilla", "Teeth Mandible", and "Neck", containing the anatomy of skull and maxilla, mandible, maxillary teeth, mandibular teeth, and neck, respectively. Only the cortical bone was segmented in the Materialise Mimics Viewer, not the cancellous bone. The inferior alveolar canal was not segmented.

The file was opened with Mimics (Version 25.0) and the parts were transformed into masks using the "Mask from Object" tool. The mask containing the mandible and the one containing the mandibular teeth were combined, and the holes inside the mandible were filled manually with the "Smart Fill" tool in order to make volumetric comparisons possible. In the cases where there were some holes in the surface of the model, we filled them without intervening in the segmentation of the cortical bone. Afterward, the part was transformed into an object using the "Calculate Part tool", smoothed for 4 iterations with the "Smooth" tool at a factor of 0.4, and finally exported as an STL file.

With Mimics Viewer, we encountered problems in 2 of the 20 test DICOMs during the segmentation process regarding image orientation and cropping. All transmitted problems were solved by the support team.

#### 2.3.3. Diagnocat

Diagnocat (Figure 6) is an established start-up that offers fully automated segmentation for CBCT images and prediagnosis for 2D dental X-rays. The segmented anatomical structures are the toothless mandible, the mandibular teeth (each tooth individually), the inferior alveolar canal, the toothless maxilla, the maxillary teeth (each tooth individually), the cranium, the airways, and the soft tissue. The bone segmentations include cortical and cancellous structures. Diagnocat has a CE mark.

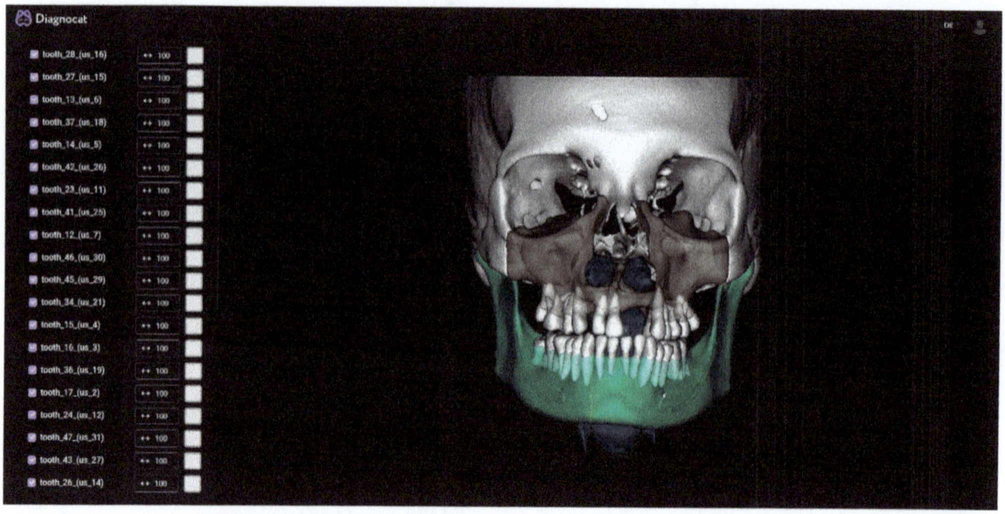


**Figure 6.** Diagnocat's user interface (CT w/A 1 displayed).

For the segmentation of the mandible, the anonymized DICOM files were uploaded onto the cloud system and the segmentations requested (all the structures as separated files option). After the segmentation was completed, the mandible, the inferior alveolar canal, and the mandibular teeth were downloaded and combined into a single file using Materialise 3-Matic (Version 17.0, Materialise NV, Leuven, Belgium). This was then imported into Mimics (Version 25.0) and transformed into an object using the "Calculate Part tool", smoothed for 4 iterations with the "Smooth" tool at a factor of 0.4, and finally exported as an STL file.

With Diagnocat, we encountered problems in all of the CT images and one CBCT image out of the twenty test DICOMs during the segmentation process. All these images had non-isotropic voxel spacing (CBCTs generally have isotropic voxel spacing, as shown in Annex S1–S5), which needed to be adapted. All transmitted problems were solved by the support team.

#### 2.3.4. Brainlab

The Brainlab Elements application (Figure 7) consists of multiple applications and backend services for image processing of medical data (data transfer and exchange, image co-registration, automatic image segmentation, manual contouring, object manipulation, trajectory planning, etc.). The anatomical structures that can be automatically segmented are the optic nerve, eye, midface, skull base, skull base anterior, skull base central, skull base posterior, orbit volume, skull, ethmoid bone, LeFort I Template, LeFort II Template, LeFort III Template, LeFort III-I Template, mandible, mandible body, mandible ramus, frontal bone, maxilla, nasal bone, orbit, orbit floor, orbit wall medial, zygomatic bone, occipital bone, parietal bone, sphenoid bone, and temporal bone. For all bony structures, the cortical and cancellous bones are segmented by Brainlab. Teeth are not part of the segmentation model.

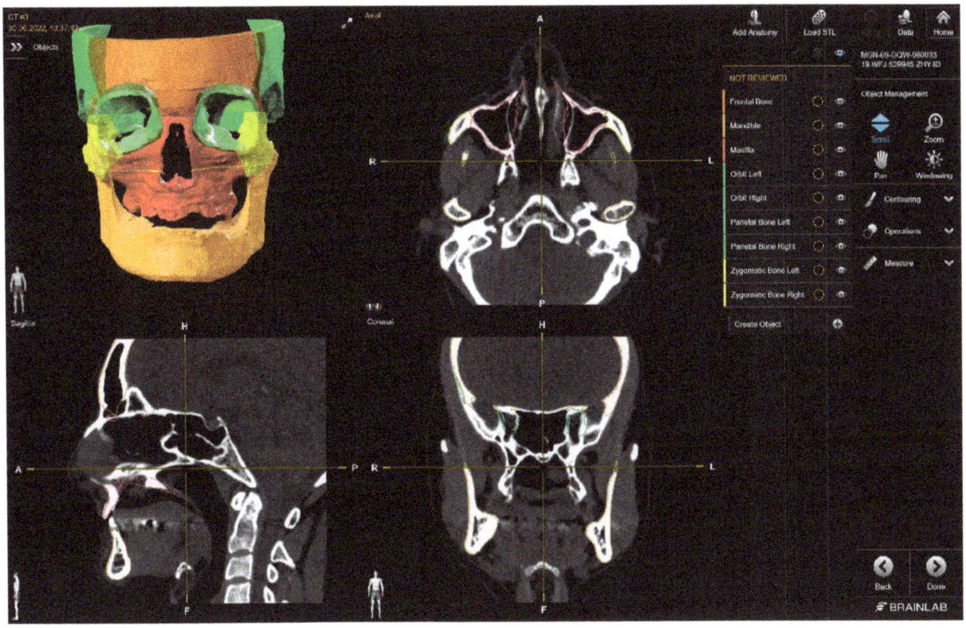


**Figure 7.** Brainlab's user interface (CT w/A 1 displayed).

The mandible was downloaded as an STL file and was then imported into Mimics (Version 25.0) and transformed into a mask, which was then manually filled with the "Smart Fill" tool. Afterward, the part was transformed into an object using the "Calculate Part tool", smoothed for 4 iterations with the "Smooth" tool at a factor of 0.4, and finally exported as an STL file.

With Brainlab, no problems were encountered during the segmentation process.

*2.4. Mandible Cutting*

The following three comparisons were made: one of the mandible with teeth, one of just the mandibular bone, and the last of just the mandibular teeth (as shown in Figure 8). In order to split the mandible into the mandibular teeth and the mandibular bone, 3-Matic was used. For each of the 20 mandibles in the test dataset, the ground truth was used to manually insert three cutting planes (one horizontal and two vertical planes), which were used to automatically cut and split the segmented mandibles for each company using the 3-Matic scripting tool. Two different STL files were obtained, one containing the mandibular bone and one containing the mandibular teeth.

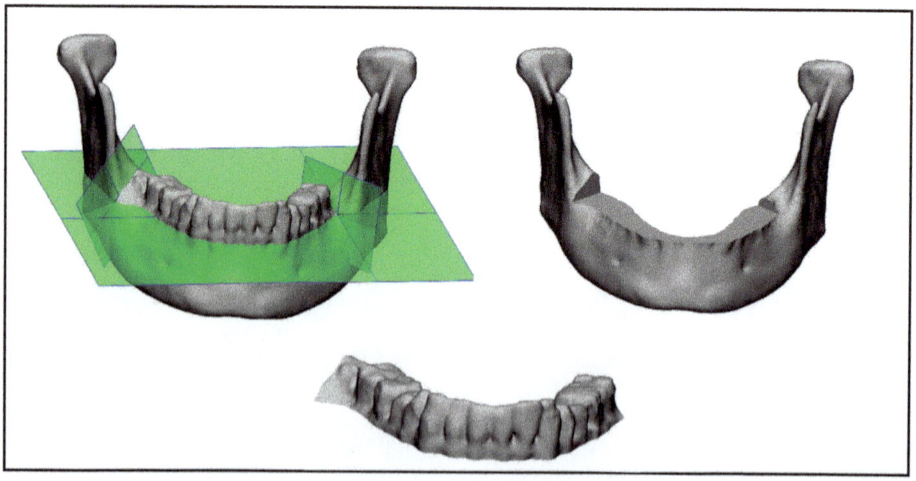

**Figure 8.** Cutting planes on mandible with teeth (**left**), mandibular bone (**right**), and mandibular teeth (**bottom**).

## 3. Results

The main results after all the assessments were made are as follows:

- Overall, Relu performed best if the mean DSC for the mandible with teeth (mean DSC of 0.938) and bone (mean DSC of 0.949) is taken into consideration, which was closely followed by Diagnocat and then Materialise, as displayed in Tables 4 and 6.
- Brainlab was only included for the assessment of the mandibular bone, as it does not offer teeth segmentation (mean DSC of 0.912), as displayed in Table 6.
- Materialise performed best over the other software in the assessment of the mandibular teeth (mean DSC of 0.864), as displayed in Table 5.
- We could observe that in all assessments, our in-house-developed software performed worst, obtaining the closest result in the mandibular bone comparison (mean DSC of 0.894), but achieved an accuracy of 94.24% in comparison to the best-performing software, as displayed in Tables 4–6.
- The segmentation performed by the inexperienced user with good anatomical understanding (CMF surgeon) had, for all assessments, the best mean DSC, as displayed in Tables 4–6.

For better visualization and understanding of the results, we chose to display in each category (CT with artifacts (Figure 9), CT without artifacts (Figure 10), CBCT with artifacts (Figure 11), and CBCT without artifacts (Figure 12)) the first segmented mandible. For that, we used the color mapping of the surface distance between the segmentation and the ground truth (where the segmentation is visible and the ground truth is hidden), with minimum and maximum ranges of −1.0 mm and +1.0 mm.

**Table 4.** Mean dice similarity coefficient (DSC) of the **mandible with teeth** comparison.

| | Manual (Beginner) | In-House | Relu | Materialise | Diagnocat | Brainlab |
|---|---|---|---|---|---|---|
| Mean CT w/A | 0.961 | 0.885 | 0.939 | 0.914 | 0.927 | - |
| Mean CT w/o A | 0.968 | 0.891 | 0.935 | 0.903 | 0.921 | - |
| Mean CBCT w/A | 0.951 | 0.863 | 0.938 | 0.947 | 0.941 | - |
| Mean CBCT w/o A | 0.958 | 0.899 | 0.939 | 0.956 | 0.947 | - |
| Mean | **0.960** | **0.884** | **0.938** | **0.930** | **0.934** | - |

**Table 5.** Mean dice similarity coefficient (DSC) of the **mandibular teeth** comparison.

| | Manual (Beginner) | In-House | Relu | Materialise | Diagnocat | Brainlab |
|---|---|---|---|---|---|---|
| Mean CT w/A | 0.923 | 0.787 | 0.814 | 0.838 | 0.817 | - |
| Mean CT w/o A | 0.953 | 0.818 | 0.792 | 0.847 | 0.797 | - |
| Mean CBCT w/A | 0.838 | 0.762 | 0.858 | 0.837 | 0.853 | - |
| Mean CBCT w/o A | 0.935 | 0.841 | 0.889 | 0.935 | 0.903 | - |
| Mean | **0.912** | **0.802** | **0.838** | **0.864** | **0.842** | - |

**Table 6.** Mean dice similarity coefficient (DSC) of the **mandibular bone** comparison.

| | Manual (Beginner) | In-House | Relu | Materialise | Diagnocat | Brainlab |
|---|---|---|---|---|---|---|
| Mean CT w/A | 0.968 | 0.898 | 0.958 | 0.925 | 0.943 | 0.948 |
| Mean CT w/o A | 0.969 | 0.900 | 0.952 | 0.909 | 0.936 | 0.943 |
| Mean CBCT w/A | 0.963 | 0.873 | 0.944 | 0.959 | 0.948 | 0.852 |
| Mean CBCT w/o A | 0.962 | 0.905 | 0.943 | 0.958 | 0.950 | 0.903 |
| Mean | **0.966** | **0.894** | **0.949** | **0.938** | **0.944** | **0.912** |

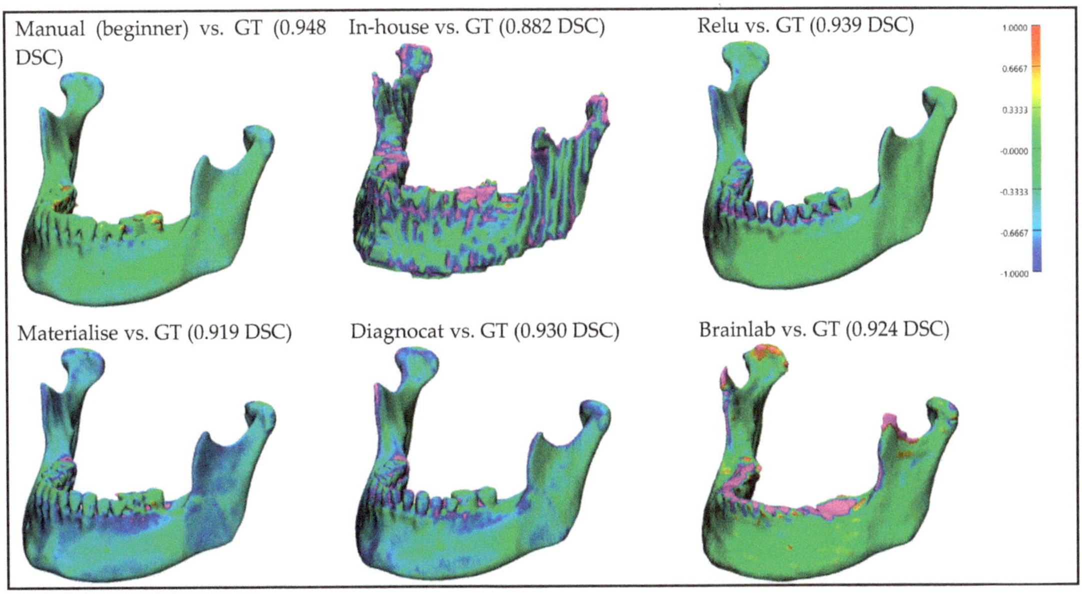


**Figure 9. CT with artifacts:** Color mapping of the surface distance between the segmented mandibles of the CT w/A 1 image and the ground truth (GT).

Timing: We calculated the mean values of the segmentation times for CT and CBCT with/without artifacts (Figure 13). We have shown that our in-house model performed best with the lowest mean time (2′03″), followed by Brainlab (3′54″) and Diagnocat (4′52″). The manually segmented mandibles (those from the expert and the inexperienced user) showed similar timings (26′09″ and 22′54″, respectively). Materialise showed the highest mean value (85′54″).

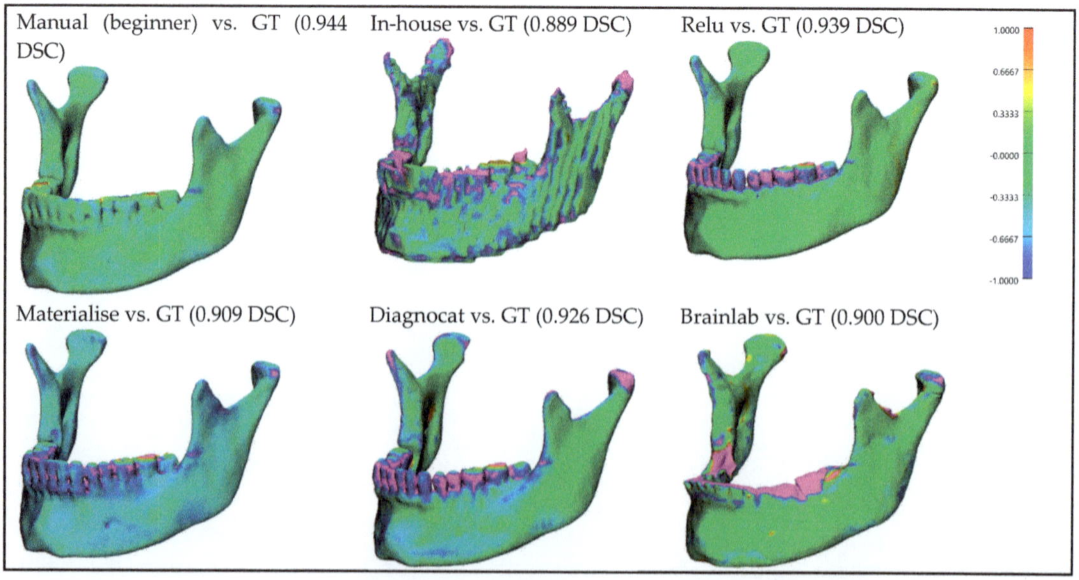


**Figure 10. CT without artifacts:** Color mapping of the surface distance between the segmented mandibles of the CT w/o A 1 image and the ground truth (GT).

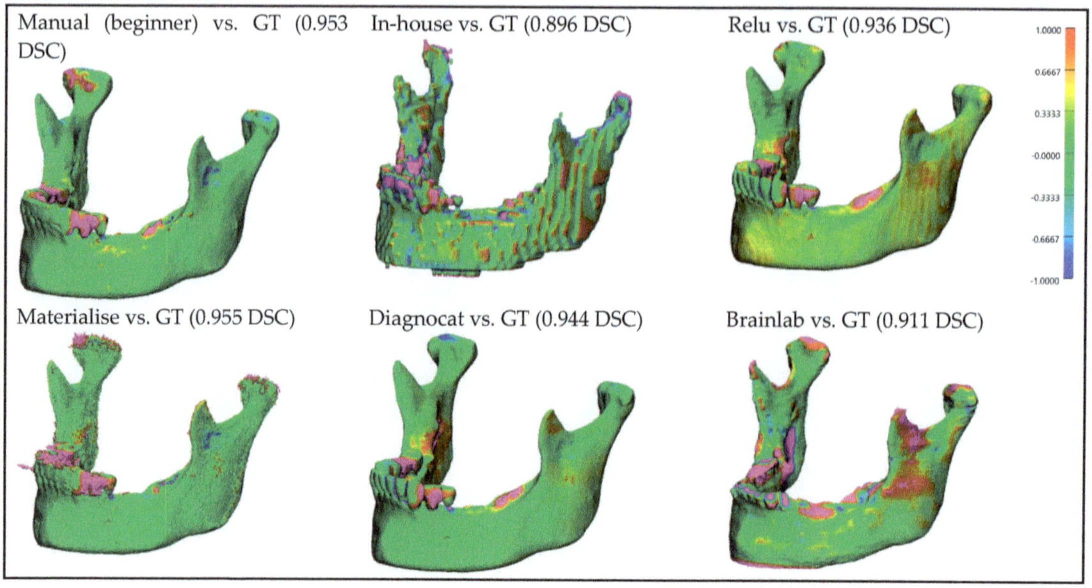


**Figure 11. CBCT with artifacts:** Color mapping of the surface distance between the segmented mandibles of the CBCT w/A 1 image and the ground truth (GT).

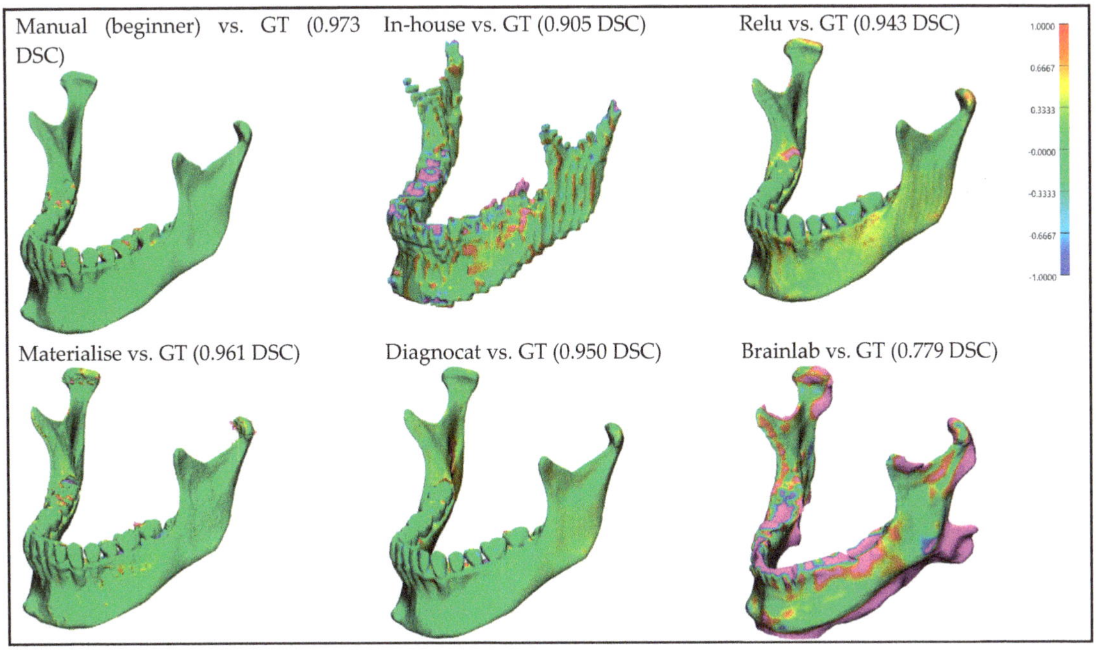


**Figure 12. CBCT <u>without</u> artifacts:** Color mapping of the surface distance between the segmented mandibles of the CBCT w/o A 1 image and the ground truth (GT).

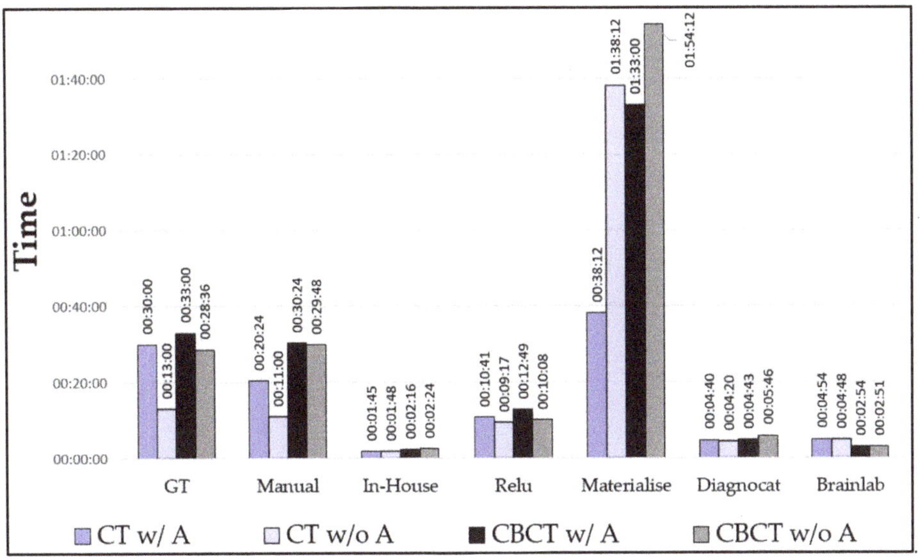


**Figure 13.** Graph of the mean timing for the segmentations.

## 4. Discussions

In a clinical routine, three important factors stand out: segmentation accuracy, cost, and time. The segmentation accuracy result was best for manual segmentation in all comparisons, followed by Relu, Diagnocat, and Materialise, which all performed very similarly to one another. Brainlab could only be included in the comparison of the mandibular bone because the segmentation did not include the teeth, as its main activity offers intraoperative navigation solutions. Our in-house-developed CNN performed worst in all of the comparisons. We encountered the problem that the segmented mandibles of our in-house CNN had a cubical surface, which was probably due to a too high voxel spacing parameter. This problem could not be fixed and will require further training and improvements to the model. The advantage of our system is that it has higher stability than the other software included in our study. We could upload all the DICOM files without any modifications and obtain a complete segmentation. The other software encountered some problems with DICOMs containing not only the skull but also, e.g., the thorax, and needed preprocessing (cropping) in order to obtain the segmentation. A further problem was with the handling of CT images, because some systems were only trained on CBCT images, and in many cases, images without isotropic voxel spacing were not supported and had to be modified. Additionally, it is worth mentioning that not all the DICOM file orientations were supported. Figures 9 and 10 show that for CT images, the segmented mandibles from Materialise and Diagnocat had a slight inaccuracy in the segmentation of the mandibular bone compared to those from Relu or Brainlab, which was probably due to different thresholds used for the clipping during the training. Finally, the manual segmentation may have performed better than other automatic systems due to a similar segmentation protocol as the one for the ground truth. The same could apply to our in-house-developed CNN, which may have performed better because it was trained with a dataset prepared by following the same segmentation protocol. Using Mimics, which is developed by Materialise, for the manual segmentation (training and test data) and the filling process, could have had a positive influence on the final outcomes. Furthermore, the filling process of the mandibles, which was performed manually and was needed due to the different segmentation approaches, could be subject to bias. Pricing is also a relevant factor that needs to be considered. As we were offered the segmentations by the companies for research purposes, pricing was not further investigated in this study. The timing may vary due to the fact that most of the companies offer a cloud service, which, depending on the server load and internet connection, affects the segmentation time. Additionally, our ground truth implies that a manual segmentation process can differ from the anatomical specimen ground truth, which implies a scanning process. Other studies are necessary to compare the segmentations with laser-scanned mandibles (anatomical specimens) as the ground truth to improve accuracy.

## 5. Conclusions

In our study, we wanted to find out if non-professional medical personnel could become close to segmentation software developed by established companies, following a clearly defined research protocol. The results showed that our in-house-developed model achieved an accuracy of 94.24% compared to the best-performing software. We also conclude that the segmentation performed by an inexperienced user with good anatomical understanding achieved the best result compared to all the other companies included in the study.

The timing required to automatically segment a mandible was, for almost all of the software, lower than the manual segmentation.

We can deduce that in order to obtain better quality segmentations, the CNN has to be trained with a dataset containing a large number of highly variable images (e.g., older and newer DICOM files, different types of DICOMs (CT and CBCT), and different image sizes, including different regions of interest and from different centers) that is constantly updated and enlarged due to the constantly improving image technologies.

To fulfill today's expectations of personalized medicine, digital workflows, including segmentation, need to offer stable solutions. Answers must be found for the current problems that are often encountered during the segmentation process: artifacts, amount of noise, voxel spacing, the size of the image, DICOM type, and image orientation. All these problems were reported to the companies so that solutions could be elaborated in the future.

For the future, the first step for implementing fully automated digital workflows is to generate accurate segmentations of the patient's anatomy, which will be possible after solving the above-mentioned issues.

Once the above-mentioned issues are solved, these software can be implemented in fully automated digital workflows, allowing new clinical applications, such as intraoperatively 3D-printed patient-specific implants, even in emergency situations.

**Supplementary Materials:** The following supporting information can be downloaded at: https://www.mdpi.com/article/10.3390/bioengineering10050604/s1, Annex S1: Test data DICOM properties; Annex S2: Dice similarity coefficient (DSC) of the mandible with teeth comparison; Annex S3: Dice similarity coefficient (DSC) of the mandibular bone comparison; Annex S4: Dice similarity coefficient (DSC) of the mandibular teeth comparison; Annex S5: Mean values for the comparison of the mandible with teeth segmentations, mandibular bone and mandibular teeth to the ground truth by using the dice similarity coefficient (DSC), average surface distance (ASD), Hausdorff distance (HD), relative volume difference (RVD), volumetric overlap error (VOE), false positive rate (FPR), and false negative rate (FNR).

**Author Contributions:** Conceptualization, R.R.I. and M.B.; methodology R.R.I. and M.B.; software M.B. and R.R.I., validation, R.R.I. and M.B.; formal analysis, R.R.I. and M.B.; investigation, R.R.I. and M.B.; resources, R.R.I. and M.B.; data curation, M.B. and R.R.I.; writing—original draft preparation, R.R.I. and M.B.; writing—review and editing, R.R.I., M.B., C.K. and F.M.T.; visualization, R.R.I. and M.B.; supervision, C.K. and F.M.T.; project administration, R.R.I. and M.B.; All authors have read and agreed to the published version of the manuscript.

**Funding:** This research was funded by the Werner Siemens Foundation (MIRACLE II/Smart Implants) and the Innovation Focus Regenerative Surgery, University Hospital Basel.

**Institutional Review Board Statement:** Not applicable.

**Informed Consent Statement:** Informed consent was obtained from all subjects involved in the study.

**Data Availability Statement:** Data are contained within the article. Additional information can be obtained from the corresponding author upon reasonable request.

**Conflicts of Interest:** The authors declare no conflict of interest.

## Abbreviations

| | |
|---|---|
| 3D | Three-dimensional |
| AI | Artificial Intelligence |
| AR | Augmented Reality |
| ASD | Average Surface Distance |
| CBCT | Cone-Beam Computed Tomography |
| CMF | Cranio-Maxillofacial |
| CNN | Convolutional Neural Network |
| CT | Computed Tomography |
| DICOM | Digital Imaging and Communications in Medicine |
| DSC | Dice Similarity Coefficient |
| FDG-PET | Fluorodeoxyglucose-Positron Emission Tomography |
| GT | Ground Truth |
| FNR | False Negative Rate |
| FPR | False Positive Rate |
| HD | Hausdorff distance |
| MIScnn | Medical Image Segmentation with Convolutional Neural Networks |
| RAS | Right, Anterior Superior |
| RVD | Relative Volume Difference |

| | |
|---|---|
| SD | Standard Deviation |
| STL | Standard Tessellation Language |
| VOE | Volumetric Overlap Error |
| VR | Virtual Reality |
| VSP | Virtual Surgical Planning |

## References


1. Ganry, L.; Quilichini, J.; Bandini, C.M.; Leyder, P.; Hersant, B.; Meningaud, J.P. Three-dimensional surgical modelling with an open-source software protocol: Study of precision and reproducibility in mandibular reconstruction with the fibula free flap. *Int. J. Oral Maxillofac. Surg.* **2017**, *46*, 946–957. [CrossRef] [PubMed]
2. Weinstock, P.; Prabhu, S.P.; Flynn, K.; Orbach, D.B.; Smith, E. Optimizing cerebrovascular surgical and endovascular procedures in children via personalized 3D printing. *J. Neurosurg. Pediatr.* **2015**, *16*, 584–589. [CrossRef] [PubMed]
3. Enciso, R.; Memon, A.; Mah, J. Three-dimensional visualization of the craniofacial patient: Volume segmentation, data integration and animation. *Orthod. Craniofac. Res.* **2003**, *6* (Suppl. 1), 66–182. [CrossRef]
4. Marschall, J.S.; Dutra, V.; Flint, R.L.; Kushner, G.M.; Alpert, B.; Scarfe, W.; Azevedo, B. In-House Digital Workflow for the Management of Acute Mandible Fractures. *J. Oral Maxillofac. Surg.* **2019**, *77*, 2084.e1–2048.e9. [CrossRef] [PubMed]
5. Sigron, G.R.; Barba, M.; Chammartin, F.; Msallem, B.; Berg, B.I.; Thieringer, F.M. Functional and Cosmetic Outcome after Reconstruction of Isolated, Unilateral Orbital Floor Fractures (Blow-Out Fractures) with and without the Support of 3D-Printed Orbital Anatomical Models. *J. Clin. Med.* **2021**, *10*, 3509. [CrossRef]
6. Valls-Ontañón, A.; Ascencio-Padilla, R.D.J.; Vela-Lasagabaster, A.; Sada-Malumbres, A.; Haas-Junior, O.L.; Masià-Gridilla, J.; Hernández-Alfaro, F. Relevance of 3D virtual planning in predicting bony interferences between distal and proximal fragments after sagittal split osteotomy. *Int. J. Oral Maxillofac. Surg.* **2020**, *49*, 1020–1028. [CrossRef] [PubMed]
7. Weissheimer, A.; de Menezes, L.M.; Sameshima, G.T.; Enciso, R.; Pham, J.; Grauer, D. Imaging software accuracy for 3-dimensional analysis of the upper airway, American. *J. Orthod. Dentofac. Orthop.* **2012**, *142*, 801–803. [CrossRef]
8. El, H.; Palomo, J.M. Measuring the airway in 3 dimensions: A reliability and accuracy study. *Am. J. Orthod. Dentofacial. Orthop.* **2010**, *137* (Suppl. 4), S50.e1–S50.e9. [CrossRef]
9. Wang, L.; Gao, Y.; Shi, F.; Li, G.; Chen, K.-C.; Tang, Z.; Xia, J.J.; Shen, D. Automated segmentation of dental CBCT image with prior-guided sequential random forests. *Med. Phys.* **2016**, *43*, 336. [CrossRef]
10. Lo Giudice, A.; Ronsivalle, V.; Grippaudo, C.; Lucchese, A.; Muraglie, S.; Lagravère, M.O.; Isola, G. One Step before 3D Printing-Evaluation of Imaging Software Accuracy for 3-Dimensional Analysis of the Mandible: A Comparative Study Using a Surface-to-Surface Matching Technique. *Materials* **2020**, *13*, 2798. [CrossRef]
11. Yang, W.F.; Su, Y.X. Artificial intelligence-enabled automatic segmentation of skull CT facilitates computer-assisted craniomaxillofacial surgery. *Oral. Oncol.* **2021**, *118*, 105360. [CrossRef] [PubMed]
12. Verhelst, P.J.; Smolders, A.; Beznik, T.; Meewis, J.; Vandemeulebroucke, A.; Shaheem, E.; Van Gerven, A.; Willems, H.; Politis, C.; Jacobs, R. Layered deep learning for automatic mandibular segmentation in cone-beam computed tomography. *J. Dent.* **2021**, *114*, 103786. [CrossRef] [PubMed]
13. Heimann, T.; Meinzer, H.-P. Statistical shape models for 3D medical image segmentation: A review. *Med. Image Anal.* **2009**, *13*, 543–563. [CrossRef] [PubMed]
14. Litjens, G.; Kooi, T.; Bejnordi, B.E.; Setio, A.A.A.; Ciompi, F.; Ghafoorian, M.; van der Laak, J.A.W.M.; van Ginneken, B.; Sánchez, C.I. A survey on deep learning in medical image analysis. *Med. Image Anal.* **2017**, *42*, 60–88. [CrossRef] [PubMed]
15. Minnema, J.; van Eijnatten, M.; Kouw, W.; Diblen, F.; Mendrik, A.; Wolff, J. CT image segmentation of bone for medical additive manufacturing using a convolutional neural network. *Comput. Biol. Med.* **2018**, *103*, 130–139. [CrossRef]
16. Ezhov, M.; Gusarev, M.; Golitsyna, M.; Yates, J.M.; Kushnerev, E.; Tamimi, D.; Aksoy, S.; Shumilov, E.; Sanders, A.; Oehan, K. Clinically applicable artificial intelligence system for dental diagnosis with CBCT. *Sci. Rep.* **2021**, *11*, 15006. [CrossRef] [PubMed]
17. Orhan, K.; Shamshiev, M.; Ezhov, M.; Plaksin, A.; Kurbanova, A.; Unsal, G.; Gusarev, M.; Golitsyna, M.; Aksoy, S.; Misieli, M.; et al. AI-based automatic segmentation of craniomaxillofacial anatomy from CBCT scans for automatic detection of pharyngeal airway evaluations in OSA patients. *Sci. Rep.* **2022**, *12*, 11863. [CrossRef]
18. Çiçek, Ö.; Abdulkadir, A.; Lienkamp, S.; Brox, T.; Ronneberger, O. *3D U-Net: Learning Dense Volumetric Segmentation from Sparse Annotation. Medical Image Computing and Computer-Assisted Intervention (MICCAI)*; Springer: Berlin/Heidelberg, Germany, 2016; Volume 9901, pp. 424–432.
19. Ambellan, F.; Tack, A.; Ehlke, M.; Zachow, S. Automated segmentation of knee bone and cartilage combining statistical shape knowledge and convolutional neural networks: Data from the Osteoarthritis Initiative. *Med. Image Anal.* **2019**, *52*, 109–118. [CrossRef]
20. Lo Giudice, A.; Ronsivalle, V.; Spampinato, C.; Leonardi, R. Fully automatic segmentation of the mandible based on convolutional neural networks (CNNs). *Orthod. Craniofac. Res.* **2021**, *24* (Suppl. 2), 100–107. [CrossRef]
21. Li, Q.; Chen, K.; Han, L.; Zhuang, Y.; Li, J.; Lin, J. Automatic tooth roots segmentation of cone beam computed tomography image sequences using U-net and RNN. *J. X-ray Sci. Technol.* **2020**, *28*, 905–922. [CrossRef]
22. Kwak, G.H.; Kwak, E.J.; Song, J.M.; Park, H.R.; Jung, Y.-H.; Cho, B.-H.; Hui, P.; Hwang, J.J. Automatic mandibular canal detection using a deep convolutional neural network. *Sci. Rep.* **2020**, *10*, 5711. [CrossRef] [PubMed]

23. Musatian, S.A.; Lomakin, A.V.; Sartasov SYu Popyvanov, L.K.; Monakhov, I.B.; Chizhova, A.S. Medical Images Segmentation Operations. *Trudy ISP RAN/Proc. ISP RAS* **2018**, *30*, 183–194. [CrossRef] [PubMed]
24. Vallières, M.; Kay-Rivest, E.; Perrin, L.J.; Liem, X.; Furstoss, C.; Aerts, H.J.W.L.; Khaouam, N.; Nguyen-Tan, P.F.; Wang, C.-S.; Sultanem, K.; et al. Radiomics strategies for risk assessment of tumour failure in head-and-neck cancer. *Sci. Rep.* **2017**, *7*, 10117. [CrossRef] [PubMed]
25. Müller, D.; Kramer, F. MIScnn: A framework for medical image segmentation with convolutional neural networks and deep learning. *BMC Med. Imaging* **2021**, *21*, 12. [CrossRef]
26. Ronneberger, O.; Fischer, P.; Brox, T. U-Net: Convolutional Networks for Biomedical Image Segmentation. In *Medical Image Computing and Computer-Assisted Intervention—MICCAI 2015*; Navab, N., Hornegger, J., Wells, W., Frangi, A., Eds.; Lecture Notes in Computer Science; Springer: Cham, Switzerland, 2015; Volume 9351. [CrossRef]
27. Isensee, F.; Jaeger, P.F.; Kohl, S.A.A.; Petersen, J.; Maier-Hein, K.H. nnU-Net: A self-configuring method for deep learning-based biomedical image segmentation. *Nat. Methods* **2021**, *18*, 203–211. [CrossRef]

bioengineering

Article

# A Lightweight Deep Learning Network on a System-on-Chip for Wearable Ultrasound Bladder Volume Measurement Systems: Preliminary Study

Hyunwoo Cho [1], Ilseob Song [2,3], Jihun Jang [2,3,*] and Yangmo Yoo [1,3,4,*]

[1] Department of Electronic Engineering, Sogang University, Seoul 04107, Republic of Korea; hyunwoocho@sogang.ac.kr
[2] Medical Solutions Institute, Sogang University, Seoul 04107, Republic of Korea; ilseob@sogang.ac.kr
[3] Edgecare Inc., TE1103, 35 Baekbeom-ro, Mapo-gu, Seoul 04107, Republic of Korea
[4] Department of Biomedical Engineering, Sogang University, Seoul 04107, Republic of Korea
* Correspondence: jhjang@sogang.ac.kr (J.J.); ymyoo@sogang.ac.kr (Y.Y.)

**Abstract:** Bladder volume assessments are crucial for managing urinary disorders. Ultrasound imaging (US) is a preferred noninvasive, cost-effective imaging modality for bladder observation and volume measurements. However, the high operator dependency of US is a major challenge due to the difficulty in evaluating ultrasound images without professional expertise. To address this issue, image-based automatic bladder volume estimation methods have been introduced, but most conventional methods require high-complexity computing resources that are not available in point-of-care (POC) settings. Therefore, in this study, a deep learning-based bladder volume measurement system was developed for POC settings using a lightweight convolutional neural network (CNN)-based segmentation model, which was optimized on a low-resource system-on-chip (SoC) to detect and segment the bladder region in ultrasound images in real time. The proposed model achieved high accuracy and robustness and can be executed on the low-resource SoC at 7.93 frames per second, which is 13.44 times faster than the frame rate of a conventional network with negligible accuracy drawbacks (0.004 of the Dice coefficient). The feasibility of the developed lightweight deep learning network was demonstrated using tissue-mimicking phantoms.

**Keywords:** deep learning; semantic segmentation; automatic volume measurement; ultrasound bladder scanner; edge computing; urinary disease

**Citation:** Cho, H.; Song, I.; Jang, J.; Yoo, Y. A Lightweight Deep Learning Network on a System-on-Chip for Wearable Ultrasound Bladder Volume Measurement Systems: Preliminary Study. *Bioengineering* **2023**, *10*, 525. https://doi.org/10.3390/bioengineering10050525

Academic Editors: Paolo Zaffino and Maria Francesca Spadea

Received: 29 March 2023
Revised: 17 April 2023
Accepted: 25 April 2023
Published: 26 April 2023



## 1. Introduction

Bladder volume measurements are commonly used in managing urinary diseases, such as urinary incontinence and benign prostate enlargement. Urinary catheterization is often used for measuring bladder volume in many cases, e.g., postoperative urinary retention [1], but it yields a high risk of urinary tract infection. To minimize unnecessary urinary catheterization, several studies have been conducted to analyze the impact and proper cycles of urinary catheterization [2,3]. Measuring post-void residual urine (PVR) is regarded as an effective way to reduce unnecessary catheterization. Additionally, PVR is a useful predictor of various diseases, such as prostatism and urinary tract infection [4,5]. To maximize the advantages of PVR measurements, a fast and accurate PVR measurement method is needed.

Ultrasound imaging (US) is a noninvasive, cost-effective, and real-time imaging modality that has been shown to be one of the most accurate and effective methods for measuring PVR [6–9]. Several studies have also demonstrated that US can potentially be used in point-of-care (POC) settings [10–12]. Recently, the development of portable US imaging devices for measuring bladder volume has been proposed [13,14]. Despite its usefulness, US has several limitations for measuring PVR in POC settings. One of the most challenging problems is its high operator dependency, which makes interpreting ultrasound images without professional experience and expertise difficult. Additionally, in POC settings, limited resources such as computing power and less experienced clinicians (e.g., nurses and care providers) can be problematic. As a result, the ultrasound image quality may be degraded, leading to misinterpretation or difficulties in PVR analysis.

To decrease operator dependency, an automatic bladder volume measurement method is needed. Traditionally, mechanical ultrasound scanning systems, such as a wobbling probe, have been used for PVR measurements. However, these methods require a prescan process to allocate the probe to the proper location before an actual volume measurement. Additionally, with mechanical scanning systems, ultrasound images or bladder volume measurements cannot be carried out in real time, resulting in inefficient repetitions of measurements. Moreover, patient motion during long scanning times may cause errors in PVR measurements. To address these issues, the need for a real-time image-based bladder measurement system has emerged.

To measure bladder volumes in real time, several studies have introduced image-based bladder volume measurements using various image analysis techniques, such as segmentation. Recent advances in deep learning and computer vision techniques have shown promising results for various tasks, including segmentation of regions of interest (e.g., organs and masses) in ultrasound images. In addition, deep learning techniques have been applied for analyzing urine in ultrasound images [15–17]. While these studies have shown that deep learning models can accurately segment the bladder and measure PVR volume, these tasks were primarily conducted on highly complex computing resources such as graphic processing units (GPUs). Additionally, in previous studies, ultrasound images were acquired by commercial cart-based ultrasound systems. In contrast, in POC settings, ultrasound images are collected by portable ultrasound systems so the imaging quality may be degraded due to the compactness and low computational power of these systems. This may reduce the accuracy of PVR measurements with deep learning models.

In this study, to address this issue, a lightweight deep-learning model for a portable bladder volume measurement system is proposed. Our proposed system was designed to detect PVR and segment the bladder region in ultrasound images with much fewer parameters; subsequently, an algorithm was employed to automatically measure the bladder volume using the segmentation results. Additionally, considering system integration, to improve its execution time in portable settings, the developed deep learning model was optimized with a fixed-point quantization technique. As a result, the optimized model could measure the bladder volume accurately with fewer than 1 million parameters on a low-resource SoC at high frame rates. The feasibility of our proposed automatic bladder volume measurement system was demonstrated by using various tissue-mimicking phantoms.

## 2. Materials and Methods

### 2.1. Data Acquisition from the Portable Ultrasound System

In this study, a system-on-chip (SoC)-based portable ultrasound system (EdgeFlow UH10w, Edgecare Inc., Seoul, Republic of Korea) was used to acquire ultrasound bladder images for training and validating the designed deep learning model. As shown in Figure 1, the commercial ultrasound system includes a SoC, a front-end processing module, and a power module and uses two 32-channel high-voltage (HV) pulsers and a transmit/receive (T/R) switch to control the cross-array probe. Front-end processing involves low noise amplification, time gain compensation, programmable gain amplification, and analog-to-digital conversion. Back-end processing is performed using the programmable logics (PL) on the SoC, with data transfer to the processing system (PS) via a direct memory access (DMA) engine. The signal is then reconstructed into an image using digital scan conversion. To acquire sagittal and transverse images simultaneously, a T-shaped array consisting of two phased array probes was used with the portable ultrasound system. The received radio-frequency signals were processed in PL in the SoC (Zynq Ultrascale+, Xilinx Inc., San Jose, CA, USA) by performing receive beamforming, quadrature demodulation, envelope detection, and log compression. The processed signal was then reconstructed into an image with a height of 330 pixels and a width of 570 pixels by the PS.

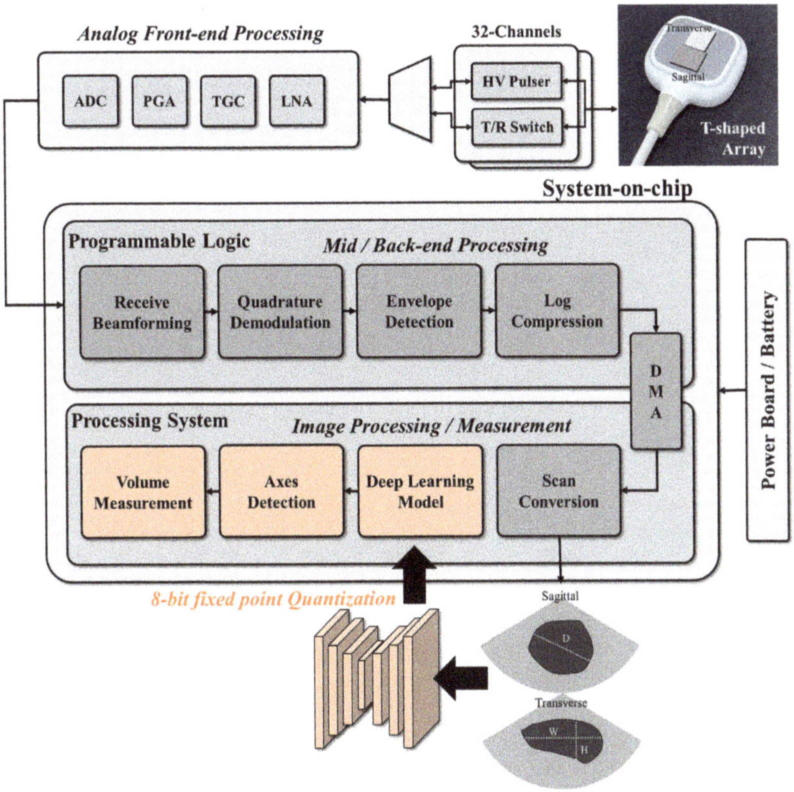


**Figure 1.** Block diagram showing the processing chains of the integrated system for ultrasound image acquisition and analysis. The gray boxes indicate the original processing blocks of the device, and the orange boxes denote the integrated processing blocks of this study. The deep learning model, trained on biplane images, is implemented on the SoC for bladder segmentation and post-void residual (PVR) detection, enabling automatic bladder volume measurement. To optimize performance, the deep learning model is quantized.

The proposed bladder measurement method, based on deep learning, was developed to be integrated into a portable ultrasound system using a system-on-chip (SoC). As depicted in Figure 1, the deep learning network was designed to perform segmentation and classification on the ultrasound images after digital scan conversion (DSC), identifying regions of interest (ROIs) and detecting the bladder. Once the bladder is detected on the image, the bladder volume is estimated by using the length of the axes.

To collect a dataset with high variability, various gain and depth settings were used. The ultrasound bladder images were obtained from two tissue-mimicking phantoms: an intravesical urine volume measurement phantom (US-16, Kyoto Kagaku, Kyoto, Japan) with urine volumes of 50 mL, 150 mL, and 300 mL, and a multimodality pelvic phantom (Model 048A, CIRS, Norfolk, VA, USA). A total of 1306 images with a bladder and 2095 images without a bladder were collected. The bladder images were randomly divided into 1044 images for training and 262 images for validation, with each image labeled with a corresponding mask for the segmentation task. The images without a bladder were divided into 1675 images for training and 420 images for validation for the classification task. To capture ROIs of various sizes, the dataset was collected by randomly selecting locations with a free hand on a static phantom. To validate the size and distribution of the dataset, the accuracies on the training phase and validation phase are compared. Examples of the dataset are shown in Figure 2.

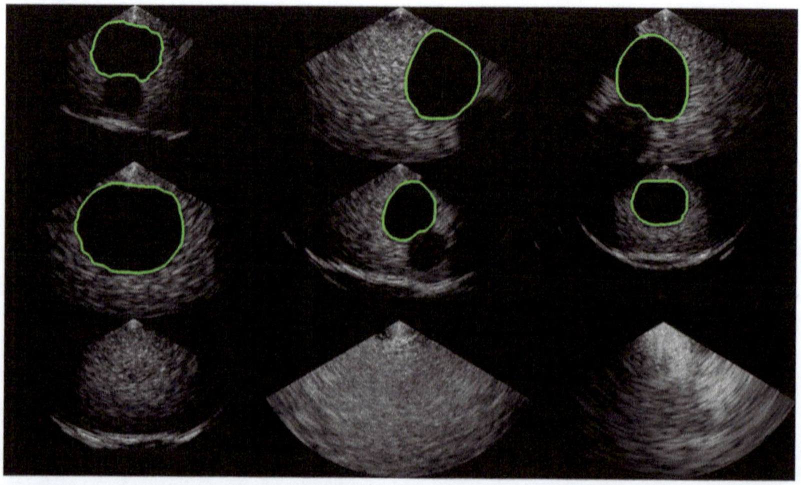

**Figure 2.** Examples of the acquired dataset. The first and second rows show images with a bladder, and the green line indicates the boundary of the mask label from human labelers. The third row represents images without a bladder, indicating that the bladder was not observed.

### 2.2. Multitask Deep Learning-Based Bladder Volume Measurement

The aim of this study was to design a convolutional neural network (CNN)-based deep learning model that is simple yet efficient for bladder volume measurement systems. The model was designed to perform multiple tasks, including classification and segmentation, as shown in Figure 3. This multitask approach can improve the model's efficiency in SoC environments and prevent unexpected measurement results from images without bladder regions. The classification path of the model detects a bladder on the ultrasound image by classifying the image into two classes, indicating the existence of the bladder in the image. The segmentation path of the model aims to find the pixelwise accurate ROI of the bladder in the image. The architecture of the model, including parameters such as the kernel size, was optimized on the collected dataset. Starting from the large-size model, parameters were gradually reduced by comparing the accuracy to the validation dataset.

To reduce the complexity of calculation and memory consumption, the input image was resized to a height and width of 192 pixels. For the feature extraction stage, MobileNetv2 [18], known for its lightweight network architecture and efficiency with regard to portable devices, was used to generate features with dimensions of 12 × 12 × 96. For the classification path, the extracted features were optimized by global average pooling, a dense layer, and classification head layers. For the segmentation path, the features were further processed by quantizable squeeze-and-excitation (QSE) blocks, depthwise separable convolution (DWC) blocks [19], upscaling layers, and a segmentation head. The squeeze-and-excitation (SE) [20] block has been widely used to embed channel weights into features. However, the SE block is not suitable for quantized networks due to elementwise multiplications. Therefore, in this study, a QSE block was designed with a channel weight operation using concatenation and convolution instead of elementwise multiplication. The QSE block was used to capture the larger context of the image. In the SE block, the feature was reduced to a small size vector (1 × 1 × 96) by averaging, and then the expanded feature was weighted by the reduced vector. To merge detailed information with the features from the QSE blocks, 3 × 3 convolution layers were placed parallel to the SE blocks. The features from the QSE blocks and convolution layers were merged by DWC blocks. The merged features were reduced into a smaller channel by convolution layers and then upscaled into the resized input size (192 × 192) with two channels (i.e., the number of pixel classes). Then, the segmentation head classified each pixel into two classes (i.e., background and bladder). Finally, the segmentation result was resized to the original size of the image.

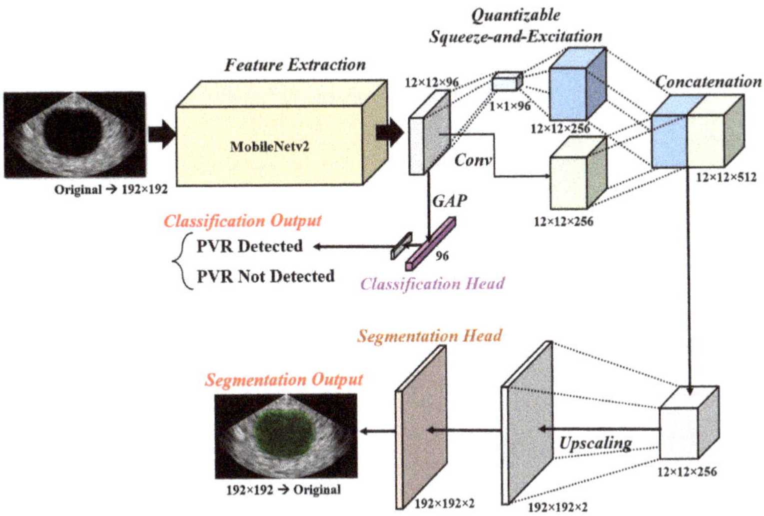


**Figure 3.** Overall architecture of the deep learning network designed for ultrasound image analysis. The original image is resized to a height and width of 192 pixels before feature extraction using MobileNetv2. The extracted feature is processed through two paths: segmentation and classification. The segmentation path uses squeeze-and-excitation and convolution to expand the features to a dimension of 12 × 12 × 256, which are then concatenated to 12 × 12 × 512 dimensions. A depthwise separable convolution is then applied to merge the gathered feature to a dimension of 12 × 12 × 256 before upscaling it to the 192 × 192 dimension using bilinear interpolation. The channel of the upscaled feature is then reduced to match the number of classes before finalizing segmentation with SoftMax. The classification path uses global average pooling to gather features that are reduced into logits according to each class. The classification is then finalized using SoftMax function.

Bladder volume is typically estimated based on shape coefficients and measurements of height, width, and depth on two different planes (i.e., sagittal and transverse). In this study, as illustrated in Figure 4, depth was estimated on the sagittal plane, while height and width were estimated on the transverse plane. The bladder volume was then calculated using Equation (1), where c is a constant determined by the shape of the bladder region (e.g., 0.52 for a spherical shape, 0.7 for an unknown shape [21]).

$$\text{Volume} \approx c \times \text{Depth} \times \text{Height} \times \text{Width} \tag{1}$$

### 2.3. Network Compression and System Implementation

To train the multitask architecture model without any degradation in accuracy, the classification path and the segmentation path were trained separately. Figure 5 illustrates the three distinct training stages of the proposed network. In the first stage, the segmentation path was trained with the initial weight of the model while the classification path was kept frozen. Next, in the second stage, the segmentation path was frozen, and the classification path was trained. Once the training of the classification path was completed, the weights from the first and second stages were merged into a single model.

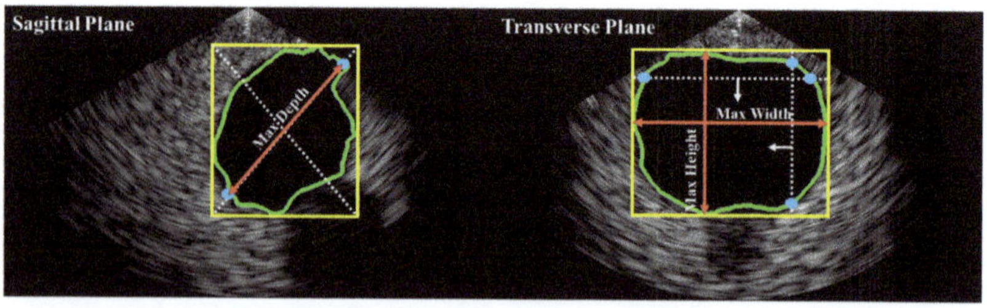


**Figure 4.** Automated axes detection for estimating bladder height, width, and depth from biplane ultrasound images. The bladder ROI dimensions were obtained by calculating the minimum enclosing bounding box (yellow line). In the sagittal plane, the greater distance between two intersections (blue dots) of the bounding box's diagonal line (white dashed lines) with the bladder ROI contour (green line) was used to estimate depth. The red arrow lines represent the estimated depth. In the transverse plane, the maximum distances of the vertical and horizontal intersections were used to estimate height and width, respectively.

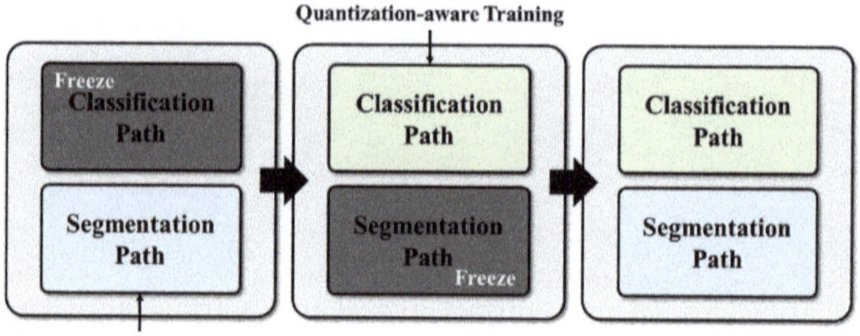


**Figure 5.** Training process for the designed network. The segmentation path is first trained in a quantization-aware training (QAT) setting with the classification path frozen. Once the segmentation path training is complete, the classification path is trained in the same QAT setting with the segmentation path frozen. Finally, the trained weights from both segmentation and classification paths are combined into a single model.

Additionally, during the training process, the model was also subjected to quantization-aware training (QAT) [22] to enhance execution speed while minimizing any drop in accuracy. The combo loss function [23], which combines the Dice loss and cross-entropy loss functions, was used to train the segmentation path. Meanwhile, the classification path was trained using the cross-entropy loss function. To avoid overfitting, data augmentation techniques, such as random intensity shift and random left–right flip, were applied. Furthermore, early stopping criteria were implemented, with a patience of 20 epochs. The Adam optimizer [24] was employed to train the network. To avoid local minimum and overfitting problem, the learning rate scheduling and early stopping criterion were used.

The deep learning model is trained using the TensorFlow (Google Inc., Mountain View, CA, USA) framework. After training, the model is compressed to enhance execution time on low-resource SoC settings. Model weights are quantized into 8-bit fixed-point using TensorFlow Lite (Google Inc., Mountain View, CA, USA). Inference is performed using the C++ programming language. To handle the entire system, the Linux operating system is utilized on the SoC with the Vitis (Xilinx Inc., San Jose, CA, USA) framework.

## 3. Results

### 3.1. Evaluation of the Trained Deep Learning Model

For the evaluation of the model's performance, the segmentation was assessed using the Dice coefficient metric, while accuracy was used to evaluate the classification performance. To compare the segmentation results with conventional models, U-Net [25], Attention U-Net [26], and BiSeNetv2 [27] were implemented at the original image resolution of 570 × 330. U-Net and Attention U-Net were implemented with 64, 128, 256, and 512 channels. The same optimizer and loss function were used for both the proposed and conventional methods. The conventional methods were trained and implemented using a 32-bit floating point, while the proposed method was trained and implemented using an 8-bit fixed point, as previously mentioned. The models were compared on the same validation data. The results are presented in Table 1, where F and Q represent 32-bit floating point and 8-bit fixed point implementation, respectively. The throughput in Table 1 was measured in the integrated SoC setting.

**Table 1.** Comparison of the Segmentation Results.

| | Dice Coefficient | # of Parameters | Throughput (FPS) |
|---|---|---|---|
| U-Net | $0.913 \pm 0.124$ | 8.56 M | 1.33 |
| Attention U-Net | $0.944 \pm 0.075$ | 7.91 M | 0.15 |
| BiSeNetv2 | $0.958 \pm 0.034$ | 3.12 M | 0.59 |
| Ours | $0.954 \pm 0.045$ | 0.97 M | 7.93 |

Table 1 shows the comparison of the proposed segmentation path with conventional CNN-based segmentation networks. U-Net, which is commonly used as a baseline for medical image segmentation, achieved an average Dice coefficient of 0.913 with a standard deviation of 0.124. The implemented U-Net had 8.56 million parameters and a throughput of 1.33 frames per second (FPS) in the SoC environment. In comparison, the Attention U-Net achieved a much higher Dice coefficient of 0.944 on average with a standard deviation of 0.075 but had a slower throughput than the U-Net, despite having fewer parameters. The recently introduced BiSe-Netv2 showed even higher Dice coefficients than both U-Net and Attention U-Net (i.e., an average of 0.958 and a standard deviation of 0.034) with even fewer parameters. However, BiSeNetv2 was slower than U-Net, running at less than 1 FPS. In contrast, the proposed method had significantly fewer parameters (i.e., 0.97 million) and could be executed at a much faster rate of approximately 8 FPS, which was 5.96x , 52.87x, and 13.44x faster than the U-Net, Attention U-Net, and BiSeNetv2, respectively. Although the proposed method had a slightly lower Dice coefficient than BiSeNetv2 (i.e., $0.954 \pm 0.045$ vs. $0.958 \pm 0.034$, respectively), the segmentation results from the proposed network are promising, as shown in Figure 6. The validation accuracy of the classification path was over 0.99, indicating high accuracy in the classification results, as shown in the confusion matrix in Figure 7.

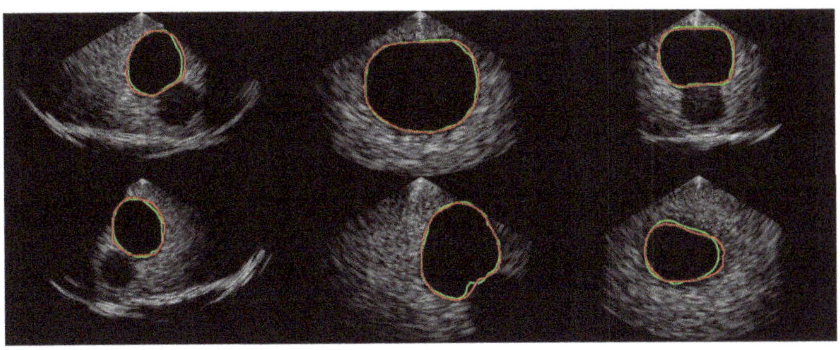

**Figure 6.** Examples of the segmentation result. The ground truth is represented by the green line, while the prediction from our proposed network is represented by the red line.

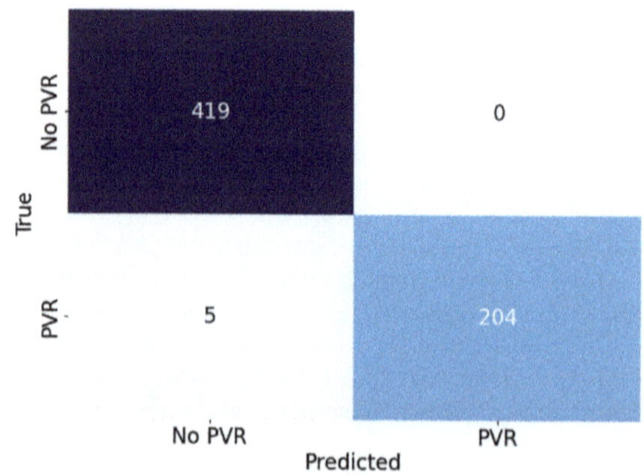


**Figure 7.** The confusion matrix shows the classification results, with only 5 out of 628 samples being incorrectly predicted, resulting in a 99.2% accuracy in predicting the correct classes.

### 3.2. Evaluation of the Bladder Volume Measurement

The bladder volume measurement results using the integrated system are depicted in Figure 8. To evaluate the system, the volume of an intravesical urine volume measurement phantom (US-16, Kyoto Kagaku, Kyoto, Japan) was measured using 30 separately acquired sagittal and transverse images for 50 mL, 150 mL, and 300 mL targets. The segmentation and classification paths were evaluated using the Dice coefficient and accuracy metrics, respectively. To quantitatively evaluate bladder volume measurements, automatic measurements were conducted five times on each phantom. The results in Table 2 show that the volumes could be calculated with less than 10% error when appropriate shape coefficients were selected. The coefficient of 0.72 was found to provide the most accurate measurement for both the 50 mL and 150 mL phantoms, while for the 300 mL phantom, the coefficient of 0.66 was most accurate. However, it is worth noting that when using the coefficient of 0.72 for the 300 mL phantom, a 5.78% error was observed, which is comparable to the 3.45% error observed when using the coefficient of 0.66. Therefore, using the coefficient of 0.72 resulted in measurements within tolerable error for all cases.

**Table 2.** Quantitative Evaluation Results of the Volume Measurement using Integrated System.

| | Coefficient | 50 mL | 150 mL | 300 mL |
|---|---|---|---|---|
| Unknown | 0.72 | 50.89 | 157.45 | 317.34 |
| Triangular prism | 0.66 | 46.57 | 141.86 | 289.66 |
| Cylinder | 0.81 | 56.85 | 171.06 | 350.57 |
| Cuboid | 0.89 | 61.69 | 184.34 | 402.86 |
| Spherical | 0.52 | 37.84 | 112.29 | 233.06 |

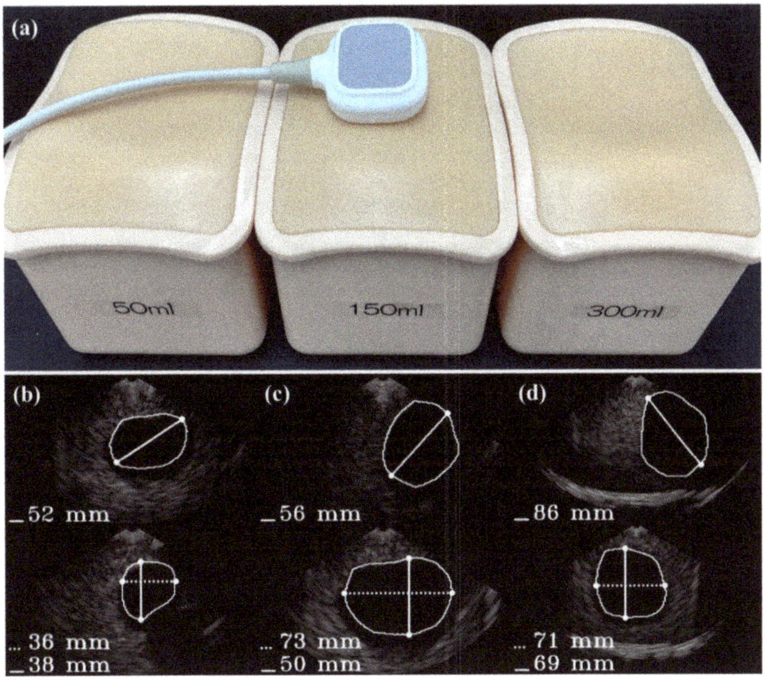


**Figure 8.** (**a**) Overall experimental setup for bladder volume measurement. Bladder phantoms with volumes of 50 mL, 150 mL, and 300 mL were measured using the integrated system, which combines the proposed method with a T-shaped array portable ultrasound system. (**b**–**d**) Measurement results for the 50 mL, 150 mL, 300 mL phantoms, respectively.

## 4. Discussion

An ultrasound-based bladder volume measurement is an effective method for detecting and managing PVR. To enhance its efficacy, an automatic bladder volume measurement method based on image analysis is necessary. Recent studies have demonstrated that deep learning-based image analysis techniques can be employed on ultrasound images for the management of urinary diseases [15,28]. However, conventional methods have high computational complexity, making them unsuitable for adoption on portable devices. As a result, it is challenging to integrate deep learning-based bladder volume measuring algorithms into portable systems.

In this study, a lightweight deep learning network was developed as a multitask network that performed classification and segmentation simultaneously. The multitask network has several advantages compared to networks that perform only segmentation or classification. In terms of computational complexity, the integrated device can save power or resources by using the classification node. For example, when the classification results show that images do not have ROIs, the post-processing algorithm for measurement does not need to be executed. In terms of accuracy and user experience, the classification node is also helpful. For instance, if the classification result indicates that an image does not have an ROI, the segmentation result is invalid and inaccurate and should not be presented to the users. These advantages make the multitask network more suitable for real clinical situations.

The segmentation accuracy of this network was validated by comparing it with conventional segmentation networks (U-Net, Attention U-Net, BiSeNetv2) using the Dice coefficient as the evaluation metric. Compared to the baseline U-Net, the Attention U-Net exhibited a significantly higher Dice coefficient of 0.944 with fewer parameters (7.91 M vs.

8.56 M). Although the recently introduced BiSeNetv2 had an even higher Dice coefficient with a smaller number of parameters, it still exhibited slow execution times in low-resource SoC environments (0.59 FPS). The slow execution time of these conventional networks could be problematic, particularly since two images (sagittal and transverse) were used to measure bladder volume automatically.

Compared to conventional networks, the developed network had fewer parameters (<1 M), and its Dice coefficient (0.954) was comparable to that of BiSeNetv2 (0.958) and higher than that of Attention U-Net and U-Net at 0.944 and 0.913, respectively. The drop in accuracy can be considered negligible, as the annotation process is performed by humans and may have labeling noise. Moreover, the accuracy of the proposed network is sufficient to be used for automatic volume calculation, as demonstrated in the previous section.

To efficiently implement the developed network on SoCs, further optimization based on quantization was carried out. After quantization, the proposed network could execute on SoCs at 7.93 FPS, enabling it to measure bladder volume approximately four times per second. Moreover, the classification accuracy of the proposed network was over 0.99, making it an efficient network with high accuracy for both PVR detection and bladder segmentation. The proposed network was also used in an end-to-end automatic algorithm to measure bladder volume. The algorithm accurately identified the axes (i.e., height, width, and depth) of the bladder, and with the proper shape coefficient, the bladder volume could be estimated within ±6% error of the actual volume. In addition, since the model was trained on dataset with various size and location of ROIs, the model can estimate ROIs regardless of their size or location.

While this study demonstrated the potential of the deep learning-based automatic bladder volume measurements on an SoC, there are still several limitations that need to be addressed. First, this study was focused on phantom studies and was not validated in in vivo cases. In further works, data from in vivo cases will be collected, and the clinical impact of the proposed method will be evaluated. Additionally, this study estimated the bladder volume using shape coefficients, but as shown in Table 2, improper shape coefficients can result in significant errors in volume estimations. To address this issue, future studies may investigate automatic estimation of bladder volume or shape coefficient. For example, a study may be conducted on deep learning methods that utilize prior knowledge of bladder shape to perform end-to-end volume prediction.

## 5. Conclusions

In this study, a lightweight deep learning network was developed to measure bladder volume on portable bladder ultrasound devices. The designed network showed comparable accuracy to conventional deep learning methods in terms of the Dice coefficient. Additionally, the execution of the designed network was much faster than that of conventional methods. An automatic axis detection algorithm was also utilized, and the bladder volume could be end-to-end automatically measured with under ±6% error with proper shape coefficients. Finally, the proposed network and algorithm were successfully integrated into a low-resource SoC-based portable bladder ultrasound system. However, further validation on in vivo cases and automatic estimation of shape coefficients may be necessary for future studies.

**Author Contributions:** Conceptualization, H.C. and Y.Y.; methodology, H.C., I.S., J.J. and Y.Y.; validation, H.C., I.S. and J.J.; writing-original draft preparation, H.C, I.S. and J.J.; writing—review and editing, I.S., J.J. and Y.Y.; supervision, J.J. and Y.Y.; funding acquisition, J.J. and Y.Y. All authors have read and agreed to the published version of the manuscript.

**Funding:** This work was supported by the Korea Medical Device Development Fund grant funded by the Korea government (the Ministry of Science and ICT, the Ministry of Trade, Industry and Energy, the Ministry of Health & Welfare, Republic of Korea, the Ministry of Food and Drug Safety) (Project Number: 202011A01). This work was supported by the National Research Foundation of Korea (NRF) grant funded by the Korea government (MSIT) (2021R1A2C3006264).

**Institutional Review Board Statement:** Not applicable.

**Informed Consent Statement:** Not applicable.

**Data Availability Statement:** The data that support the findings of this study are available from the corresponding author upon reasonable request.

**Conflicts of Interest:** I.S., J.J. and Y.Y. work for Edgecare Inc., the start-up building the wearable patient monitoring ultrasound. H.C. declare no competing interests.

## References


1. Darrah, D.M.; Griebling, T.L.; Silverstein, J.H. Postoperative urinary retention. *Anesthesiol. Clin.* **2009**, *27*, 465–484. [CrossRef]
2. Meddings, J.; Rogers, M.A.; Krein, S.L.; Fakih, M.G.; Olmsted, R.N.; Saint, S. Reducing unnecessary urinary catheter use and other strategies to prevent catheter-associated urinary tract infection: An integrative review. *BMJ Qual. Saf.* **2014**, *23*, 277–289. [CrossRef] [PubMed]
3. Schweiger, A.; Kuster, S.P.; Maag, J.; Züllig, S.; Bertschy, S.; Bortolin, E.; John, G.; Sax, H.; Limacher, A.; Atkinson, A. Impact of an evidence-based intervention on urinary catheter utilization, associated process indicators, and infectious and non-infectious outcomes. *J. Hosp. Infect.* **2020**, *106*, 364–371. [CrossRef] [PubMed]
4. Bruskewitz, R.; Iversen, P.; Madsen, P. Value of postvoid residual urine determination in evaluation of prostatism. *Urology* **1982**, *20*, 602–604. [CrossRef]
5. May, M.; Brookman-Amissah, S.; Hoschke, B.; Gilfrich, C.; Braun, K.-P.; Kendel, F. Post-void residual urine as a predictor of urinary tract infection—Is there a cutoff value in asymptomatic men? *J. Urol.* **2009**, *181*, 2540–2544. [CrossRef] [PubMed]
6. Asimakopoulos, A.D.; De Nunzio, C.; Kocjancic, E.; Tubaro, A.; Rosier, P.F.; Finazzi-Agrò, E. Measurement of post-void residual urine. *Neurourol. Urodyn.* **2016**, *35*, 55–57. [CrossRef]
7. Goode, P.; Locher, J.; Bryant, R.; Roth, D.; Burgio, K. Measurement of postvoid residual urine with portable transabdominal bladder ultrasound scanner and urethral catheterization. *Int. Urogynecol. J.* **2000**, *11*, 296–300. [CrossRef]
8. Alnaif, B.; Drutz, H. The accuracy of portable abdominal ultrasound equipment in measuring postvoid residual volume. *Int. Urogynecology J.* **1999**, *10*, 215–218. [CrossRef] [PubMed]
9. Stevens, E. Bladder ultrasound: Avoiding unnecessary catheterizations. *Medsurg Nurs.* **2005**, *14*, 249.
10. Teng, C.-H.; Huang, Y.-H.; Kuo, B.-J.; Bih, L.-I. Application of portable ultrasound scanners in the measurement of post-void residual urine. *J. Nurs. Res.* **2005**, *13*, 216–224. [CrossRef]
11. Park, Y.H.; Ku, J.H.; Oh, S.J. Accuracy of post-void residual urine volume measurement using a portable ultrasound bladder scanner with real-time pre-scan imaging. *Neurourol. Urodyn.* **2011**, *30*, 335–338. [CrossRef] [PubMed]
12. Chen, S.-C.; Chen, P.-Y.; Chen, G.-C.; Chuang, S.-Y.; Tzeng, I.-S.; Lin, S.-K. Portable bladder ultrasound reduces incidence of urinary tract infection and shortens hospital length of stay in patients with acute ischemic stroke. *J. Cardiovasc. Nurs.* **2018**, *33*, 551. [CrossRef] [PubMed]
13. Cho, M.K.; Noh, E.J.; Kim, C.H. Accuracy and precision of a new portable ultrasound scanner, the Biocon-700, in residual urine volume measurement. *Int. Urogynecol. J.* **2017**, *28*, 1057–1061. [CrossRef] [PubMed]
14. Majima, T.; Oota, Y.; Matsukawa, Y.; Funahashi, Y.; Kato, M.; Mimata, H.; Gotoh, M. Feasibility of the Lilium α-200 portable ultrasound bladder scanner for accurate bladder volume measurement. *Investig. Clin. Urol.* **2020**, *61*, 613. [CrossRef]
15. Matsumoto, M.; Tsutaoka, T.; Yabunaka, K.; Handa, M.; Yoshida, M.; Nakagami, G.; Sanada, H. Development and evaluation of automated ultrasonographic detection of bladder diameter for estimation of bladder urine volume. *PLoS ONE* **2019**, *14*, e0219916. [CrossRef]
16. Zheng, Q.; Tastan, G.; Fan, Y. Transfer learning for diagnosis of congenital abnormalities of the kidney and urinary tract in children based on ultrasound imaging data. In Proceedings of the 2018 IEEE 15th International Symposium on Biomedical Imaging (ISBI 2018), Washington, DC, USA, 4–7 April 2018; pp. 1487–1490.
17. Kuo, C.-C.; Chang, C.-M.; Liu, K.-T.; Lin, W.-K.; Chiang, H.-Y.; Chung, C.-W.; Ho, M.-R.; Sun, P.-R.; Yang, R.-L.; Chen, K.-T. Automation of the kidney function prediction and classification through ultrasound-based kidney imaging using deep learning. *NPJ Digit. Med.* **2019**, *2*, 29. [CrossRef]
18. Sandler, M.; Howard, A.; Zhu, M.; Zhmoginov, A.; Chen, L.-C. Mobilenetv2: Inverted residuals and linear bottlenecks. In Proceedings of the IEEE Conference on Computer Vision and Pattern Recognition, Salt Lake City, UT, USA, 18–22 June 2018; pp. 4510–4520.
19. Chollet, F. Xception: Deep learning with depthwise separable convolutions. In Proceedings of the IEEE Conference on Computer Vision and Pattern Recognition, Honolulu, HI, USA, 21–26 July 2017; pp. 1251–1258.
20. Hu, J.; Shen, L.; Sun, G. Squeeze-and-excitation networks. In Proceedings of the IEEE Conference on Computer Vision and Pattern Recognition, Salt Lake City, UT, USA, 18–22 June 2018; pp. 7132–7141.
21. Bih, L.-I.; Ho, C.-C.; Tsai, S.-J.; Lai, Y.-C.; Chow, W. Bladder shape impact on the accuracy of ultrasonic estimation of bladder volume. *Arch. Phys. Med. Rehabil.* **1998**, *79*, 1553–1556. [CrossRef]

22. Jacob, B.; Kligys, S.; Chen, B.; Zhu, M.; Tang, M.; Howard, A.; Adam, H.; Kalenichenko, D. Quantization and training of neural networks for efficient integer-arithmetic-only inference. In Proceedings of the IEEE Conference on Computer Vision and Pattern Recognition, Salt Lake City, UT, USA, 18–22 June 2018; pp. 2704–2713.
23. Taghanaki, S.A.; Zheng, Y.; Zhou, S.K.; Georgescu, B.; Sharma, P.; Xu, D.; Comaniciu, D.; Hamarneh, G. Combo loss: Handling input and output imbalance in multi-organ segmentation. *Comput. Med. Imaging Graph.* **2019**, *75*, 24–33. [CrossRef]
24. Kingma, D.P.; Ba, J. Adam: A method for stochastic optimization. *arXiv* **2014**, arXiv:1412.6980.
25. Ronneberger, O.; Fischer, P.; Brox, T. U-net: Convolutional networks for biomedical image segmentation. In Proceedings of the International Conference on Medical Image Computing and Computer-Assisted Intervention, Munich, Germany, 5–9 October 2015; pp. 234–241.
26. Oktay, O.; Schlemper, J.; Folgoc, L.L.; Lee, M.; Heinrich, M.; Misawa, K.; Mori, K.; McDonagh, S.; Hammerla, N.Y.; Kainz, B. Attention u-net: Learning where to look for the pancreas. *arXiv* **2018**, arXiv:1804.03999.
27. Yu, C.; Gao, C.; Wang, J.; Yu, G.; Shen, C.; Sang, N. Bisenet v2: Bilateral network with guided aggregation for real-time semantic segmentation. *Int. J. Comput. Vis.* **2021**, *129*, 3051–3068. [CrossRef]
28. Song, S.H.; Han, J.H.; Kim, K.S.; Cho, Y.A.; Youn, H.J.; Kim, Y.I.; Kweon, J. Deep-learning segmentation of ultrasound images for automated calculation of the hydronephrosis area to renal parenchyma ratio. *Investig. Clin. Urol.* **2022**, *63*, 455–463. [CrossRef] [PubMed]

bioengineering

MDPI

*Article*

# U-Net Architecture for Prostate Segmentation: The Impact of Loss Function on System Performance

**Maryam Montazerolghaem [1], Yu Sun [1], Giuseppe Sasso [2,3] and Annette Haworth [1,*]**

1 School of Physics, The University of Sydney, Sydney, NSW 2006, Australia
2 Radiation Oncology Department, Auckland City Hospital, Auckland 1023, New Zealand
3 Faculty of Medical and Health Sciences, University of Auckland, Auckland 1010, New Zealand
* Correspondence: annette.haworth@sydney.edu.au

**Abstract:** Segmentation of the prostate gland from magnetic resonance images is rapidly becoming a standard of care in prostate cancer radiotherapy treatment planning. Automating this process has the potential to improve accuracy and efficiency. However, the performance and accuracy of deep learning models varies depending on the design and optimal tuning of the hyper-parameters. In this study, we examine the effect of loss functions on the performance of deep-learning-based prostate segmentation models. A U-Net model for prostate segmentation using T2-weighted images from a local dataset was trained and performance compared when using nine different loss functions, including: Binary Cross-Entropy (BCE), Intersection over Union (IoU), Dice, BCE and Dice (BCE + Dice), weighted BCE and Dice (W (BCE + Dice)), Focal, Tversky, Focal Tversky, and Surface loss functions. Model outputs were compared using several metrics on a five-fold cross-validation set. Ranking of model performance was found to be dependent on the metric used to measure performance, but in general, W (BCE + Dice) and Focal Tversky performed well for all metrics (whole gland Dice similarity coefficient (DSC): 0.71 and 0.74; 95HD: 6.66 and 7.42; Ravid 0.05 and 0.18, respectively) and Surface loss generally ranked lowest (DSC: 0.40; 95HD: 13.64; Ravid −0.09). When comparing the performance of the models for the mid-gland, apex, and base parts of the prostate gland, the models' performance was lower for the apex and base compared to the mid-gland. In conclusion, we have demonstrated that the performance of a deep learning model for prostate segmentation can be affected by choice of loss function. For prostate segmentation, it would appear that compound loss functions generally outperform singles loss functions such as Surface loss.

**Keywords:** prostate cancer; prostate segmentation; U-Net; mp-MRI; loss function; medical imaging

**Citation:** Montazerolghaem, M.; Sun, Y.; Sasso, G.; Haworth, A. U-Net Architecture for Prostate Segmentation: The Impact of Loss Function on System Performance. *Bioengineering* **2023**, *10*, 412. https://doi.org/10.3390/bioengineering10040412

Academic Editors: Paolo Zaffino and Maria Francesca Spadea

Received: 6 February 2023
Revised: 19 March 2023
Accepted: 23 March 2023
Published: 26 March 2023



## 1. Introduction

Multiparametric magnetic resonance imaging (mp-MRI) is increasingly being used in the computer-aided diagnosis, computer-assisted surgery and radiation therapy planning for prostate cancer [1,2]. Accurate prostate segmentation for radiation therapy treatment planning is necessary to ensure the prostate receives an adequate amount of radiation for tumor control whilst minimizing the amount of dose received by other organs, such as the bladder and rectum [3]. Manual segmentation has been shown to demonstrate a high degree of intra- and inter-variability, particularly at the base and apex of the prostate [4]. Additionally, manual segmentation is subjective, time-consuming and can be affected by level of experience. In comparison, automatic segmentation is fast and can decrease human bias and errors [5–7].

The U-Net [8] architecture has been successfully applied in prostate segmentation in several studies [9–11]. However, applying deep neural networks for this task can result in variable outcomes, as multiple factors can influence the model outcome. Firstly, performance of auto-segmentation models are highly dependent on training dataset features, quality and number of samples [9]. In particular, the small sample size typically used in prostate segmentation models makes automatic segmentation very challenging. Prostate shape and texture may vary widely between different patients and the heterogeneity of the prostate tissue presents additional challenges for automated segmentation. In addition, design and configuration of a deep-learning-based segmentation model requires careful consideration. There are many parameters and hyper-parameters that need to be optimized to achieve acceptable model performance. These include network architectures, training schedules, data pre-processing, data augmentation (if used), data post-processing, and several essential hyper-parameter tuning steps such as learning rate, batch size, number of epochs, or class sampling [2]. In addition, hardware availability for training and inference of these models should be considered in advance [12,13]. Model performance varies substantially with the training dataset's properties and its size. Therefore, the applicability of trained public models for unseen datasets is limited [2], and training a model from scratch or retraining other models are popular solutions in medical image segmentation tasks.

One of the key parameters of deep-learning-based models that plays an important role in model training and success of the segmentation model is the *loss function*, also known as the cost function. The loss function is ultimately responsible for how the model's weights are adjusted for optimization goals, such as minimizing region mismatches between predicted and ground truth segmentations. Various domain-specific loss functions have been proposed and applied for segmentation of the prostate and other organs to improve results for their datasets [9,14]. It can be challenging to know which loss function meets the requirements of the task, and whether the right function for a specific dataset has been chosen [14]. In the past ten years many loss functions have been proposed. Jadon [15], for example, reported the performance of thirteen well known loss functions designed for fast model convergence, and proposed a new loss function for skull segmentation from CT data. Ma et al. [14] provided a comprehensive review of twenty loss functions based on four CT-based publicly available data sets. For our study, we have chosen to complement these works with a focus on nine loss functions, applied to a single MRI-based data set sourced from an in-house study. This data set provides ground-truth prostate-gland segmentations based on whole-mount histology (rather than clinician generated segmentations which form the basis of many segmentation models). These nine loss functions are commonly used in medical image segmentation models and are intended to be representative of the many loss functions reported in the literature, and in particular, form a sub-set of those reported by Ma et al. and Jadon [14,15] with at least one loss function from each of the four categories defined in both studies and excluding those relevant to multi-class solutions that are not relevant here. Whilst there are many applications of segmentation models, our study was motivated by the need to develop a segmentation algorithm to analyze data collected as part of a clinical trial investigating the ability of quantitative multiparametric MRI to assess response to radiation therapy (ANZCTR UTN U1111-1221-9589). Our longitudinal data set generated a large amount of data that required an objective delineation of the prostate gland prior to the extraction of radiomic features to develop treatment response predictive models. As part of this study we identified a lack of comprehensive comparisons of prostate segmentation model performance using different loss functions. In this study, we compared deep-learning-based prostate segmentations of T2-weighted (T2w) MR images, using nine different loss functions for 2D U-Net with our locally acquired dataset.

## 2. Materials and Methods

**Dataset:** In vivo mp-MRI data were collected from 70 patients prior to radical prostatectomy as part of a Human Research Ethics Committee (HREC)-approved project called "BiRT" (HREC/15/PMCC125). These images were acquired using a 3T Siemens Trio Tim machine (Siemens Medical Solutions, Erlangen, Germany). The first 37 cases imaged using a standardized imaging protocol and free of major artifacts were available for analysis at the time this study was performed [16]. Prostate segmentations were generated from the whole-mount histology slides and subsequently co-registered with the mpMRI using a sophisticated co-registration framework [16]. For quality control, the co-registered prostate masks were checked against an independent annotation by an experienced radiation oncologist (GS) on the in vivo 3D T2w images using RayStation v.8 (RaySearchLabs Stockholm, Sweden). These contours were used as ground truth for automatic segmentation. Following segmentation of the entire prostate gland, each prostate volume was mathematically divided into sub-regions by thirds in the craniocaudal axis, with the most superior volume labelled "base", the inferior volume "apex" and the central volume "mid-gland". T2w images were acquired using a turbo spin echo sequence with two sets of resolutions. For the first set, the in-plane resolution was 0.6 mm × 0.6 mm, the inter-plane distance was 6 mm. The volumes of the first set contained between 80 and 96 slices each, with each slice resolution being 384 × 384 pixels. For the second set, the in-plane resolution was 0.8 mm × 0.8 mm, and the inter-plane distance was 0.8 mm. The volumes of the second set contained between 80 and 88 slices each, with each slice resolution being 256 × 256 pixels.

Pre-processing of the input data included bias field correction, resampling, and image normalization. The intensity range of each image was normalized using minimum and maximum intensity values of each single image before incorporation into the network. The datasets were resampled into 128 × 128 × 64 voxels. A flow chart indicating the image processing pipeline is shown in Figure 1. The full pelvic field of view was used without cropping.

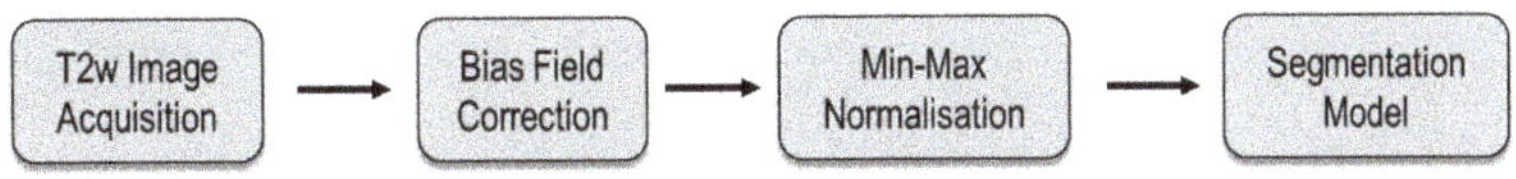


**Figure 1.** The image pre-processing for T2-weighted images. After acquisition, bias field correction was applied using the N4 algorithm to correct for the magnetic field inhomogeneity. The images were then normalized using the min–max approach before entering the segmentation network.

**U-Net architecture and Loss Functions:** The effect of various loss functions on the performance of a basic 2D U-Net architecture [8] was investigated using the T2-weighted MR images. Loss functions were selected from traditional distribution-based and region-based categories, as well as more recent compound and boundary-based loss functions. Most of the loss functions used in this study were selected based on their suitability for use with strongly and mildly imbalanced data sets in segmentation tasks and those commonly used in medical image segmentation models [14]. These include Binary Cross-Entropy (BCE), Intersection over Union Loss (IoU), Dice Loss, combination of Dice and BCE loss functions (BCE + Dice), weighted BCE and Dice Loss (W (BCE + Dice)), Focal Loss, Tversky Loss, Focal Tversky Loss and Surface [9,14,17]. Table 1 summarizes the loss functions used in this paper with loss function definitions based on those of Ma et al. [14], with details included in Appendix A.

The U-Net architecture contains two main components: the encoder or contracting path, which extracts the features of the image by applying a stack of convolutional and max pooling layers (Figure 2, left), and the decoder or expanding path (Figure 2, right). The U-Net architecture is an end-to-end fully convolutional network (FCN) and contains only convolutional layers without any dense layers. This allowed the network to accept images of any size.

The encoder of the network used in the current study had five convolutional layers to extract high-level feature maps. In each convolutional layer, the input feature map was convolved with a set of trainable filters, kernels of size 3 × 3 and a 2 × 2 max pooling operation with a stride of 2. Max pooling operations or down-sampling reduced the feature map size by a factor of 2 in each dimension. Then, a batch normalization operation was applied, followed by rectified linear unit (ReLU) activation functions. ReLU performed the thresholding operation (*max* (*x*,0)), used to introduce nonlinearity to the trained network. The number of feature channels started at 16 for the first stage, and doubled after each stage of the decoder to 32, 64, 128, and finally 256.

A decoder reverses the operations of the encoder to recover the original input size and enable the network to perform a voxel-wise classification. Each stage of the decoder included two types of operations. Firstly, layers were up-sampled to increase the size of the feature map gradually until it reached the size of the original input image. Secondly, deconvolutional layers reduced the number of feature channels to half at each stage of the decoder to match the number of channels with the corresponding encoder layers. Features extracted from earlier stages were added to the encoder side (Figure 2) using short-circuit layers to help recover the spatial information from the convolutions in the encoder.

The U-Net model applied in this study had nine convolutional layers. Model parameters, except the loss function, were fixed for all models. The Adam optimizer [18] was selected as the optimization algorithm, with an initial learning rate $\alpha = 0.0001$, a learning rate drop factor of 0.1, and a patience of 10 (meaning that the learning rate dropped by a factor of 0.1 when the validation loss did not improve for 10 epochs). The training was performed for 10,000 epochs with an early stopping strategy and a batch size of 2 to avoid overfitting. Model training was stopped when the validation loss did not improve for 10 epochs. Dropout was applied for each convolutional layer at a rate of 10% to avoid overfitting. Batch normalization was applied after each convolution layer to prevent gradient vanishing/exploding [19]. The results for each model reported the best epoch based on the validation set. The number of model parameters was 1,189,264, of which 1,187,792 were trainable. Sigmoid activation was used as the output layer for binary predictions. A threshold value of 0.5 for the probability was applied to obtain the segmentation mask, this value was found to be the optimal value that gave the highest Dice coefficient (DSC) and fewer false positives.

Five-fold cross-validation was used to validate the results [20]. For model selection, the best model was determined based on performance of the validation datasets [14]. Our proposed network was implemented in Keras v2.3.1 [21], using TensorFlow v2.0.0 [22] backend with Python. For each loss function, the network was trained by performing a five-fold cross-validation using all 37 cases from the BiRT dataset. All calculations were performed using the University of Sydney's HPC service and GPU access, NVIDIA V100 SXM2.

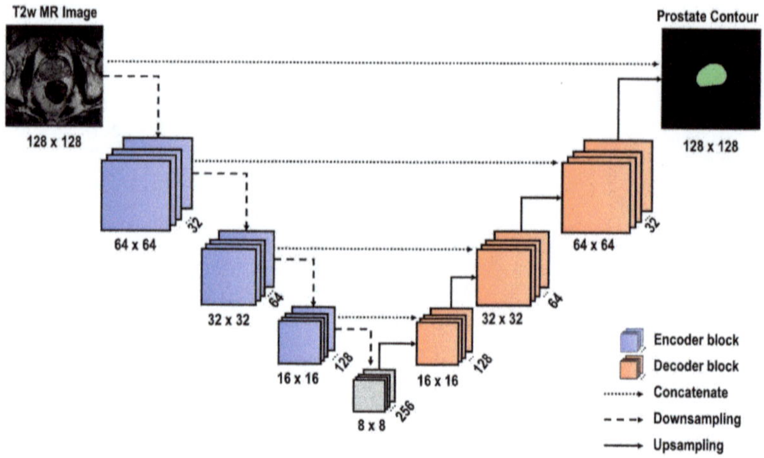


**Figure 2.** The U-Net architecture used in this study. The encoder contains four convolution layers with pooling. The decoder is symmetrical as the encoder, expanding the in-plane resolution back to the input image.

**Table 1.** The family and individual loss functions used in this study.

| Category | Loss Functions/Use Case |
|---|---|
| Distribution-based | • **Binary CrossEntropy (BCE) Loss:**<br>○ Balanced dataset<br>○ Bernoulli distribution-based loss function<br>• **Focal Loss:** Suitable for highly imbalanced datasets<br>○ Enables models to learn hard examples by down-weighting simple samples |
| Region-based | • **Intersection over Union (IoU) Loss**<br>○ Inspired from Jaccard similarity coefficient, a metric for segmentation validation<br>• **Dice Loss:**<br>○ Based on Dice coefficient<br>• **Tversky Loss:**<br>○ Variant of Dice coefficient<br>○ Adds weights to false positive and false negative<br>• **Focal Tversky Loss:**<br>○ Suitable for highly imbalanced dataset<br>○ Enables models to learn hard examples by down-weighting simple samples |
| Boundary-based | • **Surface (Boundary) Loss** |
| Compound | • **Weighted BCE and Dice W(BCE + Dice) Loss:**<br>○ Combination of Dice Loss and Binary CrossEntropy Loss<br>○ Used for lightly class imbalance<br>○ Benefits from both BCE and Dice Loss properties<br>• **BCE and Dice:** (BCE + Dice) Loss |

**Evaluation Metrics:** Models were compared and evaluated using commonly used metrics for medical image segmentation [23]. These include the DSC, 95% Hausdorff Distance (95HD), relative absolute volume difference (Ravd), precision, and sensitivity. These metrics were selected to cover evaluations for region-based, contour-based and volume-based similarities between the ground truth and auto-segmentation output. A DSC score of 1 shows perfect agreement. The Hausdorff Distance measures the distance between the borders of the ground truth and the auto-segmentation output. Lower values of 95HD indicate a better performance of segmentation. Ravd is the difference between the total volume of the segmentation and the ground truth divided by the total volume of the ground truth. The Ravd value for a perfect segmentation is equal to zero.

## 3. Results

Table 2 provides a summary of the results of the different loss functions applied to the nine models used in this study. Figure 3 shows box plots for each of the nine models and evaluation metrics for the whole prostate. Supplementary material Figures S1–S3 contain boxplots for these models for the prostate mid-gland, apex, and base, respectively. Figure 4 shows DSC box plots for different parts of the prostate. The mid-gland (Figure 4C) shows a consistently high performance (except for Surface loss), followed by the base and the apex (Figure 4B,D, respectively). Table 2 shows that the Focal Tversky loss function had the highest average of DSC scores for the whole gland and the lowest standard deviation

(0.74 ± 0.09). Models with IoU, Dice, Tversky and W (BCE +Dice) and BCE + Dice loss functions obtained similar DSC scores of 0.73, 0.73, 0.72, 0.71, and 0.71, respectively. A high DSC score was expected for these loss functions as they are variates of the Dice coefficient and aim to minimize this metric during the training process. Additionally, the Dice loss function and its variates perform better in class-imbalanced problems such as prostate segmentation. Models with surface and BCE loss functions had the lowest whole gland DSC, with values of 0.40 and 0.58, respectively. The maximum difference in DSC score across all models' performance was approximately 34%.

In considering DSC scores shown in Figure 4, it can be seen that all models achieved the highest DSC score for the mid-gland (Figure 4C), which had a 20% (up to 93–94%) higher accuracy compared to the whole prostate (Figure 4A), most likely because the whole gland resembles the mid-gland, and it accounts for the majority of the prostate volume. Model performance was lower in the apex (Figure 4B) when considering all parts of the prostate and the prostate as a whole. Higher standard deviations of the DSC scores were observed for the apex from all models (Table 2).

Regarding 95HD, the best performance was achieved by W (BCE + Dice), with a value of 6.66 ± 2.82 for the whole prostate gland, followed by Tversky and Focal Tversky with values of 7.17 ± 4.21 and 7.42 ± 5.81, respectively (Table 2). The worst performing model was Surface, with a value of 13.64 ± 4.38, approximately double that of the best performing model (W (BCE + Dice). When considering the base, mid-gland, and apex, as expected, the mid-gland reported lower 95HD values, followed by the apex, with the best performance achieved by W (BCE + Dice) and Dice, respectively.

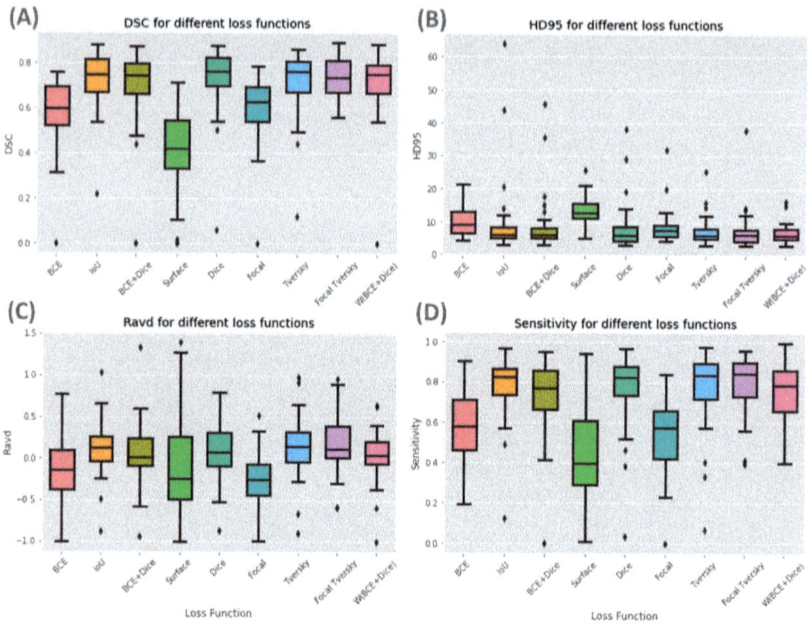


**Figure 3. Each** box plot (**A–D**) represents metrics DSC, HD95, Ravid and sensitivity respectively for the whole prostate on validation data from the five-fold cross-validation for models with different loss functions. DSC: Dice similarity coefficient; HD95: 95% Housdorff Distance.

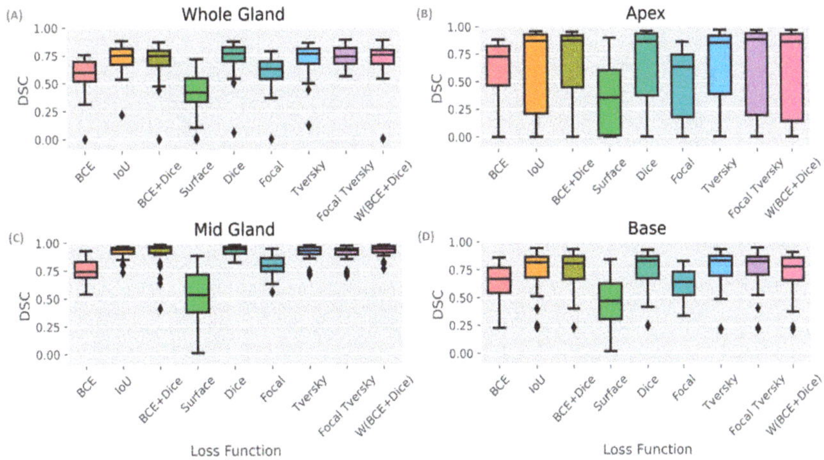


**Figure 4.** Boxplots showing the Dice similarity coefficient (DSC) scores for different parts of the prostate. The mid-gland (**C**) shows a consistent high performance (except for Surface loss), followed by the base (**D**) and the apex (**B**). The whole gland's performance resembles the mid-gland (**A**,**C**), as it accounts for the majority of the prostate volume. Results are from the model trained using the Dice loss.

Ravd is an appropriate metric for applications with an interest in accurate volume estimation and similarity. An absolute value of Ravd approaching zero shows a better model performance. The lowest absolute values of Ravd for the whole prostate were obtained from W (BCE + Dice), BCE + Dice, Surface and Dice (0.05, 0.07, 0.09 and 0.09, respectively) and the largest deviation from a score of zero was Focal with a value of −0.25 ± 0.31 (Table 2). The standard deviations of Ravd for models with W (BCE + Dice) and BCE + Dice were small, with values of 0.31 and 0.37, respectively.

The highest sensitivity value was achieved for the whole prostate gland using Focal Tversky (80%), and the lowest using the Surface loss function (44%). Similar values of precision were achieved for all loss functions for the whole gland (69–73%), with the exception of Surface (51%). Focal Tversky, W (BCE + Dice) and Focal each have parameters which can control trade-off between false positives and false negatives (FP and FN). These parameters can be optimized based on segmentation task needs and data properties.

The surface loss function had the lowest DSC score and a higher 95HD. This model had the lowest performance considering the majority of metrics used in this study. Furthermore, models with a surface loss function required longer training times and higher numbers of iterations.

There was a pattern of improved DSC score in slices that covered a larger area of prostate, mainly in the mid-gland with cross sectional areas greater than 600 mm$^2$ and less than 2100 mm$^2$. This is represented in Figure 5, where the data shown is based on the prediction from the model using W (BCE + Dice) on the validation data. The same pattern is seen in all models. Figure 6 presents the box plots for all loss functions.

**Table 2.** Mean value of the five-fold cross validation for each metric used in the current study are shown for the whole gland, base, mid-gland, and apex regions. Values shown in **bold** represent best performing results.

| Loss Function | Part | BCE | BCE + Dice | Dice | Focal | Focal Tversky | IoU | Surface | Tversky | W (BCE + Dice) |
|---|---|---|---|---|---|---|---|---|---|---|
| DSC (mean± std) | Whole | 0.58 ± 0.15 | 0.71 ± 0.16 | **0.73** ± 0.15 | 0.60 ± 0.14 | **0.74** ± 0.09 | **0.73** ± 0.12 | 0.40 ± 0.18 | **0.72** ± 0.14 | 0.71 ± 0.15 |
| | Base | 0.65 ± 0.16 | 0.73 ± 0.19 | **0.74** ± 0.18 | 0.62 ± 0.15 | **0.75** ± 0.19 | **0.74** ± 0.18 | 0.46 ± 0.22 | **0.74** ± 0.18 | 0.72 ± 0.19 |
| | Mid | 0.75 ± 0.11 | 0.90 ± 0.12 | **0.93** ± 0.05 | 0.79 ± 0.09 | 0.90 ± 0.07 | **0.92** ± 0.06 | 0.52 ± 0.24 | 0.90 ± 0.07 | **0.93** ± 0.05 |
| | Apex | 0.59 ± 0.31 | **0.65** ± 0.36 | **0.65** ± 0.36 | 0.50 ± 0.31 | 0.63 ± 0.39 | **0.64** ± 0.38 | 0.38 ± 0.32 | 0.63 ± 0.37 | 0.62 ± 0.39 |
| 95HD (mean± std) | Whole | 10.41 ± 4.51 | 8.63 ± 8.29 | **7.99** ± 7.05 | 8.54 ± 4.87 | **7.42** ± 5.81 | 9.48 ± 11.37 | 13.64 ± 4.38 | **7.17** ± 4.21 | **6.66** ± 2.82 |
| | Base | 4.88 ± 1.62 | 4.51 ± 2.47 | **4.41** ± 2.99 | 4.69 ± 1.51 | 5.11 ± 6.25 | 7.83 ± 12.71 | 9.47 ± 4.33 | **4.22** ± 2.05 | 4.64 ± 2.39 |
| | Mid | 3.30 ± 1.13 | 1.65 ± 1.11 | **1.49** ± 0.65 | 2.73 ± 0.76 | 1.85 ± 0.89 | 1.54 ± 0.73 | 5.52 ± 2.26 | 1.74 ± 0.85 | 1.51 ± 0.66 |
| | Apex | 4.12 ± 2.32 | 3.01 ± 2.43 | **2.89** ± 2.45 | 4.04 ± 1.96 | **2.97** ± 2.63 | 3.06 ± 2.56 | 5.63 ± 2.61 | 3.25 ± 2.79 | 3.13 ± 2.54 |
| Ravd | Whole | −0.13 ±0.39 | **0.07** ± 0.37 | 0.09 ± 0.31 | −0.25 ± 0.31 | 0.18 ± 0.35 | 0.13 ± 0.32 | −0.09 ± 0.62 | 0.15 ± 0.36 | **0.05** ± 0.31 |
| | Base | **−0.06** ± 0.93 | 0.69 ± 1.50 | 0.58 ± 1.47 | **0.00** ± 0.98 | 0.73 ± 1.74 | 0.64 ± 1.51 | 0.21 ± 1.91 | 0.65 ± 1.66 | 0.70 ± 1.63 |
| | Mid | −0.32 ± 0.27 | **0.03** ± 0.25 | **0.04** ± 0.15 | −0.28 ± 0.21 | 0.15 ± 0.26 | 0.06 ± 0.20 | −0.55 ± 0.26 | 0.09 ± 0.26 | 0.06 ± 0.17 |
| | Apex | 1.63 ± 4.64 | 1.23 ± 4.05 | 1.30 ± 3.72 | **0.70** ± 2.98 | 1.89 ± 5.14 | 1.33 ± 3.97 | **−0.31** ± 0.94 | 2.06 ± 5.76 | 1.7 ± 5.0 |
| Sensitivity | Whole | 0.58 ± 0.17 | 0.74 ± 0.21 | 0.77 ± 0.20 | 0.54 ± 0.19 | **0.80** ± 0.17 | 0.78 ± 0.23 | 0.44 ± 0.28 | 0.78 ± 0.20 | 0.76 ± 0.05 |
| | Base | 0.60 ± 0.001 | 0.87 ± 0.01 | 0.85 ± 0.01 | 0.59 ± 0.001 | **0.88** ± 0.01 | 0.86 ± 0.01 | 0.48 ± 0.01 | 0.86 ± 0.01 | 0.84 ± 0.01 |
| | Mid | 0.64 ± 0.17 | 0.92 ± 0.17 | 0.95 ± 0.06 | 0.68 ± 0.14 | **0.96** ± 0.04 | 0.94 ± 0.06 | 0.41 ± 0.24 | 0.94 ± 0.06 | 0.95 ± 0.05 |
| | Apex | 0.74 ± 0.17 | 0.87 ± 0.21 | 0.87 ± 0.20 | 0.66 ± 0.19 | **0.90** ± 0.17 | 0.84 ± 0.23 | 0.45 ± 0.28 | 0.87 ± 0.20 | **0.95 ± 0.05** |
| Precision | Whole | 0.69 ± 0.14 | 0.71 ± 0.19 | 0.72 ± 0.14 | **0.73** ± 0.17 | 0.71 ± 0.14 | 0.72 ± 0.13 | 0.51 ± 0.16 | 0.71 ± 0.13 | **0.73** ± 0.13 |
| | Base | **0.83** ± 0.22 | 0.69 ± 0.24 | 0.73 ± 0.24 | 0.79 ± 0.23 | 0.72 ± 0.25 | 0.72 ± 0.25 | 0.64 ± 0.33 | 0.72 ± 0.24 | 0.70 ± 0.26 |
| | Mid | **0.96** ± 0.09 | 0.91 ± 0.09 | 0.92 ± 0.08 | **0.96** ± 0.07 | 0.86 ± 0.13 | 0.91 ± 0.10 | 0.92 ± 0.11 | 0.89 ± 0.13 | 0.91 ± 0.09 |
| | Apex | 0.77 ± 0.34 | 0.77 ± 0.27 | **0.79** ± 0.30 | 0.78 ± 0.31 | 0.75 ± 0.32 | 0.78 ± 0.30 | 0.77 ± 0.26 | 0.75 ± 0.32 | 0.71 ± 0.30 |

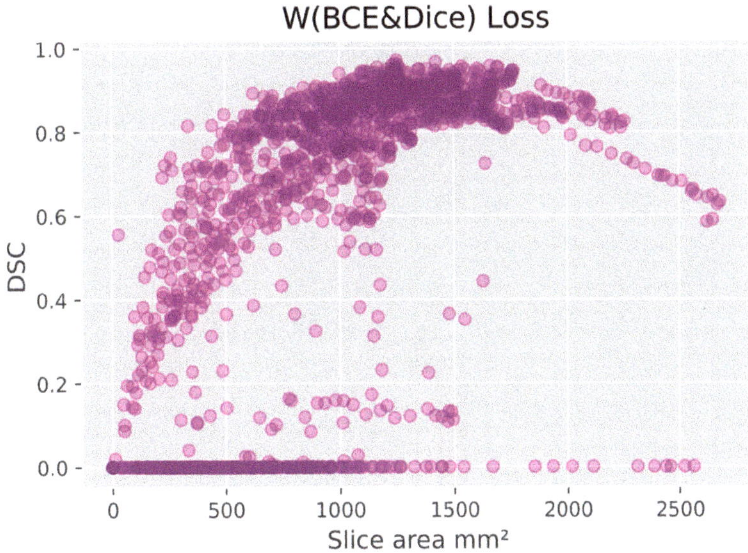


**Figure 5.** Dice score as a function of prostate area A reverse U shape is observed, indicating the prediction at the mid-gland (500–2000 mm$^2$) outperformed those at the base (>2000 mm$^2$) and apex (<500 mm$^2$). The zeros at the bottom correspond to cases where the model totally missed the prostate region (Dice score = 0). Data is based on the prediction from the model using W (BCE + Dice) on the validation data. The same pattern is seen in all models. W (BCE + Dice): weighted binary cross-entropy with Dice.

Within the Supplementary material, Figure S4 shows the DSC scores for individual patients for each model. Box plots of the DSC scores of all the models for each patient on the validation datasets in the five-fold cross validation are shown in Figure 7. DSC scores of models varied between patients, but for each patient the results were generally consistent across all three models (Tversky, Focal Tversky and W (BCE + Dice) (Table S1).

Model performance was generally lower in the apex and base compared with the mid-gland. This was not surprising, as inter-observer variability has been reported to be higher in these regions [4]. However, this may be an effect of the small cross-sectional areas (Figures 5 and 6). Additionally, the DSC score was lower for the slices that covered small areas or very large areas. We investigated the relationship between DSC score and prostate volume (Figure 4). No clear trend was identified, possibly due to the limited number of samples. However, in general, the model showed lower performance in DSC scores for smaller volumes in comparison to the average volume.

*Qualatative Comparison*

A selection of cases representing high and low performance are shown in Figures 8 and 9 for the models' outputs using two different loss functions, Focal Tversky and W (BCE + Dice). Samples with DSC scores higher than 0.80 were considered high-performance cases, and lower than 0.70 were considered low-performance cases. Higher DSC scores were achieved, for example in patients (cases) #2, # 16, #21, and #33. Cases #3, #8, and #22 are examples of lower performance.

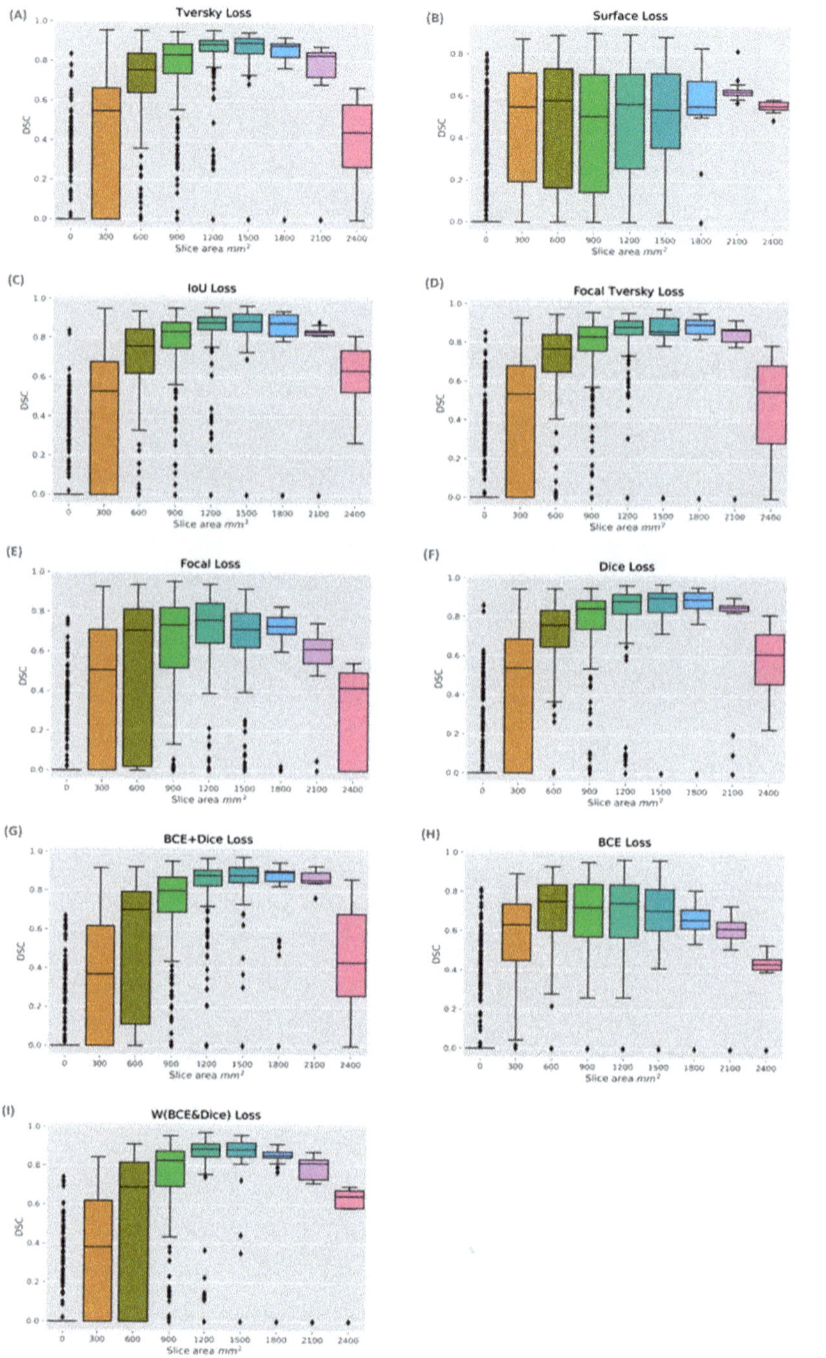


**Figure 6.** Box plots of DSC score vs. prostate area for each of the nine loss functions (**A–I**) listed in Table 1. DSC: Dice similarity coefficient.

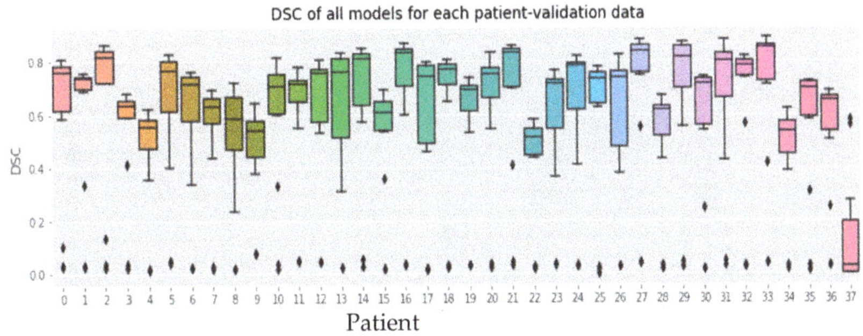


**Figure 7.** Box plots of Dice similarity coefficient (DSC) scores for all models for each patient in the validation data set.

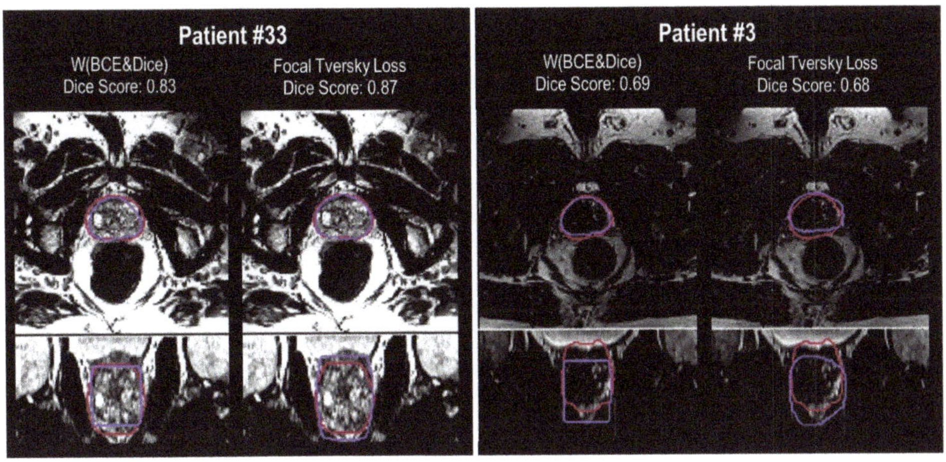


**Figure 8.** Outputs of two models (Focal Tversky and W (BCE + Dice)) in the axial (top image) and coronal views (bottom image for each patient), demonstrating both high and low performances, measured in Dice scores. The ground truth is represented by the red contour, the model's prediction contour is shown in purple.

Segmentation results show higher DSC scores from the model with Focal Tversky for case # 33 compared to W (BCE + Dice), with values of 0.83 and 0.87, respectively. Both models failed to capture the shape of the prostate at the apex and base. However, the output of the model with Focal Tversky had a greater similarity to the shape of the prostate than W (BCE + Dice) (Figure 9, case #2). This indicates that the model using Focal Tversky was more effective in defining the prostate boundaries.

Both models failed to define the prostate boundary for cases #15 and #22, especially in the apex and base regions. From the rectum shape in case #22, it is possible that there is some gas in the rectum which can reduce the quality of the MRI image.

For case #16, the model using W (BCE + Dice), with a DSC score of 0.79, had a worse performance compared with Focal Tversky (DSC score 0.85). The segmentation output of the model with W (BCE + Dice) was rectangular in shape, which can be seen in the coronal and sagittal views (Figure 9). Shapes of the segmentation outputs from Focal Tversky had a closer shape to the prostate than those from models with W (BCE + Dice) loss function.

In general, the W (BCE + Dice) model under-estimated the prostate volume and the Focal Tversky over-estimated the volume. Examples are cases #22, #15, #8, #16, and #2 (Figure 9).

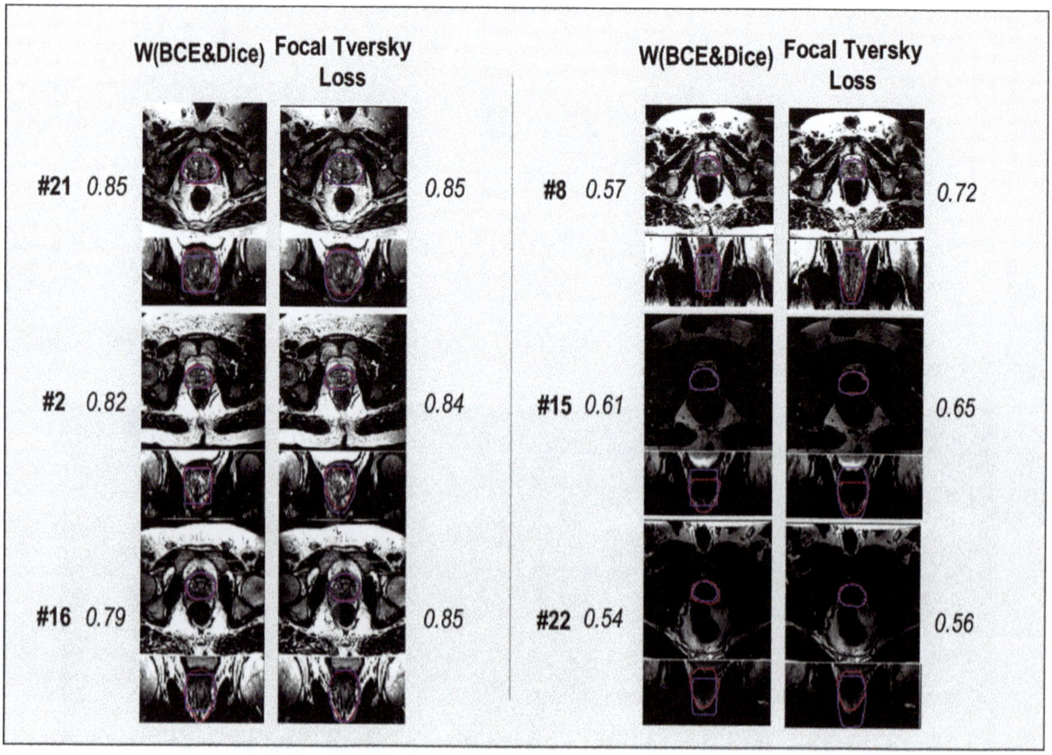


**Figure 9.** Outputs of two models (Focal Tversky and W (BCE + Dice)) for 6 patients (each patient identified by a number, e.g., #21 represents patient 21) in the axial (top image) and coronal views (bottom image for each patient), demonstrating both high and low performances, measured in Dice scores (shown adjacent to each image set). The ground truth is represented by the red contour, the model's prediction contour is shown in purple.

## 4. Discussion

Finding the most appropriate loss function for prostate segmentation is challenging. In this study we compared the performance of nine loss functions in a 37-patient data set. These nine loss functions were chosen as they are commonly used in medical image segmentation tasks [14]. The 37-patient data set included locally acquired data with a common imaging protocol (two resolutions) and a single MRI scanner [16] to avoid variations due to image acquisition. These data were co-registered with whole mount pathology to provide ground truth delineations of the prostate [16] in contrast to many publicly available datasets that rely on clinician-generated segmentations which are subject to interobserver variation [4]. A limitation of the generalizability of our study is the small sample size and homogeneity in the methods used to acquire the MRI data. We therefore recommend that future studies that intend to use data from a variety of sources and scanning protocols confirm the findings of our study using the methodology we describe, and consider the most appropriate metric for their evaluation. Publicly available data can be sourced from a variety of locations such as those described by Ma et al. [14], however, the purpose of our study was to remove uncertainties due to heterogeneity in data source and clinician contouring, and focus only on the relative performance of the loss functions selected for our study and a range of metrics for their evaluation. Our study found the proposed architecture performed with notable variations when different loss functions were applied. As the base and the apex of the prostate are particularly challenging to segment manually due to the lack of a clear boundary [1,17], we therefore also evaluated the performance at the mid-gland, apex, and base of the prostate independently.

Focal Tversky had the highest scores for the whole gland in terms of DSC score and sensitivity. However, W (BCE + Dice) outperformed all competing methods in precision, followed by 95HD, Ravd, and Tversky. With performance measured by the median and standard deviation, the best performance was achieved by applying W (BCE + Dice), Tversky, and Focal Tversky loss functions. However, the performance of models with Focal Tversky, Tversky, W (BCE + Dice), Dice, and IoU loss functions were very close for our dataset. Lower performance was observed using Surface loss, BCE loss and Focal loss functions. Focal Tversky and Tversky loss functions have been recommended by other researchers as returning optimal results when their parameters are set to the correct values [15]. However, for challenging medical segmentation tasks, we suggest using Focal Tversky and W (BCE + Dice), and by optimizing their parameters, the best solution can be achieved in accordance with the application requirements. The loss function parameters of W (BCE + Dice) allow the user to define the best trade-off between FNs and FPs. Additionally, Focal Tversky and W (BCE + Dice) have the advantage of adjustable parameters, which make it possible to tune the loss function based on the application requirements. For example, Focal Tversky and W (BCE + Dice) have parameters which can be tuned to address under- and over-segmentation issues that may arise with other loss functions. As a result, in the future, we plan to investigate the effectiveness of a combination of Tversky and BCE loss functions for prostate segmentation.

Lower performance was observed using Surface loss, BCE loss and Focal loss functions. All models achieved higher performance for mid-gland and lower performance in the apex and base regions. When considering model performance for individual data sets, we observed that all models had a similar performance for each image, but performance varied across the patient cohort. This may be related to patient-specific image quality, however, all models generalized the average shape of images and failed to perform well for outlier shapes.

Intuitively, it can be expected that model performance will be affected by the choice of the metric used to measure performance and the principal components driving the loss function. For example, DSC measures the overlap between two regions. If the Dice loss is used, the training process is exactly guided as the final metric, which theoretically should achieve a good performance. This can be seen in Table 2; the Dice loss achieved a consistently high DSC in the whole prostate gland (0.73) as well as the sub-volumes (0.65–0.93). In addition, the close variants of the Dice loss, including Tversky, Focal Tversky, and IoU loss, also obtained high performances (0.63–0.92), but slightly inferior to the Dice loss. For losses that are not region-based, compound losses such as BCE + Dice and W (BCE + Dice) showed relatively higher DSC (0.62–0.93) as they consist of a Dice loss component. In contrast, Surface loss (boundary-based) and BCE (distribution-based) demonstrated the lowest DSC (0.38–0.75). However, this pattern is not shown between all metrics and categories. For example, HD95 is a boundary-based metric and it was expected that Surface loss would achieve a high performance. However, as shown in Table 2, Dice loss has the lowest HD95, while Surface loss had the highest. One possible reason is that the Surface loss is relatively hard to train, requiring more epoches for it to converge. Since the training process was consistent across all loss functions, this may explain why some functions did not perform as well as expected.

To overcome variability in performance of individual loss functions, compound loss functions can be considered. For example, in the case of prostate segmentation, data imbalance is a major problem, and loss functions, such as BCE, that are suitable for balanced data are not suitable for this task. However, as shown in our study, weighted BCE combined with Dice can improve model performance significantly.

Tuning hyper-parameters of U-Net, such as the learning rate and number of iterations, requires significant computational time. To address this, we defined the best learning rate for Dice and BCE loss functions, as most of the other loss functions are variations of these loss functions. We used a grid search for optimization of the learning rate and defined the optimal value of loss function parameters in Focal, W (Dice + BCE) and Focal Tversky loss functions on the validation data set. The optimal learning rate was selected as $\alpha = 0.0001$, from 0.001, 0.0001, 0.00001. The parameters of the W (Dice + BCE) loss function allocated a higher contribution to the cross-entropy term, $\alpha$ equal to 0.6, in comparison to the Dice term with a weight of 0.4. The optimum value of $\beta$ for the weighted cross-entropy term was found to be 0.7, which penalizes false negatives more. This aligned with other recommendations for segmentation problems on MRI data [24]. Different values of $\alpha$ and $\beta$ can be applied to obtain the best model result and handle the imbalance problem of each dataset appropriately.

Models were trained using the T2w axial data and performed better visually in the axial view. Training a model using axial, sagittal, and coronal (or a 3D data set) might improve the model performance. However, adding more inputs will also add complexity and extra computation cost. In this study, we used the 2D U-Net model, which has a lower number of components, to optimize in comparison to a 3D U-Net. In addition, 3D U-Net models underfit when trained on a small number of datasets [6]. Furthermore, it is easier to identify the loss function contribution to the model performance where there is less model complexity. It has been shown that a simple network with a proper loss function can outperform more complex architectures, including networks with specific up-sampling or with skip connection [24].

Regarding implementation, Keras offers a number of tools to construct a U-Net with its sequential and functional interface. Hence, the model itself can be constructed and set up for training in a straightforward approach. However, for the loss function, a potential challenge is to carefully choose the exact equation to implement. This is because even for the same loss function, there are slight variations. For example, the denominator of a Dice loss can be the sum of squared signal intensities, while another form will leave out the square operation. Such subtle differences can add to confounding factors when comparing model performance reported in the literature.

A model's output can improve using post-processing methods that reduce false positives and false negatives in segmented images [25]. CNN segmentation results improve using energy-based refinement post-processing steps [26]. We applied threshold-based refinement to cope with false positives [27]. A threshold value of 0.5 was found to be the optimal value to return the highest Dice score with the least number of false positives.

## 5. Conclusions

The performance of a 2D U-Net model with nine different loss functions for prostate gland segmentation was compared. Ranking of model performance was found to depend on the metric used to measure performance. Performance was also found to vary based on the region within the prostate being considered, with the base and apex generally being less compared with the mid-glad and entire prostate gland. There was some evidence that performance was also affected by cross-sectional area of the image, with peak performance in the range of 600–2100 $mm^2$. The performance of models using different loss functions varied by approximately 34% using the DSC score metric. Focal Tversky, Tversky, and W (Dice + BCE) loss functions achieve better performance considering majority of metrics. However, performance of models with Focal Tversky, Tversky, W (Dice + BCE), Dice, and IoU were close. Lower performance was observed using the distribution-based and boundary-based loss functions (Surface, BCE, and Focal loss functions). Based on this 37-patient data set, it is suggested that the Focal Tversky and W (Dice + BCE) loss functions are most suitable for the task of prostate segmentation as their parameters allow the user to modify the loss function for a specific dataset.

**Supplementary Materials:** The following supporting information can be downloaded at: https://www.mdpi.com/article/10.3390/bioengineering10040412/s1, Figure S1: Box plots of all metrics used in this study for the *prostate mid-gland* on validation data from the five-fold cross-validation for models with different loss functions. Figure S2: Box plots of all metrics used in this study for the *apex region* on validation data from the five-fold cross-validation for models with different loss functions. Figure S3: Box plots of all metrics used in this study for the prostate base region on validation data from the five-fold cross-validation for models with different loss functions. Figure S4: Dice similarity coefficient (DSC) score of all the models for each patient. Figure S5: DSC score vs. prostate volume ($mm^3$) for model using Focal Tversky loss. Table S1: W (BCE + Dice), Tversky and Focal Tversky performances, DSC score, for each patient.

**Author Contributions:** Conceptualization, M.M.; methodology, M.M.; software, M.M.; validation, M.M. and Y.S.; formal analysis, M.M.; investigation, M.M., Y.S., G.S., A.H.; resources, A.H.; data curation, Y.S. and M.M.; writing—original draft preparation, M.M.; writing—review and editing, M.M., Y.S., G.S., A.H.; supervision, A.H.; project administration, A.H.; funding acquisition, A.H. All authors have read and agreed to the published version of the manuscript.

**Funding:** This research was funded by National Health and Medical Research Council (Grant No.: 1126955), Cancer Institute of New South Wales: TPG 182165 and SW-TCRC Partner Program grant 2019.

**Institutional Review Board Statement:** The study was conducted in accordance with the Declaration of Helsinki, and approved by the Peter MacCallum Cancer Centre Human Research Ethics Committee Ethics Committee (HREC/15/PMCC125).

**Informed Consent Statement:** Informed consent was obtained from all subjects involved in the study.

**Data Availability Statement:** The data presented in this study and the code used to perform the calculations with these data are available on request from the corresponding author. The data are not publicly available due to privacy restrictions.

**Acknowledgments:** The authors acknowledge the Sydney Informatics Hub and the University of Sydney's high-performance computing cluster Artemis for providing the high-performance computing resources, and the technical assistance of Nathaniel Butterworth who contributed to the research results reported within this paper. We would also like to thank Nym Vandenberg for their critical review of the manuscript.

**Conflicts of Interest:** The authors declare no conflict of interest. The funders had no role in the design of the study; in the collection, analyses, or interpretation of data; in the writing of the manuscript, or in the decision to publish the results. G.S. has received consultancy fees and/or travel support paid to his institution from MSD Ltd. New Zealand, Siemens HealthcareGmb, RaySearch Laboratories, Elekta New Zealand Ltd., AccuRay Ltd. G.S. has received scientific grant support from Bayer and Astella pharmaceuticals for clinical trials administered through the Australia and New Zealand Urogenital and Prostate (ANZUP) clinical trials group. G.S. also has research supported by Bristol-Myers Squibb, Prostate Cancer Foundation (USA), Prostate Cancer Foundation Australia, the PeterMac Foundation, Cancer Council Victoria and Victoria Cancer Agency. None of these relate to the current work. A.H. has a non-financial research agreement with Siemens Healthcare which does not relate to the current work. G.S. has a non-financial research agreement with MIM Software Inc. and RaySearch Laboratories; their software was used for the purposes of this work.

## Appendix A Definition of Loss Functions Used in This Study

Loss functions are important key drivers in determining the success of neural network models. They define how neural network models calculate the overall error between the prediction and the ground truth. During training, the loss is calculated for each batch and minimized using optimization algorithms. Selecting an appropriate loss function has a larger effect on model performance than using a complex architecture [17]. Loss functions can generally be classified into four groups: distribution-based, region-based, boundary-based, and compound loss [14]. Compound loss is the combination of different types of loss functions. The main role of loss functions is to quantify the mismatch region between ground truth and segmentation. The main differences between them are the weighting methods [14].

The following equations use these generic notations. Specific parameters will be explained otherwise.

$g_i$, $s_i$: voxels $i$ in ground truth and segmentation output, respectively;

$C$: the number of classes;

$c$: notation for an individual class. If class $c$ is the correct classification for voxel $i$, ${g_i}^c$ is equal to 1 and ${s_i}^c$ is the corresponding predicted probability;

$N$: the total number of samples.

**Distribution-based loss functions:** Distribution-based loss functions aim to minimize dissimilarity between two distributions. We used binary cross-entropy (BCE) and Focal loss from distribution-based loss functions. The fundamental function in this category is cross-entropy and all functions were derived from cross-entropy function.

**Cross-entropy loss:** Cross-entropy (CE) loss is the most commonly used loss function for training deep learning models. It measures dissimilarity between two distributions using CE. Data distribution comes from the training set properties. The formulation for the CE loss function is:

$$Loss_{CE} = -\frac{1}{N}\sum_{i=1}^{N}\sum_{c}^{C}{g_i}^c\log\left({s_i}^c\right)$$

In this study the segmentation task was a binary classification, therefore, the loss function is a binary cross-entropy (BCE).

A CE loss function can control output imbalance, false positive, and false negative rates. However, model performance with a cross-entropy loss function is not optimal for segmentation tasks with highly class-imbalanced input images [28]. There are several different loss-function-based techniques using weighted cross-entropy [29].

A variation is the weighted cross-entropy (WCE):

$$Loss_{WCE} = -\frac{1}{N}\sum_{i=1}^{N}\sum_{c}^{C}w_c{g_i}^c\log\left({s_i}^c\right)$$

where $w_c$ is the weight for each class. This loss function penalizes majority classes by weighting them inversely proportional to the class frequencies.

**Focal loss:** The focal loss function is one of the WCE loss functions shown to better manage unbalanced classes in a dataset [30]. The Focal loss function reduces the loss function corresponding to well-classified examples. It uses a scaling method to allocate higher weights on the examples that are difficult to classify over easier cases.

$$Loss_{focal} = -\frac{1}{N}\sum_{i=1}^{N}\sum_{c}^{C}\left(1-s_i\right)^{\gamma}{g_i}^c\log\left({s_i}^c\right)$$

where $\gamma$ is a hyperparameter called focusing parameter.

**Region-based loss:** Region-based loss functions aim to minimize mismatch by maximizing the overlap regions between the output of segmentation ($Ss$) and ground truth ($Gg$). Dice loss is the key element of this category.

**Dice loss:** Dice loss aims to directly maximize the Dice coefficient, which is the most commonly used segmentation evaluation metric [31]. Segmentation models with Dice loss functions have shown superior performance for binary segmentation [29,31,32]. The loss function is formulated as the negative DSC:

$$Loss_{Dice} = -\frac{2\sum_{i=1}^{N}s_i g_i}{\sum_{i=1}^{N}{s_i}^2+\sum_{i=1}^{N}{g_i}^2+\varepsilon}$$

where $\varepsilon$ is a small number to avoid division by zero. In this study, $\varepsilon = 1$ was used for all models.

**IoU loss:** The IoU loss function aims to maximize the intersection-over-union coefficient, known as the Jaccard coefficient. IoU is an evaluation metric for segmentation similar to Dice loss [33]:

$$Loss_{IoU} = 1 - \frac{\sum_{i=1}^{N} s_i g_i}{\sum_{i=1}^{N} (s_i + g_i - s_i g_i)}$$

**Tversky:** The Tversky loss function reshapes Dice loss and prioritizes false negatives to achieve a better trade-off between precision and recall [33]. Background voxels that are labelled as the target object are false positives. False negatives refer to the voxels of a target object that are misclassified as background. Segmentation with fewer false positives and false negatives are ideal, but it is not easy to decrease both at the same time.

$$Loss_{Tversky} = \frac{\sum_{i=1}^{N} s_i g_i}{\sum_{i=1}^{N} s_i g_i + \alpha \sum_{i=1}^{N} s_i (1 - g_i) + \beta \sum_{i=1}^{N} g_i (1 - s_i)}$$

where $\alpha$ and $\beta$ are weighting factors to weight the contribution of false positives and false negatives. For certain applications, reducing the false positive (FP) rate is more important than reducing the false negative (FN) rate or vice versa [34].

**Focal Tversky:** Focal Tversky applies the concept of focal loss to improve model performance for cases with low probabilities [35]:

$$L_{FTL} = (1 - L_{Tversky})^{1/\gamma}$$

where $\gamma$ varies in the range [1, 3].

**Boundary-based loss functions:** Boundary-based loss functions are a new type of loss function that aims to minimize the distance between two boundaries of the ground truth and segmentation output.

**Boundary (BD) loss (Surface loss):** A boundary (BD) loss (or surface loss) function aims to minimize the mean surface distance, Dist $(\partial G, \partial S)$, between two boundaries (surfaces) of the ground truth $G$ and segmentation output $S$. The boundary of the ground truth $(G)$ is denoted as $\partial G$, and $\partial S$ represents the boundary of segmentation $(S)$. This means that BD loss minimizes the mean of the distance between surface voxels in $S$ and the closest voxels in $G$.

Boundary loss uses an integral over the boundary between regions instead of integrals within the regions.

$$\text{Dist}(\partial G, \partial S) = \int_{\partial G} ||y_{as}(p) - p||^2 dp$$

where $p$ is a point on boundary $\partial G$ and $y_{as}(p)$ is the corresponding point on segmentation boundary $\partial S$.

**Compound loss:** Compound loss functions are a combination of different types of loss functions, mostly cross-entropy and Dice similarity coefficient. This loss function comes from both the WCE and the Dice loss functions.

$$Loss_{Combo} = \alpha \left( -\frac{1}{N} [\sum_{i=1}^{N} \beta (g_i \log s_i) + (1 - \beta)(1 - g_i) log(1 - s_i)] \right) - (1 - \alpha) \left( \frac{2\sum_{i=1}^{N} s_i g_i + \varepsilon}{\sum_{i=1}^{N} s_i^2 + \sum_{i=1}^{N} g_i^2 + \varepsilon} \right)$$

where $\alpha$ controls the contribution of the WCE loss and the Dice terms; $\beta$ controls the contribution from positive voxels within WCE. Values of $\alpha$ and $\beta$ can be defined from a grid search. In this study, two configurations are used. One has equal weights on BCE and Dice, referred to as BCE + Dice. The other uses grid search to determine the best combination ($\alpha$ = 0.6, $\beta$ = 0.7), known as weighted BCE and Dice, or W (BCE + Dice). The latter applies more penalty to false negatives. This aligns with the observation that under-segmentation (false negative) is a common problem for MRI data [23].

## References


1. Litjens, G.; Toth, R.; van de Ven, W.; Hoeks, C.; Kerkstra, S.; van Ginneken, B.; Vincent, G.; Guillard, G.; Birbeck, N.; Zhang, J.; et al. Evaluation of Prostate Segmentation Algorithms for MRI: The PROMISE12 Challenge. *Med. Image Anal.* **2014**, *18*, 359–373. [CrossRef] [PubMed]
2. Litjens, G.; Kooi, T.; Bejnordi, B.E.; Setio, A.A.A.; Ciompi, F.; Ghafoorian, M.; van der Laak, J.A.W.M.; van Ginneken, B.; Sánchez, C.I. A Survey on Deep Learning in Medical Image Analysis. *Med. Image Anal.* **2017**, *42*, 60–88. [CrossRef] [PubMed]
3. Dowling, J.A.; Fripp, J.; Chandra, S.; Pluim, J.P.W.; Lambert, J.; Parker, J.; Denham, J.; Greer, P.B.; Salvado, O. Fast Automatic Multi-Atlas Segmentation of the Prostate from 3D MR Images. In Proceedings of the Lecture Notes in Computer Science, Toronto, ON, Canada, 22 September 2011; pp. 10–21.
4. Becker, A.S.; Chaitanya, K.; Schawkat, K.; Muehlematter, U.J.; Hötker, A.M.; Konukoglu, E.; Donati, O.F. Variability of Manual Segmentation of the Prostate in Axial T2-Weighted MRI: A Multi-Reader Study. *Eur. J. Radiol.* **2019**, *121*, 108716. [CrossRef] [PubMed]
5. Xu, X.; Lu, Q.; Hu, Y.; Yang, L.; Hu, S.; Chen, D.; Shi, Y. Quantization of Fully Convolutional Networks for Accurate Biomedical Image Segmentation. In Proceedings of the IEEE Computer Society Conference on Computer Vision and Pattern Recognition, Salt Lake City, UT, USA, 13 March 2018; pp. 8300–8308.
6. Yu, L.; Yang, X.; Chen, H.; Qin, J.; Heng, P.A. Volumetric ConvNets with Mixed Residual Connections for Automated Prostate Segmentation from 3D MR Images. *Proc. AAAI Conf. Artif. Intell.* **2017**, *31*, 66–72. [CrossRef]
7. Mahapatra, D.; Buhmann, J.M. Prostate MRI Segmentation Using Learned Semantic Knowledge and Graph Cuts. *IEEE Trans. Biomed. Eng.* **2014**, *61*, 756–764. [CrossRef] [PubMed]
8. Ronneberger, O.; Fischer, P.; Brox, T. U-Net: Convolutional Networks for Biomedical Image Segmentation. In Proceedings of the Lecture Notes in Computer Science (including subseries Lecture Notes in Artificial Intelligence and Lecture Notes in Bioinformatics), Munich, Germany, 5–9 October 2015; pp. 234–241.
9. Aldoj, N.; Biavati, F.; Michallek, F.; Stober, S.; Dewey, M. Automatic Prostate and Prostate Zones Segmentation of Magnetic Resonance Images Using DenseNet-like U-Net. *Sci. Rep.* **2020**, *10*, 14315. [CrossRef] [PubMed]
10. Huang, G.; Liu, Z.; Van Der Maaten, L.; Weinberger, K.Q. Densely Connected Convolutional Networks. In Proceedings of the 2017 IEEE Conference on Computer Vision and Pattern Recognition (CVPR), Honolulu, HI, USA, 21–26 July 2017; IEEE: Manhattan, NY, USA, 2017; pp. 2261–2269.
11. Isensee, F.; Jäger, P.F.; Kohl, S.A.A.; Petersen, J.; Maier-Hein, K.H. nnU-Net: A Self-Configuring Method for Deep Learning-Based Biomedical Image Segmentation. *Nat. Methods* **2021**, *18*, 203–211. [CrossRef] [PubMed]
12. Elsken, T.; Metzen, J.H.; Hutter, F. Neural Architecture Search. In *Automated Machine Learning*; The Springer Series on Challenges in Machine Learning; Springer: Cham, Switzerland, 2019; pp. 63–77.
13. LeCun, Y. 1.1 Deep Learning Hardware: Past, Present, and Future. In Proceedings of the 2019 IEEE International Solid-State Circuits Conference—(ISSCC), San Francisco, CA, USA, 17–21 February 2019; IEEE: Manhattan, NY, USA, 2019; pp. 12–19.
14. Ma, J.; Chen, J.; Ng, M.; Huang, R.; Li, Y.; Li, C.; Yang, X.; Martel, A.L. Loss Odyssey in Medical Image Segmentation. *Med. Image Anal.* **2021**, *71*, 102035. [CrossRef] [PubMed]
15. Jadon, S. A Survey of Loss Functions for Semantic Segmentation. In Proceedings of the 2020 IEEE Conference on Computational Intelligence in Bioinformatics and Computational Biology, Via del Mar, Chile, 26 June 2020; pp. 1–7.
16. Reynolds, H.M.; Williams, S.; Zhang, A.; Chakravorty, R.; Rawlinson, D.; Ong, C.S.; Esteva, M.; Mitchell, C.; Parameswaran, B.; Finnegan, M.; et al. Development of a Registration Framework to Validate MRI with Histology for Prostate Focal Therapy. *Med. Phys.* **2015**, *42*, 7078–7089. [CrossRef] [PubMed]
17. Asgari Taghanaki, S.; Abhishek, K.; Cohen, J.P.; Cohen-Adad, J.; Hamarneh, G. Deep Semantic Segmentation of Natural and Medical Images: A Review. *Artif. Intell. Rev.* **2021**, *54*, 137–178. [CrossRef]
18. Kingma, D.P.; Ba, J. Adam: A Method for Stochastic Optimization. In Proceedings of the International Conference on Learning Representations, San Diego, CA, USA, 22 December 2015.
19. Singh, S.; Krishnan, S. Filter Response Normalization Layer: Eliminating Batch Dependence in the Training of Deep Neural Networks. In Proceedings of the IEEE Computer Society Conference on Computer Vision and Pattern Recognition, San Juan, PR, USA, 21 November 2020; pp. 11234–11243.
20. Kohavi, R. A Study of Cross-Validation and Bootstrap for Accuracy Estimation and Model Selection. In Proceedings of the 14th International Joint Conference on Artificial Intelligence, Montreal, QC, Canada, 20–25 August 1995; Morgan Kaufmann Publishers Inc.: San Francisco, CA, USA, 1995; Volume 2, pp. 1137–1143.
21. Gulli, A.; Pal, S. *Deep Learning with Keras*; Packt Publishing: Birmingham, UK, 2017; ISBN 9781787128422.
22. Abadi, M.; Barham, P.; Chen, J.; Chen, Z.; Davis, A.; Dean, J.; Devin, M.; Ghemawat, S.; Irving, G.; Isard, M.; et al. TensorFlow: A System for Large-Scale Machine Learning. In Proceedings of the 12th USENIX Symposium on Operating Systems Design and Implementation, Savannah, GA, USA, 2–4 November 2016; ISBN 978-1-931971-33-1.
23. Taha, A.A.; Hanbury, A. Metrics for Evaluating 3D Medical Image Segmentation: Analysis, Selection, and Tool. *BMC Med. Imaging* **2015**, *15*, 29. [CrossRef] [PubMed]
24. Taghanaki, S.A.; Zheng, Y.; Kevin Zhou, S.; Georgescu, B.; Sharma, P.; Xu, D.; Comaniciu, D.; Hamarneh, G. Combo Loss: Handling Input and Output Imbalance in Multi-Organ Segmentation. *Comput. Med. Imaging Graph.* **2019**, *75*, 24–33. [CrossRef] [PubMed]

25. Yang, D.; Xu, D.; Zhou, S.K.; Georgescu, B.; Chen, M.; Grbic, S.; Metaxas, D.; Comaniciu, D. Automatic Liver Segmentation Using an Adversarial Image-to-Image Network. In Proceedings of the Lecture Notes in Computer Science (including subseries Lecture Notes in Artificial Intelligence and Lecture Notes in Bioinformatics), Quebec City, QC, Canada, 25 July 2017; pp. 507–515.
26. Hu, P.; Wu, F.; Peng, J.; Bao, Y.; Chen, F.; Kong, D. Automatic Abdominal Multi-Organ Segmentation Using Deep Convolutional Neural Network and Time-Implicit Level Sets. *Int. J. Comput. Assist. Radiol. Surg.* **2017**, *12*, 399–411. [CrossRef] [PubMed]
27. Gibson, E.; Giganti, F.; Hu, Y.; Bonmati, E.; Bandula, S.; Gurusamy, K.; Davidson, B.R.; Pereira, S.P.; Clarkson, M.J.; Barratt, D.C. Towards Image-Guided Pancreas and Biliary Endoscopy: Automatic Multi-Organ Segmentation on Abdominal CT with Dense Dilated Networks. In Proceedings of the Lecture Notes in Computer Science (including subseries Lecture Notes in Artificial Intelligence and Lecture Notes in Bioinformatics), Quebec City, QC, Canada, 25 July 2017; pp. 728–736.
28. Badrinarayanan, V.; Kendall, A.; Cipolla, R. SegNet: A Deep Convolutional Encoder-Decoder Architecture for Image Segmentation. *IEEE Trans. Pattern Anal. Mach. Intell.* **2017**, *39*, 2481–2495. [CrossRef] [PubMed]
29. Sudre, C.H.; Li, W.; Vercauteren, T.; Ourselin, S.; Jorge Cardoso, M. Generalised Dice Overlap as a Deep Learning Loss Function for Highly Unbalanced Segmentations. In Proceedings of the Deep Learn Med Image Anal Multimodal Learn Clin Decis Support, Quebec City, QC, Canada, 14 September 2017; pp. 240–248. [CrossRef]
30. Lin, T.-Y.; Goyal, P.; Girshick, R.; He, K.; Dollár, P. Focal Loss for Dense Object Detection. *IEEE Trans. Pattern Anal. Mach. Intell.* **2017**, *42*, 318–327. [CrossRef]
31. Milletari, F.; Navab, N.; Ahmadi, S.-A. V-Net: Fully Convolutional Neural Networks for Volumetric Medical Image Segmentation. In Proceedings of the 2016 Fourth International Conference on 3D Vision (3DV), Stanford, CA, USA, 25–28 October 2016; IEEE: Manhattan, NY, USA, 2016; pp. 565–571.
32. Brosch, T.; Tang, L.Y.W.; Yoo, Y.; Li, D.K.B.; Traboulsee, A.; Tam, R. Deep 3D Convolutional Encoder Networks With Shortcuts for Multiscale Feature Integration Applied to Multiple Sclerosis Lesion Segmentation. *IEEE Trans. Med. Imaging* **2016**, *35*, 1229–1239. [CrossRef] [PubMed]
33. Rahman, M.A.; Wang, Y. Optimizing Intersection-Over-Union in Deep Neural Networks for Image Segmentation. In Proceedings of the Lecture Notes in Computer Science (including subseries Lecture Notes in Artificial Intelligence and Lecture Notes in Bioinformatics), Las Vegas, NV, USA, 12–14 December 2016; pp. 234–244.
34. Salehi, S.S.M.; Erdogmus, D.; Gholipour, A. Tversky Loss Function for Image Segmentation Using 3D Fully Convolutional Deep Networks. In Proceedings of the Lecture Notes in Computer Science (including subseries Lecture Notes in Artificial Intelligence and Lecture Notes in Bioinformatics), Quebec City, QC, Canada, 18 June 2017; pp. 379–387.
35. Abraham, N.; Khan, N.M. Multimodal Segmentation with MGF-Net and the Focal Tversky Loss Function. In Proceedings of the Lecture Notes in Computer Science, Shenzhen, China, 17 October 2020; pp. 191–198.

bioengineering

MDPI

*Article*

# Radiomics-Based Machine Learning Model for Predicting Overall and Progression-Free Survival in Rare Cancer: A Case Study for Primary CNS Lymphoma Patients

**Michela Destito [1,*,†], Aldo Marzullo [2,†], Riccardo Leone [3,†], Paolo Zaffino [1], Sara Steffanoni [4,†], Federico Erbella [4], Francesco Calimeri [2], Nicoletta Anzalone [3,5], Elena De Momi [6], Andrés J. M. Ferreri [4], Teresa Calimeri [4,‡] and Maria Francesca Spadea [1,7,‡]**

1 Department of Experimental and Clinical Medicine, University of Catanzaro, 88100 Catanzaro, Italy
2 Department of Mathematics and Computer Science, University of Calabria, 87036 Rende, Italy
3 Neuroradiology Unit, IRCCS San Raffaele Scientific Institute, 20132 Milan, Italy
4 Lymphoma Unit, IRCCS San Raffaele Scientific Institute, 20132 Milan, Italy
5 Neuroradiology Unit and CERMAC, San Raffaele Scientific Institute, Vita-Salute San Raffaele University, 20132 Milan, Italy
6 Department of Electronics, Information and Bioengineering, Politecnico of Milan, 20133 Milan, Italy
7 Institute of Biomedical Engineering, Karlsruhe Institute of Technology (KIT), 76131 Karlsruhe, Germany
* Correspondence: michela.destito@unicz.it; Tel.: +39-0961-3694196
† These authors contributed equally to this work.
‡ These authors contributed equally to this work.

**Abstract:** Primary Central Nervous System Lymphoma (PCNSL) is an aggressive neoplasm with a poor prognosis. Although therapeutic progresses have significantly improved Overall Survival (OS), a number of patients do not respond to HD–MTX-based chemotherapy (15–25%) or experience relapse (25–50%) after an initial response. The reasons underlying this poor response to therapy are unknown. Thus, there is an urgent need to develop improved predictive models for PCNSL. In this study, we investigated whether radiomics features can improve outcome prediction in patients with PCNSL. A total of 80 patients diagnosed with PCNSL were enrolled. A patient sub-group, with complete Magnetic Resonance Imaging (MRI) series, were selected for the stratification analysis. Following radiomics feature extraction and selection, different Machine Learning (ML) models were tested for OS and Progression-free Survival (PFS) prediction. To assess the stability of the selected features, images from 23 patients scanned at three different time points were used to compute the Interclass Correlation Coefficient (ICC) and to evaluate the reproducibility of each feature for both original and normalized images. Features extracted from Z-score normalized images were significantly more stable than those extracted from non-normalized images with an improvement of about 38% on average ($p$-value $< 10^{-12}$). The area under the ROC curve (AUC) showed that radiomics-based prediction overcame prediction based on current clinical prognostic factors with an improvement of 23% for OS and 50% for PFS, respectively. These results indicate that radiomics features extracted from normalized MR images can improve prognosis stratification of PCNSL patients and pave the way for further study on its potential role to drive treatment choice.

**Keywords:** rare tumor; PCNSL; radiomics; image normalization; MRI

**Citation:** Destito, M.; Marzullo, A.; Leone, R.; Zaffino, P.; Steffanoni, S.; Erbella, F.; Calimeri, F.; Anzalone, N.; De Momi, E.; Ferreri, A.J.M.; et al. Radiomics-Based Machine Learning Model for Predicting Overall and Progression-Free Survival in Rare Cancer: A Case Study for Primary CNS Lymphoma Patients. *Bioengineering* **2023**, *10*, 285. https://doi.org/10.3390/bioengineering10030285

Academic Editors: Zhen Cheng and Larbi Boubchir

Received: 19 January 2023
Revised: 15 February 2023
Accepted: 20 February 2023
Published: 22 February 2023



## 1. Introduction

Primary diffuse large B-cell lymphoma (DLBCL) of the central nervous system (CNS) (PCNSL) is a rare form of aggressive extranodal non-Hodgkin's lymphoma limited to the CNS and, thus, potentially involving the brain, spinal cord, meninges, and eyes [1,2]. Magnetic resonance imaging (MRI) before and after contrast injection is the recommended imaging modality in the case of PCNSL suspicion and for disease staging after diagnosis confirmation by histopathological examination of a tumor biopsy [3]. The modern treatment of PCNSL is based on two phases, induction and consolidation [3,4]. The first one typically consists of high-dose methotrexate (MTX)-based chemotherapy, while the second one may include several options, among which high-dose chemotherapy, followed by autologous stem cell transplantation (HCT–ASCT), is presently the golden standard [5–7]. Although new therapeutic approaches have improved overall survival [5,8], about 30% of patients <70 years are primary refractory to HD–MTX-based chemotherapy and nearly 25% of patients relapse after consolidation [9]. Unfortunately, the reasons underlying this poor response to therapy are not known. Nevertheless, being able to identify, in advance, patients who are going to respond to the current treatment would be of the uttermost importance, as it may help in driving clinical decision making and in tailoring treatment accordingly.

Radiomics is a computational technique to extract high-dimensional quantitative features from medical images [10], which embed information about shape, intensity, and texture of a particular Volume of Interest (VoI). It assumes that medical images reflect underlying characteristics of disease-specific pathological processes and quantitative analysis can objectively capture and describe such mechanisms [11]. In recent years, the application of Artificial Intelligence (AI) techniques in the biomedical field[12,13] has been rapidly expanding. Advanced analytical and machine learning (ML) tools with radiomics features [14] have been used to improve diagnosis [15], or to allow prognostic stratification [16] and customization of therapy in oncology [17]. In contrast to a traditional biopsy, which is limited to the analysis of a small amount of tissue sample, one of the advantages of Radiomics is the possibility to characterize the whole tumor volume, and, thus, capturing extended lesion properties, such as size, shape and heterogeneity, or changes over time on image series [18]. Several radiomics studies have so far been conducted for highly prevalent common cancer types, such as lung [19], breast [20], and colon [21]. However, for rarer cancer types, especially for PCNSL, the literature is still very limited. In this context, studies have mainly focused on differentiating PCNSL from glioblastoma (GBM) [22–27] starting from multi-parametric MRI [22,28]. On the other hand, the correlation between radiomics features and therapy response or outcome has been barely investigated for PCNSL [29]. Chen et al. [30] evaluated the prognostic value of radiomics features for predicting Overall Survival (OS) in 52 PCNSL patients. However, the study was limited only to the analysis of textural features on contrast enhanced MRI. Ale et al. [31] carried out a predictive analysis on OS and Progression-Free Survival (PFS) considering a population of 47 patients, respectively. Promising results were achieved, although few details about the methodology and the patient cohort were provided. A schematic overview about the State of Art (SoA) of PCNSL and Radiomics Analysis is given in Table S1 in the Supplementary Data. A common problem for studies related to PCNSL is that recruiting patients with such a disease in a single center may be difficult, due to the relatively low incidence of the tumor [32]. Nonetheless, some issues must be taken into account for radiomics data deriving from multiple institutions. Inter- and intra-scanner variability is a common problem for multicenter MRI studies and, for this reason, the normalization of the intensity of the gray level becomes of fundamental importance in radiomics analyses.

Herein, we report a machine learning-based approach for predicting one-year OS and PFS in patients with PCNSL undergoing treatment with a high-dose methotrexate-based chemotherapy regimen. The proposed method relies on extracting robust and stable radiomics features from MRI scans. Such robustness and stability was assessed by comparing different intensity normalization methods on patient images acquired at different time points. To our knowledge, only a few studies have investigated the importance of image normalization in radiomics studies, despite it constituting an important challenge when using MRI data. In fact, the definition of a protocol is still missing [33–37]. Moreover, to date, the role of image normalization for radiomics analysis of PCNSL tumors has not yet been evaluated.

## 2. Materials and Methods

### 2.1. Dataset Description

Clinical and MRI data from 80 patients with histological or cytological diagnosis of PCNSL, as well as absence of extra-CNS disease as per international guidelines [38], treated at San Raffaele Scientific Institute of Milano, Italy, between January, 2010, and November, 2019, were retrospectively collected. MRIs were acquired in different centers and with different scanners. Patients were considered eligible for subsequent analyses based on the following criteria (see Figure 1): (1) availability of T1-W, T2-W, Fluid Attenuated Inversion Recovery (FLAIR) and T1-W with gadolinium (T1 gd) pulse sequences on MR scans obtained before the start of therapy; (2) tumor contours clearly distinguishable for manual segmentation. Overall, 56 patients were included for the OS classification (Group A) and 47 patients (Group A2) for PFS. From Group A, 23 patients (Group A1) were imaged at 3 different time points (before, during and after the treatment) and with different scanners (described for each group in Table S2 in Supplementary Data) were selected for feature stability analysis. The demographics and clinical features of the patient cohort are summarized in Table 1. This observational study was approved by the Ethical Committee of San Raffaele Hospital in Milan (Italy) with number 22/INT/2021 and conducted in accordance with all international laws and rules, and in accordance with the national laws, as well as in accordance with all applicable guidelines. Due to the retrospective nature of this study and anonymized clinical data, ad hoc informed consent was waived.

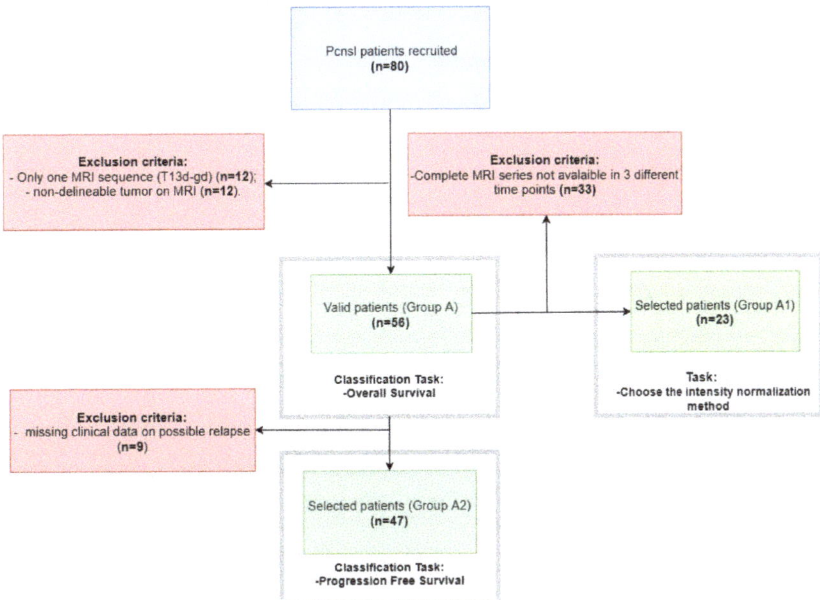


**Figure 1.** Flowchart of the patient enrolment process. In the blue box, the initial number of patients available for this study. In the red box, the reasons for exclusion of some patients (unavailability of complete MRI sequences or missing clinical data). In the green box, the number of patients selected for the specific task.

**Table 1.** Description of the patient dataset (Group A).

| **Eligible Patients (#)** | 56/80 (70%) |
|---|---|
| Male:Female | 0.56 |
| Median Age | 69 (41–85) |
| Multiple lesions | 32 (58%) |
| Involvement of deep areas § | 45 (80%) |
| Lactic dehydrogenase serum level >ULN | 35 (52%) |
| Cerebrospinal-fluid protein concentration >ULN * | 34(60%) |
| ECOG—Performance Status >2 | 30 (53%) |
| **IELSG risk score** | |
| -Low | 5 (9%) |
| -Intermediate | 28 (50%) |
| -High | 23 (41%) |
| **Sites of disease** | |
| -Brain parenchyma | 56 (100%) |
| **Treatment details** | |
| **Induction** | |
| MATRix | 37 (66%) |
| MAT | 2 (3%) |
| HD-MTX + HD-ARAC | 10 (17%) |
| HD-MTX + Alkylators | 4 (7%) |
| WBRT ± TMZ | 4 (7%) |
| Rituximab | 43 (77%) |
| **Consolidations** | |
| ASCT | 15 (27%) |
| WBRT | 6 (11%) |
| DeVIC | 5 (9%) |
| Oral Maintenance | 3 (5%) |
| None | 26 (46%) |
| Unknown | 1(2%) |
| Treatment delay >20 gg | 40 (71%) |
| Refractory to first line @ | 22 (39%) |
| 1-year PFS | 24/47 (51%) |
| 1-year OS | 30/56 (54%) |

* Lumbar puncture was contraindicated in 3 patients; CSF protein concentration was considered an unfavorable feature in IELSG risk score in these patients. § At least one of the following brain structures: periventricular regions, basal ganglia, corpus callosum, brainstem, and cerebellum. @ PD < 6 months from the end of first line treatment; HD-ARAC: high dose Cytarabine; ASCT: autologous stem cell transplantation; DeVIC: Dexamethasone, Etoposide, Ifosfamide and Carboplatin; ECOG—PS: Eastern Cooperative Oncology Group—Performance Status; HD-MTX: High dose Methotrexate; IELSG: International Extranodal Lymphoma Study Group; LDH: Lactic dehydrogenase serum level; MATRix: High dose Methotrexate, high dose Cytarabine, Thiotepa and Rituximab; pCSF: Cerebrospinal-fluid protein concentration; PFS: Progression free survival; OS: Overall Survival; TMZ: Temozolomide; ULN: upper limit normal; WBRT: Whole brain radiation therapy.

### 2.2. *Image Pre-Processing*

All images were pre-processed according to the steps described below (see Figure 2), in order to improve their quality and to increase the reproducibility of radiomics features [39]:

1. to correct the non-homogeneous intensity of the magnetic field present in MR images, the module "N4ITK MRI bias correction" available in 3D Slicer [40] was used [41];
2. for each patient, all available MRI acquisitions were registered on the T1-gd image (sequence where segmentation was performed);

3. skull stripping [42] was performed from images to remove extra brain tissue from the brain volume and to increase the accuracy of subsequent MRI processing. The "Swiss skull stripper" module of 3D Slicer was used[43];
4. normalization methods were applied for MRI intensities normalization (described in detail in Section 2.2.1);
5. all sequences were resampled (voxels 1 $mm^3$) [44].

#### 2.2.1. Intensity Normalization of MR Images

Three gray level intensity normalization methods were tested on the MR images: Z-score, WhiteStripe and Nyul.

The Z-score method normalizes the image $I(x)$ by subtracting the mean of the image $\mu_{brain}$ and dividing by the standard deviation of all the voxel intensities $\sigma_{brain}$:

$$I_{Zscore}(x) = \frac{(I(x) - \mu_{brain})}{\sigma_{brain}} \quad (1)$$

The WhiteStripe method [45] was developed to bring raw image intensities to a biologically interpretable intensity scale. The method applies a z-score transformation to the whole brain using parameters estimated from a latent subdistribution of normal-appearing white matter (NAWM). In detail, this method normalizes the image $I(x)$ intensities by subtracting $\mu_{ws}$, which corresponds to the mean intensity value of the (NAWM), from each voxel intensity $I(x)$ and dividing the result by the standard deviation of the NAWM $\sigma_{ws}$:

$$I_{ws}(x) = \frac{(I(x) - \mu_{ws})}{\sigma_{ws}} \quad (2)$$

The method developed from Nyul et al. [46], also called piecewise linear histogram matching normalization, learns a standard image histogram from a set of images, and then linearly maps the intensities of each image to this standard image histogram. MRI intensities are not standardized. For this reason, before carrying out Radiomics analyses, the intensity normalization of the gray levels of images is essential.

The code used for this implementation is available at https://github.com/jcreinhold/intensity-normalization (accessed on 19 February 2023).

### *2.3. Segmentation VOI (Volume of Interest) and Features Extraction*

The hyperintense tumor lesion on post-contrast T1-W images was manually segmented for each patient resulting in volume of interest (VOI). The same VOI was reported in the other sequences for each patient applying the linear transformation identified by the registration process. All segmentations were performed by R.L., a radiologist with 4 years of experience, at the time of the study. Radiomics features were extracted from the VOI using Pyradiomics 3.0.1 (https://pyradiomics.readthedocs.io/en/latest/features.html, (accessed on 19 February 2023) [47]: 19 First Order (F0) features, 14 Shape features, 23 Gray Level Co-I Matrix (GLCM) features, 16 Gray Level Run Length Matrix (GLRLM) features, 16 Gray Level Size Zone Matrix (GLSZM) features and 14 Gray Level Dependence Matrix (GLDM) features [48]. In total, 120 features (including radiological features) were extracted from the tumor region of each MRI sequence from both non-normalized images and normalized images with the chosen method.

### *2.4. Machine Learning Model Building*

Given as input a set of radiomics features extracted from processed MRIs (Group A), the goal was to train a machine learning model to predict the probability of survival of a patient with PCNSL. Since the prediction task had only two possible outcomes (survive/not survive after 1 year), the task was modeled as a binary classification problem. A first selection of the features was performed, using a high correlation filter to remove variables having large absolute correlation. To overcome the curse of dimensionality issues and reduce overfitting, the Min–Max Normalization method was applied to linearly transform radiomics features by using scikit-learn library (https://scikit-learn.org/stable/modules/generated/sklearn.preprocessing.MinMaxScaler.html, accessed on 19 February 2023). Only relevant features were selected in cross validation according to an ensemble of four selection methods: (i) SelectKBest for the chi-square test method; (ii) the Recursive Feature Elimination (RFE) using the Logistic Regression model; (iii) least absolute shrinkage and selection operator (Lasso), and (iv) Select From Model using RandomForestClassifier model. In detail, each method extracted $k = 15$ candidate features and only the ones selected by at least three algorithms over four were chosen to feed the classification algorithm. Five classifiers were tested, namely: Extra Tree Classifier (ETC), Support Vector Machine (SVM), Logistic Regression (LR), Random Forest (RF), kNeighbors (KN). Feature selection methods and ML classifiers were implemented, based on the scikit-learn library version 0.23. The whole process, from the normalization of features to the selection and classification, was performed in a repeated five-fold stratified cross-validation (https://scikit-learn.org/stable/modules/generated/sklearn.model_selection.RepeatedStratifiedKFold.html, accessed on 19 February 2023) (10 repetitions) was adopted to assess overfitting and to evaluate the stability of the results. The workflow of this study is described in Figure 2.

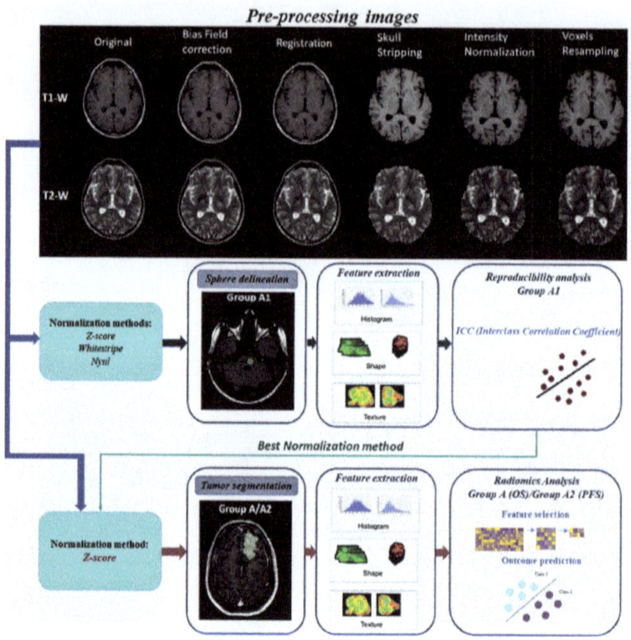


**Figure 2.** The workflow of the study was divided into two main sections: (1) Reproducibility analysis of features extracted from pathological tissue of MR images normalized with three different methods (Z-score, WhiteStripe and Nyul); (2) Radiomics Analysis for predictive OS and PFS of PCNSL patients (features extracted from segmentation tumor). For both sections, the first step was to pre-process MRI sequences. From the results of the reproducibility of features, the Z-score method was selected for application to the MRI sequences.

### 2.5. Experiments

#### 2.5.1. Feature Robustness

To determine which normalization method was best suited for our dataset, we studied the effect of image intensity normalization on the reproducibility of the radiomics features. To this aim, Group A1 (subgroup of Group A, as described in Section 2.1 and shown in Figure 1) was considered. Notice that the selected subgroup of patients was not considered during the survival prediction analysis, in order to avoid any bias in the classification results.

Given a patient, all his/her longitudinal T1-W and T2-W sequences were, in turn, normalized using three methods: Z-score, WhiteStripe and Nyul (described in Section 2.2.1). Then, a region of the pons, where no pathological modifications were observed, was identified on the patient's FLAIR image. From this region, a 1 cm diameter spherical segmentation was extracted using the segmentation tool of 3D Slicer software. The segmentation was reported for all the longitudinal sequences of the patient by applying the linear transformation of the registration between the images made previously. A total of 94 radiomics features were extracted with the Pyradiomics library. Shape features were excluded as the selected spherical VOI was equal for all patients. For the three longitudinal acquisitions of each patient, we extracted features from images normalized with three methods previously described and from the non-normalized images for sequences T1-W and T2-W.

The Interclass Correlation Coefficient (ICC) was calculated to evaluate the reproducibility of each feature for each normalization method. Formally, the ICC is a descriptive statistic that can be used when quantitative measurements are made on units organized into groups [49]. It ranges between 0 and 1, indicating null and perfect reproducibility. ICCs were calculated with IBM's SPSS statistical software, using the two-way random mean measurement ICC (2,k). We defined a matrix nxk, with n number of features extracted for each patient and k, number of observers (i.e., MRI acquired with different scanners). Given MSr the average square for rows, MSe the residual average, and MSc the average square for columns:

$$ICC(2,k) = \frac{(MS_r - MS_e)}{MS_r + \frac{(MS_c - MS_e)}{k}} \quad (3)$$

ICCs were computed to assess the stability of first-order and textural features across the three acquisitions before and after normalization. The Kruskal–Wallis test and its post hoc were used to compare the obtained ICCs for T1-w and T2-w sequences, under the assumption that data were not normally distributed. The best normalization method was applied to images of groups A/A2 for subsequent Radiomics analysis.

2.5.2. Overall and Progression Free Survival Prediction

Patients were dichotomized, based on OS or PFS greater than, or lower than, 12 months, respectively. OS was defined as time from diagnosis until death due to any cause or date of last follow-up visit, and PFS was defined as time from diagnosis until progression, relapse, death or date of last follow-up visit [50].

Each of the selected ML algorithms was trained at classifying OS for patients in Group A. Classification performances were evaluated in terms of F1-score (https://scikit-learn.org/stable/modules/generated/sklearn.metrics.f1_score.html, accessed on 19 February 2023) and Area Under ROC curve (AUC) (https://scikit-learn.org/stable/modules/generated/sklearn.metrics.auc.html, accessed on 19 February 2023). It is worth noticing that machine learning model validation is a crucial step, especially in the biomedical domain. We also compared the performance of the classifiers using both radiomics features alone as well as combined with clinical features. Age > 60, PS > 2, LDH > ULN, protein CSF > ULN and deep lesion were considered as clinical features, these being considered as available and validating PCNSL risk scores [51,52].

To better evaluate the impact of normalization on survival prediction, each algorithm was trained and tested using radiomics features obtained either from raw or normalized images. Only the Z-score method was used in these experiments since, as shown in Section 3.1, it provided the most stable features.

## 3. Results

### 3.1. Impact of the Intensity Normalization Method on Radiomics Feature

Figure 3 shows median ± quartiles of ICCs computed on both original and normalized images in T1-W and T2-W Sequences (Group A1). Z-score normalization determined the highest increase in ICC for features extracted from T1-W (30% average increase compared with non-normalized sequences, $p < 10^{-9}$). No statistically significant differences were observed when comparing non-normalized T1-W sequences with Nyul or WhiteStripe normalized sequences. All three normalization methods showed a clear increase of ICC values in T2-W sequence (Kruskal–Wallis test, $p < 10^{-12}$ (Z-score), $p < 10^{-13}$ (WhiteStripe), $p < 10^{-15}$ (Nyul).

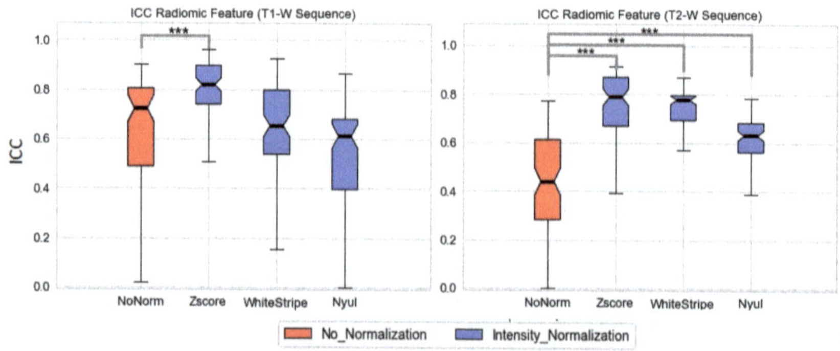


**Figure 3.** The distribution of ICC values computed from extracted features for non-normalized images and for normalized images with Z-score, WhiteStripe and Nyul methods. *** (significant statical difference).

## 3.2. *Performance Comparison of Classification Models*

The results of median and quartiles of the F1-scores obtained from the five selected machine learning models for the OS and PFS prediction classification tasks are reported in Table 2. For both tasks, we performed the classification with radiomics features alone, radiomics features + clinical features and clinical features alone.

**Table 2.** median ± quartiles of the F1-Score (T1-W and T2-W and combination T1-W and T2-W), obtained using all 5 test folds with 10 repeated of the cross-validation and 5 machine learning models. The difference between the quartiles provided information on the distribution of results. Each result was compared with features extracted from non-normalized images and normalized images using only radiomics features, radiomics plus clinical features and only clinical features.

| OS | Radiomics Features | ETC | SVM | LR | RF | KN |
|---|---|---|---|---|---|---|
| T1-W | No Normalizazion | 0.67 (0.61–0.79) | 0.71 (0.70–0.71) | 0.71 (0.67–0.71) | 0.67 (0.61–0.72) | 0.67 (0.61–0.73) |
| | Intensity Normalization | 0.75 (0.67–0.83) | 0.77 (0.68–0.83) | 0.77 (0.73–0.83) | 0.73 (0.67–0.83) | 0.73 (0.63–0.80) |
| T2-W | No Normalization | 0.67 (0.55–0.73) | 0.67 (0.57–0.71) | 0.71 (0.67–0.71) | 0.59 (0.50–0.71) | 0.57 (0.44–0.70) |
| | Intensity Normalization | 0.79 (0.73–0.86) | 0.80 (0.77–0.86) | 0.80 (0.75–0.86) | 0.73 (0.67–0.830) | 0.77 (0.72–0.80) |
| T1-W/T2-W | No Normalization | 0.67 (0.57–0.72) | 0.67 (0.55–0.76) | 0.67 (0.60–0.76) | 0.61 (0.54–0.71) | 0.61 (0.54–0.70) |
| | Intensity Normalization | 0.80 (0.77–0.86) | 0.80 (0.72–0.83) | 0.80 (0.73–0.83) | **0.83 (0.77–0.86)** | 0.80 (0.72–0.83) |
| **OS** | **Radiomics + Clinical Features** | **ETC** | **SVM** | **LR** | **RF** | **KN** |
| T1-W | No Normalizazion | 0.72 (0.67–0.80) | 0.73 (0.60–0.80) | 0.73 (0.60–0.80) | 0.67 (0.61–0.75) | 0.73 (0.60–0.77) |
| | Intensity Normalization | 0.80 (0.73–0.83) | 0.79 (0.68–0.83) | 0.80 (0.68–0.83) | 0.80 (0.71–0.83) | **0.82 (0.73–0.86)** |
| T2-W | No Normalization | 0.73 (0.66–0.80) | 0.72 (0.60–0.825) | 0.72 (0.66–0.77) | 0.72 (0.60–0.77) | 0.67 (0.60–0.77) |
| | Intensity Normalization | 0.77 (0.66–0.86) | 0.77 (0.68–0.83) | 0.77 (0.66–0.83) | 0.73 (0.68–0.80) | 0.77 (0.67–0.77) |
| T1-W/T2-W | No Normalization | 0.77 (0.67–0.86) | 0.73 (0.66–0.83) | 0.73 (0.66–0.83) | 0.67 (0.60–0.72) | 0.73 (0.60–0.80) |
| | Intensity Normalization | 0.80 (0.73–0.86) | 0.80 (0.73–0.83) | 0.80 (0.72–0.83) | 0.77 (0.68–0.83) | 0.80 (0.72–0.83) |
| **OS** | **Clinical Features** | **ETC** | **SVM** | **LR** | **RF** | **KN** |
| | | 0.60 (0.44–0.67) | 0.71 (0.66–0.79) | 0.71 (0.66–0.77) | 0.60 (0.54–0.67) | 0.67 (0.60–0.77) |

**Table 2.** *Cont.*

| PFS | Radiomics Features | ETC | SVM | LR | RF | KN |
|---|---|---|---|---|---|---|
| T1-W | No Normalizazion<br>Intensity Normalization | 0.67 (0.54–0.72)<br>0.60 (0.50–0.66) | 0.68 (0.58–0.75)<br>0.68 (0.60–0.68) | 0.71 (0.66–0.79)<br>0.68 (0.66–0.73) | 0.60 (0.50–0.72)<br>0.60 (0.50–0.66) | 0.60 (0.54–0.73)<br>0.67 (0.55–0.66) |
| T2-W | No Normalization<br>Intensity Normalization | 0.67 (0.55–0.75)<br>0.68 (0.57–0.76) | 0.67 (0.61–0.68)<br>**0.80 (0.67–0.88)** | 0.67 (0.61–0.68)<br>0.80 (0.67–0.86) | 0.60 (0.50–0.73)<br>0.68 (0.55–0.76) | 0.67 (0.51–0.73)<br>0.73 (0.67–0.83) |
| T1-W/T2-W | No Normalization<br>Intensity Normalization | 0.67 (0.50–0.74)<br>0.63 (0.50–0.75) | 0.67 (0.60–0.73)<br>0.70 (0.60–0.80) | 0.67 (0.58–0.73)<br>0.73 (0.62–0.80) | 0.60 (0.46–0.67)<br>0.67 (0.58–0.75) | 0.67 (0.58–0.73)<br>0.68 (0.60–0.75) |
| **PFS** | **Radiomics + Clinical Features** | **ETC** | **SVM** | **LR** | **RF** | **KN** |
| T1-W | No Normalizazion<br>Intensity Normalization | 0.60 (0.44–0.67)<br>0.60 (0.45–0.67) | 0.67 (0.60–0.71)<br>0.68 (0.61–0.67) | 0.67 (0.55–0.71)<br>0.68 (0.60–0.67) | 0.60 (0.51–0.73)<br>0.60 (0.50–0.72) | 0.60 (0.47–0.72)<br>0.60 (0.45–0.67) |
| T2-W | No Normalization<br>Intensity Normalization | 0.60 (0.48–0.71)<br>0.68 (0.50–0.76) | 0.61 (0.55–0.67)<br>0.72 (0.60–0.80) | 0.67 (0.55–0.76)<br>0.69 (0.60–0.75) | 0.60 (0.50–0.70)<br>0.70 (0.60–0.77) | 0.62 (0.55–0.67)<br>0.69 (0.60–0.73) |
| T1-W/T2-W | No Normalization<br>Intensity Normalization | 0.64 (0.55–0.68)<br>0.61 (0.44–0.68) | 0.67 (0.66–0.71)<br>0.69 (0.60–0.73) | 0.67 (0.61–0.77)<br>0.65 (0.55–0.68) | 0.61 (0.50–0.73)<br>0.60 (0.50–0.67) | 0.61 (0.50–0.67)<br>0.60 (0.45–0.62) |
| **PFS** | **Clinical Features** | **ETC** | **SVM** | **LR** | **RF** | **KN** |
| | | 0.55 (0.41–0.60) | 0.62 (0.51–0.67) | 0.67 (0.63–0.71) | 0.57(0.47–0.65) | 0.55 (0.40–0.61) |

### 3.2.1. OS Classification Task

For features extracted from T1-W, T2-W and the combination of T1-W and T2-W features (T1-W/T2-W), classification results obtained from images normalized with Z-score are presented in this section (providing the best results in terms of reproducibility and stability compared to the other normalization methods, as reported in Section 3.1).

Considering only radiomics features, the best performances of T1-W sequence were obtained from classifiers SVM and LR with the median and quartiles, respectively, F1 = 0.77 (0.68–0.83) and F1 = 0.77 (0.73–0.83). For T2-W sequence, the best performances were obtained by the SVM classifier with F1 = 0.80 (0.77–0.86) and LR with F1 = 0.80 (0.75–0.86). For T1-W/T2-W, the performance improved and we obtained a median of F1-score equal to 0.83 (0.77–0.86) with RF classifier. The best results were obtained from normalized images, with a significant statistical difference from the results obtained using features extracted from non-normalized images.

When introducing clinical features, the results did not significantly change. In this case, the best performances were obtained with T1-W (KN = 0.82 (0.73–0.86) and T1-W/T2-W (ETC = 0.80 (0.73–0.86)). Instead, The F1-score for predicting OS using only clinical features was 0.71 (0.66–0.79) with the SVM classifier.

Figure 4 shows the ROC curves of the best performances of classifiers. The AUC values of radiomics features alone, radiomics + clinical features and clinical features alone for predicting OS were $0.86 \pm 0.09$, $0.83 \pm 0.11$ and $0.70 \pm 0.14$, respectively. In comparing the best performance for OS prediction with clinical features, and with radiomics features a significant statistical difference ($p < 10^{-9}$) was found. There was no significant statistical difference between performance with radiomics features alone, and with Radiomics plus clinical features ($p = 0.38$).

### 3.2.2. PFS Classification Task

Patients of Group A2 were considered to assess PFS classification task. The patients' characteristics are summarized in Section 2.1. Considering radiomics features alone, the best performances were obtained from the sequence T2-W (SVM = 0.80 (0.67–0.88) and LR = 0.80 (0.67–0.86)) and from T1-W/T2-W (LR = 0.73 (0.62–0.80)). For the PFS, the combination of T1-W and T2-W sequences did not improve the performance of the model (compared to single sequence) .

The addition of clinical features for PFS did not improve performances and, considering only clinical features, the best result was LR = 0.67 (0.63–0.71). Compared to OS prediction, in this case also the best performance was obtained with normalized images for the sequence T2-W and T1-W/T2-W with a statistical difference with non-normalized images.

ROC curves of the best performances for PFS classification (Figure 4) also showed the prediction of radiomics features (AUC = 0.84 ± 0.13) in respect to clinical features (AUC= 0.56 ± 0.18) with a significant statistical difference ($p$-value < $10^{-12}$). There was also a statistical difference between the prediction with radiomics features alone and with the addition of clinical features ($p$ = 0.002).

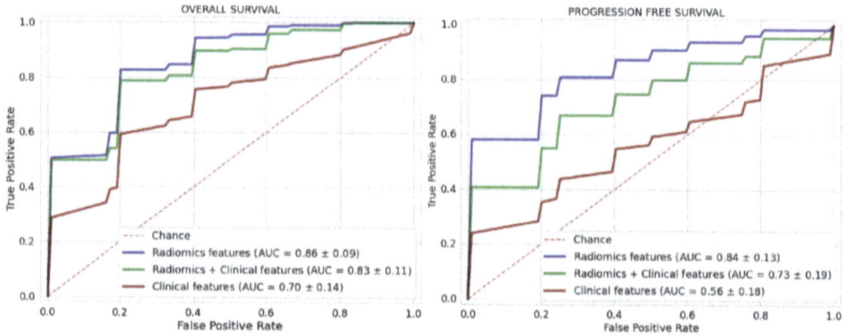


**Figure 4.** Roc curves of the best classifiers for each feature category: only radiomics features in blue, radiomics + clinical features in green and only clinical features in red (currently validated).

### 3.3. Feature Importance

Beyond the classification scores, further analyses were conducted to better understand the role of the features in the classification process. The study was performed for each imaging modality, with and without intensity normalization. We considered the RF classifier, where the feature importance was computed as the mean and standard deviation of accumulation of the impurity decreased within each tree of the forest. In more detail, for each independent training in the cross-validation procedure, we ranked the features according to importance score and selected the top 15 (top-15). Then, for each feature, we calculated the frequency with which that feature was selected as top-15 and, from the resulting distribution, we selected the top 13 features for analysis. Simply put, we selected the top 13 features most often ranked as "most important" in each independent training of the cross-validation procedure.

Figures 5 and 6 represent the selected clinical and radiomics features for T1-W, T2-W sequences, and T1-W/T2-W sequences. As per the OS classification task, the most selected clinical features were Age and Performance status (PS) (Figure 5), while for the PFS classification task, LDH>ULN, deep lesion, and Age were almost always selected (Figure 6). Considering the feature importance score, radiomics features seemed to give a greater contribution to the outcome than clinical features.

For T1-W and T2-W sequences (without intensity normalization) in the OS classification, the most important contribution was given by shape features (Elongation and Sphericity) and first order features (https://pyradiomics.readthedocs.io/en/1.1.1/features.html#radiomics-firstorder-label, accessed on 19 February 2023) (Minimum, Maximum and Skeweness). For T1-W and T2-W sequences (with intensity normalization), GLCM features (Cluster Shade, Joint Average) and GLRLM features (Long Run Low Gray Level Emphasis, Run Length Non-Uniformity and High Gray Level Run Emphasis ) received the highest scores.

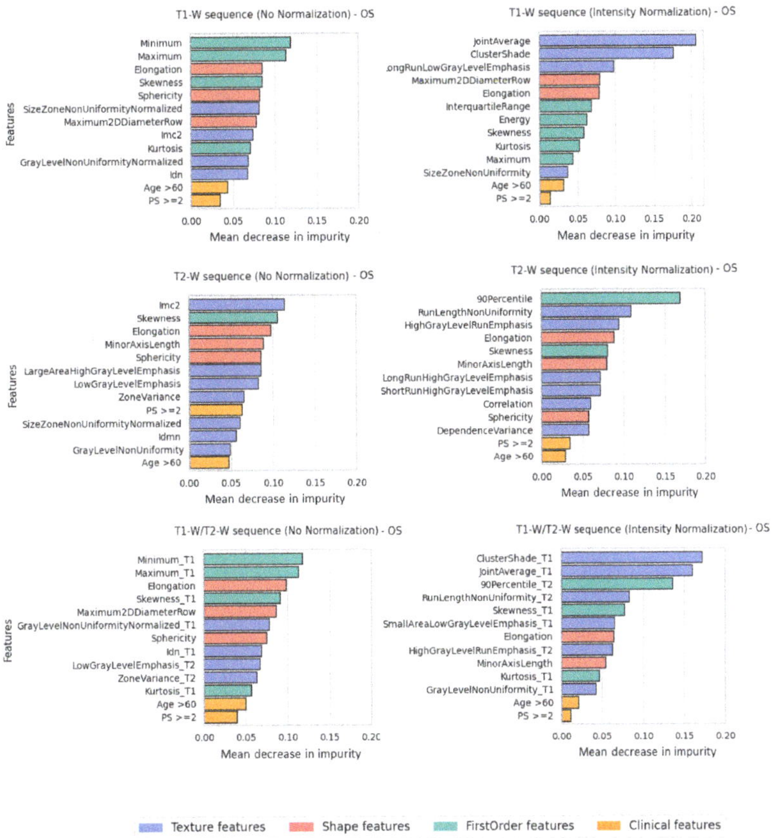


**Figure 5.** Feature importance for all MRIsequences with and without normalization (OS classification task). Features were grouped using different colors for shape features, texture features, first order features and clinical features.

Considering the PFS classification task, an important role seemed to be played by Elongation (shape feature), that shows the relationship between the two largest principal components in the ROI shape, and its value, ranging from 0 (line-like object) to 1 (circle-like object).

$$Elongation = \sqrt{\frac{\lambda_{minor}}{\lambda_{mayor}}} \tag{4}$$

Here, $\lambda_{mayor}$ and $\lambda_{minor}$ were the lengths of the largest and second largest principal component axes. Amongst the selected, we also found Zone Percentage (GLSZM) and Imc2 (GLCM) for non-normalized images, and, concerning normalized images, Large Dependence High Gray Level Emphasis (GLDM) for T1-W and Gray Level Emphasis (GLSZM) for T2-W.

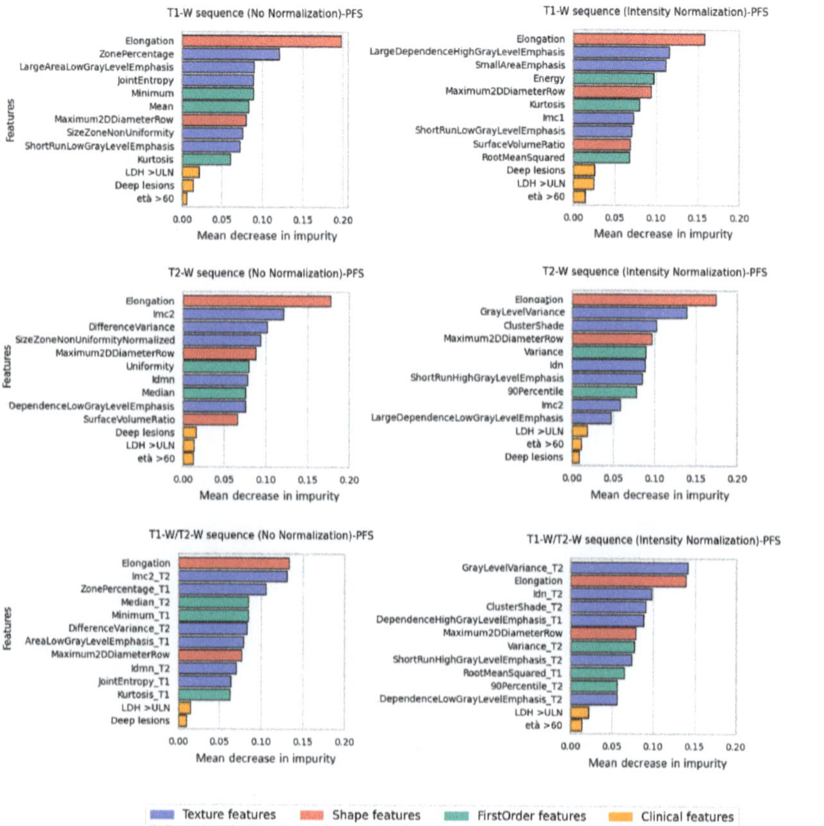


**Figure 6.** Feature importance for all MRI sequences with and without normalization (PFS classification task). Features were grouped using different colors for shape features, texture features, first order features and clinical features.

## 4. Discussion

To the best of our knowledge, this is the first study investigating the capability of radiomics features as outcome predictors in patients with newly diagnosed PCNSL, while also evaluating the impact of MR image normalization [53] on feature stability. To overcome the curse of dimensionality issues and to reduce overfitting, feature selection was performed by using multiple approaches and reaching consensus by a voting procedure. A post-hoc analysis of the most salient features learned by the selected ML models was performed, with the aim of trying to collect more insight about the pathology and to partially explain the classification process.

Significant results were obtained for both OS and PFS prediction using all the selected classifiers with a statistically significant difference ($p$-value $< 10^{-4}$) between image intensity normalization and no normalization (best median F1-score 0.83 vs. 0.71 for OS and 0.80 vs. 0.71 for PFS, respectively). Interestingly, it was observed that combining features from both T1-W and T2-W sequences improved results in the OS classification task compared to using features from a single sequence. On the other hand, the best performance for PFS was obtained using only the T2-W sequence (median F1-score T1-W/T2-W = 0.73 (0.62–0.80) vs. T1-W = 0.68 (0.66–0.73) vs. T2-W = 0.80 (0.67–0.88)). Noteworthy was the fact that introduction of clinical features commonly used to calculate the IELSG score (age, PS, deep lesions, CSF protein, and LDH) marginally improved the performance of some classifiers only in OS analysis. However, their contribution did not have a significant impact.

AUC scores achieved by the best classifiers (RF for OS and SVM for PFS) were observed to be significantly higher compared with scores obtained using only clinical features ($p$-value < $10^{-9}$ and $p$-value < $10^{-12}$, respectively), showing that radiomics features better contributed to the outcome prediction than clinical features. This work has some limitations that are worth mentioning. First, the relatively small number of patients, mainly due to the low incidence rate of PCNSL [32], which could highly impact the learning process and might cause sub-optimal prediction performances and overfitting. Obviously, we resorted to numerous techniques to mitigate the effect of the low number of patients, but, in any case, our future goal is to increase the dataset in order to validate these promising results. Furthermore, images from multiple centers were collected to mitigate the issue and a repeated cross-validation approach was used to evaluate the robustness of our results. Furthermore, up to 30% of the initial study population could not be considered eligible for this study because of lack of MR sequences or delineable tumor. However, we believe this number could be reduced in future radiomics studies in the PCNSL setting, given increasing use of stereotactic biopsy instead of surgery for diagnosis, as well as the potential availability of pre-biopsy MRI scans which could also reduce other technical problems, such as bleeding. Moreover, the recent IPCG (International Primary CNS Lymphoma Collaborative Group) recommendations for MRI imaging should, potentially, also impact on the homogeneity of future studies [54]. Second, our models processed the radiomics features representing the tumor, excluding the possible prediction capability of extra lesion tissues as well as the association between radiomics features and pathological/molecular characteristics, which might reveal hidden relations useful to better understand the history of the disease. Third, information about the performed treatment was not included in the prediction process of the final analysis, as it differed from the main focus of this study. However, up to 93% of patients received an HD–MTX based treatment with a subsequent consolidation/maintenance in nearly 50%, unless there was progression or death due to lymphoma or other causes and, overall, all patients received the best available treatment based on clinical stratification. Further investigation is needed to use this integrated clinical and radiomics approach to stratify patients for therapy response prediction. This would allow not only the division of patients into risk groups, but also definition of the better potential treatment to be studied in future clinical trials.

Furthermore, some aspects of this trial merit discussion. The analyses were performed on the features extracted from T1-W and T2-W sequence and not from the T1 as contrast, as we did not want the radiomics features to be affected by the contrast. All analyses were also carried out on the FLAIR sequence, but the data were not reported in this paper as the results were not satisfactory. We plan to consider it again in future work where deep learning-based models will be explored.

Almost all the work related to this rare tumor has been focused, to date, on the differentiation of PCNSL from atypical glioblastoma [23,24,28]. Instead, in the present study, we evaluated the prognostic value of images normalization to use radiomics features for predicting OS and PFS in PCNSL. Indeed, for rare tumors, one of the limitations is to collect a sufficient quantity of patient data to analyze; thus, assembling data from different centers is usually a valid solution. However, in the case of MRI acquired in a multicenter setting, inter- and intra-scanner variability can be an important limitation in the radiomics analysis. Thus, the study of the effect of normalization on both task prediction and reproducibility of radiomics features is of important value. To this end, a subgroup of patients with three longitudinal acquisitions over time was selected and the ICC for each radiomics features was computed in non-pathological tissue. Three state-of-the-art normalization methods were tested (Z-score, WhiteStripe and Nyul), according to many MR image harmonization studies [33,34,55]. While a similar study performed for Glioblastoma [53] found the Nyul method to be the most robust for radiomics analysis, for MRI of PCNSL patients we found that the Z-score normalization gave the highest number of reproducible features (median and quartile values of all ICCs = 0.8 (0.74–0.90)) for both the T1-W and the T2-W sequences, as shown in Figure 4. Furthermore, in contrast with [53], we performed a feature stability analysis on a portion of healthy tissue so that the results were unaffected by disease progression or regression.

The normalization step had a significant impact on the learning process for both OS and PFS (all results summarized in Table 2). Figures 5 and 6 show the feature importance for each sequence at inference time. As is observable, first order features had the highest importance among the features extracted from non-normalized images. By contrast, when using normalized images, the classifiers seemed to rely more on textural features (GLCM and GLRLM). First order statistics describe the distribution of individual voxel values without concern for spatial relationships. Instead, textural features are obtained calculating the statistical inter-relationships between neighbouring voxels (hence, they provide a measure of intra-lesion heterogeneity) [17]. We speculate that the latter may contain more robust and informative content for the survival prediction, therefore explaining the better classification results. Indeed, textural analysis derived from conventional sequences reflects histopathology features in solid cancer and has been proposed as a novel noninvasive modality to further characterize tumors in clinical oncology [56,57]. Furthermore, it is worth noticing that shape features may also act as confounding factors. If spurious correlation exists (e.g., between tumor size and disease progression) the learning process may be biased. In this case, elongation was the most important feature for almost all sequences and there seemed to be no difference between normalized and non-normalized images, but that was because the shape features were not affected by intensity normalization and depended only on tumor segmentation. Furthermore, the performance improved significantly for the prediction classification task, especially for the T2-W sequence. Probably, the other textural features made the difference. Finally, for the OS survival classification, features of both sequences (T1-W and T2-W) were equally important. The PFS features of the T2-W sequence provided a greater contribution and, in fact, the performance results were better than for the T1-W sequence.

## 5. Conclusions

This work presented the effect of normalization of MR images on a radiomic-based approach to predict OS and PFS in PCNSL patients. Despite the limited number of cases (mainly due to the rarity of the tumor), the proposed method made a breakthrough in radiomics-based precision medicine for PCNSL patients.

**Supplementary Materials:** The following supporting information can be downloaded at:https://www.mdpi.com/article/10.3390/bioengineering10030285/s1. Table S1: The SoA of Radiomics Analysis on PCNSL. Table S2: The scanner characteristics for groups of patients examined (All scanners are 1.5 T). For each group, we reported the Ripetition Time (TR), Echo Time (TE), and Flip Angle (FA) for both sequences T1-W and T2-W.

**Author Contributions:** Conceptualization: T.C. and M.F.S.; Methodology: M.D., A.M., P.Z. and E.D.M.; Formal analysis: M.D., A.M. and P.Z.; Validation: M.D., A.M., P.Z., R.L. and N.A.; Data annotation: R.L., S.S., F.E. and T.C.; Data curation: R.L., S.S., F.E., T.C., N.A. and A.J.M.F.; Resources: E.D.M., F.C., N.A., A.M., A.J.M.F. and M.F.S.; Supervision: E.D.M., N.A., T.C. and M.F.S.; Writing—original draft preparation: M.D. and A.M.; Figures: M.D.; Writing-review and editing: all Authors. All authors have read and agreed to the published version of the manuscript.

**Funding:** This research received no external funding.

**Institutional Review Board Statement:** The study was conducted in accordance with the Declaration of Helsinki, and approved by the Ethics Committee of San Raffaele Hospital in Milan (Italy) (code 22/INT/2021).

**Informed Consent Statement:** Informed consent was obtained from all subjects involved in the study.

**Data Availability Statement:** The dataset analyzed in the current study is available from the corresponding author on reasonable request.

**Acknowledgments:** The authors are indebted to patients and their families for their generous commitment. We appreciate the excellent technical assistance and sustained scientific collaboration of

Chiara Tarquini (Data managment office of the Lymphoma Unit, San Raffaele Scientific Institute, Milano, Italy Giuseppina D'Elia (Research Nurse of the Lymphoma Unit, San Raffaele Scientific Institute, Milano, Italy), Stefano Orezzi (Neuroradiology Unit, San Raffaele Scientific Institute, Milano, Italy), Anna Chiara (Radiotherapy and Tomotherapy Unit, San Raffaele Scientific Institute, Milano, Italy), Maria Rosa Terreni (Pathology Unit, San Raffaele Scientific Institute, Milano, Italy), Filippo Gagliardi (Neurosurgery Unit, San Raffaele Scientific Institute, Milano, Italy), Elisabetta Miserocchi and Giulio Modorati (Ophthalmology Unit, San Raffaele Scientific Institute, Milano, Italy). We also acknowledge Fabio Ciceri, all the hematologists and collaborators of Hematology and BMT Unit, San Raffaele Scientific Institute, Milano, Italy for their excellent clinical assistance.

**Conflicts of Interest:** The authors declare no conflict of interest.

## References


1. Kluin, P. Primary diffuse large B-cell lymphoma of the CNS. In *World Health Organization: Pathology and Genetics of Tumors of Haematopoietic and Lymphoid Tissues*; World Health Organization: Geneva, Switzerland, 2008; pp. 240–241.
2. Ferreri, A.J.; Holdhoff, M.; Nayak, L.; Rubenstein, J.L. Evolving Treatments for Primary Central Nervous System Lymphoma. In *American Society of Clinical Oncology Educational Book*; American Society of Clinical Oncology: Alexandria, VA, USA, 2019; Volume 39, pp. 454–466.
3. Grommes, C.; Rubenstein, J.L.; DeAngelis, L.M.; Ferreri, A.J.; Batchelor, T.T. Comprehensive approach to diagnosis and treatment of newly diagnosed primary CNS lymphoma. *Neuro-Oncology* **2019**, *21*, 296–305. [CrossRef]
4. Calimeri, T.; Steffanoni, S.; Gagliardi, F.; Chiara, A.; Ferreri, A. How we treat primary central nervous system lymphoma. *ESMO Open* **2021**, *6*, 100213. [CrossRef]
5. Ferreri, A.J.; Cwynarski, K.; Pulczynski, E.; Fox, C.P.; Schorb, E.; La Rosée, P.; Binder, M.; Fabbri, A.; Torri, V.; Minacapelli, E.; et al. Whole-brain radiotherapy or autologous stem-cell transplantation as consolidation strategies after high-dose methotrexate-based chemoimmunotherapy in patients with primary CNS lymphoma: Results of the second randomisation of the International Extranodal Lymphoma Study Group-32 phase 2 trial. *Lancet Haematol.* **2017**, *4*, e510–e523.
6. Houillier, C.; Taillandier, L.; Dureau, S.; Lamy, T.; Laadhari, M.; Chinot, O.; Moluçon-Chabrot, C.; Soubeyran, P.; Gressin, R.; Choquet, S.; et al. Radiotherapy or autologous stem-cell transplantation for primary CNS lymphoma in patients 60 years of age and younger: Results of the intergroup ANOCEF-GOELAMS randomized phase II PRECIS study. *J. Clin. Oncol.* **2019**, *37*, 823–833. [CrossRef]
7. Batchelor, T.; Giri, S.; Ruppert, A.S.; Bartlett, N.L.; Hsi, E.D.; Cheson, B.D.; Nayak, L.; Leonard, J.P.; Rubenstein, J.L. Myeloablative versus non-myeloablative consolidative chemotherapy for newly diagnosed primary central nervous system lymphoma: Results of CALGB 51101 (Alliance). *J. Clin. Oncol.* **2021**, *39*, 7506. [CrossRef]
8. Houillier, C.; Soussain, C.; Ghesquières, H.; Soubeyran, P.; Chinot, O.; Taillandier, L.; Lamy, T.; Choquet, S.; Ahle, G.; Damaj, G.; et al. Management and outcome of primary CNS lymphoma in the modern era: An LOC network study. *Neurology* **2020**, *94*, e1027–e1039. [CrossRef]
9. Ambady, P.; Holdhoff, M.; Bonekamp, D.; Wong, F.; Grossman, S.A. Late relapses in primary CNS lymphoma after complete remissions with high-dose methotrexate monotherapy. *Cns Oncl.* **2015**, *4*, 393–398. [CrossRef]
10. Gillies, R.J.; Kinahan, P.E.; Hricak, H. Radiomics: Images are more than pictures, they are data. *Radiology* **2016**, *278*, 563–577. [CrossRef]
11. Zhou, M.; Scott, J.; Chaudhury, B.; Hall, L.; Goldgof, D.; Yeom, K.W.; Iv, M.; Ou, Y.; Kalpathy-Cramer, J.; Napel, S.; et al. Radiomics in brain tumor: Image assessment, quantitative feature descriptors, and machine-learning approaches. *Am. J. Neuroradiol.* **2018**, *39*, 208–216. [CrossRef]
12. Khemchandani, M.A.; Jadhav, S.M.; Iyer, B. Brain Tumor Segmentation and Identification Using Particle Imperialist Deep Convolutional Neural Network in MRI Images. *Int. J. Interact. Multimed. Artif. Intell.* **2022**, *7*, 7. [CrossRef]
13. Hassan, L.; Saleh, A.; Abdel-Nasser, M.; Omer, O.A.; Puig, D. Promising deep semantic nuclei segmentation models for multi-institutional histopathology images of different organs. *Int. J. Interact. Multimed. Artif. Intell.* **2021**, *6*, 6. [CrossRef]
14. Tomaszewski, M.R.; Gillies, R.J. The biological meaning of radiomic features. *Radiology* **2021**, *298*, 505–516. [CrossRef]
15. Liu, Z.; Wang, S.; Di Dong, J.W.; Fang, C.; Zhou, X.; Sun, K.; Li, L.; Li, B.; Wang, M.; Tian, J. The applications of radiomics in precision diagnosis and treatment of oncology: Opportunities and challenges. *Theranostics* **2019**, *9*, 1303. [CrossRef]
16. Luo, H.; Zhuang, Q.; Wang, Y.; Abudumijiti, A.; Shi, K.; Rominger, A.; Chen, H.; Yang, Z.; Tran, V.; Wu, G.; et al. A novel image signature-based radiomics method to achieve precise diagnosis and prognostic stratification of gliomas. *Lab. Investig.* **2021**, *101*, 450–462. [CrossRef]
17. Rizzo, S.; Botta, F.; Raimondi, S.; Origgi, D.; Fanciullo, C.; Morganti, A.G.; Bellomi, M. Radiomics: The facts and the challenges of image analysis. *Eur. Radiol. Exp.* **2018**, *2*, 1–8. [CrossRef]
18. Mayerhoefer, M.E.; Materka, A.; Langs, G.; Häggström, I.; Szczypiński, P.; Gibbs, P.; Cook, G. Introduction to radiomics. *J. Nucl. Med.* **2020**, *61*, 488–495. [CrossRef]
19. Thawani, R.; McLane, M.; Beig, N.; Ghose, S.; Prasanna, P.; Velcheti, V.; Madabhushi, A. Radiomics and radiogenomics in lung cancer: A review for the clinician. *Lung Cancer* **2018**, *115*, 34–41. [CrossRef]

20. Valdora, F.; Houssami, N.; Rossi, F.; Calabrese, M.; Tagliafico, A.S. Rapid review: Radiomics and breast cancer. *Breast Cancer Res. Treat.* **2018**, *169*, 217–229. [CrossRef]
21. Staal, F.C.; van der Reijd, D.J.; Taghavi, M.; Lambregts, D.M.; Beets-Tan, R.G.; Maas, M. Radiomics for the prediction of treatment outcome and survival in patients with colorectal cancer: A systematic review. *Clin. Color. Cancer* **2021**, *20*, 52–71. [CrossRef]
22. Kang, D.; Park, J.E.; Kim, Y.H.; Kim, J.H.; Oh, J.Y.; Kim, J.; Kim, Y.; Kim, S.T.; Kim, H.S. Diffusion radiomics as a diagnostic model for atypical manifestation of primary central nervous system lymphoma: Development and multicenter external validation. *Neuro-Oncology* **2018**, *20*, 1251–1261. [CrossRef]
23. Chen, C.; Zheng, A.; Ou, X.; Wang, J.; Ma, X. Comparison of radiomics-based machine-learning classifiers in diagnosis of glioblastoma from primary central nervous system lymphoma. *Front. Oncol.* **2020**, *10*, 1151. [CrossRef]
24. Xia, W.; Hu, B.; Li, H.; Geng, C.; Wu, Q.; Yang, L.; Yin, B.; Gao, X.; Li, Y.; Geng, D. Multiparametric-MRI-based radiomics model for differentiating primary central nervous system lymphoma from glioblastoma: Development and cross-vendor validation. *J. Magn. Reson. Imaging* **2021**, *53*, 242–250. [CrossRef]
25. Yun, J.; Park, J.E.; Lee, H.; Ham, S.; Kim, N.; Kim, H.S. Radiomic features and multilayer perceptron network classifier: A robust MRI classification strategy for distinguishing glioblastoma from primary central nervous system lymphoma. *Sci. Rep.* **2019**, *9*, 5746. [CrossRef]
26. Eisenhut, F.; Schmidt, M.A.; Putz, F.; Lettmaier, S.; Fröhlich, K.; Arinrad, S.; Coras, R.; Luecking, H.; Lang, S.; Fietkau, R.; et al. Classification of primary cerebral lymphoma and glioblastoma featuring dynamic susceptibility contrast and apparent diffusion coefficient. *Brain Sci.* **2020**, *10*, 886. [CrossRef]
27. Kunimatsu, A.; Kunimatsu, N.; Kamiya, K.; Watadani, T.; Mori, H.; Abe, O. Comparison between glioblastoma and primary central nervous system lymphoma using MR image-based texture analysis. *Magn. Reson. Med. Sci.* **2018**, *17*, 50. [CrossRef]
28. Kim, Y.; Cho, H.h.; Kim, S.T.; Park, H.; Nam, D.; Kong, D.S. Radiomics features to distinguish glioblastoma from primary central nervous system lymphoma on multi-parametric MRI. *Neuroradiology* **2018**, *60*, 1297–1305. [CrossRef]
29. Wang, H.; Zhou, Y.; Li, L.; Hou, W.; Ma, X.; Tian, R. Current status and quality of radiomics studies in lymphoma: A systematic review. *Eur. Radiol.* **2020**, *30*, 6228–6240. [CrossRef]
30. Chen, C.; Zhuo, H.; Wei, X.; Ma, X. Contrast-enhanced MRI texture parameters as potential prognostic factors for primary central nervous system lymphoma patients receiving high-dose methotrexate-based chemotherapy. *Contrast Media Mol. Imaging* **2019**, *2019*, 5481491. [CrossRef]
31. Ali, O.M.; Nalawade, S.S.; Xi, Y.; Wagner, B.; Mazal, A.; Ahlers, S.; Rizvi, S.M.; Awan, F.T.; Kumar, K.A.; Desai, N.B.; et al. A Radiomic Machine Learning Model to Predict Treatment Response to Methotrexate and Survival Outcomes in Primary Central Nervous System Lymphoma (PCNSL). *Blood* **2020**, *136*, 29–30. [CrossRef]
32. Villano, J.; Koshy, M.; Shaikh, H.; Dolecek, T.; McCarthy, B. Age, gender, and racial differences in incidence and survival in primary CNS lymphoma. *Br. J. Cancer* **2011**, *105*, 1414–1418. [CrossRef]
33. Scalco, E.; Belfatto, A.; Mastropietro, A.; Rancati, T.; Avuzzi, B.; Messina, A.; Valdagni, R.; Rizzo, G. T2w-MRI signal normalization affects radiomics features reproducibility. *Med. Phys.* **2020**, *47*, 1680–1691. [CrossRef]
34. Isaksson, L.J.; Raimondi, S.; Botta, F.; Pepa, M.; Gugliandolo, S.G.; De Angelis, S.P.; Marvaso, G.; Petralia, G.; De Cobelli, O.; Gandini, S.; et al. Effects of MRI image normalization techniques in prostate cancer radiomics. *Phys. Medica* **2020**, *71*, 7–13. [CrossRef]
35. Hoebel, K.V.; Patel, J.B.; Beers, A.L.; Chang, K.; Singh, P.; Brown, J.M.; Pinho, M.C.; Batchelor, T.T.; Gerstner, E.R.; Rosen, B.R.; et al. Radiomics Repeatability Pitfalls in a Scan-Rescan MRI Study of Glioblastoma. *Radiol. Artif. Intell.* **2020**, *3*, e190199. [CrossRef]
36. Schwier, M.; van Griethuysen, J.; Vangel, M.G.; Pieper, S.; Peled, S.; Tempany, C.; Aerts, H.J.; Kikinis, R.; Fennessy, F.M.; Fedorov, A. Repeatability of multiparametric prostate MRI radiomics features. *Sci. Rep.* **2019**, *9*, 9441. [CrossRef]
37. Crombé, A.; Kind, M.; Fadli, D.; Le Loarer, F.; Italiano, A.; Buy, X.; Saut, O. Intensity harmonization techniques influence radiomics features and radiomics-based predictions in sarcoma patients. *Sci. Rep.* **2020**, *10*, 15496. [CrossRef]
38. Shenkier, T.N.; Blay, J.Y.; O'Neill, B.P.; Poortmans, P.; Thiel, E.; Jahnke, K.; Abrey, L.E.; Neuwelt, E.; Tsang, R.; Batchelor, T.; et al. Primary CNS lymphoma of T-cell origin: A descriptive analysis from the international primary CNS lymphoma collaborative group. *J. Clin. Oncol.* **2005**, *23*, 2233–2239. [CrossRef]
39. Moradmand, H.; Aghamiri, S.M.R.; Ghaderi, R. Impact of image preprocessing methods on reproducibility of radiomic features in multimodal magnetic resonance imaging in glioblastoma. *J. Appl. Clin. Med. Phys.* **2020**, *21*, 179–190. [CrossRef]
40. Fedorov, A.; Beichel, R.; Kalpathy-Cramer, J.; Finet, J.; Fillion-Robin, J.C.; Pujol, S.; Bauer, C.; Jennings, D.; Fennessy, F.; Sonka, M.; et al. 3D Slicer as an image computing platform for the Quantitative Imaging Network. *Magn. Reson. Imaging* **2012**, *30*, 1323–1341. [CrossRef]
41. Sled, J.G.; Zijdenbos, A.P.; Evans, A.C. A nonparametric method for automatic correction of intensity nonuniformity in MRI data. *IEEE Trans. Med. Imaging* **1998**, *17*, 87–97. [CrossRef]
42. Kalavathi, P.; Prasath, V.S. Methods on skull stripping of MRI head scan images—A review. *J. Digit. Imaging* **2016**, *29*, 365–379. [CrossRef]
43. Bauer, S.; Fejes, T.; Reyes, M. A skull-stripping filter for ITK. *Insight J.* **2013**, *2012*, 1–7. [CrossRef]
44. Aganj, I.; Yeo, B.T.T.; Sabuncu, M.R.; Fischl, B. On removing interpolation and resampling artifacts in rigid image registration. *IEEE Trans. Image Process* **2012**, *22*, 816–827. [CrossRef]

45. Shinohara, R.T.; Sweeney, E.M.; Goldsmith, J.; Shiee, N.; Mateen, F.J.; Calabresi, P.A.; Jarso, S.; Pham, D.L.; Reich, D.S.; Crainiceanu, C.M.; et al. Statistical normalization techniques for magnetic resonance imaging. *Neuroimage Clin.* **2014**, *6*, 9–19. [CrossRef]
46. Nyúl, L.G.; Udupa, J.K. On standardizing the MR image intensity scale. *Magn. Reson. Med. Off. J. Int. Soc. Magn. Reson. Med.* **1999**, *42*, 1072–1081. [CrossRef]
47. Van Griethuysen, J.J.; Fedorov, A.; Parmar, C.; Hosny, A.; Aucoin, N.; Narayan, V.; Beets-Tan, R.G.; Fillion-Robin, J.C.; Pieper, S.; Aerts, H.J. Computational radiomics system to decode the radiographic phenotype. *Cancer Res.* **2017**, *77*, e104–e107. [CrossRef]
48. Liang, Z.G.; Tan, H.Q.; Zhang, F.; Rui Tan, L.K.; Lin, L.; Lenkowicz, J.; Wang, H.; Wen Ong, E.H.; Kusumawidjaja, G.; Phua, J.H.; et al. Comparison of radiomics tools for image analyses and clinical prediction in nasopharyngeal carcinoma. *Br. J. Radiol.* **2019**, *92*, 20190271. [CrossRef]
49. Shrout, P.E.; Fleiss, J.L. Intraclass correlations: Uses in assessing rater reliability. *Psychol. Bull.* **1979**, *86*, 420. [CrossRef]
50. Kasenda, B.; Ferreri, A.J.; Marturano, E.; Forst, D.; Bromberg, J.; Ghesquieres, H.; Ferlay, C.; Blay, J.Y.; Hoang-Xuan, K.; Pulczynski, E.; et al. First-line treatment and outcome of elderly patients with primary central nervous system lymphoma (PCNSL)—A systematic review and individual patient data meta-analysis. *Ann. Oncol.* **2015**, *26*, 1305–1313. [CrossRef]
51. Ferreri, A.J.; Blay, J.Y.; Reni, M.; Pasini, F.; Spina, M.; Ambrosetti, A.; Calderoni, A.; Rossi, A.; Vavassori, V.; Conconi, A.; et al. Prognostic scoring system for primary CNS lymphomas: The International Extranodal Lymphoma Study Group experience. *J. Clin. Oncol.* **2003**, *21*, 266–272. [CrossRef]
52. Abrey, L.E.; Ben-Porat, L.; Panageas, K.S.; Yahalom, J.; Berkey, B.; Curran, W.; Schultz, C.; Leibel, S.; Nelson, D.; Mehta, M.; et al. Primary central nervous system lymphoma: The Memorial Sloan-Kettering Cancer Center prognostic model. *J. Clin. Oncol.* **2006**, *24*, 5711–5715. [CrossRef]
53. Carré, A.; Klausner, G.; Edjlali, M.; Lerousseau, M.; Briend-Diop, J.; Sun, R.; Ammari, S.; Reuzé, S.; Alvarez-Andres, E.; Estienne, T.; et al. Standardization of Brain MRI across Machines and Protocols: Bridging the Gap for MRI-Based Radiomics. In Proceedings of the Radiotherapy and Oncology, Online, 28 November–1 December 2020; Elsevier Ireland Ltd. Elsevier House: East Park Shannon, UK, 2020; Volume 152, p. S294.
54. Barajas, R.F., Jr.; Politi, L.S.; Anzalone, N.; Schöder, H.; Fox, C.P.; Boxerman, J.L.; Kaufmann, T.J.; Quarles, C.C.; Ellingson, B.M.; Auer, D.; et al. Consensus recommendations for MRI and PET imaging of primary central nervous system lymphoma: Guideline statement from the International Primary CNS Lymphoma Collaborative Group (IPCG). *Neuro-Oncology* **2021**, *23*, 1056–1071. [CrossRef]
55. Li, Y.; Ammari, S.; Balleyguier, C.; Lassau, N.; Chouzenoux, E. Impact of Preprocessing and Harmonization Methods on the Removal of Scanner Effects in Brain MRI Radiomic Features. *Cancers* **2021**, *13*, 3000. [CrossRef]
56. Fujima, N.; Homma, A.; Harada, T.; Shimizu, Y.; Tha, K.K.; Kano, S.; Mizumachi, T.; Li, R.; Kudo, K.; Shirato, H. The utility of MRI histogram and texture analysis for the prediction of histological diagnosis in head and neck malignancies. *Cancer Imaging* **2019**, *19*, 5. [CrossRef]
57. Meyer, H.J.; Schob, S.; Höhn, A.K.; Surov, A. MRI texture analysis reflects histopathology parameters in thyroid cancer–a first preliminary study. *Transl. Oncol.* **2017**, *10*, 911–916. [CrossRef]

bioengineering

MDPI

Article

# Synthetic CT in Carbon Ion Radiotherapy of the Abdominal Site

Giovanni Parrella [1,*], Alessandro Vai [2], Anestis Nakas [1], Noemi Garau [1], Giorgia Meschini [1], Francesca Camagni [1], Silvia Molinelli [2], Amelia Barcellini [3,4], Andrea Pella [5], Mario Ciocca [2], Viviana Vitolo [3], Ester Orlandi [6], Chiara Paganelli [1] and Guido Baroni [1]

1 Department of Electronics, Information and Bioengineering, Politecnico di Milano, Piazza Leonardo da Vinci 32, 20133 Milan, Italy
2 Medical Physics Unit, National Center of Oncological Hadrontherapy (CNAO), Strada Campeggi, 53, 27100 Pavia, Italy
3 Radiotherapy Unit, National Center of Oncological Hadrontherapy (CNAO), Strada Campeggi, 53, 27100 Pavia, Italy
4 Department of Internal Medicine and Medical Therapy, University of Pavia, 27100 Pavia, Italy
5 Bioengineering Unit, National Center of Oncological Hadrontherapy (CNAO), Strada Campeggi, 53, 27100 Pavia, Italy
6 Clinical Unit, National Center of Oncological Hadrontherapy (CNAO), Strada Campeggi, 53, 27100 Pavia, Italy
* Correspondence: giovanni.parrella@polimi.it; Tel.: +39-02-2399-18-9022

**Abstract:** The generation of synthetic CT for carbon ion radiotherapy (CIRT) applications is challenging, since high accuracy is required in treatment planning and delivery, especially in an anatomical site as complex as the abdomen. Thirty-nine abdominal MRI-CT volume pairs were collected and a three-channel cGAN (accounting for air, bones, soft tissues) was used to generate sCTs. The network was tested on five held-out MRI volumes for two scenarios: (i) a CT-based segmentation of the MRI channels, to assess the quality of sCTs and (ii) an MRI manual segmentation, to simulate an MRI-only treatment scenario. The sCTs were evaluated by means of similarity metrics (e.g., mean absolute error, MAE) and geometrical criteria (e.g., dice coefficient). Recalculated CIRT plans were evaluated through dose volume histogram, gamma analysis and range shift analysis. The CT-based test set presented optimal MAE on bones (86.03 ± 10.76 HU), soft tissues (55.39 ± 3.41 HU) and air (54.42 ± 11.48 HU). Higher values were obtained from the MRI-only test set ($MAE_{BONE}$ = 154.87 ± 22.90 HU). The global gamma pass rate reached 94.88 ± 4.9% with 3%/3 mm, while the range shift reached a median (IQR) of 0.98 (3.64) mm. The three-channel cGAN can generate acceptable abdominal sCTs and allow for CIRT dose recalculations comparable to the clinical plans.

**Keywords:** synthetic CT; MRI guidance; MRI-only; image-guided radiotherapy; carbon ion radiotherapy; particle therapy; deep learning

**Citation:** Parrella, G.; Vai, A.; Nakas, A.; Garau, N.; Meschini, G.; Camagni, F.; Molinelli, S.; Barcellini, A.; Pella, A.; Ciocca, M.; et al. Synthetic CT in Carbon Ion Radiotherapy of the Abdominal Site. *Bioengineering* **2023**, *10*, 250. https://doi.org/10.3390/bioengineering10020250

Academic Editors: Paolo Zaffino and Maria Francesca Spadea

Received: 13 January 2023
Revised: 6 February 2023
Accepted: 9 February 2023
Published: 14 February 2023



## 1. Introduction

In the treatment of abdominal tumors such as those of the liver and pancreas, carbon ion radiotherapy (CIRT) is considered a promising therapeutic option, thanks to its excellent geometrical selectivity and radiobiological effectiveness [1,2]. However, tumors subject to respiratory motion may suffer from inter- and intra-fraction motion that needs to be properly accounted for during planning and delivery. So far, the repeated acquisition of computed tomography (CT), in the form of respiratory-correlated 4DCT, is the clinical routine for motion management, but concurrently exposes the patient to additional non-therapeutic radiation [3].

The growing interest in magnetic resonance imaging (MRI) in recent years has fostered the development of MRI-only workflows that would guarantee the absence of ionizing radiation while exploiting the superior soft tissue contrasts of MRI scans. In this regard, online MR-guided radiotherapy (MRgRT) is a clinical reality in conventional radiotherapy [4–6], while no integrated system is clinically available for online MR guidance in particle therapy (PT), with only a few feasibility studies addressing protons but not carbon ions [7,8]. Nonetheless, the off-line use of MRI in CIRT may support treatment planning and adaptation through the generation of synthetic CT (sCT), used as alternatives to verification CTs, and avoid non-therapeutic radiation along with CT-MRI registration errors that are typically relevant in the abdominal site [9,10]. Different methods (i.e., bulk density override, atlas based, voxel based) have been studied to generate sCTs starting from MRI scans [8,11]. As an alternative, deep learning (DL) methods have been largely investigated [11–17], relying on their capability of autonomously learning hidden relationships among data [18]. In particular, deep convolutional neural networks (DCNN) showed promising results for the generation of sCTs of the head and neck with respect to conventional methods, as they are able to catch complex nonlinear relationships among MRI and CT [12,15]. Generative adversarial networks (GAN), especially in the conditional form (conditional GAN—cGAN) have been instead considered for more complex anatomical districts such as the abdomen, thanks to the adversarial learning process, which provides great efficacy in image-to-image translation applications [19–21]. The main limitation of these networks is the need for paired MRI-CT samples to perform the training, which is a critical aspect when dealing with multi-modal volumes. In recent years, new architectures have been proposed to overcome the need for paired CT-MRI training datasets, with cycleGAN and fully bidirectional networks showing great results in end-to-end synthetic image generation tasks [22]. Clearly, the use of such complex and multi-network architectures requires high dimensional training datasets, which still remain an evident concern in most clinical realities. The use of multiple MRI sequences (e.g., Dixon in-phase (IP), opposed-phase (OP), fat, and water) has been also investigated to increase the quality of critical anatomical structures, but no relevant improvements were shown [23].

The dosimetric evaluation of DL-based sCTs has been investigated for X-ray and proton radiotherapy in most anatomical districts [20–28], while a similar analysis for CIRT applications is still missing in the literature.

In the treatment of abdominal targets, the presence of inter- and intra-acquisition motions, which mainly translate into the different dispositions of the air fillings in the organs, highly affects the dose distribution for carbon ions plans [29]. An MRI-only workflow for CIRT would thus require more stringent constraints and tolerances in terms of accuracy in the definition of HU values, if compared to conventional radiotherapy, since a precise knowledge of particle stopping power, estimated from HU values within the patient anatomy, is essential for accurate treatment planning, to limit any range shift and damage to healthy tissues [30,31]. Up to now, to the authors' best knowledge, the application of sCT to CIRT plans has been investigated only by Knäusl et al. [32] for head and neck imaging, while no study has been performed on CIRT plans for abdominal tumors.

In this study, we focused on the generation of abdominal sCTs through a cGAN, with the final aim of simulating an MRI-only workflow for CIRT, based on recalculation of the clinical CIRT treatment plans for generated abdominal sCT volumes. This study, therefore, was conducted to evaluate, for the first time, the feasibility of sCT in CIRT for abdominal sites.

## 2. Materials and Methods

### *2.1. Patient Cohort*

Image datasets were collected from 24 patients affected by liver or pancreatic cancer and treated with CIRT at the National Centre for Oncological Hadrontherapy (CNAO, Pavia, Italy) between 2017 and 2021. Standard clinical workflow comprised the acquisition of a 4DCT followed by a 3D MRI on the same day, for contouring and planning preparation. The same immobilization setup, consisting of customized pillows (MOLDCARE Cushion, QFix, Avondale, PA, USA) and non-perforated body thermoplastic masks (Klarity Medical Products, Heath, OH, USA), was used both for CT and MRI acquisitions and treatment delivery. Acquisition in two different scanners along with re-positioning of the patient with the thermoplastic mask caused inter-acquisition motion. For 15 patients, re-evaluative images (both CT and MRI) were acquired during the treatment course with the same immobilizations setup and were considered independent from the first acquisition, leading to a total of 39 CT-MRI volume pairs collected. The study was approved by the local ethical committee, and all patients signed the informed consent (CNAO 37-2019 4D-MRI).

The 4DCTs were acquired during patient free breathing on a Siemens SOMATOM Sensation Open CT scanner (resolution $0.98 \times 0.98 \times 2$ mm$^3$). Clinical plans were optimized at end-exhale for gated treatments; as such, only this phase was used in this work to derive sCTs. CT acquisitions had a variable number of slices, resulting in a volume size of $512 \times 512 \times [96 - 145]$ voxels. MRI acquisitions were performed with a Siemens Magnetom Verio 3T scanner. Three-dimensional breath-hold T1-weighted volumetric interpolated breath-hold examination (VIBE) sequences were acquired at end-exhale with $1.06 \times 1.06 \times 3$ mm$^3$ resolution (repetition time TR = 3.87 ms, echo time TE = 1.92 ms). For two patients, MRI acquisitions had a voxel size of $1.25 \times 1.25 \times 3$ mm$^3$ and $1.125 \times 1.125 \times 3$ mm$^3$. Most of the MRI acquisitions had $320 \times 260 \times 64$ voxels, except for one having 88 transversal slices. Two CT-MRI volume pairs were discarded from the study because of the low quality of the acquired images. Therefore, 37 volume pairs were used: 32 pairs were exploited for cross-validation (CV) and training, while five pairs were randomly selected and held out for testing (Table 1) for more details on the treatment plans. All treatment plans were optimized with the RayStation (Raysearch Laboratories, Stockholm, Sweden—version 10.B) Treatment Planning System (TPS) on the end-exhale reconstructed CT phase and clinically approved. Corresponding organs at risk (OARs), gross tumor volume (GTV) and clinical target volume (CTV) were segmented by radiation oncologists. The relevant OARs included were kidneys, aorta, colon, duodenum, stomach and spinal cord.

**Table 1.** CIRT plan details for the patients used in the test set.

| Patient | N of Beams | Prescribed Dose [Gy(RBE)] | Fractions | Position | Tumor Location |
|---|---|---|---|---|---|
| P17 | 1 | 43 | 10 | Prone | Pancreas |
| P20 | 1 | 38.4 | 8 | Prone | Pancreas |
| P21 | 2 | 57.6 | 12 | Supine | Pancreas |
| P27 | 1 | 38.4 | 8 | Prone | Pancreas |
| P31 | 1 | 48 | 10 | Prone | Pancreas |

### 2.2. Pre-Processing

Although CT and MRI scans were acquired the same day and with an immobilization setup, inter-acquisition motion (i.e., anatomical changes between CT and MRI scans) was present. A manual rigid registration was firstly performed to align CT and MRI scans. The application of deformable image registration (DIR) was investigated but did not show relevant improvements on the quality of sCTs; therefore, it was not applied (Supplementary Material S1). Indeed, in most cases, the multi-modal DIR could hardly compensate for air filling mismatches and caused large bone deformations and artefacts that reduced the size of the training dataset, making it a less effective approach. Nonetheless, given the lack of a real ground truth (i.e., a CT with MRI anatomical configuration), DIR was applied for the generation of pseudo ground truths (see Section 2.4).

CT slices were resampled to the corresponding MRI spacing to guarantee the voxel consistency among the two volumes and clipped to [−1000,+1047] HU as previously performed by Maspero et al. [14] to reduce the discretization step, while background values were set to −1000 HU. For MRI volumes, the pre-processing consisted of: (i) bias field correction to reduce low frequency noise due to magnetic field inhomogeneities [33]; (ii) reduction of Gaussian noise through a non-linear bilateral filter; (iii) contrast enhancement through histogram clipping to 99th percentile of intensity [14,21,24]: (iv) setting of background values to zero; and (v) histogram matching to similarly distribute grey levels across all MRI volumes [34].

CT and MRI preprocessed volumes were all resized to 256 × 256 to match the network input dimensions [19]. Then, the input MRI and target CT transversal slices were segmented in three channels by means of CT-based masks (i.e., air [−1000,−800] HU, bone [150,1047] HU, soft tissues [−800,150] HU) and then linearly scaled to [−1,1].

As a final step, eight transformations were applied to each CT/MRI channel triplet used for training, including horizontal or vertical flip, Gaussian noise adding, shear, rotation and cropping. Thus, the initial training dataset composed of 2014 CT-MRI slice pairs was enlarged to 16,112. Pre-processing was performed through Python scripts implemented using SimpleITK modules (version 2.0.2).

### 2.3. Neural Network

The neural network used in this work consisted of a cGAN derived from the open-source network "Pix2Pix" by Isola et al. [19], adapted to work on three channels (air, bone, soft tissues) to better account for the anatomical complexity of the abdomen [35] and to deal with the limited dataset of 39 volume pairs.

The net was trained on transversal slices and composed of a 256 × 256 × 3 U-net generator (Figure 1a), used to generate sCTs, and a 70 × 70 × 3 PatchGAN discriminator (Figure 1b), used to judge the quality of the output with high resolution during the training [19]. The U-net is composed of eight encoder blocks, each comprising convolution, batch normalization and leaky rectified linear unit (ReLU) layers, and seven decoder blocks, each composed of transposed convolution, batch normalization, dropout and ReLU layers. The PatchGAN architecture is made of four encoding blocks, such that each pixel of the 30 × 30 output classifies a 70 × 70 pixel patch of the input image. For the detailed architecture, refer to Figure 1a,b.

The loss function (Equation (3)) combined $L_1$ norm (Equation (1)) and cGAN adversarial cross-entropy loss (Equation (2)) to reduce blurring effects and artefacts [19]:

$$L_1 = E_{x,y,z}[\|x - G(y,z)\|_1] \tag{1}$$

$$L_{cGAN} = -[E_{y,x}[\log D(y,x)] + E_{y,z}[\log(1 - D(y, G(y,z)))]] \tag{2}$$

$$L_{tot} = L_{cGAN}(\theta_g, \theta_d) + \lambda \cdot L_1(\theta_g) \tag{3}$$

where $x$ is the target CT, $y$ is the input MRI, $z$ the noise, $G(y,z)$ the sCT, $D(y,x)$ the PatchGAN output, $\theta g$, $\theta d$ are the trainable parameters of generator and discriminator, and $\lambda = 100$ is a weight for the $L_1$ norm, as from Isola et al. [19]. The role of this additional loss is to have the generator not only mislead the discriminator, but also generate synthetic images that mimic the target CT in an $L_1$ sense, reducing the blurring and improving the representation of structures. The cGAN loss was evaluated on the channel triplets, while the $L_1$ norm was evaluated after reassembling the sCT slice.

The MRI represents the conditional input of the net, with the noise z being provided in the form of dropout on several layers of the U-net generator during training and testing. The training was performed by alternating one gradient descent step on the discriminator and one on the generator, using ADAM optimizer with momentum parameters $\beta 1 = 0.5$, $\beta 2 = 0.999$. The training was stopped after 20 epochs, which were sufficient to achieve convergence thanks to data augmentation.

The cGAN was optimized through a six-fold cross-validation (CV), setting the batch size to 1 and discriminator learning rate to $2 \times 10^{-7}$ (Supplementary Material S2). This confirmed that the instance normalization (i.e., the use of batch-normalization layers with batch = 1) is well-suited for image generation tasks [36]. The network was trained on the full dataset (i.e., 32 volumes) and tested on the remaining five volumes (Figure 1c). Before stacking all of the output slices to rebuild the synthetic volumes, the values of each channel were re-scaled to the corresponding HU range and re-assembled by means of the same masks used for the channels' segmentation. The stacked volumes were finally resized to the original MRI [$320 \times 260 \times N_{slices}$] size.

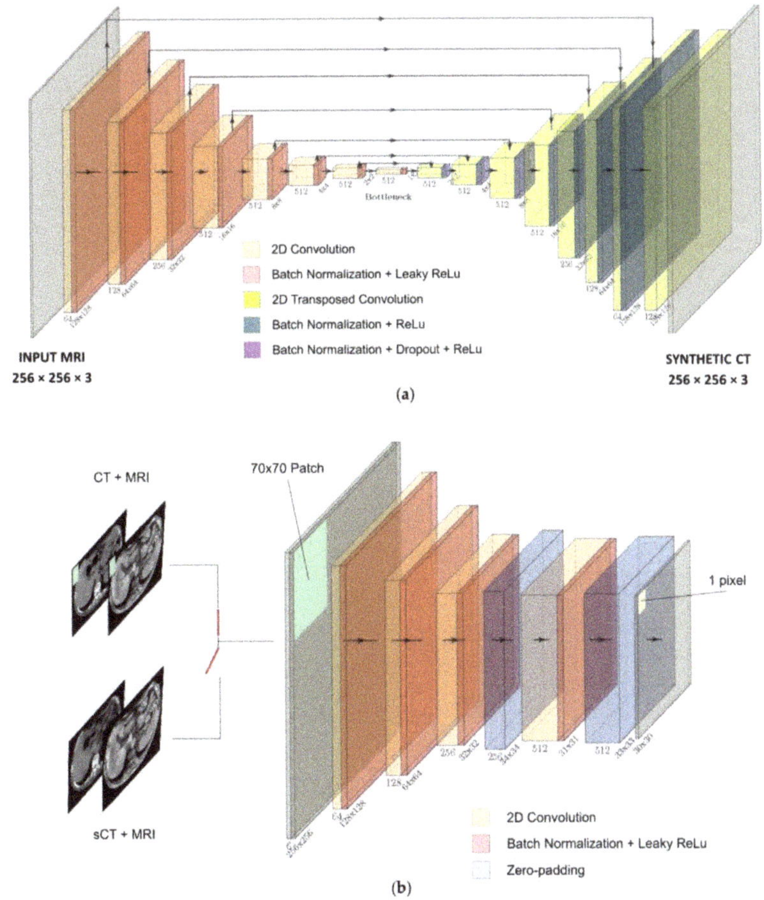


**Figure 1.** (**a**) U-net generator. (**b**) PatchGAN discriminator.

### 2.4. Experiments

The results were evaluated by means of similarity metrics such as mean absolute error (MAE), root mean squared error (RMSE), normalized cross correlation (NCC), structural similarity index (SSIM) and peak signal-to-noise ratio (PSNR) between sCT and target CT (Supplementary Material S3), with the exclusion of the background, both within CV and testing scenarios. All metrics were evaluated on the basis of reassembled volumes, applying the corresponding tissue mask. Specifically, the five held-out patients were used (i) to create the test set, where CT-based masks were used for the analysis, and (ii) to build an MRI-only simulation set, where the use of CT-based masks was replaced by manual segmentation of the three channels directly on MRI (Figure 1c).

Within the MRI-only workflow, the evaluation of similarity metrics was performed between sCTs and reference volumes obtained by applying DIR between the target CT and MRI (i.e., pseudo ground truths, $CT_{PGT}$) to compensate for MRI-CT inter-acquisition motion (Figure 2a).

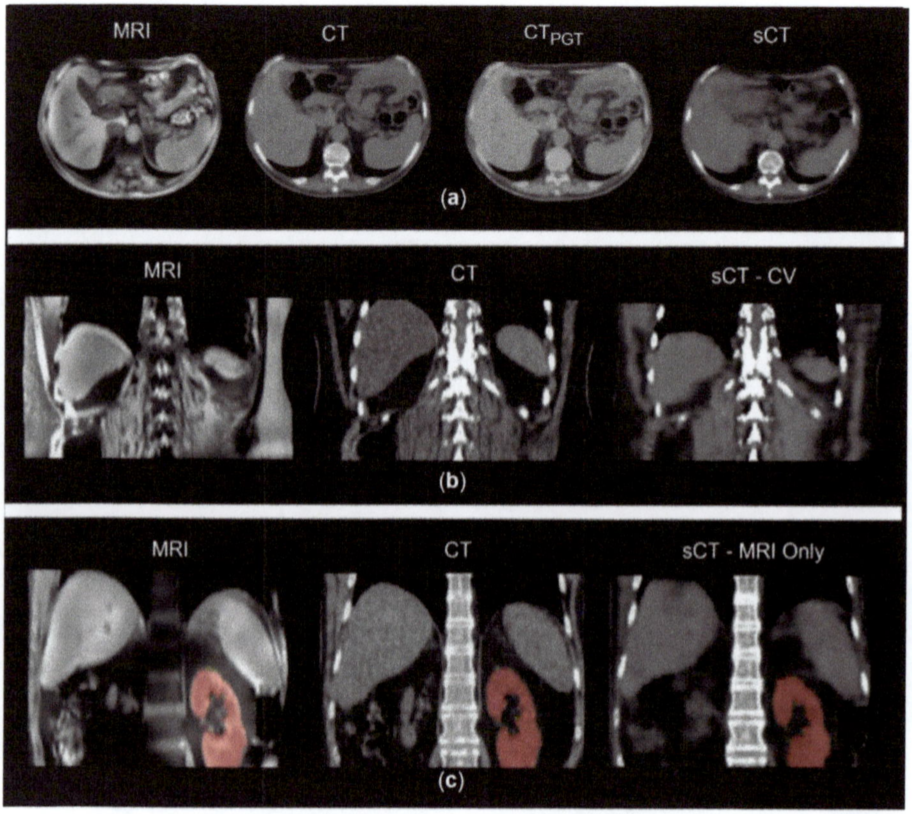


**Figure 2.** (**a**) Example of MRI, CT, sCT and pseudo ground truth ($CT_{PGT}$) on axial plane. $CT_{PGT}$ still shows visible discrepancies with MRI anatomical condition. (**b**) Example of inter-acquisition motion in a CT-MRI pair and the resulted sCT in CV. (**c**) Example of MRI, planning CT and synthetic CT from MRI-only scenario. In red, the segmentation of kidneys used for the geometrical analysis.

The MRI-only scenario was evaluated also through geometrical criteria: dice coefficient (DSC), 95th percentile Hausdorff distance (HD) and the center of mass distance (CoMD) were calculated on kidney segmentations of MRI, CT and sCT, to assess the quality of the sCT in terms of correct reproduction of soft tissues with respect to the MRI anatomy. HD, DSC and CoMD were assessed for each couple of segmented volumes (CT-MRI, sCT-CT, sCT-MRI). sCTs were compared to MRIs since we expected the sCT to be representative of the MRI anatomical condition; CT was compared to MRI to determine the initial mismatch, while sCTs were compared to target CTs to confirm that sCT was far from matching CT structures.

Due to the lack of a real ground truth, the net was also validated on a CT-MRI volume pairs obtained through a computational phantom (i.e., XCAT [37] for CT and correspondent ComBAT [38] for MRI) that guaranteed an improved match of anatomical structures between the two volumes, avoiding any inter-acquisition motion (Supplementary Material S4).

The clinical CIRT plans were recalculated on the MRI-only sCT for each patient through the TPS and evaluated on the basis of DVH-based metrics (GTV $D_{95\%}$, CTV $D_{95\%}$, $D_{2\%}$ on OARs) and dose difference maps with respect to the original plan. A two one-sided test of equivalence for paired samples (TOST-P) was used to compare the DVH metrics, considering a confidence interval of 95% and an equivalence interval of $\pm 0.5\%$. The global gamma analysis was also performed with 1 mm/1%, 2 mm/2%, 3 mm/3% as tolerance criteria, and three dose thresholds: 10%, 50%, and 90% on the prescription dose. A range shift (*RS*) analysis was performed to take into account possible dose shifts, which are generally averaged in gamma analysis [15,39]. In this regard, both range shift (*RS*) and relative range shift (*RRS*) were evaluated on a beam-by-beam basis, considering an acceptability threshold set to 5 mm, in accordance with clinical margins used at CNAO [40]. RS and RRS are defined as:

$$RS = (R_{sCT80} - R_{CT80}) \tag{4}$$

$$RRS = \left( {}^{RS} / {R_{CT80}} \right) \tag{5}$$

where $R_{sCT80}$ and $R_{CT80}$ are the beam ranges computed at 80% of the dose peak. This evaluation was performed on the dose profile along the central axis of each beam, considering 10 transversal slices, for a total of 60 RSs.

All computational steps were performed on a Precision 5820 Tower DELL workstation equipped with a 16 GB RAM Nvidia GPU (QUADRO P5000). A full training procedure took around 12 h, while the generation of a synthetic volume took ~5 s.

### 3. Results

The similarity metrics from CV, testing and MRI-only simulation are presented in Table 2. Comparable results were found between CV and test performance, with MAE on soft tissues and air channels being lower than that on bone. Notwithstanding inter-acquisition discrepancies (Figure 2b), the air channel showed an error of 54.42 ± 11.48 HU, the soft tissues 55.39 ± 3.41 HU, and the bone structures 86.03 ± 10.76 HU.

**Table 2.** Average results for CV, test and MRI-Only procedures, compared to the literature. Average (St. Dev.). * Soft tissue. ** Lungs. *** Vertebral bodies. **** Bidirectional network.

| | | MAE_Body [HU] | RMSE [HU] | SSIM | PSNR [dB] | NCC | MAE_Air [HU] | MAE_Bone [HU] | MAE_Soft [HU] |
|---|---|---|---|---|---|---|---|---|---|
| Our work | CV | 56.52 (8.31) | 97.24 (17.56) | 0.651 (0.043) | 27.73 (1.23) | 0.857 (0.054) | 46.19 (6.30) | 90.76 (7.86) | 54.79 (8.98) |
| | TEST | 57.08 (2.79) | 99.69 (4.90) | 0.67 (0.06) | 27.64 (0.68) | 0.92 (0.02) | 54.42 (11.48) | 86.03 (10.76) | 55.39 (3.41) |
| | MRI-ONLY | 88.22 (9.88) | 181.10 (11.84) | 0.59 (0.08) | 20.99 (1.49) | 0.76 (0.10) | 279.01 (142.46) | 154.87 (22.90) | 75.00 (8.12) |
| Literature | [20] | 78.71 (18.46) | - | - | - | - | - | 152.71 (30.14) | 53.89 (10.7) |
| | [24] | 62(13) | - | - | 30.0 (1.8) | - | 104(38) ** | 167 (22) | 36 (8) * |
| | [25] | 72.48 (18.16) | - | - | 22.65 (3.63) | 0.92 (0.04) | 108.06 (49.45) | 216.81 (63.0) | 58.62 (30.61) |
| | [26] | 55.56 (2.27) | 106.43 (11.45) | - | - | 0.87 (0.03) | | - | - |
| | [27] | - | - | - | - | - | - | - | 90 (29) |
| | [28] | - | - | - | - | - | - | 110.09 (29.23) *** | - |
| | [21] | 89.8 (18.7) | - | - | 27.4 (1.6) | - | - | - | - |
| | [23] | 60.42 (2.27) | - | - | - | 0.88 (0.03) | - | - | - |
| | [22] | 6.30 (0.56) **** | - | 0.90 (0.42) | - | - | - | - | - |

The results relative to the MRI-only evaluation were characterized by larger values on all metrics. In particular, the errors grew to 279.01 ± 142.46 HU on the air channel, 154.87 ± 22.90 HU in the case of bone structures, and 88.22 ± 9.88 HU within the body.

The geometrical analysis of the MRI-only workflow (Figure 2c) showed the lowest discrepancies between CT and MRI segmentation, confirming good accuracy in replicating MRI anatomy. As an example, the average CoMD in sCT-MRI was 5.85 ± 4.87 mm versus 7.75 ± 5.92 mm in CT-MRI and 13.30 ± 10.42 mm for sCT-CT. Detailed results are reported in Supplementary Material S5.

The validation performed on the phantom showed an MAE of 73.3 HU on the whole volume, while the MAEs on the three channels were of 66.11 HU, 77.93 HU and 167.36 HU for soft tissues, air, and bones, respectively (see Supplementary Material S4).

As for dose accuracy, Figure 3 shows the DVH comparison for patient and dose recalculations on P17 and P27 (complete results reported in Supplementary Materials S6–S8). The GTV and CTV $D_{95\%}$ as well as the $D_{2\%}$ on the organs at risk (OARs) are displayed in Figure 4, expressed in terms of dose difference Δ(sCT-CT), MAE and relative error (E[%]) with respect to the prescribed dose. For all five patients, good reproducibility was shown relative to the dose to the GTV, with patients P20 and P21 being characterized by errors of −2.04 Gy[RBE] and −1.88 Gy[RBE] respectively, corresponding to −5.3% and −3.3% of the prescribed dose. The MAE on the GTV $D_{95\%}$ was 0.86 ± 0.90 Gy[RBE].

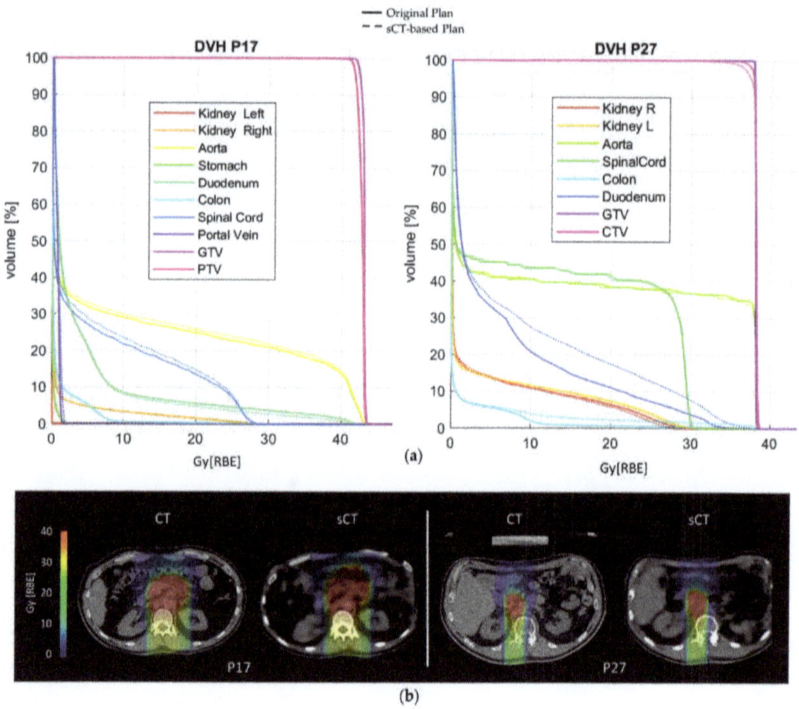


**Figure 3.** (**a**) DVH comparison on patients P17 and P27; (**b**) original CIRT plan (RBE) and sCT-based recalculation for patients P17 and P27.

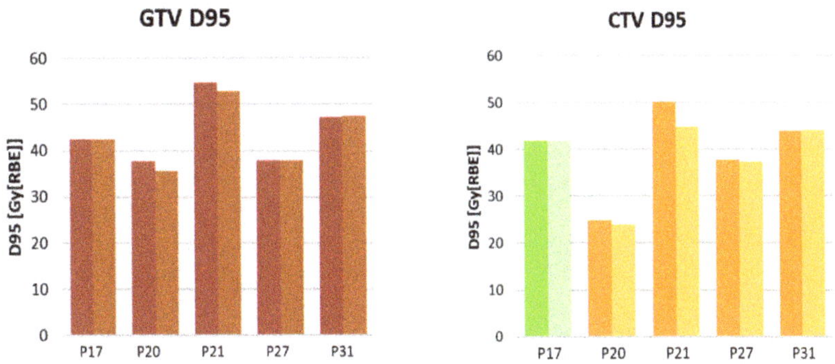


| GTV | P17 | P20 | P21 | P27 | P31 |
|---|---|---|---|---|---|
| CT [Gy[RBE]] | 42.50 | 37.62 | 54.69 | 37.86 | 47.22 |
| sCT [Gy[RBE]] | 42.45 | 35.58 | 52.81 | 37.94 | 47.46 |
| Δ [Gy[RBE]] | -0.05 | -2.04 | -1.88 | 0.08 | 0.24 |
| E[%] | -0.12 | -5.31 | -3.26 | 0.21 | 0.50 |

| CTV | P17 | P20 | P21 | P27 | P31 |
|---|---|---|---|---|---|
| CT [Gy[RBE]] | 41.85 | 24.87 | 49.98 | 37.77 | 43.92 |
| sCT [Gy[RBE]] | 41.80 | 23.87 | 44.88 | 37.39 | 44.08 |
| Δ [Gy[RBE]] | -0.05 | -1.00 | -5.10 | -0.38 | 0.16 |
| E[%] | -0.12 | -2.60 | -8.85 | -0.99 | 0.34 |

(a)

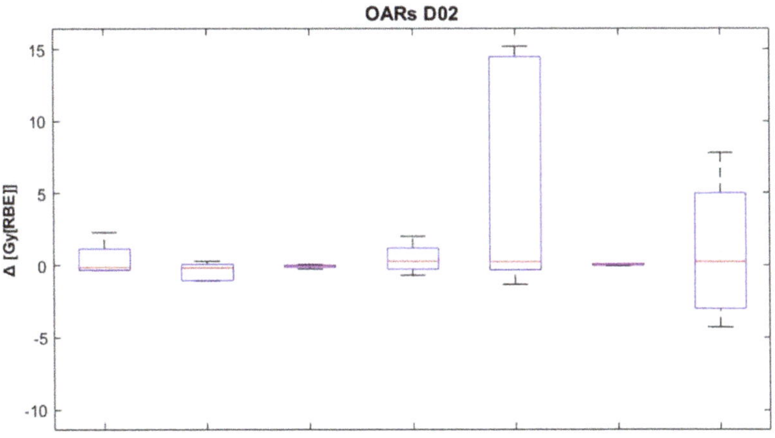


| Δ [Gy[RBE]] | Kidney R | Kidney L | Aorta | Stomach | Colon | Spinal Cord | Duodenum |
|---|---|---|---|---|---|---|---|
| P17 | -1.08 | 0.01 | -0.01 | 0.15 | 0.24 | 0.09 | -4.27 |
| P20 | -0.17 | 2.27 | -0.26 | 0.37 | -1.34 | -0.01 | -1.72 |
| P21 | 0.28 | - | 0.05 | -0.69 | 15.18 | 0.09 | - |
| P27 | -1.03 | -0.30 | -0.10 | - | 14.22 | -0.06 | 2.19 |
| P31 | 0.00 | -0.34 | -0.03 | 1.99 | 0.02 | -0.01 | 7.8 |
| MAE [Gy[RBE]] | 0.512(0.45) | 0.84(1.09) | 0.09(0.09) | 0.80(0.71) | 6.20(6.96) | 0.05(0.04) | 3.99(2.40) |

(b)

**Figure 4.** (**a**) D95% values for GTV and CTV in the original plan (CT) and the recalculated one (sCT). For P17, PTV was considered in this comparison (shown in Green). The table contains the dose values and the dose difference Δ[Gy[RBE]], as well as the error relative to the prescribed dose E [%]. (**b**) D2% difference (sCT-CT) for the main OARs on each patient, and the D2% MAE over each OAR. The red mark indicates the median, and edges of the box show the 25th and 75th percentiles.

Similarly, the values relative to the CTV showed the two dose distributions to have comparable results, with a maximum error of −5.10 Gy[RBE] for P21 (−8.9% of the prescribed dose). The MAE on the $D_{95\%}$ CTV was 1.34 ± 1.33 Gy[RBE]. The $D_{2\%}$ relative errors on OARs lay in an interquartile range (IQR) of [−0.24,0.22]%, although the colon reached a peak relative error of 37.03% (14.22 Gy[RBE]), and the duodenum of 16.25% (7.8 Gy[RBE]) (Supplementary Material S6). The voxel-wise dose difference maps are shown in Figure S5 in Supplementary Material S7, highlighting the worst-case slice for each test patient. Regions with a high dose difference can be mainly seen with correspondence of air pocket mismatches, while the overall distribution of dose to the body is comparable. Indeed, as in Table S6 in Supplementary Material S7, all test patients showed a median dose difference close to zero, with the widest IQR being 0.158 Gy[RBE] on patient P21. The maximum errors were in the range [22.64,42.36] Gy[RBE]. In this regard, an incomplete reproduction of the kidney affected the dose distribution in patient P27 (Figure S6), while the limited field of view of MRI introduced dose artefacts on recalculation for patient P21 (Figure S7).

A peak gamma pass rate of 94.88% was obtained in the 3%/3 mm analysis (Supplementary Material S9).

The range shift analysis showed a median (IQR) RS of 0.98 (3.64) mm and RRS of 0.61 (2.14)% (Figure 5). Considering each beam individually, as shown in Table S9 in Supplementary Material S10, patient P27 was the one showing the highest errors, with a median RS of 5.69 (6.97) mm.

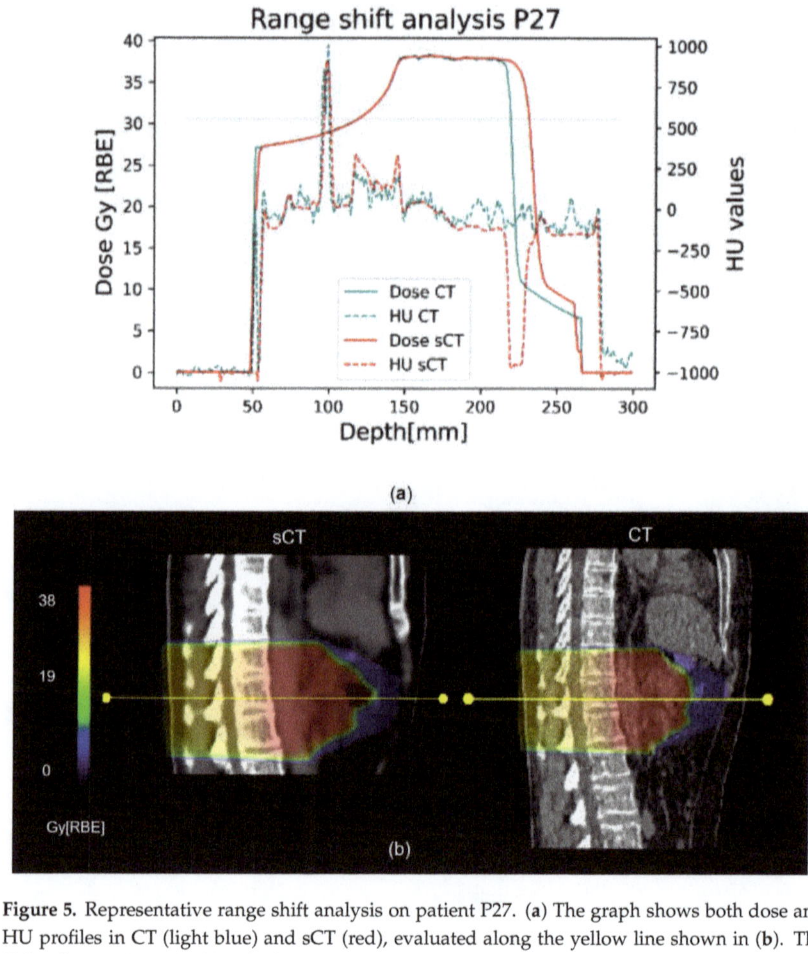


**Figure 5.** Representative range shift analysis on patient P27. (**a**) The graph shows both dose and HU profiles in CT (light blue) and sCT (red), evaluated along the yellow line shown in (**b**). The 80% reference is marked by the horizontal line. (**b**) Corresponding CT and sCT sagittal views are compared. The yellow line is one of the 10 considered for each beam, on different transversal slices.

## 4. Discussion

In this work, we investigated for the first time the feasibility of a cGAN in generating sCTs of the abdominal site for applications in CIRT. The cGAN, trained with transversal CT-MRI slice pairs, was optimized to work on three channels corresponding to air, bone and soft tissues to better account for the anatomical complexity of the abdomen.

The performance of the network was firstly evaluated on the basis of similarity metrics of the test set built with CT-based segmentation. Despite MRI to CT inter-acquisition motion, the MAE on the body (57.08 $\pm$ 2.79 HU) was comparable to results obtained by other works on the abdominal site that used multiple Dixon sequences and more complex architectures (55.56 $\pm$ 2.27 HU [26] and 60.42 $\pm$ 2.27 HU [23]), as well as U-nets trained on T1w–T2w MRI acquisitions (62 $\pm$ 13 HU) [24]. In addition, this work favorably compares to other studies exploiting cGAN or cycleGAN in terms of $MAE_{BODY}$ [20,21,25], even if the bidirectional network from Xu et al. achieved outstanding results, although working on a much wider unpaired dataset [22]. The NCC (0.92 $\pm$ 0.02) was shown to be consistent with those of other works [23,25,26], while the PSNR (27.64 $\pm$ 0.68 dB) was comparable to the work by Fu et al. [21]. Optimal MAE metrics were obtained in the generation of bone structures (86.03 $\pm$ 10.76 HU) and soft tissues (55.39 $\pm$ 3.4 HU), outperforming other works in the literature [20,24,25,28].

The MAE on the air channel and bones was also low with respect to other approaches with cycleGAN [25]. This could be due to the use of CT-derived masks on the test set, which may have aided the replication of CT air pockets and bone structures on the sCTs.

In order to cope with this and to simulate a real-case scenario, the network was then tested on an MRI-only simulation set, where the MRI segmentation was completely independent from the use of CT. Given the limited performance of multi-modal DIR on the considered dataset, manual segmentation of the three channels on MRI was considered the most accurate approach, despite the time-consuming task for clinical purposes. Nonetheless, DIR was applied between the target CT and MRI to overcome the lack of a ground truth CT representative of MRI anatomical condition, notwithstanding the minor contribution of such registration. Indeed, this process did not fully compensate for different air cavities and inter-acquisition motion; therefore, the volumes (i.e., $CT_{PGT}$) used as reference were still showing visible discrepancies with respect to the MRI, biasing the evaluation of the performance of the network (Figure 2a). A similar consideration applied by Florkow et al. highlighted errors in HU intensities caused by inter-acquisition variations that were not compensated by the deformable registration [24]. Moreover, the manual segmentation of MRI represented a complex step, especially for bones, since they are not clearly visible on VIBE volumes. Nonetheless, the results were shown to still be acceptable when compared to the literature, with MAE on soft tissues (75.00 $\pm$ 8.12 HU) being aligned to results obtained with GAN (90 $\pm$ 29 HU) or cycleGAN (58.62 $\pm$ 30.61 HU) [25,27].

The geometrical analysis, performed on the MRI-only test set, was conducted to support the good geometrical agreement between sCT and MRI: in general, sCT and MRI segmentations showed the best match, confirming the expected performance of the net in reproducing the MRI anatomy, while sCTs showed higher deviations with respect to the target CT scans. This evaluation was performed on segmentations of the kidneys, which are well contrasted organs, and can be considered representative of the geometrical accuracy in the reproduction of soft tissues.

Due to the lack of a proper ground truth on patients' data, we supported our results with a validation performed on a single CT-MRI volume pair obtained through a computational phantom (i.e., XCAT [37] for CT and correspondent ComBAT [38] for MRI). This approach, which guaranteed the perfect anatomical match between the two volumes, provided promising outcomes (i.e., MAE of 73.3 HU on the whole volume) aligned with the presented results and the literature. The MAE on the three channels described a discrete reproduction of the air fillings and soft tissues, with poorer performance on bones, as noticed also in the literature. NCC reached 0.89, describing an acceptable reproduction of the anatomical structures (see Supplementary Material S4).

For dose analysis, most patients (Figure 4a and Supplementary Materials S6–S8) presented DVH metrics within clinical tolerances (i.e., below 3% of the prescribed dose) and aligned with studies in the literature on protons (relative error < 3% on PTV [25] and relative error < 2% on ITV—the internal target volume [24]). Moreover, *p*-values gave statistical evidence of the equivalence between the DVH metrics from sCT and CT ($p < 0.05$, Table S7 in Supplementary Material S8); however, a larger test dataset would make these considerations more solid.

Concerning DVH metrics, Liu Y. et al. for protons [25] as well as Liu L. for photons [28] obtained promising results, with errors on the maximum dose on the PTV being in the range [−1,+1] Gy[RBE], also thanks to the use of lateral beams that avoided most of the air-filled organs. Florkow et al. [24] obtained acceptable errors on the OARs, with $D_{2\%}$ being in the range [−2.7,3.7]% of the prescribed dose. Notwithstanding the use of DIR, the work by Florkow et al. suffered from inter-scan variations (i.e., air fillings), that may have caused an overestimation of the actual differences between the planning CT and sCT [24]. In our work, the IQR on the $D_{2\%}$ relative error was [−0.24,0.22]%, but high errors were highlighted on the colon (37.03%) and the duodenum (16.25%, Figure 4c), which were highly affected by inter-acquisition motion of air fillings between sCT (representative of MRI anatomy) and planning CT. In this regard, the work by Knäusl et al. [32] showed that the constraints on OARs are very challenging for compliance, presenting errors of up to 28%, mostly due to the incorrect representation of bones or air cavities. Discrepancies in the dose distributions were also confirmed by the gamma analysis, which in our work showed a peak value of 94.88 ± 4.9% against the 99.37 ± 0.99% reported in the literature [25]. Table S8 in Supplementary Material S9 shows a comparison of gamma pass rates from relevant studies on proton and photon applications, which were, therefore, not fully comparable to our application and highlighted the need for more studies on CIRT. The range shift analysis presented median RS within the clinical threshold (i.e., 5 mm), but with critical results for patients P27 and P31 (median (IQR), 5.69 (6.97) mm and 3.09 (2.60) mm, respectively), because of the presence of unmatched air pockets (Figure 5b). In proton plans, a maximum RS value of 5.6 mm (5.68%) was reported [25], whereas in our case, the inconsistencies of air cavities led to RS values of up to 15.37 mm (i.e., RRS = 9.09%, patient P27). This aspect may be critical for CIRT application, and needs to be analyzed on a wider population.

Similarly, the highest dose discrepancies were mostly found in correspondence of the different disposition of air pockets between sCT and planning CT (Figure S5 in Supplementary Material S7). The maximum error for patient P27 was due to an incomplete reproduction of the kidney (Figure S6), which caused an overdose for duodenum (+5.7% $D_{02}$) with respect to the prescribed dose (Supplementary Material S6). The dose artefacts on recalculation for patient P21 reported regions of high dose differences (Figure S7), but this issue was not correlated with the quality of the sCT and could be easily overcome by acquiring wider volumes.

The main limitation of the study was the lack of a proper ground truth to validate the proposed approach, which could not be fully compensated due to the poor performance of multi-modal DIR in the abdominal site. As such, the use of computational phantoms [37,38] to ensure the correspondence between CT and MRI scans will be considered in the future as an effective approach to validate the proposed network, as anticipated in this work (Supplementary Material S4), and enlarge the training dataset. In addition, although the three channel implementation allowed good performance of the net in such a complex anatomical site with limited data, manual segmentation can be demanding, especially for not well-contrasted structures in VIBE acquisitions, such as bone, and is definitely not suited for clinical application. Further steps could be, therefore, to include the acquisition of specific MRI sequences (e.g., ultrashort echo time, UTE) to facilitate bone segmentation or avoid channel separation in an improved version of the net [26]. We also expect that our results could be improved by increasing the dataset; in future analyses, we intend to include different respiratory phases to (i) achieve higher accuracy and limit errors in the reproduction of tissues, (ii) eliminate separation in the three channels, and (iii) derive synthetic respiratory-correlated 4DCT [10,41]. Finally, although our results were mainly affected by air-filling effects, the absence of the thermoplastic mask on the sCT could also have an impact [32]; as such, a uniform and pre-defined outline to the sCT could be applied [32], although it would not be an optimal countermeasure.

Despite the above-mentioned limitations, this work showed that the three-channel cGAN can generate accurate sCTs of the abdominal site that can support treatment planning, evaluation, and adaptation in CIRT. To the authors' best knowledge, this work is the first analysis applied to the abdomen for CIRT, and thus represents a starting point for future in-depth analyses of the feasibility of MRI-only workflows in CIRT.

**Supplementary Materials:** The following supporting information can be downloaded at: https://www.mdpi.com/article/10.3390/bioengineering10020250/s1, Figure S1: Average MAE values on the three channels, for each batch tested; Figure S2: Coronal view of the XCAT phantom (A. original, B. with gaussian noise), ComBAT phantom (C) and the resulting synthetic CT (D); Figure S3: A. Histogram representing the HD 95th percentile. B. Histogram relative to the CoM distances. C. Histogram of Dice coefficient; Figure S4: DVH graphs comparing original plan (solid line) with sCT-based recalculations (dotted line); Figure S5: CT, sCT and dose difference maps for the test patients. The transversal slice shown contains the pixel with maximum error (black dot). Red arrows show the possible causes; Figure S6: Coronal view of P27, showing an error in the reproduction of the kidney, that affects the dose distribution; Figure S7: Dose recalculation on patient P21, coronal view. The discrepancy in the MRI-CT FoV causes dose artefacts in the recalculation; Figure S8: Representative range shift analysis on patient P17. The graph shows the Dose and HU profiles evaluated along the white line, as shown on CT and sCT below. The horizontal black line represents the 80% of the peak of dose; Table S1: DIR vs No-DIR average metrics results on the test set (St. Dev.); Table S2: Cross validation results, Table S3: Results of the network validation on the phantom; Table S4: Geometrical evaluation of kidney's segmentations; Table S5: $D_{2\%}$ Relative errors on each OAR. The values are referred to the prescribed dose; Table S6: Dose difference median values [IQR] for the test patients on the whole plan; Table S7: TOST-P test of equivalence P-values; Table S8: Average Gamma pass rates (St. Dev.) of this work and literature references; Table S9: Median (IQR) RS and RRS evaluated over 10 transversal slices for each beam. Ap = Anterior-posterior. L = Lateral.

**Author Contributions:** Conceptualization, G.P., A.V., C.P., S.M., G.M. and G.B.; methodology, G.P., C.P., G.M. and N.G.; software, G.P.; validation, G.P. and F.C.; formal analysis, G.P., C.P., A.N. and N.G.; investigation, G.P.; resources, G.B.; data curation, G.P., V.V., A.B., S.M., A.V. and M.C.; writing—original draft preparation, G.P. and C.P.; writing—review and editing, A.N., N.G., M.C., A.P., S.M. and E.O.; visualization, G.P. and C.P.; supervision, C.P., G.B. and A.V.; project administration, G.B., C.P. and E.O.; funding acquisition, G.B. All authors have read and agreed to the published version of the manuscript.

**Funding:** This research was funded by Marie Skłodowska-Curie grant RAPTOR—Real-Time Adaptive Particle Therapy of Cancer, grant agreement No. 955956 and partially supported by AIRC (Associazione Italiana per la Ricerca sul Cancro), grant number IG2020-24946.

**Institutional Review Board Statement:** The study was conducted in accordance with the Declaration of Helsinki, and approved by the Institutional Review Board (or Ethics Committee) of CENTRO NAZIONALE DI ADROTERAPIA ONCOLOGICA (CNAO) (protocol code CNAO 37-2019 4D-MRI, 2019).

**Informed Consent Statement:** Informed consent was obtained from all subjects involved in the study.

**Data Availability Statement:** Data are unavailable due to privacy and ethical restrictions.

**Acknowledgments:** The authors would like to thank Luca Anemoni and Alice Mancin for patients' data collection. Anestis Nakas would like to express thanks for the support of the European Union's Horizon 2020 research and innovation program under the Marie Skłodowska-Curie grant RAPTOR—Real-Time Adaptive Particle Therapy of Cancer.

**Conflicts of Interest:** The authors declare no conflict of interest.

## References


1. Durante, M.; Orecchia, R.; Loeffler, J.S. Charged-Particle Therapy in Cancer: Clinical Uses and Future Perspectives. *Nat. Rev. Clin. Oncol.* **2017**, *14*, 483–495. [CrossRef]
2. Liermann, J.; Shinoto, M.; Syed, M.; Debus, J.; Herfarth, K.; Naumann, P. Carbon Ion Radiotherapy in Pancreatic Cancer: A Review of Clinical Data. *Radiother. Oncol.* **2020**, *147*, 145–150. [CrossRef] [PubMed]
3. Li, H.; Dong, L.; Bert, C.; Chang, J.; Flampouri, S.; Jee, K.; Lin, L.; Moyers, M.; Mori, S.; Rottmann, J.; et al. AAPM Task Group Report 290: Respiratory Motion Management for Particle Therapy. *Med. Phys.* **2022**, *49*, e50–e81. [CrossRef]
4. Jaffray, D.A. Image-Guided Radiotherapy: From Current Concept to Future Perspectives. *Nat. Rev. Clin. Oncol.* **2012**, *9*, 688–699. [CrossRef] [PubMed]
5. Otazo, R.; Lambin, P.; Pignol, J.P.; Ladd, M.E.; Schlemmer, H.P.; Baumann, M.; Hricak, H. MRI-Guided Radiation Therapy: An Emerging Paradigm in Adaptive Radiation Oncology. *Radiology* **2021**, *298*, 248–260. [CrossRef] [PubMed]
6. Keall, P.J.; Brighi, C.; Glide-Hurst, C.; Liney, G.; Liu, P.Z.Y.; Lydiard, S.; Paganelli, C.; Pham, T.; Shan, S.; Tree, A.C.; et al. Integrated MRI-Guided Radiotherapy—Opportunities and Challenges. *Nat. Rev. Clin. Oncol.* **2022**, *19*, 458–470. [CrossRef]
7. Kurz, C.; Buizza, G.; Landry, G.; Kamp, F.; Rabe, M.; Paganelli, C.; Baroni, G.; Reiner, M.; Keall, P.J.; van den Berg, C.A.T.; et al. Medical Physics Challenges in Clinical MR-Guided Radiotherapy. *Radiat. Oncol.* **2020**, *15*, 93. [CrossRef] [PubMed]
8. Hoffmann, A.; Oborn, B.; Moteabbed, M.; Yan, S.; Bortfeld, T.; Knopf, A.; Fuchs, H.; Georg, D.; Seco, J.; Spadea, M.F.; et al. MR-Guided Proton Therapy: A Review and a Preview. *Radiat. Oncol.* **2020**, *15*, 129. [CrossRef]
9. Paganelli, C.; Meschini, G.; Molinelli, S.; Riboldi, M.; Baroni, G. Patient-Specific Validation of Deformable Image Registration in Radiation Therapy: Overview and Caveats. *Med. Phys.* **2018**, *45*, e908–e922. [CrossRef] [PubMed]
10. Meschini, G.; Vai, A.; Paganelli, C.; Molinelli, S.; Maestri, D.; Fontana, G.; Pella, A.; Vitolo, V.; Valvo, F.; Ciocca, M.; et al. Investigating the Use of Virtual 4DCT from 4DMRI in Gated Carbon Ion Radiation Therapy of Abdominal Tumors. *Z. Med. Phys.* **2022**, *32*, 98–108. [CrossRef]
11. Johnstone, E.; Wyatt, J.J.; Henry, A.M.; Short, S.C.; Sebag-Montefiore, D.; Murray, L.; Kelly, C.G.; McCallum, H.M.; Speight, R. Systematic Review of Synthetic Computed Tomography Generation Methodologies for Use in Magnetic Resonance Imaging–Only Radiation Therapy. *Int. J. Radiat. Oncol. Biol. Phys.* **2018**, *100*, 199–217. [CrossRef]
12. Han, X. MR-Based Synthetic CT Generation Using a Deep Convolutional Neural Network Method. *Med. Phys.* **2017**, *44*, 1408–1419. [CrossRef]
13. Dinkla, A.M.; Florkow, M.C.; Maspero, M.; Savenije, M.H.F.; Zijlstra, F.; Doornaert, P.A.H.; van Stralen, M.; Philippens, M.E.P.; van den Berg, C.A.T.; Seevinck, P.R. Dosimetric Evaluation of Synthetic CT for Head and Neck Radiotherapy Generated by a Patch-Based Three-Dimensional Convolutional Neural Network. *Med. Phys.* **2019**, *46*, 4095–4104. [CrossRef]
14. Maspero, M.; Savenije, M.H.F.; Dinkla, A.M.; Seevinck, P.R.; Intven, M.P.W.; Jurgenliemk-Schulz, I.M.; Kerkmeijer, L.G.W.; van den Berg, C.A.T. Dose Evaluation of Fast Synthetic-CT Generation Using a Generative Adversarial Network for General Pelvis MR-Only Radiotherapy. *Phys. Med. Biol.* **2018**, *63*, 185001. [CrossRef]
15. Spadea, M.F.; Pileggi, G.; Zaffino, P.; Salome, P.; Catana, C.; Izquierdo-Garcia, D.; Amato, F.; Seco, J. Deep Convolution Neural Network (DCNN) Multiplane Approach to Synthetic CT Generation From MR Images—Application in Brain Proton Therapy. *Int. J. Radiat. Oncol. Biol. Phys.* **2019**, *105*, 495–503. [CrossRef]
16. Kazemifar, S.; Montero, A.M.B.; Souris, K.; Rivas, S.T.; Timmerman, R.; Park, Y.K.; Jiang, S.; Geets, X.; Sterpin, E.; Owrangi, A. Dosimetric Evaluation of Synthetic CT Generated with GANs for MRI-Only Proton Therapy Treatment Planning of Brain Tumors. *J. Appl. Clin. Med. Phys.* **2020**, *21*, 76–86. [CrossRef]
17. Koerkamp, M.L.G.; de Hond, Y.J.M.; Maspero, M.; Kontaxis, C.; Mandija, S.; Vasmel, J.E.; Charaghvandi, R.K.; Philippens, M.E.P.; van Asselen, B.; van den Bongard, H.J.G.D.; et al. Synthetic CT for Single-Fraction Neoadjuvant Partial Breast Irradiation on an MRI-Linac. *Phys. Med. Biol.* **2021**, *66*, 085010. [CrossRef]
18. Spadea, M.F.; Maspero, M.; Zaffino, P.; Seco, J. Deep Learning Based Synthetic-CT Generation in Radiotherapy and PET: A Review. *Med. Phys.* **2021**, *48*, 6537–6566. [CrossRef]
19. Isola, P.; Zhu, J.; Efros, A.A.; Ai, B.; Berkeley, U.C. Image-to-Image Translation with Conditional Adversarial Networks. In Proceedings of the IEEE Conference on Computer Vision and Pattern Recognition, Honolulu, HI, USA, 21–26 July 2017.
20. Cusumano, D.; Lenkowicz, J.; Votta, C.; Boldrini, L.; Placidi, L.; Catucci, F.; Dinapoli, N.; Antonelli, M.V.; Romano, A.; de Luca, V.; et al. A Deep Learning Approach to Generate Synthetic CT in Low Field MR-Guided Adaptive Radiotherapy for Abdominal and Pelvic Cases. *Radiother. Oncol.* **2020**, *153*, 205–212. [CrossRef]
21. Fu, J.; Singhrao, K.; Cao, M.; Yu, V.; Anand, P.S.; Yang, Y.; Guo, M.; Raldow, A.C.; Ruan, D.; Lewis, J.H. Generation of Abdominal Synthetic CTs from 0.35T MR Images Using Generative Adversarial Networks for MR-Only Liver Radiotherapy. *Biomed. Phys. Eng. Express* **2020**, *6*, 015033. [CrossRef]
22. Xu, L.; Zeng, X.; Zhang, H.; Li, W.; Lei, J.; Huang, Z. BPGAN: Bidirectional CT-to-MRI Prediction Using Multi-Generative Multi-Adversarial Nets with Spectral Normalization and Localization. *Neural Netw.* **2020**, *128*, 82–96. [CrossRef]
23. Xu, K.; Cao, J.; Xia, K.; Yang, H.; Zhu, J.; Wu, C.; Jiang, Y.; Qian, P. Multichannel Residual Conditional GAN-Leveraged Abdominal Pseudo-CT Generation via Dixon MR Images. *IEEE Access* **2019**, *7*, 163823–163830. [CrossRef]
24. Florkow, M.C.; Guerreiro, F.; Zijlstra, F.; Seravalli, E.; Janssens, G.O.; Maduro, J.H.; Knopf, A.C.; Castelein, R.M.; van Stralen, M.; Raaymakers, B.W.; et al. Deep Learning-Enabled MRI-Only Photon and Proton Therapy Treatment Planning for Paediatric Abdominal Tumours. *Radiother. Oncol.* **2020**, *153*, 220–227. [CrossRef] [PubMed]

25. Liu, Y.; Lei, Y.; Wang, Y.; Wang, T.; Ren, L.; Lin, L.; McDonald, M.; Curran, W.J.; Liu, T.; Zhou, J.; et al. MRI-Based Treatment Planning for Proton Radiotherapy: Dosimetric Validation of a Deep Learning-Based Liver Synthetic CT Generation Method. *Phys. Med. Biol.* **2019**, *64*, 145015. [CrossRef] [PubMed]
26. Qian, P.; Xu, K.; Wang, T.; Zheng, Q.; Yang, H.; Baydoun, A.; Zhu, J.; Traughber, B.; Muzic, R.F. Estimating CT from MR Abdominal Images Using Novel Generative Adversarial Networks. *J. Grid Comput.* **2020**, *18*, 211–226. [CrossRef]
27. Olberg, S.; Chun, J.; Choi, B.S.; Park, I.; Kim, H.; Kim, T.; Kim, J.S.; Green, O.; Park, J.C. Abdominal Synthetic CT Reconstruction with Intensity Projection Prior for MRI-Only Adaptive Radiotherapy. *Phys. Med. Biol.* **2021**, *66*, 204001. [CrossRef]
28. Liu, L.; Johansson, A.; Cao, Y.; Dow, J.; Lawrence, T.; Balter, J. Abdominal Synthetic CT Generation from MR Dixon Images Using a U-Net Trained with 'Semi-Synthetic' CT Data. *Phys. Med. Biol.* **2020**, *65*, 125001. [CrossRef] [PubMed]
29. Kumagai, M.; Hara, R.; Mori, S.; Yanagi, T.; Asakura, H.; Kishimoto, R.; Kato, H.; Yamada, S.; Kandatsu, S.; Kamada, T. Impact of Intrafractional Bowel Gas Movement on Carbon Ion Beam Dose Distribution in Pancreatic Radiotherapy. *Int. J. Radiat. Oncol. Biol. Phys.* **2009**, *73*, 1276–1281. [CrossRef] [PubMed]
30. Rietzel, E.; Schardt, D.; Haberer, T. Range Accuracy in Carbon Ion Treatment Planning Based on CT-Calibration with Real Tissue Samples. *Radiat. Oncol.* **2007**, *2*, 14. [CrossRef] [PubMed]
31. Witt, M.; Weber, U.; Kellner, D.; Engenhart-Cabillic, R.; Zink, K. Optimization of the Stopping-Power-Ratio to Hounsfield-Value Calibration Curve in Proton and Heavy Ion Therapy. *Z. Med. Phys.* **2015**, *25*, 251–263. [CrossRef]
32. Knäusl, B.; Kuess, P.; Stock, M.; Georg, D.; Fossati, P.; Georg, P.; Zimmermann, L. Possibilities and Challenges When Using Synthetic Computed Tomography in an Adaptive Carbon-Ion Treatment Workflow. *Z. Med. Phys.* **2022**. [CrossRef]
33. Tustison, N.J.; Avants, B.B.; Cook, P.A.; Zheng, Y.; Egan, A.; Yushkevich, P.A.; Gee, J.C. N4ITK: Improved N3 Bias Correction. *IEEE Trans. Med. Imaging* **2010**, *29*, 1310–1320. [CrossRef]
34. Udupa, J.K. On Standardizing the MR Image Intensity Scale. *Magn. Reson. Med. Off. J. Int. Soc. Magn. Reson. Med.* **1999**, *42*, 1072–1081.
35. Gupta, D.; Kim, M.; Vineberg, K.A.; Balter, J.M. Generation of Synthetic CT Images from MRI for Treatment Planning and Patient Positioning Using a 3-Channel U-Net Trained on Sagittal Images. *Front. Oncol.* **2019**, *9*, 964. [CrossRef]
36. Ulyanov, D.; Vedaldi, A.; Lempitsky, V. Instance Normalization: The Missing Ingredient for Fast Stylization. *arXiv* **2016**, arXiv:1607.08022.
37. Segars, W.P.; Sturgeon, G.; Mendonca, S.; Grimes, J.; Tsui, B.M.W. 4D XCAT Phantom for Multimodality Imaging Research. *Med. Phys.* **2010**, *37*, 4902–4915. [CrossRef]
38. Paganelli, C.; Summers, P.; Gianoli, C.; Bellomi, M.; Baroni, G.; Riboldi, M. A Tool for Validating MRI-Guided Strategies: A Digital Breathing CT/MRI Phantom of the Abdominal Site. *Med. Biol. Eng. Comput.* **2017**, *55*, 2001–2014. [CrossRef]
39. Pileggi, G.; Speier, C.; Sharp, G.C.; Izquierdo Garcia, D.; Catana, C.; Pursley, J.; Amato, F.; Seco, J.; Spadea, M.F. Proton Range Shift Analysis on Brain Pseudo-CT Generated from T1 and T2 MR. *Acta Oncol.* **2018**, *57*, 1521–1531. [CrossRef]
40. Vitolo, V.; Cobianchi, L.; Brugnatelli, S.; Barcellini, A.; Peloso, A.; Facoetti, A.; Vanoli, A.; Delfanti, S.; Preda, L.; Molinelli, S.; et al. Preoperative Chemotherapy and Carbon Ions Therapy for Treatment of Resectable and Borderline Resectable Pancreatic Adenocarcinoma: A Prospective, Phase II, Multicentre, Single-Arm Study. *BMC Cancer* **2019**, *19*, 922. [CrossRef]
41. Meschini, G.; Vai, A.; Paganelli, C.; Molinelli, S.; Fontana, G.; Pella, A.; Preda, L.; Vitolo, V.; Valvo, F.; Ciocca, M.; et al. Virtual 4DCT from 4DMRI for the Management of Respiratory Motion in Carbon Ion Therapy of Abdominal Tumors. *Med. Phys.* **2020**, *47*, 909–916. [CrossRef]

bioengineering

MDPI

Article

# Comparing 3D, 2.5D, and 2D Approaches to Brain Image Auto-Segmentation

Arman Avesta [1,2,3], Sajid Hossain [2,3], MingDe Lin [1,4], Mariam Aboian [1], Harlan M. Krumholz [3,5] and Sanjay Aneja [2,3,6,*]

[1] Department of Radiology and Biomedical Imaging, Yale School of Medicine, New Haven, CT 06510, USA
[2] Department of Therapeutic Radiology, Yale School of Medicine, New Haven, CT 06510, USA
[3] Center for Outcomes Research and Evaluation, Yale School of Medicine, New Haven, CT 06510, USA
[4] Visage Imaging, Inc., San Diego, CA 92130, USA
[5] Division of Cardiovascular Medicine, Yale School of Medicine, New Haven, CT 06510, USA
[6] Department of Biomedical Engineering, Yale University, New Haven, CT 06510, USA
[*] Correspondence: sanjay.aneja@yale.edu; Tel.: +1-203-200-2100; Fax: +1-203-737-1467

**Abstract:** Deep-learning methods for auto-segmenting brain images either segment one slice of the image (2D), five consecutive slices of the image (2.5D), or an entire volume of the image (3D). Whether one approach is superior for auto-segmenting brain images is not known. We compared these three approaches (3D, 2.5D, and 2D) across three auto-segmentation models (capsule networks, UNets, and nnUNets) to segment brain structures. We used 3430 brain MRIs, acquired in a multi-institutional study, to train and test our models. We used the following performance metrics: segmentation accuracy, performance with limited training data, required computational memory, and computational speed during training and deployment. The 3D, 2.5D, and 2D approaches respectively gave the highest to lowest Dice scores across all models. 3D models maintained higher Dice scores when the training set size was decreased from 3199 MRIs down to 60 MRIs. 3D models converged 20% to 40% faster during training and were 30% to 50% faster during deployment. However, 3D models require 20 times more computational memory compared to 2.5D or 2D models. This study showed that 3D models are more accurate, maintain better performance with limited training data, and are faster to train and deploy. However, 3D models require more computational memory compared to 2.5D or 2D models.

**Keywords:** auto-segmentation; deep learning; neuroimaging; magnetic resonance imaging

**Citation:** Avesta, A.; Hossain, S.; Lin, M.; Aboian, M.; Krumholz, H.M.; Aneja, S. Comparing 3D, 2.5D, and 2D Approaches to Brain Image Auto-Segmentation. *Bioengineering* **2023**, *10*, 181. https://doi.org/10.3390/bioengineering10020181

Academic Editors: Paolo Zaffino and Maria Francesca Spadea

Received: 4 November 2022
Revised: 9 January 2023
Accepted: 9 January 2023
Published: 1 February 2023



## 1. Introduction

Segmentation of brain magnetic resonance images (MRIs) has widespread applications in the management of neurological disorders [1–3]. In patients with neurodegenerative disorders, segmenting brain structures such as the hippocampus provides quantitative information about the amount of brain atrophy [4]. In patients undergoing radiotherapy, segmentation is used to demarcate important brain structures that should be avoided to limit potential radiation toxicity [5]. Pre-operative or intra-operative brain MRIs are often used to identify important brain structures that should be avoided during neurosurgery [6,7]. Manual segmentation of brain structures on these MR images is a time-consuming task that is prone to intra- and inter-observer variability [8]. As a result, deep learning auto-segmentation methods have been increasingly used to efficiently segment important anatomical structures on brain MRIs [9].

Compared to two-dimensional (2D) auto-segmentation tasks, the three-dimensional (3D) nature of brain MRIs makes auto-segmentation considerably more challenging. There have been three proposed approaches to handling auto-segmentation of 3D images: (1) analyze and segment a two-dimensional slice of the image at a time (2D), [10] (2) analyze five consecutive two-dimensional slices at a time to generate a segmentation of the middle slice (2.5D), [11] and (3) analyze and segment the image volume in three-dimensional space (3D) [10]. Although each approach has shown some promise in medical image segmentation, a comprehensive comparison and benchmarking of these approaches for auto-segmentation of brain MRIs is lacking. Prior studies on comparing these auto-segmentation approaches have often not evaluated their efficacy in segmenting brain MRIs, or have limited their comparison narrowly to one deep learning architecture [10,12–14]. Additionally, previous studies have focused primarily on segmentation accuracy and failed to evaluate more practical metrics such as computational efficiency or accuracy in data-limited settings. As a result, it is difficult for clinicians and researchers to easily choose the appropriate auto-segmentation method for a desired clinical task. There is a need to compare and benchmark these three approaches for brain MRI auto-segmentation across different models and using comprehensive performance metrics.

In this study, we comprehensively compared 3D, 2.5D, and 2D approaches to brain MRI auto-segmentation across three different deep learning architectures and used metrics of accuracy and computational efficiency. We used a multi-institutional cohort of 3430 brain MRIs to train and test our models, and evaluated the efficacy of each approach across three clinically-relevant anatomical structures of the brain.

## 2. Methods

### 2.1. Dataset

This study used a dataset of 3430 T1-weighted brain MR images belonging to 841 patients from 19 institutions enrolled in the Alzheimer's Disease Neuroimaging Initiative (ADNI) study [15]. The inclusion and exclusion criteria of ADNI have been previously described [16]. On average, each patient underwent four MRI acquisitions. Each patient underwent MR imaging using a single scanner at each site. However, the diversity of scanners in all study sites included nine different types of MR scanners. Supplementary Material S1 describes the details of MRI acquisition parameters. We downloaded the anonymized MRIs of these patients from Image and Data Archive, which is a data-sharing platform [15]. The patients were randomly split into training (3199 MRIs, 93% of data), validation (117 MRIs, 3.5% of data), and test (114 MRIs, 3.5% of data) sets at the patient level. Therefore, all images belonging to a patient were assigned to either the training, validation, or test set. Table 1 summarizes patient demographics. For external validation, we additionally trained and tested a subset of our models on a dataset that contains 400 images of right and left hippocampi. The details of these experiments are provided in Supplementary Material S2.

**Table 1.** Study participants tabulated by the training, validation, and test sets.

| Data Partitions | Number of MRIs | Number of Patients | Age (Mean ± SD) | Gender † | Diagnosis †† |
|---|---|---|---|---|---|
| Training set | 3199 | 841 | $76 \pm 7$ | 42% F, 58% M | 29% CN, 54% MCI, 17% AD |
| Validation set | 117 | 30 | $75 \pm 6$ | 30% F, 70% M | 21% CN, 59% MCI, 20% AD |
| Test set | 114 | 30 | $77 \pm 7$ | 33% F, 67% M | 27% CN, 47% MCI, 26% AD |

† F: female; M: male. †† CN: cognitively normal; MCI: mild cognitive impairment; AD: Alzheimer's disease.

### 2.2. Anatomic Segmentations

We trained our models to segment three representative structures of the brain: the third ventricle, thalamus, and hippocampus. These structures represent varying degrees of segmentation difficulty: the third ventricle is an easy structure to segment because it is filled with cerebrospinal fluid (CSF) with a distinct image contrast compared to surrounding structures; the thalamus is a medium-difficulty structure because it is bounded by CSF on one side and is bounded by white matter on the other side, and the hippocampus is a difficult structure because it has a complex shape and is neighbored by multiple brain structures with different image contrasts. Preliminary ground-truth segmentations were initially generated by FreeSurfer [4,17,18], and were manually corrected by a board-eligible radiologist (AA).

### 2.3. Image Pre-Processing

MRI preprocessing included corrections for B1-field variations as well as intensity inhomogeneities [19,20]. The 3D brain image was cropped around the brain after removing the skull, face, and neck tissues [21]. The input to the 3D capsule networks and 3D UNets were image patches sized 64 × 64 × 64 voxels. The inputs to the 2.5D capsule networks and 2.5D UNets were five consecutive slices of the image. The inputs to the 2D capsule networks and 2D UNets were one slice of the image. The inputs to the 3D and 2D nnUNet models were respectively 3D and 2D patches of the images with self-configured patch sizes that were automatically set by the nnUNet paradigm [22]. Supplementary Material S3 describes the details of pre-processing.

### 2.4. Auto-Segmentation Models

We compared the 3D, 2.5D, and 2D approaches (Figure 1) across three segmentation models: capsule networks (CapsNets) [23], UNets [24], and nnUNets [22]. These models are considered the highest-performing auto-segmentation models in the biomedical domain [9,22,23,25–29]. The 3D models process a 3D patch of the image as input, all feature maps and parameter tensors in all layers are 3D, and the model output is the segmented 3D patch of the image. Conversely, 2D models process a 2D slice of the image as input, all feature maps and parameter tensors in all layers are 2D, and the model output is the segmented 2D slice of the image. The 2.5D models process five consecutive slices of the image as input channels. The remaining parts of the 2.5D model, including the feature maps and parameter tensors, are 2D, and the model output is the segmented 2D middle slice among the five slices. We did not develop 2.5D nnUNets, because the self-configuring paradigm of nnUNets was developed for 3D and 2D inputs but not for 2.5D inputs. Notably, the aim of training and testing nnUNets (in addition to UNets) was to ensure that our choices of hyperparameters did not cause one approach (such as 3D) to perform better than other approaches. The nnUNet can self-configure the best hyperparameters for the 3D and 2D approaches but not for the 2.5D approach. As a result, we did not train or test 2.5D nnUNets. The model architectures are described in Supplementary Material S4.

### 2.5. Training

We trained the CapsNet and UNet models for 50 epochs using Dice loss and the Adam optimizer [30]. Initial learning rate was set at 0.002. We used dynamic paradigms for learning rate scheduling, with a minimal learning rate of 0.0001. The hyperparameters for our CapsNet and UNet models were chosen based on the model with the lowest Dice loss over the validation set. The hyperparameters for the nnUNet model were self-configured by the model [22]. Supplementary Material S5 describes the training hyperparameters for CapsNet and UNet.

### 2.6. Performance Metrics

For each model (CapsNet, UNet, and nnUNet), we compared the performance of 3D, 2.5D, and 2D approaches using the following metrics: (1) Segmentation accuracy: we used the Dice score to quantify the segmentation accuracy of the fully trained models over the test set.31 We compared Dice scores between the three approaches for three representative anatomic structures of the brain: the third ventricle, thalamus, and hippocampus. The mean Dice scores for the auto-segmentation of these brain structures are reported together with their 95% confidence interval. To compute the 95% confidence interval for each Dice score, we used bootstrapping to sample the 114 Dice scores over the test set, with replacement, 1000 times. We then calculated the mean Dice score for each of the 1000 samples, giving us 1000 mean Dice scores. We then sorted these mean Dice scores and found the range that covered 95% of them, which is equivalent to the 95% confidence interval for each Dice score. (2) Performance when training data is limited: we trained the models using the complete training set and random subsets of the training set with 600, 240, 120, and 60 MR images. The models trained on these subsets were then evaluated over the test set. (3) Computational speed during training: we compared the time needed to train the 3D, 2.5D, and 2D models per training example per epoch until the model converged. (4) Computational speed for segmenting an MR image: we compared how quickly each of the 3D, 2.5D, and 2D models segment one brain MRI volume. (5) Computational memory: we compared how much GPU memory is required, in units of megabytes, to train and deploy each of the 3D, 2.5D, and 2D models.

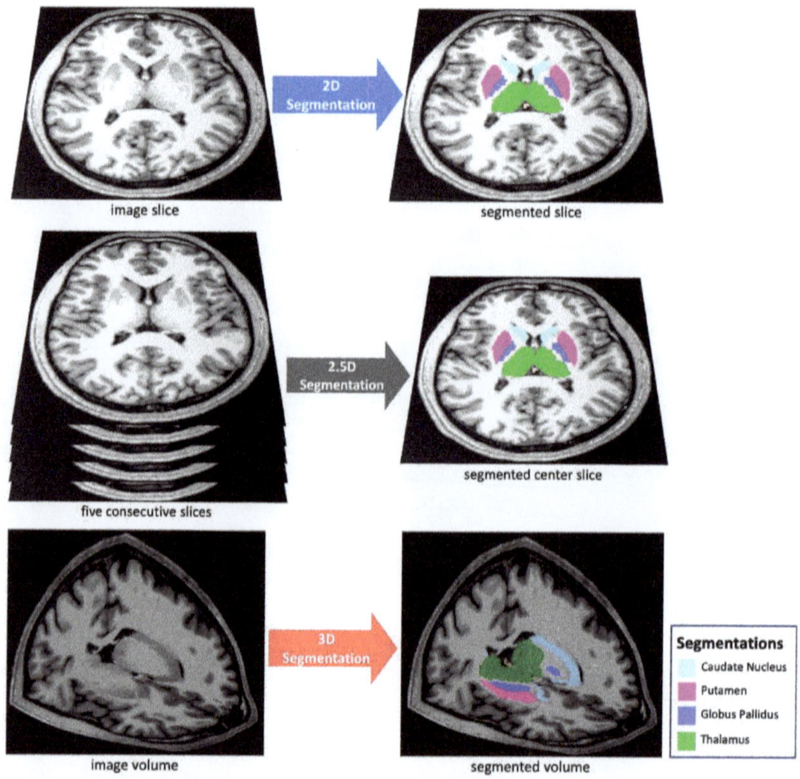


**Figure 1.** We compared three segmentation approaches: 3D, 2.5D, and 2D. The 2D approach analyzes and segments one slice of the image, the 2.5D approach analyzes five consecutive slices of the image to segment the middle slice, and the 3D approach analyzes and segments a 3D volume of the image.

### 2.7. Implementation

Image pre-processing was carried out using Python (version 3.10) and FreeSurfer (version 7). PyTorch (version 1.12) was used for model development and testing. Training and testing of the models were run on GPU-equipped servers (4 vCPUs, 16 GB RAM, 16 GB NVIDIA GPU). The code used to train and test our models is available on our lab's GitHub page: https://github.com/Aneja-Lab-Yale/Aneja-Lab-Public-3D2D-Segmentation (accessed on 4 November 2022).

## 3. Results

The segmentation accuracy of the 3D approach across all models and all anatomic structures of the brain was higher than that of the 2.5D or 2D approaches, with Dice scores of the 3D models above 90% for all anatomic structures (Table 2). Within the 3D approach, all models (CapsNet, UNet, and nnUNet) performed similarly in segmenting each anatomic structure, with their Dice scores within 1% of each other. For instance, the Dice scores of 3D CapsNet, UNet, and nnUNet in segmenting the hippocampus were respectively 92%, 93%, and 93%. Figure 2 shows auto-segmented brain structures in one patient using the three approaches. Likewise, our experiments using the external hippocampus dataset showed that 3D nnUNets achieved higher Dice scores compared to 2D nnUNets. Supplementary Material S2 details the results of our experiments with the external hippocampus dataset.

3D models maintained higher accuracy, compared to 2.5D and 2D models, when training data were limited (Figure 3). When we trained the 3D, 2.5D, and 2D CapsNets using the full training set containing 3199 MRIs, their Dice scores in segmenting the third ventricle were respectively 95%, 90%, and 90%. When we trained the same models on smaller subsets of the training set containing 600, 240, 120, and 60 MRIs, the performance of 3D, 2.5D, and 2D CapsNets gradually decreased down to 90%, 88%, and 87% for the 3D, 2.5D, and 2D CapsNets, respectively (Figure 3). Importantly, the 3D CapsNet maintained higher Dice scores (over the test set) compared to 2.5D or 2D CapsNets in all these experiments. Similarly, when we trained 3D, 2.5D, and 2D UNets using the full training set, their Dice scores in segmenting the third ventricle were respectively 96%, 91%, and 90%. Decreasing the size of the training set down to 60 MRIs resulted in Dice scores of 90%, 88%, and 87% for the 3D, 2.5D, and 2D UNets, respectively. Again, the 3D UNet maintained higher Dice scores compared to 2.5D or 2D UNets in all these experiments. Lastly, when we trained 3D and 2D nnUNets using the full training set, their Dice scores in segmenting the third ventricle were respectively 96% and 90%. Decreasing the size of the training set down to 60 MRIs resulted in Dice scores of 92% and 87% for the 3D and 2D nnUNets, respectively. Once more, the 3D nnUNet maintained higher Dice scores compared to the 2D nnUNet in all these experiments (Figure 3).

The 3D models trained faster compared to 2.5D or 2D models (Figure 4). The 3D, 2.5D, and 2D CapsNets respectively took 0.8, 1, and 1 s per training example per epoch to converge during training. The 3D, 2.5D, and 2D UNets respectively took 1.6, 2.2 and 2.9 s per training example per epoch to converge during training. The 3D and 2D nnUNets respectively took 2 and 2.9 s per training example per epoch to converge during training. Therefore, 3D models converged 20% to 40% faster compared to 2.5D or 2D models. Supplementary Material S6 also compares total convergence times between the 3D, 2.5D, and 2D approaches.

**Table 2.** Comparing the segmentation accuracy of 3D, 2.5D, and 2D approaches across three auto-segmentation models to segment brain structures. The three auto-segmentation models included CapsNet, UNet, and nnUNet. These models were used to segment three representative brain structures: third ventricle, thalamus, and hippocampus, which respectively represent easy, medium, and difficult structures to segment. The segmentation accuracy was quantified by Dice scores over the test (114 brain MRIs).

| | **CapsNet** | | |
|---|---|---|---|
| **Brain Structure** | **3D Dice (95% CI)** | **2.5D Dice (95% CI)** | **2D Dice (95% CI)** |
| 3rd ventricle | 95% (94 to 96) | 90% (89 to 91) | 90% (88 to 92) |
| Thalamus | 94% (93 to 95) | 76% (74 to 78) | 75% (72 to 78) |
| Hippocampus | 92% (91 to 93) | 73% (71 to 75) | 71% (68 to 74) |
| | **UNet** | | |
| **Brain Structure** | **3D Dice (95% CI)** | **2.5D Dice (95% CI)** | **2D Dice (95% CI)** |
| 3rd ventricle | 96% (95 to 97) | 92% (91 to 93) | 91% (89 to 91) |
| Thalamus | 95% (94 to 96) | 92% (91 to 93) | 90% (88 to 92) |
| Hippocampus | 93% (92 to 94) | 86% (84 to 88) | 88% (86 to 90) |
| **nnUNet** | **nnUNet** | **nnUNet** | **nnUNet** |
| **Brain Structure** | **Brain Structure** | **Brain Structure** | **Brain Structure** |
| 3rd ventricle | 3rd ventricle | 3rd ventricle | 3rd ventricle |
| Thalamus | Thalamus | Thalamus | Thalamus |
| Hippocampus | Hippocampus | Hippocampus | Hippocampus |

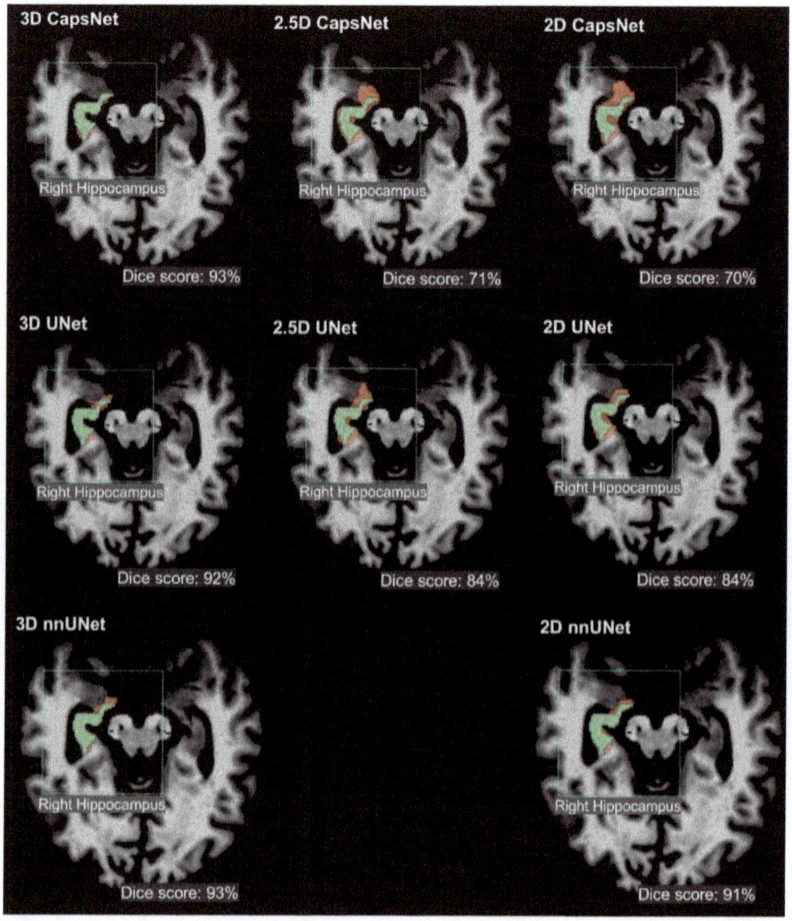


**Figure 2.** Examples of 3D, 2.5D, and 2D segmentations of the right hippocampus by CapsNet, UNet, and nnUNet. Target segmentations and model predictions are respectively shown in green and red. Dice scores are provided for the entire volume of the right hippocampus in this patient (who was randomly chosen from the test set).

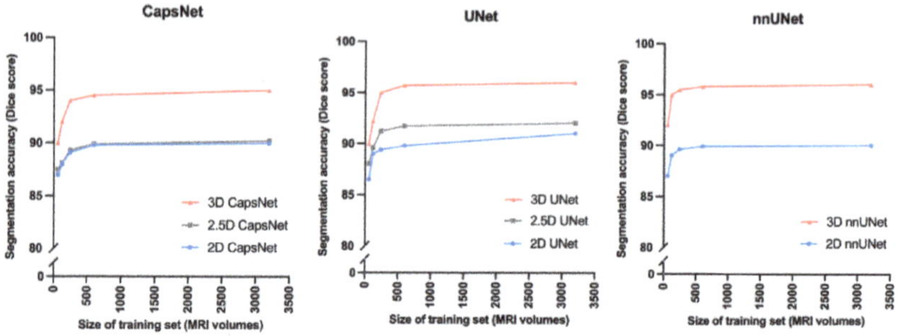


(a) (b) (c)

**Figure 3.** Comparing 3D, 2.5D, and 2D approaches when training data is limited. As we decreased the size of the training set from 3000 MRIs down to 60 MRIs, the CapsNet (**a**), UNet (**b**), and nnUNet (**c**) models maintained higher segmentation accuracy (measured by Dice scores).

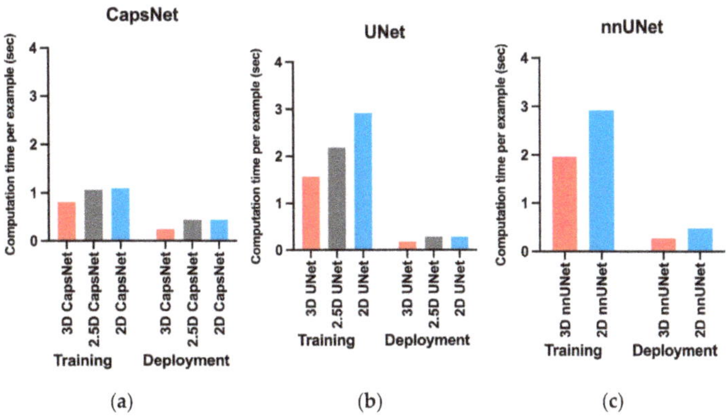


**Figure 4.** Comparing computational time required by 3D, 2.5D, and 2D approaches to train and deploy auto-segmentation models. The training times represent how much time it would take per training example per epoch for the model to converge. The deployment times represent how much time each model would require to segment one brain MRI volume. The 3D approach trained and deployed faster across all auto-segmentation models, including CapNet (**a**), UNet (**b**), and nnUNet (**c**).

Fully-trained 3D models could segment brain MRIs faster during deployment compared to 2.5D or 2D models (Figure 4). Fully-trained 3D, 2.5D, and 2D CapsNets could respectively segment a brain MRI in 0.2, 0.4, and 0.4 s. Fully-trained 3D, 2.5D, and 2D UNets could respectively segment a brain MRI in 0.2, 0.3, and 0.3 s. Lastly, fully-grained 3D and 2D nnUNets could respectively segment a brain MRI in 0.3 and 0.5 s. Therefore, fully-trained 3D models segmented a brain MRI 30% to 50% faster compared to fully-trained 2.5D or 2D models.

The 3D models needed more computational memory to train and deploy as compared to the 2.5D or 2D models (Figure 5). The 3D, 2.5D, and 2D CapsNets respectively required 317, 19, and 19 megabytes of GPU memory during training. The 3D, 2.5D, and 2D UNets respectively required 3150, 180, and 180 megabytes of GPU memory. The 3D and 2D nnUNets respectively required 3200 and 190 megabytes of GPU memory. Therefore, 3D models required about 20 times more GPU memory compared to 2.5D or 2D models. Notably, CapsNets required 10 times less GPU memory compared to UNets or nnUNets. Therefore, 3D CapsNets only required two times more GPU memory compared to 2.5D or 2D UNets or nnUNets (Figure 5).

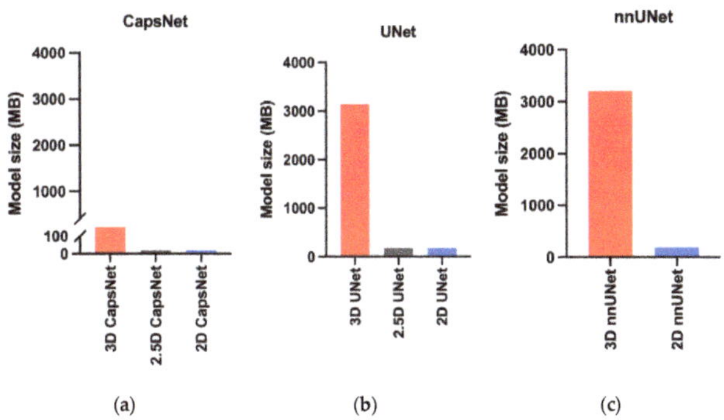


**Figure 5.** Comparing the memory required by the 3D, 2.5D, and 2D approaches. The bars represent the computational memory required to accommodate the total size of each model, including the parameters plus the cumulative size of the forward- and backward-pass feature volumes. Within each auto-segmentation model including the CapsNet (**a**), UNet (**b**), and nnUNet (**c**), the 3D approach required 20 times more computational memory compared to the 2.5D or 2D approaches.

## 4. Discussion

In this study, we compared the 3D, 2.5D, and 2D approaches of auto-segmentation across three different deep learning architectures, and found that the 3D approach is more accurate, faster to train, and faster to deploy. Moreover, the 3D auto-segmentation approach maintained better performance in the setting of limited training data. We found the major disadvantage of 3D auto-segmentation approaches to be increased computational memory requirement compared to similar 2.5D and 2D auto-segmentation approaches.

Our results extend the prior literature [10,12,13,31–34] in key ways. We provide the first comprehensive benchmarking of 3D, 2.5D, and 2D approaches in auto-segmenting of brain MRIs, measuring both accuracy and computational efficiency. We compared 3D, 2.5D, and 2D approaches across three of the most successful auto-segmentation models to date, namely capsule networks, UNets, and nnUNets [22,23,26,30,33–36]. Our findings provide a practical comparison of these three auto-segmentation approaches that can provide insight when attempting auto-segmentation in settings where computational resources are bounded or when the training data are limited.

We found that the 3D approach to auto-segmentation trains faster and deploys more quickly. Previous studies that compared the computational speed of 3D and 2D UNets have concluded conflicting results: some suggested that 2D models converge faster, [10,13,32], whereas others suggested that 3D models converge faster [22]. Notably, one training iteration of 2.5D or 2D models is faster than 3D models because 2.5D and 2D models have 20 times fewer trainable parameters compared to 3D models. However, feeding a 3D image volume into a 2.5D or 2D model requires a for loop that iterates through multiple slices of the image, which slows down 2.5D and 2D models. Additionally, 3D models can converge faster during training because they can use the contextual information in the 3D image volume to segment each structure [10]. Conversely, 2.5D models can only use the contextual information in a few slices of the image [11], and 2D models can only use the contextual information in one slice only [12]. Since the 3D approach provides more contextual information for each segmentation target, the complex shape of structures such as the hippocampus can be learned faster, and, as a result, the convergence of 3D models can become faster. Lastly, each training iteration through a 3D model can be accelerated by larger GPU memory, since the training of learnable parameters can be parallelized. However, each training iteration through a 2.5D or 2D model cannot be accelerated by larger GPU memory because iterations through the slices of the image (for loop) cannot be parallelized. We propose that our findings, that 3D models converge faster, resulted from using state-of-the-art GPUs and efficient 3D models that learn contextual information in the 3D volume of the MR image faster. Our results also show that the 3D models are faster during deployment since they can process the 3D volume of the image at once, while 2.5D or 2D models must loop through 2D image slices.

Our results do highlight one of the drawbacks of 3D auto-segmentation approaches. Specifically, we found that within each model, the 3D approach requires 20 times more computational memory compared to the 2.5D or 2D approaches. Previous studies that compared 3D and 2D UNets have found similar results [10,31]. This seems to be the only downside of the 3D approach compared to the 2.5D or 2D approaches. Notably, the 2.5D approach was initially developed to achieve segmentation accuracy similar to the 3D approach while requiring computational resources similar to the 2D approach. In brain image segmentation, however, our results show that the 2.5D approach could not achieve the segmentation accuracy of the 3D approach. This raises the question of which approach to use when computational memory is limited. Our results show that *3D CapsNets* outperformed all 2.5D and 2D models while only requiring twice more computational memory than the 2.5D or 2D UNets or nnUNets. Conversely, 3D UNets and nnUNets required 20 times more computational memory compared to 2.5D or 2D UNets and nnUNets. Therefore, 3D CapsNets may be preferred in settings where computational memory is limited.

Our results corroborate previous studies showing that deep learning is accurate in biomedical image auto-segmentation [9,22,26–29]. Prior studies have shown that capsule networks, UNets, and nnUNets are the most accurate models to auto-segment biomedical images [9,11,22,23,25,26,28,33,34,36–38]. Prior studies have also shown that the 3D, 2.5D, and 2D versions of these models can auto-segment medical images [9,11,22,23,28,29,34]. However, evidence was lacking about which of the 3D, 2.5D, or 2D approaches would be more accurate in auto-segmenting brain structures on MR images. Our results also provide practical benchmarking of computational efficiency between the three approaches, which is often under-reported.

Our study has several notable limitations. First, we only compared the 3D, 2.5D, and 2D approaches to the auto-segmentation of brain structures on MR images. The results of this study may not generalize to other imaging modalities or other body organs. Second, there are multiple ways to develop a 2.5D auto-segmentation model [11,39,40]. While we did not implement all of the different versions of 2.5D models, we believe that our implementation of 2.5D models (using five consecutive image slices as input channels) is the best approach to segment the neuroanatomy on brain images. Third, our results about the relative deployment speed of 3D models as compared to 2.5D or 2D models might change as computational resources change. If the GPU memory is large enough to accommodate large 3D patches of the image, 3D models can segment the 3D volume faster. However, in settings where the GPU memory is limited, the 3D model should loop through multiple smaller 3D patches of the image, eroding the faster performance of the 3D models during deployment. However, we used a 16 GB GPU to train and deploy our models, which is commonplace in modern computing units used for deep learning. Finally, we compared 3D, 2.5D, and 2D approaches across three auto-segmentation models only: CapsNets, UNets, and nnUNets. While multiple other auto-segmentation models are available, we believe that our study has compared 3D, 2.5D, and 2D approaches across the most successful deep-learning models for medical image auto-segmentation. Further studies comparing the three approaches across other auto-segmentation models can be an area of future research.

## 5. Conclusions

In this study, we compared 3D, 2.5D, and 2D approaches to brain image auto-segmentation across different models and concluded that the 3D approach is more accurate, achieves better performance in the context of limited training data, and is faster to train and deploy. Our results hold across various auto-segmentation models, including capsule networks, UNets, and nnUNets. The only downside of the 3D approach is that it requires 20 times more computational memory compared to the 2.5D or 2D approaches. Because 3D capsule networks only need twice the computational memory that 2.5D or 2D UNets and nnUNets need, we suggest using 3D capsule networks in settings where computational memory is limited.

**Supplementary Materials:** The following are available online at https://www.mdpi.com/article/10.3390/bioengineering10020181/s1, S1: MRI Acquisition Parameters, S2: Comparing 3D and 2D Segmentation using the Hippocampus Dataset, S3: Pre-Processing, S4: Segmentation Models, S5: Training Hyperparameters for CapsNet and UNet Models, S6: Comparison of Total Convergence Times.

**Author Contributions:** Conceptualization, methodology, validation, formal analysis, investigation, and visualization: A.A. and S.A.; software: A.A., S.H. and S.A.; resources: A.A., M.L., M.A., H.M.K. and S.A.; data curation: A.A. and M.A.; writing—original draft preparation: A.A., H.M.K. and S.A.; writing—review and editing: all co-authors; supervision: M.A., H.M.K. and S.A.; project administration: A.A. and S.A.; funding acquisition: A.A. and S.A. All authors have read and agreed to the published version of the manuscript.

**Funding:** Arman Avesta is a PhD Student in the Investigative Medicine Program at Yale, which is supported by CTSA Grant Number UL1 TR001863 from the National Center for Advancing Translational Science, a component of the National Institutes of Health (NIH). This work was also directly supported by the National Center for Advancing Translational Sciences grant number KL2 TR001862 as well as by the Radiological Society of North America's (RSNA) Fellow Research Grant Number RF2212. The contents of this article are solely the responsibility of the authors and do not necessarily represent the official views of NIH or RSNA.

**Institutional Review Board Statement:** This study was conducted in accordance with the Declaration of Helsinki, and approved by the Institutional Review Board of Yale School of Medicine (IRB number 2000027592, approved 20 April 2020).

**Informed Consent Statement:** Informed consent was obtained from all subjects involved in the ADNI study. We used the ADNI study data that is publicly available.

**Data Availability Statement:** The data used in this study were obtained from the Alzheimer's Disease Neuroimaging Initiative (ADNI) database (adni.loni.usc.edu). We obtained T1-weighted MRIs of 3430 patients in the Alzheimer's Disease Neuroimaging Initiative study from this data-sharing platform. The investigators within the ADNI contributed to the design and implementation of ADNI but did not participate in the analysis or writing of this article.

**Conflicts of Interest:** The funders had no role in the design of the study; in the collection, analyses, or interpretation of data; in the writing of the manuscript; or in the decision to publish the results.

## Abbreviations

| | |
|---|---|
| 2D segmentation | two-dimensional segmentation |
| 2.5D segmentation | enhanced two-dimensional segmentation |
| 3D segmentation | three-dimensional segmentation |
| ADNI | Alzheimer's disease neuroimaging initiative |
| CapsNet | capsule network |
| CPU | central processing unit |
| CT | computed tomography |
| GB | giga-byte |
| GPU | graphics processing unit |
| MRI | magnetic resonance imaging |

## References

1. Feng, C.H.; Cornell, M.; Moore, K.L.; Karunamuni, R.; Seibert, T.M. Automated contouring and planning pipeline for hippocampal-avoidant whole-brain radiotherapy. *Radiat. Oncol.* **2020**, *15*, 251. [CrossRef] [PubMed]
2. Dasenbrock, H.H.; See, A.P.; Smalley, R.J.; Bi, W.L.; Dolati, P.; Frerichs, K.U.; Golby, A.J.; Chiocca, E.A.; Aziz-Sultan, M.A. Frameless Stereotactic Navigation during Insular Glioma Resection using Fusion of Three-Dimensional Rotational Angiography and Magnetic Resonance Imaging. *World Neurosurg.* **2019**, *126*, 322–330. [CrossRef] [PubMed]
3. Dolati, P.; Gokoglu, A.; Eichberg, D.; Zamani, A.; Golby, A.; Al-Mefty, O. Multimodal navigated skull base tumor resection using image-based vascular and cranial nerve segmentation: A prospective pilot study. *Surg. Neurol. Int.* **2015**, *6*, 172. [CrossRef] [PubMed]
4. Clerx, L.; Gronenschild, H.B.M.; Echavarri, C.; Aalten, P.; Jacobs, H.I.L. Can FreeSurfer Compete with Manual Volumetric Measurements in Alzheimer's Disease? *Curr. Alzheimer Res.* **2015**, *12*, 358–367. [CrossRef]
5. Bousabarah, K.; Ruge, M.; Brand, J.-S.; Hoevels, M.; Rueß, D.; Borggrefe, J.; Hokamp, N.G.; Visser-Vandewalle, V.; Maintz, D.; Treuer, H.; et al. Deep convolutional neural networks for automated segmentation of brain metastases trained on clinical data. *Radiat. Oncol.* **2020**, *15*, 87. [CrossRef]
6. Nimsky, C.; Ganslandt, O.; Cerny, S.; Hastreiter, P.; Greiner, G.; Fahlbusch, R. Quantification of, visualization of, and compensation for brain shift using intraoperative magnetic resonance imaging. *Neurosurgery* **2000**, *47*, 1070–1079. [CrossRef]
7. Gerard, I.J.; Kersten-Oertel, M.; Petrecca, K.; Sirhan, D.; Hall, J.A.; Collins, D.L. Brain shift in neuronavigation of brain tumors: A review. *Med. Image Anal.* **2017**, *35*, 403–420. [CrossRef]
8. Lorenzen, E.L.; Kallehauge, J.F.; Byskov, C.S.; Dahlrot, R.H.; Haslund, C.A.; Guldberg, T.L.; Lassen-Ramshad, Y.; Lukacova, S.; Muhic, A.; Nyström, P.W.; et al. A national study on the inter-observer variability in the delineation of organs at risk in the brain. *Acta Oncol.* **2021**, *60*, 1548–1554. [CrossRef]
9. Duong, M.; Rudie, J.; Wang, J.; Xie, L.; Mohan, S.; Gee, J.; Rauschecker, A. Convolutional Neural Network for Automated FLAIR Lesion Segmentation on Clinical Brain MR Imaging. *Am. J. Neuroradiol.* **2019**, *40*, 1282–1290. [CrossRef]

10. Zettler, N.; Mastmeyer, A. Comparison of 2D vs. 3D U-Net Organ Segmentation in abdominal 3D CT images. *arXiv* **2021**, arXiv:2107.04062.
11. Ou, Y.; Yuan, Y.; Huang, X.; Wong, K.; Volpi, J.; Wang, J.Z.; Wong, S.T.C. LambdaUNet: 2.5D Stroke Lesion Segmentation of Diffusion-weighted MR Images. *arXiv* **2021**, arXiv:2104.13917. [CrossRef]
12. Bhattacharjee, R.; Douglas, L.; Drukker, K.; Hu, Q.; Fuhrman, J.; Sheth, D.; Giger, M.L. Comparison of 2D and 3D U-Net breast lesion segmentations on DCE-MRI. In *Medical Imaging 2021: Computer-Aided Diagnosis*; SPIE: Bellingham, WA, USA, 2021; Volume 11597, pp. 81–87.
13. Kern, D.; Klauck, U.; Ropinski, T.; Mastmeyer, A. 2D vs. 3D U-Net abdominal organ segmentation in CT data using organ bounds. In *Medical Imaging 2021: Imaging Informatics for Healthcare, Research, and Applications*; SPIE: Bellingham, WA, USA, 2021; Volume 11601, pp. 192–200.
14. Kulkarni, A.; Carrion-Martinez, I.; Dhindsa, K.; Alaref, A.A.; Rozenberg, R.; van der Pol, C.B. Pancreas adenocarcinoma CT texture analysis: Comparison of 3D and 2D tumor segmentation techniques. *Abdom. Imaging* **2020**, *46*, 1027–1033. [CrossRef] [PubMed]
15. Crawford, K.L.; Neu, S.C.; Toga, A.W. The Image and Data Archive at the Laboratory of Neuro Imaging. *Neuroimage* **2016**, *124*, 1080–1083. [CrossRef] [PubMed]
16. Weiner, M.; Petersen, R.; Aisen, P. Alzheimer's Disease Neuroimaging Initiative 2014. Available online: https://clinicaltrials.gov/ct2/show/NCT00106899 (accessed on 21 March 2022).
17. Ochs, A.L.; Ross, D.E.; Zannoni, M.D.; Abildskov, T.J.; Bigler, E.D.; Alzheimer's Disease Neuroimaging Initiative. Comparison of Automated Brain Volume Measures obtained with NeuroQuant® and FreeSurfer. *J. Neuroimaging* **2015**, *25*, 721–727. [CrossRef] [PubMed]
18. Fischl, B. FreeSurfer. *NeuroImage* **2012**, *62*, 774–781. [CrossRef]
19. Fischl, B.; Salat, D.H.; Busa, E.; Albert, M.; Dieterich, M.; Haselgrove, C.; van der Kouwe, A.; Killiany, R.; Kennedy, D.; Klaveness, S.; et al. Whole Brain Segmentation: Automated Labeling of Neuroanatomical Structures in the Human Brain. *Neuron* **2002**, *33*, 341–355. [CrossRef]
20. Ganzetti, M.; Wenderoth, N.; Mantini, D. Quantitative Evaluation of Intensity Inhomogeneity Correction Methods for Structural MR Brain Images. *Neuroinformatics* **2015**, *14*, 5–21. [CrossRef]
21. Somasundaram, K.; Kalaiselvi, T. Automatic brain extraction methods for T1 magnetic resonance images using region labeling and morphological operations. *Comput. Biol. Med.* **2011**, *41*, 716–725. [CrossRef]
22. Isensee, F.; Jaeger, P.F.; Kohl, S.A.; Petersen, J.; Maier-Hein, K.H. nnU-Net: A self-configuring method for deep learning-based biomedical image segmentation. *Nat. Methods* **2021**, *18*, 203–211. [CrossRef]
23. Avesta, A.; Hui, Y.; Aboian, M.; Duncan, J.; Krumholz, H.M.; Aneja, S. 3D Capsule Networks for Brain MRI Segmentation. *medRxiv* **2021**. [CrossRef]
24. Yin, X.-X.; Sun, L.; Fu, Y.; Lu, R.; Zhang, Y. U-Net-Based Medical Image Segmentation. *J. Healthc. Eng.* **2022**, *2022*, 4189781. [CrossRef] [PubMed]
25. Rudie, J.D.; Weiss, D.A.; Colby, J.B.; Rauschecker, A.M.; Laguna, B.; Braunstein, S.; Sugrue, L.P.; Hess, C.P.; Villanueva-Meyer, J.E. Three-dimensional U-Net Convolutional Neural Network for Detection and Segmentation of Intracranial Metastases. *Radiol. Artif. Intell.* **2021**, *3*, e200204. [CrossRef] [PubMed]
26. LaLonde, R.; Xu, Z.; Irmakci, I.; Jain, S.; Bagci, U. Capsules for biomedical image segmentation. *Med. Image Anal.* **2020**, *68*, 101889. [CrossRef] [PubMed]
27. Rauschecker, A.M.; Gleason, T.J.; Nedelec, P.; Duong, M.T.; Weiss, D.A.; Calabrese, E.; Colby, J.B.; Sugrue, L.P.; Rudie, J.D.; Hess, C.P. Interinstitutional Portability of a Deep Learning Brain MRI Lesion Segmentation Algorithm. *Radiol. Artif. Intell.* **2022**, *4*, e200152. [CrossRef]
28. Rudie, J.D.; Weiss, D.A.; Saluja, R.; Rauschecker, A.M.; Wang, J.; Sugrue, L.; Bakas, S.; Colby, J.B. Multi-Disease Segmentation of Gliomas and White Matter Hyperintensities in the BraTS Data Using a 3D Convolutional Neural Network. *Front. Comput. Neurosci.* **2019**, *13*, 84. [CrossRef]
29. Weiss, D.A.; Saluja, R.; Xie, L.; Gee, J.C.; Sugrue, L.P.; Pradhan, A.; Bryan, R.N.; Rauschecker, A.M.; Rudie, J.D. Automated multiclass tissue segmentation of clinical brain MRIs with lesions. *NeuroImage Clin.* **2021**, *31*, 102769. [CrossRef]
30. Yaqub, M.; Feng, J.; Zia, M.; Arshid, K.; Jia, K.; Rehman, Z.; Mehmood, A. State-of-the-Art CNN Optimizer for Brain Tumor Segmentation in Magnetic Resonance Images. *Brain Sci.* **2020**, *10*, 427. [CrossRef]
31. Sun, Y.C.; Hsieh, A.T.; Fang, S.T.; Wu, H.M.; Kao, L.W.; Chung, W.Y.; Chen, H.-H.; Liou, K.-D.; Lin, Y.-S.; Guo, W.-Y.; et al. Can 3D artificial intelligence models outshine 2D ones in the detection of intracranial metastatic tumors on magnetic resonance images? *J. Chin. Med. Assoc. JCMA* **2021**, *84*, 956–962. [CrossRef]
32. Nemoto, T.; Futakami, N.; Yagi, M.; Kumabe, A.; Takeda, A.; Kunieda, E.; Shigematsu, N. Efficacy evaluation of 2D, 3D U-Net semantic segmentation and atlas-based segmentation of normal lungs excluding the trachea and main bronchi. *J. Radiat Res.* **2020**, *61*, 257–264. [CrossRef]
33. Tran, M.; Vo-Ho, V.-K.; Le, N.T.H. 3DConvCaps: 3DUnet with Convolutional Capsule Encoder for Medical Image Segmentation. *arXiv* **2022**, arXiv:2205.09299. [CrossRef]
34. Tran, M.; Ly, L.; Hua, B.-S.; Le, N. SS-3DCapsNet: Self-supervised 3D Capsule Networks for Medical Segmentation on Less La-beled Data. *arXiv* **2022**, arXiv:2201.05905. [CrossRef]

35. Ronneberger, O.; Fischer, P.; Brox, T. U-Net: Convolutional Networks for Biomedical Image Segmentation. In *Medical Image Computing and Computer-Assisted Intervention—MICCAI 2015*; Navab, N., Hornegger, J., Wells, W.M., Frangi, A.F., Eds.; Lecture Notes in Computer Science; Springer International Publishing: Cham, Switzerland, 2015; pp. 234–241.
36. Nguyen, T.; Hua, B.-S.; Le, N. 3D UCaps: 3D Capsule Unet for Volumetric Image Segmentation. In *Medical Image Computing and Computer Assisted Intervention—MICCAI 2021*; Springer International Publishing: Cham, Switzerland, 2021; pp. 548–558.
37. Bonheur, S.; Štern, D.; Payer, C.; Pienn, M.; Olschewski, H.; Urschler, M. Matwo-CapsNet: A Multi-label Semantic Segmentation Capsules Network. In *Medical Image Computing and Computer Assisted Intervention—MICCAI 2019*; Lecture Notes in Computer Science; Springer International Publishing: Cham, Switzerland, 2019; Volume 11768, pp. 664–672.
38. Dong, J.; Liu, C.; Yang, C.; Lin, N.; Cao, Y. Robust Segmentation of the Left Ventricle from Cardiac MRI via Capsule Neural Network. In Proceedings of the 2nd International Symposium on Image Computing and Digital Medicine, ISICDM 2018, New York, NY, USA, 23 October 2019; pp. 88–91.
39. Angermann, C.; Haltmeier, M. Random 2.5D U-net for Fully 3D Segmentation. In *Machine Learning and Medical Engineering for Cardiovascular Health and Intravascular Imaging and Computer Assisted Stenting*; Springer: Cham, Switzerland, 2019; Volume 11794, pp. 158–166.
40. Li, J.; Liao, G.; Sun, W.; Sun, J.; Sheng, T.; Zhu, K.; von Deneen, K.M.; Zhang, Y. A 2.5D semantic segmentation of the pancreas using attention guided dual context embedded U-Net. *Neurocomputing* **2022**, *480*, 14–26. [CrossRef]

bioengineering

MDPI

Article

# Improving the Segmentation Accuracy of Ovarian-Tumor Ultrasound Images Using Image Inpainting

**Lijiang Chen [1,†], Changkun Qiao [1,†], Meijing Wu [2,†], Linghan Cai [1], Cong Yin [2], Mukun Yang [2], Xiubo Sang [2] and Wenpei Bai [2,*]**

1 School of Electronic and Information Engineering, Beihang University, No. 37 Xueyuan Road, Haidian District, Beijing 100191, China
2 Department of Obstetrics and Gynecology, Beijing Shijitan Hospital, Capital Medical University, Beijing 100038, China
* Correspondence: baiwp@bjsjth.cn; Tel.: +86-153-1195-5379
† These authors contributed equally to this work.

**Abstract:** Diagnostic results can be radically influenced by the quality of 2D ovarian-tumor ultrasound images. However, clinically processed 2D ovarian-tumor ultrasound images contain many artificially recognized symbols, such as fingers, crosses, dashed lines, and letters which assist artificial intelligence (AI) in image recognition. These symbols are widely distributed within the lesion's boundary, which can also affect the useful feature-extraction-utilizing networks and thus decrease the accuracy of lesion classification and segmentation. Image inpainting techniques are used for noise and object elimination from images. To solve this problem, we observed the MMOTU dataset and built a 2D ovarian-tumor ultrasound image inpainting dataset by finely annotating the various symbols in the images. A novel framework called mask-guided generative adversarial network (MGGAN) is presented in this paper for 2D ovarian-tumor ultrasound images to remove various symbols from the images. The MGGAN performs to a high standard in corrupted regions by using an attention mechanism in the generator to pay more attention to valid information and ignore symbol information, making lesion boundaries more realistic. Moreover, fast Fourier convolutions (FFCs) and residual networks are used to increase the global field of perception; thus, our model can be applied to high-resolution ultrasound images. The greatest benefit of this algorithm is that it achieves pixel-level inpainting of distorted regions without clean images. Compared with other models, our model achieveed better results with only one stage in terms of objective and subjective evaluations. Our model obtained the best results for 256 × 256 and 512 × 512 resolutions. At a resolution of 256 × 256, our model achieved 0.9246 for SSIM, 22.66 for FID, and 0.07806 for LPIPS. At a resolution of 512 × 512, our model achieved 0.9208 for SSIM, 25.52 for FID, and 0.08300 for LPIPS. Our method can considerably improve the accuracy of computerized ovarian tumor diagnosis. The segmentation accuracy was improved from 71.51% to 76.06% for the Unet model and from 61.13% to 66.65% for the PSPnet model in clean images.

**Keywords:** ovarian tumor; 2D ultrasound image; image inpainting; lesion segmentation; attention mechanism; GAN; deep learning; medical image analysis

**Citation:** Chen, L.; Qiao, C.; Wu, M.; Cai, L.; Yin, C.; Yang, M.; Sang, X.; Bai, W. Improving the Segmentation Accuracy of Ovarian-Tumor Ultrasound Images Using Image Inpainting. *Bioengineering* **2023**, *10*, 184. https://doi.org/10.3390/bioengineering10020184

Academic Editors: Paolo Zaffino, Maria Francesca Spadea and Larbi Boubchir

Received: 18 November 2022
Revised: 5 January 2023
Accepted: 28 January 2023
Published: 1 February 2023



## 1. Introduction

Medical ultrasonography has turned out to be the preferred imaging technique for many illnesses due to the fact of its simplicity, speed, and safety [1–5]. Two-dimensional gray-scale ultrasound and coloration Doppler ultrasound has been broadly used in the diagnostic tasks of ovarian tumors. Doctors can first perceive the benign and malignant nature of tumors. With the non-stop development and improvement of deep learning [6,7], AI, as a riding pressure for intelligent healthcare, has acquired a massive range of achievements in tasks such as clinical image classification and segmentation [8–11]. The accuracy of the model additionally relies upon the quality of the dataset [12,13]. There is exceedingly little research on the current use of AI for lesion recognition and segmentation of ovarian tumor diseases. In addition, the effectiveness of AI in processing ovarian-tumor images depends on a large-scale AI dataset. Zhao et al. [14] proposed an ovarian-tumor ultrasound image dataset for lesion classification and segmentation. The dataset consists of a complete of 1469 2D ovarian ultrasound images which are divided into eight categories according to tumor types. The giant majority of the images in the dataset contain annotated symbols, which are overwhelmingly allotted to inside the lesion.

Nevertheless, hidden but crucial trouble has been recognized in practice: most 2D ovarian-tumor ultrasound images incorporate extra symbols. Actually, in clinical operations where ovarian ultrasound images are acquired, the physician will mark the location, size, and border of the tumor in the ovarian ultrasound image, and observe where the lesion is positioned (left or right ovary). Due to equipment factors and the clinical practice environments, the artificially marked component of these aids to image recognition (symbols such as fingers, crosses, dashes, and letters) cannot be separated from the original image. This phenomenon is also widespread in different medical fields [15–18]. The ideal situation would be to train and test deep learning models using clean images without any symbols in lesion areas.

We observe that these symbols are centered in ovarian tumor lesions, which negatively affects the training of the model to a positive extent, causing the network to focus more on the symbols in the lesions, which in turn reduces the recognition accuracy of ovarian tumors in the clean images and the segmentation accuracy of the lesions. The different types of images in this paper are shown in Figure 1. The original images with symbols were used as the training set, and two different test sets of clean images and original images with symbols were used as a way to discover the impact of symbols on the segmentation accuracy of the model. Figures 2 and A1 exhibit the effects of our experiments. Fewer training epochs are required to segment more accurate lesion regions in images with symbols, and the segmented regions targeted the yellow line roughly. The clean images, on the other hand, required more epochs and reached lower segmentation accuracy. The results show that the symbols in the images provide additional information to the model enhancing the accuracy of segmentation, which is unrealistic in clinical practice. There is little research on this issue, and it is certainly inappropriate to use the marked ovarian-tumor ultrasound images directly to train the segmentation model. Thus, it is critical for the corrupted areas of the images to be painted, so it is significant for healthcare professionals to use clean images for the artificial intelligence-aided diagnosis of ovarian tumors.

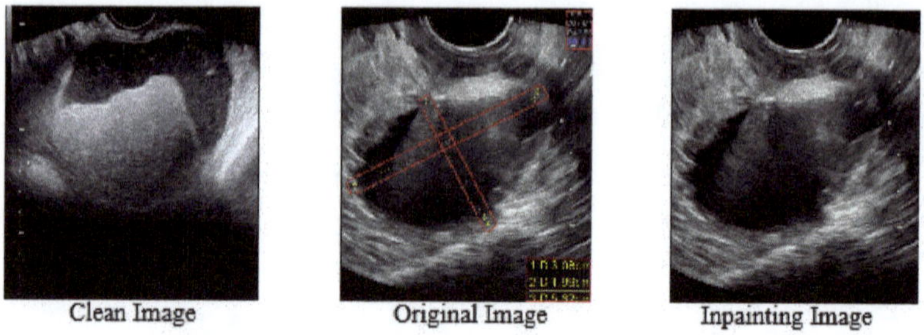


**Figure 1.** (**Clean Image**) The clean images indicate images that are not clinically labeled. (**Original Image**) The original images indicate clinical images that are labeled. The red-boxed areas show the various marker symbols used by physicians. (**Inpainting Image**) The inpainting images indicate images whose symbols are repaired.

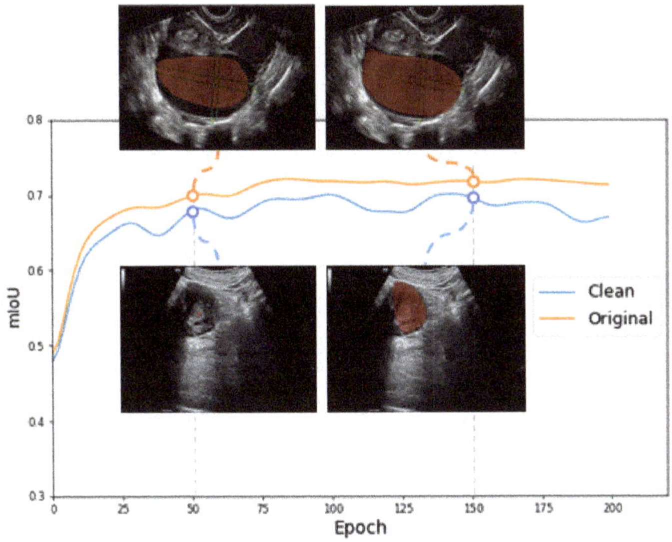


**Figure 2.** The accuracy graph of lesion segmentation of the Unet [19] model. The blue line represents the accuracy of using the clean images as the testing set. The yellow line represents the accuracy of using the original images with symbols as the testing set. The figure also shows the visualization of the segmentation results for the 50th and 150th epochs.

Currently, image inpainting in medical images is in the process of booming and has a lot of potential for development. Existing methods are primarily divided into traditional methods and deep learning-based methods. Traditional methods make use of patch-based or diffusion-based methods, the core of which is to use the redundancy of the image itself to fill in the missing areas with low-level texture features of the image. The following four methods are historically used for inpainting: interpolation [20], non-local means [21], diffusion techniques [22], and texture-dependent synthesis [23]. However, ordinary methods cannot learn the deep semantic features of medical images frequently and can not attain excellent results.

Deep-learning-based methods use convolutional neural networks to extract and learn high-level semantic features in the image to guide the model to fill the missing parts. Inspired by EdgeConnect [24], Wang et al. [25] migrated the method using edge information to medical images. This paper details the study of these methods and use of an attention mechanism, a pyramid-structured generator, to enforce the inpainting of thyroid ultrasound images, which automatically detects and reconstructs the cross symbols in ultrasound images. However, this method has some limitations: the cross symbols in the thyroid ultrasound images used in this approach are small and few, and the effect is negative for ultrasound images containing many large symbols; the detected cross symbols are labeled with rectangular boxes, and this approach does not apply to different symbols with irregular shapes; the real background is covered by these symbols, and the restoration areas have no real background, so how to guide the generative adversarial network for training and evaluation, in this case, is a very necessary issue. Wei et al. [26] proposed the MagGAN for face-attribute editing. The MagGAN does this by introducing a novel mask-guided adjustment strategy to encourage the affected regions of each target attribute to be positioned in the generator, using the corresponding attributes of the face (eyes, nose, mouth, etc.). The method is applied to the face-attribute editing task, which requires segmentation of the face's attributes, which is different from our task. However, the motivation of making the results more realistic by bootstrapping the model is similar.

In addition, various attention mechanisms have been proposed and are broadly used in image processing. These attention mechanisms have been steadily utilized in the image inpainting task. Zeng et al. [27] expanded on this by proposing a pyramidal structure for contextual attention. Yi et al. [28] proposed a contextual residual aggregation of attention for high-resolution images. The spatial attention mechanism was utilized to solve this problem. To acquire results with a clear structure and texture, the Shift-Net model proposed by Yan et al. [29] replaced the fully detailed layer in the upsampling process with a shift-connected layer, through which the features in the background region are shifted to fill in the holes.

Due to the above issues, in this paper, a one-stage generation model based on GANs is proposed, which swaps the regular convolution with fast Fourier convolutions to enhance the image-wide acceptance field of the model and includes a channel attention mechanism to minimize the model's focus on symbols to fill the holes using effective features. To the best of our knowledge, we are the first to accomplish image inpainting on 2D ovarian-tumor ultrasound images with large and irregular masks, and our approach achieves more convincing results than others.

Our contributions are as follows:

- We refined 1469 2D ovarian-tumor ultrasound images for irregular symbols and obtained binary masks to establish a 2D ovarian-tumor ultrasound image inpainting dataset.
- We introduced fast Fourier convolution to enhance the model's global perceptual field and a channel attention mechanism to enhance the model's attention to significant features, and the model uses global features and significant channel features to fill the holes.
- Our model achieved better results both subjectively and objectively compared to existing models while for the first time performing image inpainting without clean images.
- We use the restoration images for segmentation training, which significantly enhances the accuracy of the classification and segmentation of clean images.

The rest of the paper is organized as follows: Section 2 describes our dataset and model in detail. The associated experiments and results are detailed in Section 3. The conclusions are introduced in Section 4.

## 2. Methodology

### 2.1. Dataset

In recent years, research about ovarian tumors has increased, and researchers have combined ovarian tumor sonograms with deep learning for ovarian tumor classification and lesion segmentation [30–33]. Most of the 2D ovarian-tumor ultrasound images used in these studies contain symbols, which are broadly allotted to the edges or inner parts of the lesions. We experimentally confirmed the negative effect of these symbols on the classification accuracy and lesion segmentation accuracy of tumors. The MMOTU dataset [14] is a publicly available ovarian ultrasound image dataset. We obtained a 2D ovarian-tumor ultrasound image inpainting dataset based on the MMOTU dataset by refining annotation processing. As shown in Figure 3, the green dashed line in the figure is how the MMOTU dataset is annotated. We labeled the fingers and letters (brown boxes), numbers (blue boxes), and yellow lines (yellow boxes) in the figure on this basis.

With annotation, a corresponding mask for each image is generated, which masks the various symbols in the image. Figure 4 indicates our pipeline. With these annotations, the corresponding mask for each image was generated to build an inpainting dataset containing 1469 2D ultrasound images of ovarian tumors and masks. We performed experiments about image inpainting on our dataset and the effect of image inpainting on lesion segmentation accuracy in the MMOUT dataset.

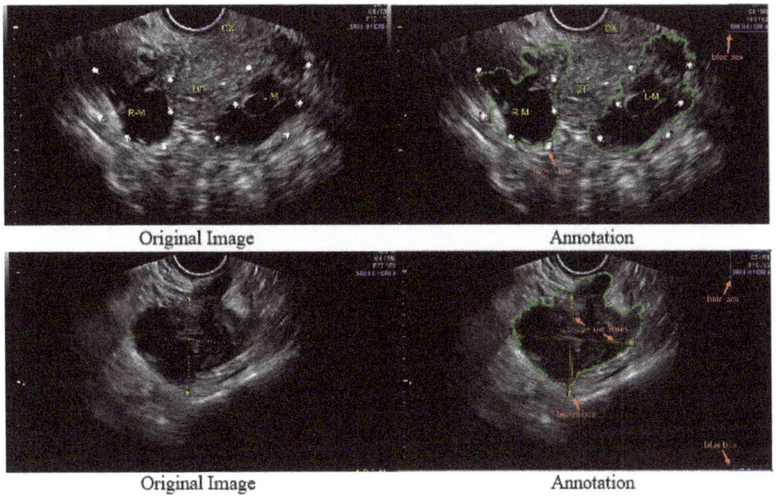


**Figure 3.** Original 2D ovarian-tumor ultrasound images and images with annotated symbols.

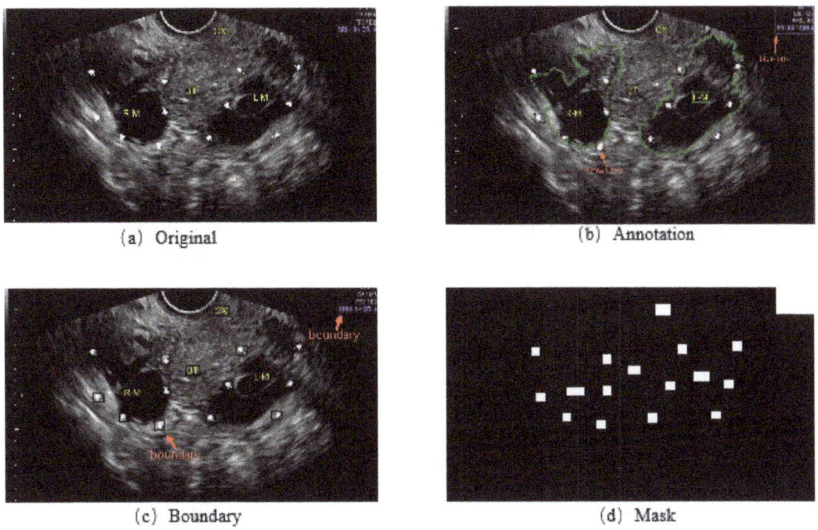


**Figure 4.** The pipeline of mask generation. (**a**) The original image. (**b**) The annotation. (**c**) The boundary. (**d**) The mask.

### *2.2. Implementation Details*

In this study, we used a complete, 2D ovarian-tumor ultrasound dataset with 1469 images that we produced, of which 1200 images were used for training and 269 images were used for testing. Arbitrarily shaped masks were used during training and testing. To make certain the equity of the experiments, we generated unique irregular masks for the images used for testing. The inputs in our experiments had two specifications: one specification was $256 \times 256$ ($h \times w$), and the other specification was $512 \times 512$ ($h \times w$). We trained and tested our model with both image specifications. The Adam optimizer was chosen to optimize the network. We set the initial learning rate to 0.0001, the batch size for training to 16, and the epoch to 1000. In addition to generating masks using our proposed mask generation strategy, we also performed data enhancement operations on the images during training. The framework was PyTorch, and the devices were two NVIDIA GeForce RTX3090Ti.

### 2.3. *Proposed Methods*

#### 2.3.1. Network Architecture

We propose an image inpainting model based on fast Fourier convolutions (FFCs) with a channel attention mechanism. Figure 5 indicates the details of our model. The images are downsampled by three convolutional layers and then encoded with the aid of nine fast Fourier Convolution Residual Network Blocks. The decoder obtains the inpainting image by predicting the output of the encoder. These inpainting and original images are fed into the discriminator for adversarial training. Traditional fully convolutional models, such as ResNet [34], suffer from slow perceptual-field growth due to a small convolutional kernel size and limited receptive fields. Due to this reason, many layers in the network lack global context, such that the result has a lack of global structural consistency. We replaced the regular convolution with the fast Fourier convolution to solve this problem. In addition, due to the presence of symbols such as yellow dashed lines in the images, we added a channel attention layer to our model to permit the model to focus more on useful features and make the results more realistic. Figure 6 suggests the specified architecture of the Fast Fourier Convolution Block.

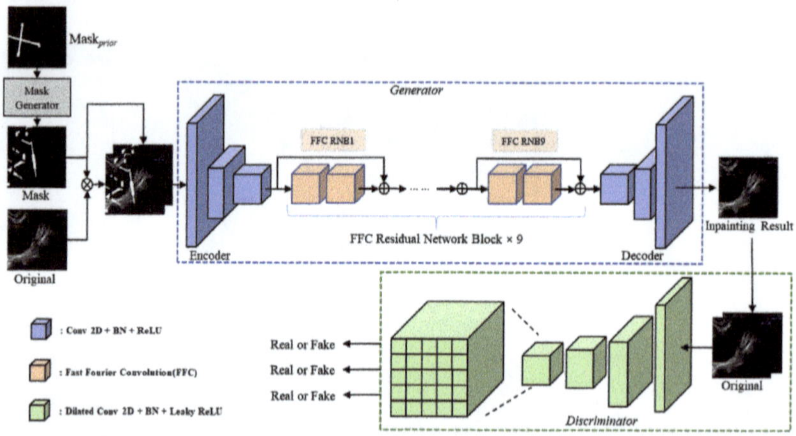


**Figure 5.** The overall architecture of our MGGAN model. The generator consists of 9 FFC Residual Network Blocks with our mask to a priori guide the generator for image inpainting.

#### 2.3.2. Fast Fourier Convolution Block

Regular convolution is mostly used in deep learning models; however, it cannot capture the global features. Fast Fourier convolutions [35] can be an appropriate solution to this problem. The FFCs divide the input channel into local and global paths: the local path uses regular convolution to capture local information; the global path uses the real fast Fourier transform to obtain information with a global receptive field. The fast Fourier change consists of the following five steps:

- Transforming the input tensor to the frequency domain using the real fast Fourier transform: $\mathbb{R}^{H\times W\times C} \rightarrow \mathbb{C}^{H\times \frac{W}{2}\times C}$.
- Concatenating the real and imaginary parts in the frequency domain: $\mathbb{C}^{H\times \frac{W}{2}\times C} \rightarrow \mathbb{R}^{H\times \frac{W}{2}\times 2C}$.
- Obtaining convolution results in the frequency domain through the ReLU layer, BatchNorm layer, and $1 \times 1$ convolution layer: $\mathbb{R}^{Hand2\times 2C} \rightarrow \mathbb{R}^{H\times \frac{W}{2}\times 2C}$.

- Separating the result of frequency domain convolution into real and imaginary parts: $\mathbb{R}^{H\times\frac{W}{2}\times 2C} \rightarrow \mathbb{C}^{H\times\frac{W}{2}\times C}$.
- Recovering its spatial structure using Fourier inverse transform: $\mathbb{C}^{H\times\frac{W}{2}\times C} \rightarrow \mathbb{R}^{H\times W\times C}$.

As shown in Figure 6, we add a squeeze-and-excitation (SE) layer after the spectral transform block, which performs the squeeze, excitation, and reweight operations in turn. The SE layer automatically acquires each feature channel's weight via learning, then boosts the beneficial features and suppresses the ones that are no longer beneficial according to the weight. By using the SE layer, we make the model focus more on the useful features rather than on the features of these symbols in the image. Finally, the output of the local and global paths are merged.

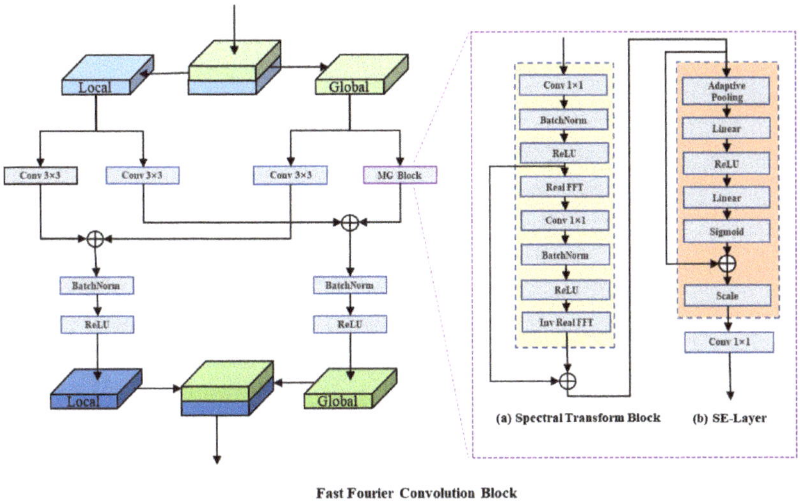


**Figure 6.** The architecture of the Fast Fourier Convolution Block (FFC Block).

### 2.3.3. Generation of masks during training

The approach of mask generation during training has been extensively mentioned in previous research, and it is crucial for the inpainting effect of the model. In early studies, the generated masks are rectangular in shape [36], centered on the geometric center of the image. Models trained with these masks have bad results for images with non-centered rectangular masks. Therefore, the method of generating masks at random locations [37] in the image during training was proposed, but this method fails to provide effective and realistic inpainting of images with irregular masks. Subsequently, the strategy of randomly generating irregular masks [38–40] at random locations in the image has emerged.

There are many symbols in the image that obscure the clean image. If these areas are repaired, the results cannot be evaluated realistically due to the fact there is no clean image. We need to guide the network to learn to use features of the non-symbolic regions to fill holes. In our task, we propose a new mask generation strategy by generating random irregular masks at random locations outside the symbolic regions in the image. The generation formula for the masks is as follows:

$$m = m_{gen} - m_{prior} \quad (1)$$

where $m_{prior}$ is the mask corresponding to the image in the dataset, $m_{gen}$ is the mask generated by the mask generator, and $m$ is the final mask.

### 2.4. Loss Function

The loss function in the generation task is essential for the training of the model, and it calculates the distinction between the ground truth and the inpainting image as the loss value. The loss values are back-propagated, and the model is penalized to update the parameters of each layer. In the end, the loss value is reduced, and the result is closer to the ground truth.

Several extraordinary loss functions were used in our task. In our model, the input uses the corrupted image $I_{in} = I_{ori} \odot (1 - m)$, where $I_{ori}$ denotes the original images and $m$ denotes the corresponding mask, for which one denotes the missing pixels and zero denotes the existing pixels. The symbol $\odot$ denotes the multiplication of the matrix. G denotes the generator, $I_{inp}$ denotes the final inpainting image generated by the model, and the expressions for the inputs and outputs are shown in Equation (2).

$$I_{inp} = G(I_{in}) \tag{2}$$

The perceptual loss [41] is derived by calculating the distance between features captured by the pre-trained network $\Psi(.)$ from the generated images and the original images. To enable the network to understand global contextual information, we compute high receptive field perceptual loss [42] using a pre-trained ResNet with global receptive fields. The calculation of $\mathcal{L}_{ResNet}$ can be expressed as follows:

$$\mathcal{L}_{ResNet}(I_{ori}, I_{inp}) = \mathcal{M}\left( [\Psi_{ResNet}(I_{ori}) - \Psi_{ResNet}(I_{inp})]^2 \right) \tag{3}$$

where $I_{ori}$ is the original image or the target image of the generator, $I_{inp}$ is the generated image, and $\mathcal{M}$ is the operation of calculating the inter-layer mean after calculating the intra-layer mean. $\Psi_{ResNet}(.)$ is a pre-trained ResNet implemented with dilated convolution.

To make the generated inline images more realistic and natural in detail, we additionally use adversarial loss. The adversarial loss function $\mathcal{L}_{adv}$ is calculated as follows:

$$\mathcal{L}_{adv}(I_{inp}, I_{ori}, I_{in}) = \max_{D} E_{x \in \mathcal{X}}[\log(D(I_{ori}, I_{in})) + \log(1 - D(I_{inp}, I_{in}))] \tag{4}$$

where $I_{ori}$ is the target image, $I_{inp}$ is the inapinting image, $I_{in}$ is the corrupted image, and $D$ is the adversarial discriminator.

In our total loss, we also use the $\mathcal{L}_1$ loss and the perceptual loss of the discriminator network $\mathcal{L}_{Disc}$ [43]. The formula for the perceptual loss of the discriminator network $\mathcal{L}_{Disc}$ is similar to Equation (2). The $\mathcal{L}_1$ loss is calculated as follows:

$$\mathcal{L}_1 = \frac{1}{N} \sum |I_{ori}(p) - I_{inp}(p)| \tag{5}$$

where $I_{ori}$ denotes the original image, $I_{inp}$ denotes the inpainting image, and $p$ represents the pixel at the same location in both images.

Our total losses are calculated as follows:

$$\mathcal{L}_{total} = \eta_1 \mathcal{L}_1 + \eta_2 \mathcal{L}_{adv} + \eta_3 \mathcal{L}_{ResNet} + \eta_4 \mathcal{L}_{Disc} \tag{6}$$

where $\eta$ is the weight of each loss function. Following [36,39,42], we set $\eta_1 = 10$, $\eta_2 = 10$, $\eta_3 = 30$, and $\eta_4 = 100$ in training.

### 2.5. Evaluation Criterion

We used the evaluation metrics of *structural similarity* (SSIM) [44], *Frechet inception distance score* (FID) [45], and *learned perceptual image patch similarity* (LPIPS) [46] to measure the performance of our model. In addition, we used the *mean intersection over union* (mIoU) evaluation metric to measure the accuracy of lesion segmentation results.

The SSIM is calculated between two windows of size H × W. The value of SSIM is between −1 and 1, where 1 means the two images are identical and −1 means the opposite. The closer the value of SSIM is to one, the better the inpainting effect is. The SSIM calculation formula is defined as follows:

$$SSIM = \frac{(2\mu_A\mu_B + c_1)(2\sigma_{AB} + c_2)}{(\mu_A{}^2 + \mu_B{}^2 + c_1)(\sigma_A{}^2 + \sigma_B{}^2 + c_2)} \tag{7}$$

where $\mu_A$ and $\sigma_A{}^2$ are the mean and variance of image $A$, $\mu_B$ and $\sigma_B{}^2$ are the mean and variance of image $B$, $\sigma_{AB}$ is the covariance of the two images, and $c_1$ and $c_2$ are the constants that maintain stability.

The Frechet inception distance score (FID) is a metric to calculate the distance between the real image and the generated image feature vectors. It uses the 2048-dimensional vector of Inception Net-V3 before full concatenation as the feature of the image to evaluate the similarity of the two sets of images. The value of FID is greater than or equal to zero. A lower score means that the two sets of images are more similar, and the FID score in the best case is 0.0, which means that the two sets of images are identical. The FID calculation formula is described as follows:

$$FID = \left\|\mu_{gt} - \mu_{pred}\right\|^2 + \mathrm{Tr}\left(\Sigma_{gt} + \Sigma_{pred} - 2\left(\Sigma_{gt}\Sigma_{pred}\right)^{1/2}\right) \tag{8}$$

where $\mu_{gt}$ and $\Sigma_{gt}$ are the mean and covariance matrices of the real image features, $\mu_{pred}$ and $\Sigma_{pred}$ are the mean and covariance matrices of the generated image features, and Tr is the operation to calculate the matrix trace.

LPIPS is used to measure the difference between two images in terms of deep-level features, and LPIPS is more consistent with human perception than traditional methods such as $\ell_2$, PSNR, and FSIM. The value of LPIPS is greater than or equal to zero. A lower value of LPIPS indicates that the two images are more similar, and vice versa. The LPIPS calculation formula is defined as follows:

$$d\left(I_{gt}, I_{pred}\right) = \sum_l \frac{1}{H_l W_l} \sum_{h,w} \left\|w_l \odot \left(\hat{y}^l_{gt-hw} - \hat{y}^l_{pred-hw}\right)\right\|_2^2 \tag{9}$$

where $l$ is the current computed layer; $H_l$ and $W_l$ are the sizes of the patches; and $\hat{y}^l_{gt-hw}$ and $\hat{y}^l_{pred-hw} \in \mathbb{R}^{H_l \times W_l \times C_l}$ are the outputs of the current layer. The feature stack is extracted from the $L$ layers and unit-normalized in the channel dimension. The vector $w_l$ is used to deflate the number of active channels and calculate the $\ell_2$ distance.

MIoU is a widely used standard metric in semantic segmentation, which calculates the mean of the ratio of intersection and merges sets of all categories. The value is between zero and one. Closer to one means better the segmentation, and closer to zero is the opposite. Its calculation formula is defined as follows:

$$MIoU = \frac{1}{k+1} \sum_{i=0}^{k} \frac{TP}{FN + FP + TP} \tag{10}$$

where $k$ is the number of categories, $TP$ is the number of true positive pixels, $FP$ is the number of false positives, and $FN$ is the number of false negatives.

## 3. Experiments and Results

### 3.1. Results

#### 3.1.1. Experiments on the Image Inpainting

Figure 7 indicates the effects of our model on the restoration of the symbolic regions in the ovarian ultrasound images. The boundary, texture, and structure have high similarity to those in the original image. The results show that we have flawlessly removed the symbols from the images. Especially in the lesion area, we removed the yellow line while reconstructing the boundary of the lesion and the content filling of the yellow line area very well. This proves the power of our model. Furthermore, we compare our approach with robust baselines that are publicly available on FID, LPIPS, and SSIM metrics. We performed statistical analysis of the inpainting results on 269 images of the test set.

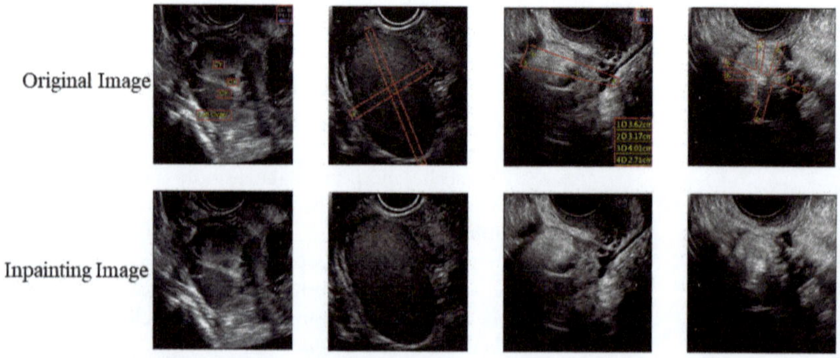


**Figure 7.** The results of our model for the inpainting of the symbolic regions in the ovarian ultrasound images.

Table 1 suggests the overall performance of each baseline on our dataset, and the values of the three metrics in the table are the means of the test samples. Smaller FID and LPIPS indicate better performance of the model, and a larger SSIM indicates better performance of the model. Table 2 presents the overall performance of each baseline on our dataset, and the values of the three metrics in the table are the variance of the test samples. The size of the input images in the experiment was 256 × 256. In the statistical analysis, we observed that our model outperformed all other comparable models in SSIM, FID, and LPIPS metrics. Our model achieved 0.9246 for SSIM, 22.66 for FID, and 0.07806 for LPIPS. Table 3 suggests that the upper and lower limits of our method surpass those of the other methods for all three metrics at a confidence level of 95%.

Figure 8 indicates the inpainting results for different models (we show more results in Appendix A). A clear distinction can be found in the blue box area. These baseline models use the learned symbol features to generate the symbol regions, resulting in yellow pixels in the restoration regions. In addition, the regions they generate show significant distortions and folds, with unsatisfactory textures and structures. We address this problem by using an attention mechanism for the model to focus on the features of the fee-symbolic region in the image. Fast Fourier convolution allows the first few layers of the network to quickly increase the receptive field, which allows the model to gain a global receptive field faster and increase the connection between global and local features. The model can better use the global and local features to fill the holes, and the results of the restoration will have the same structural and textural features as the original image, including smoother boundaries and more realistic content. By introducing the channel attention mechanism, our model pays more attention to the features of non-symbolic regions rather than the features of symbolic regions and chooses useful features for image inpainting. Thereby, the restored image is closer to the original image in terms of content, and no yellow pixels appear in the restoration region. In the qualitative comparison, our model showed the best authenticity and details in the results, including smooth edges and high similarity to the original images. Our method better reconstructed the edge structure and content of the lesion in the image, which dramatically improved lesion segmentation accuracy.

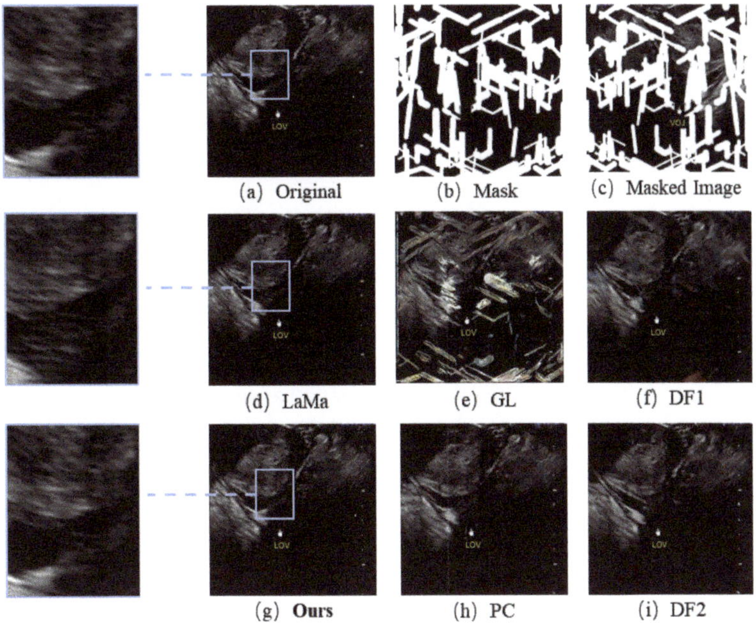


**Figure 8.** Comparison between the results of our proposed model and other models. (**a**) The original image. (**b**) The mask for the original image. (**c**) The masked image. (**d**) Results from publicly available code using the LaMa method. (**e**) Results from publicly available code using the GL method. (**f**) Results from publicly available code using the DF 1 method. (**g**) Results from publicly available code using our method. (**h**) Results from publicly available code using the PC method. (**i**) Results from publicly available code using the DF 2 method.

**Table 1.** Means of the quantitative comparison of the proposed method with already publicly available, robust baselines in FID, LPIPS, and SSIM metrics. The results of each model were derived from its public code.

| Model | SSIM | FID | LPIPS |
|---|---|---|---|
| PC [39] | 0.6847 | 79.42 | 0.13550 |
| GL [38] | 0.3026 | 170.69 | 0.29589 |
| DF 1 [37] | 0.6578 | 81.74 | 0.14090 |
| Df 2 [40] | 0.8932 | 54.38 | 0.10150 |
| LaMa [42] | 0.9209 | 25.54 | 0.08215 |
| Ours | 0.9246 | 22.66 | 0.07806 |

**Table 2.** Variances of quantitative comparison of the proposed method with already publicly available, robust baselines in FID, LPIPS, and SSIM metrics. The results of each model were derived from its public code.

| Model | SSIM | FID | LPIPS |
|---|---|---|---|
| PC [39] | $1.47 \times 10^{-5}$ | 0.2755 | $4.5 \times 10^{-8}$ |
| GL [38] | $1.81 \times 10^{-5}$ | 0.4878 | $9.7 \times 10^{-8}$ |
| DF 1 [37] | $1.39 \times 10^{-5}$ | 0.2801 | $4.3 \times 10^{-8}$ |
| Df 2 [40] | $1.10 \times 10^{-5}$ | 0.2311 | $2.1 \times 10^{-8}$ |
| LaMa [42] | $9.90 \times 10^{-6}$ | 0.1777 | $1.1 \times 10^{-8}$ |
| Ours | $9.10 \times 10^{-6}$ | 0.1373 | $8.1 \times 10^{-9}$ |

**Table 3.** The lower (left) and upper (right) limits of confidence are 95% of quantitative comparison of the proposed method with an already publicly available, robust baselines in FID, LPIPS, and SSIM metrics. The results of each model were derived from its public code.

| Model | SSIM | FID | LPIPS |
|---|---|---|---|
| PC [39] | (0.6771, 0.6923) | (78.17, 80.67) | (0.13510, 0.13590) |
| GL [38] | (0.2939, 0.3111) | (169.81, 171.57) | (0.29556, 0.29622) |
| DF 1 [37] | (0.6502, 0.6654) | (80.57, 82.40) | (0.14050, 0.14130) |
| Df 2 [40] | (0.8860, 0.9004) | (53.46, 55.30) | (0.10122, 0.10178) |
| LaMa [42] | (0.9145, 0.9273) | (24.72, 26.36) | (0.08195, 0.08235) |
| Ours | (0.9186, 0.9306) | (21.96, 23.36) | (0.07788, 0.07824) |

#### 3.1.2. Ablation Experiments

To verify that our approaches do reduce the capabilities of the model, we designed ablation experiments for the baseline model. The dataset used for the experiments was our inpainting dataset. We used solely FFCs as the baseline in this experiment.

- FFCs
  Fast Fourier convolutions have a larger and more effective field of repetition, which can effectively enhance the field of repetition of our model and improve its capability. We performed quantitative experiments on fast Fourier convolution, dilated convolution, and regular convolution. The convolution kernel size was set to 3 × 3, and the expansion rate of the dilated convolution was set to 3. Table 4 shows the scores of different types of convolution. FFC performed the best, and dilated convolution was second only to FFC; however, dilated convolution depends on the resolution of the image and has poor generalization.
- Mask generation
  The types, sizes, and positions of the mask during training impact the generative and generalization capabilities of the model. In our task, we focused on exploring the effect of mask generation location on the model. Regular, irregularly shaped masks will overlap with a variety of symbols in the image, and this part of the region was devoid of realistic background for a realistic inpainting quality assessment. Additionally, we avoided network learning to use the features of these symbols. We compare our mask generation approach with the conventional method, and Tables 5 and 6 show that our method effectively improves the SSIM, LPIPS, and FID.
- Attention mechanism
  For the network to attenuate the focus on symbolic features in the image and enhance the focus on other features in the real background, we introduced the SE layer. By introducing the channel attention mechanism, our model pays more attention to the features of non-symbolic regions rather than the features of symbolic regions and chooses useful features. By this method, the restored image is more similar to the original image in terms of content and no yellow pixels show up in the restoration region. Tables 5 and 6 show the effects of the experiments.

**Table 4.** Effects of different convolutions.

| Convs | LPIPS | FID |
|---|---|---|
| Regular | 0.92230 | 30.84 |
| Dilated | 0.08447 | 26.77 |
| Fast Fourier | 0.08215 | 25.54 |

**Table 5.** Results of experiments on input with a resolution of 256 × 256.

| Model | SSIM | FID | LPIPS |
|---|---|---|---|
| Base (only FFCs) | 0.9209 | 25.54 | 0.08215 |
| Base + Mask | 0.9240 | 23.08 | 0.08044 |
| Base + SE-Layer | 0.9238 | 23.02 | 0.07987 |
| Base + Mask + SE-Laye | 0.9246 | 22.56 | 0.07806 |

**Table 6.** Results of experiments on input with a resolution of 512 × 512.

| Model | SSIM | FID | LPIPS |
|---|---|---|---|
| Base (only FFCs) | 0.9170 | 28.58 | 0.08939 |
| Base + Mask | 0.9189 | 27.15 | 0.08842 |
| Base + SE-Layer | 0.9102 | 26.89 | 0.08769 |
| Base + Mask + SE-Layer | 0.9208 | 25.52 | 0.08300 |

3.1.3. Experiments on the Lesion Segmentation

As we noted in the introduction, our aim of inpainting of 2D ovarian-tumor ultrasound images is to enhance the accuracy of currently popular segmentation models such as Unet and PSPnet for the segmentation of ovarian lesions.

Figures 2 and A1 show the negative effect of symbols in the image on the segmentation of the lesion: they make the model focus more on these symbols. These symbols provide additional information such that the accuracy of segmentation of ovarian-tumor images that are completely clean and without symbols is substantially reduced, which is unacceptable in clinical practice. Therefore, we used the inpainting images and the original images as two training sets, and the clean images as the common test set for experiments on lesion segmentation. Figures 9 and A5 confirm that the segmentation accuracy was improved from 71.51% to 76.06% for the Unet [19] model and from 61.13% to 66.65% for the PSPnet [47] model in clean images. Figure 10 indicates the segmentation results of the Unet model using the clean images as a testing set. Our approach appreciably improves the accuracy of lesion segmentation, and the visualization of segmentation is much better for experiments on lesion segmentation with clean images. These experiments confirm our conjecture and our original aim of performing image inpainting.

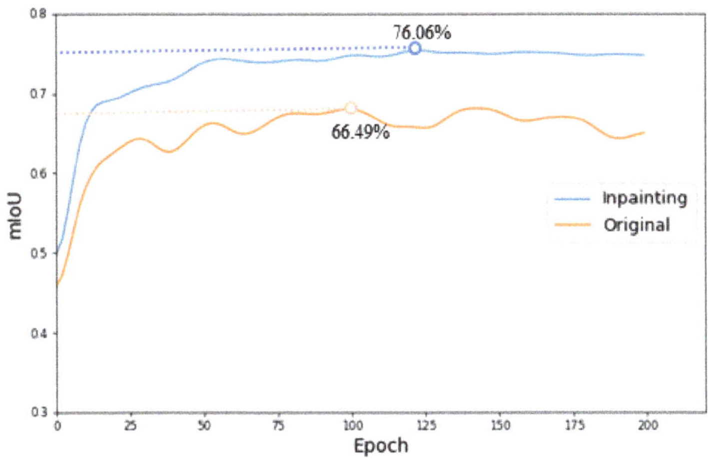


**Figure 9.** The accuracy graph of lesion segmentation of the Unet [19] model. The blue line represents the accuracy of using the inpainting images as the training set. The yellow line represents the accuracy of using the original images with symbols as the training set.

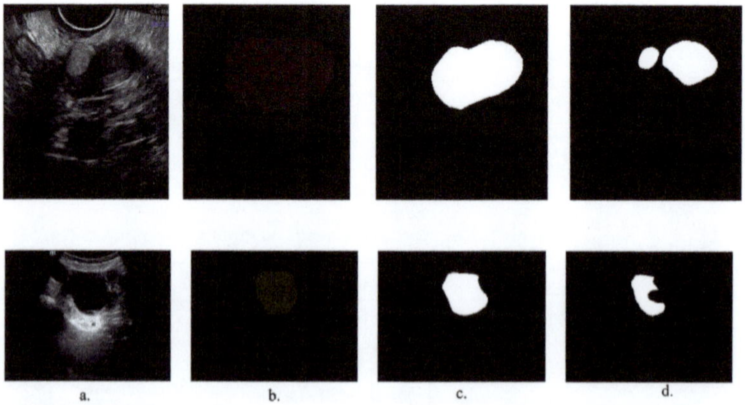


**Figure 10.** Visualization of the results of lesion segmentation of Unet. (**a**) The clean image. (**b**) The ground truth image. (**c**) Segmentation result of the Unet model using the inpainting images as the training set. (**d**) Segmentation result of the Unet model using the original images as the training set.

## 4. Conclusions

In this paper, we proposed a 2D ovarian-tumor ultrasound image inpainting dataset to investigate the effect of prevalent symbols in images on ovarian-lesion segmentation. Based on this image inpainting dataset, we proposed a 2D ovarian-tumor ultrasound image inpainting model based on fast Fourier convolution and a channel attention mechanism. Labeled images are used as a priori information to guide the model to focus on features in the non-symbolic regions of the images, and fast Fourier convolution is used to extend the receptive field of the model to make the texture and structure of the inpainting images more realistic and the boundaries smoother. Our model outperformed existing methods in both qualitative and quantitative comparisons. It received the highest scores in all three metrics, LPIPS, FID, and SSIM, which proves the effectiveness of our model. We used the inpainting images for training and validation with Unet and PSPnet models, which appreciably enhanced the accuracy of lesion segmentation in clean images. This additionally demonstrates the great significance of our study for computer-aided diagnosis of ovarian tumors.

Our study in this paper did not currently use ground truth of lesion segmentation in the dataset, which may further improve the similarity of lesion boundaries in inpainted images. In future work, we will do further exploration on how to apply the edge information of the lesion to the model to make the boundaries more similar to those in the original image and extend our model to other types of medical images—CT, MRI, etc.

**Author Contributions:** Conceptualization, C.Q. and L.C. (Lijiang Chen); methodology, C.Q.; writing—original draft preparation, C.Q.; writing—review and editing, L.C. (Linghan Cai) and W.B.; project administration, L.C. (Lijiang Chen) and W.B.; funding acquisition, W.B. and L.C (Lijiang Chen).; data curation, M.W., C.Y., X.S. and M.Y. All authors have read and agreed to the published version of the manuscript.

**Funding:** This research was partly supported by the National Natural Science Foundation of China (Grant No. 62072021), the Fundamental Research Funds for the Central Universities (Grant No. YWF-22-L-532), and the Beijing Hospitals Authority's Ascent Plan (Grant No. DFL20190701).

**Institutional Review Board Statement:** Not applicable.

**Informed Consent Statement:** Not applicable.

**Data Availability Statement:** The 2D ovarian-tumor ultrasound image inpainting dataset we created by annotation can be accessed at https://github.com/freedom423/2D-ovarian-tumor-ultrasound-image-inpainting, (accessed on 18 November 2022).

**Acknowledgments:** The authors of this paper would like to thank all subjects involved in the building these two datasets used in this study.

**Conflicts of Interest:** The authors declare no conflict of interest.

## Appendix A

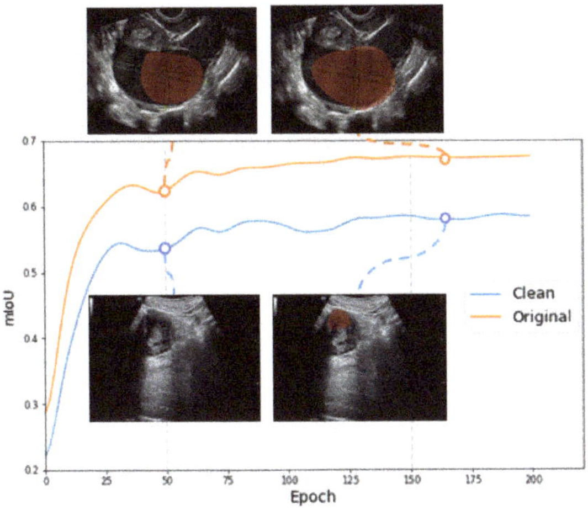


**Figure A1.** The accuracy graph of lesion segmentation of the PSPnet [47] model. The blue line represents the accuracy of using the clean images as the testing set. The yellow line represents the accuracy of using the original images with symbols as the testing set. The figure also shows a visualization of the segmentation results for the 50th and 150th epochs.

Figures A2–A4 show the results of different methods on images containing different types of symbols.

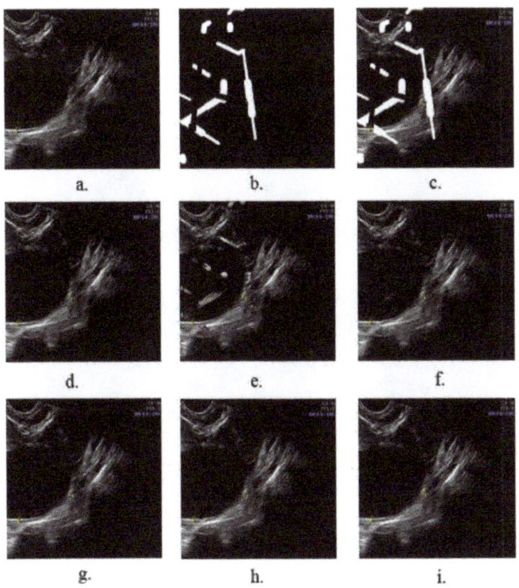


**Figure A2.** Comparison between the results of our proposed model and those of other models. (**a**) The ground truth image. (**b**) The mask for the ground truth image. (**c**) The masked image. (**d**) LaMa method. (**e**) GL method. (**f**) DF 1 method. (**g**) Our method. (**h**) PC method. (**i**) DF 2 method.

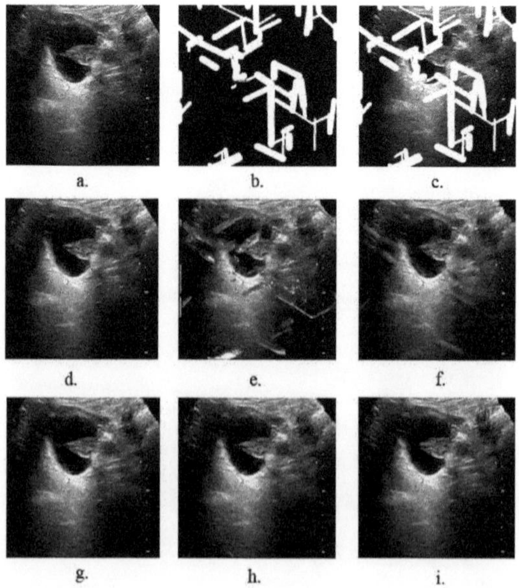


**Figure A3.** Comparison between the results of our proposed model and those of other models. (**a**) The ground truth image. (**b**) The mask for the ground truth image. (**c**) The masked image. (**d**) LaMa method. (**e**) GL method. (**f**) DF 1 method. (**g**) Our method. (**h**) PC method. (**i**) DF 2 method.

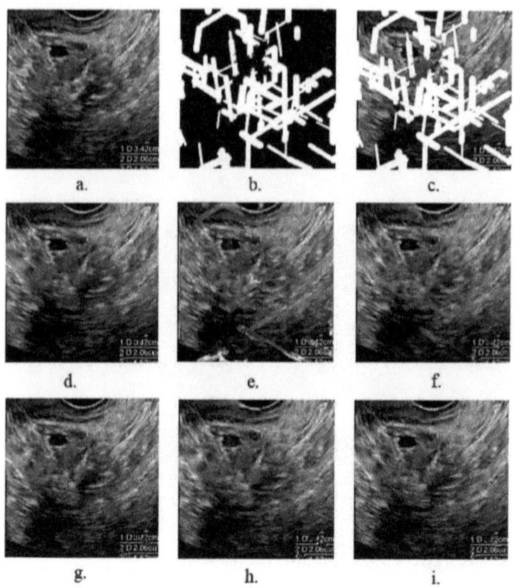


**Figure A4.** Comparison between the results of our proposed model and those of other models. (**a**) The ground truth image. (**b**) The mask for the ground truth image. (**c**) The masked image. (**d**) LaMa method. (**e**) GL method. (**f**) DF 1 method. (**g**) Our method. (**h**) PC method. (**i**) DF 2 method.

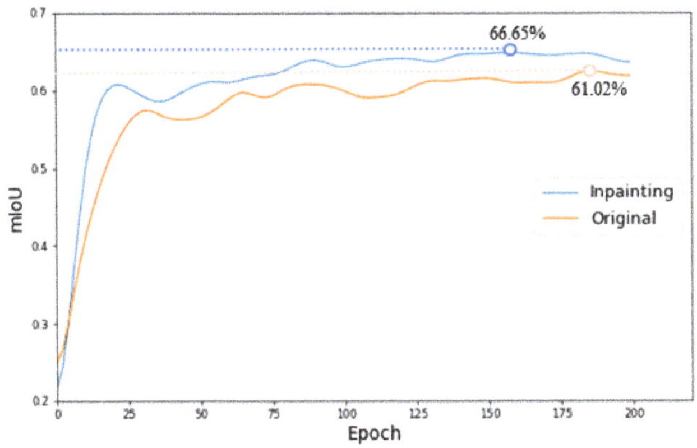


**Figure A5.** The accuracy graph of lesion segmentation of the PSPnet [47] model. The blue line represents the accuracy of using the inpainting images as the training set. The yellow line represents the accuracy of using the original images as the training set.

## References


1. Al-Dhabyani, W.; Gomaa, M.; Khaled, H.; Fahmy, A. Dataset of breast ultrasound images. *Data Brief* **2020**, *28*, 104863. [CrossRef]
2. George, M.; Anita, H. Analysis of kidney ultrasound images using deep learning and machine learning techniques: A review. *Pervasive Comput. Soc. Netw.* **2022**, *317*, 183–199.
3. Savaş, S.; Topaloğlu, N.; Kazcı, Ö.; Koşar, P.N. Classification of carotid artery intima media thickness ultrasound images with deep learning. *J. Med. Syst.* **2019**, *43*, 1–12. [CrossRef]
4. Li, H.; Weng, J.; Shi, Y.; Gu, W.; Mao, Y.; Wang, Y.; Liu, W.; Zhang, J. An improved deep learning approach for detection of thyroid papillary cancer in ultrasound images. *Sci. Rep.* **2018**, *8*, 1–12. [CrossRef]
5. Karimi, D.; Zeng, Q.; Mathur, P.; Avinash, A.; Mahdavi, S.; Spadinger, I.; Abolmaesumi, P.; Salcudean, S.E. Accurate and robust deep learning-based segmentation of the prostate clinical target volume in ultrasound images. *Med Image Anal.* **2019**, *57*, 186–196. [CrossRef]
6. Zhao, Q.; Ma, Y.; Lyu, S.; Chen, L. Embedded self-distillation in compact multibranch ensemble network for remote sensing scene classification. *IEEE Trans. Geosci. Remote Sens.* **2021**, *60*, 1–15. [CrossRef]
7. Zhao, Q.; Lyu, S.; Li, Y.; Ma, Y.; Chen, L. MGML: Multigranularity multilevel feature ensemble network for remote sensing scene classification. *IEEE Trans. Neural Netw. Learn. Syst.* **2021**. [CrossRef]
8. Cai, L.; Wu, M.; Chen, L.; Bai, W.; Yang, M.; Lyu, S.; Zhao, Q. Using Guided Self-Attention with Local Information for Polyp Segmentation. In Proceedings of the International Conference on Medical Image Computing and Computer Assisted Intervention–MICCAI 2022: 25th International Conference, Singapore, 18–22 September 2022; pp. 629–638.
9. Singh, A.; Sengupta, S.; Lakshminarayanan, V. Explainable deep learning models in medical image analysis. *J. Imaging* **2020**, *6*, 52. [CrossRef]
10. Chouhan, V.; Singh, S.K.; Khamparia, A.; Gupta, D.; Tiwari, P.; Moreira, C.; Damaševičius, R.; De Albuquerque, V.H.C. A novel transfer learning based approach for pneumonia detection in chest X-ray images. *Appl. Sci.* **2020**, *10*, 559. [CrossRef]
11. Loey, M.; Smarandache, F.M.; Khalifa, N.E. Within the lack of chest COVID-19 X-ray dataset: A novel detection model based on GAN and deep transfer learning. *Symmetry* **2020**, *12*, 651. [CrossRef]
12. Frid-Adar, M.; Diamant, I.; Klang, E.; Amitai, M.; Goldberger, J.; Greenspan, H. GAN-based synthetic medical image augmentation for increased CNN performance in liver lesion classification. *Neurocomputing* **2018**, *321*, 321–331. [CrossRef]
13. Zhao, A.; Balakrishnan, G.; Durand, F.; Guttag, J.V.; Dalca, A.V. Data augmentation using learned transformations for one-shot medical image segmentation. In Proceedings of the IEEE/CVF Conference on Computer Vision and Pattern Recognition, Long Beach, CA, USA, 15–20 June 2019; pp. 8543–8553.
14. Zhao, Q.; Lyu, S.; Bai, W.; Cai, L.; Liu, B.; Wu, M.; Sang, X.; Yang, M.; Chen, L. A Multi-Modality Ovarian Tumor Ultrasound Image Dataset for Unsupervised Cross-Domain Semantic Segmentation. *arXiv* **2022**, arXiv:2207.06799.
15. Yao, S.; Yan, J.; Wu, M.; Yang, X.; Zhang, W.; Lu, H.; Qian, B. Texture synthesis based thyroid nodule detection from medical ultrasound images: Interpreting and suppressing the adversarial effect of in-place manual annotation. *Front. Bioeng. Biotechnol.* **2020**, *8*, 599. [CrossRef]
16. Armanious, K.; Mecky, Y.; Gatidis, S.; Yang, B. Adversarial inpainting of medical image modalities. In Proceedings of the ICASSP 2019–2019 IEEE International Conference on Acoustics, Speech and Signal Processing (ICASSP), Brighton, UK, 12–17 May 2019, pp. 3267–3271.

17. Xie, E.; Ni, P.; Zhang, R.; Li, X. Limited-Angle CT Reconstruction with Generative Adversarial Network Sinogram Inpainting and Unsupervised Artifact Removal. *Appl. Sci.* **2022**, *12*, 6268. [CrossRef]
18. Kwon, H.J.; Lee, S.H. A Two-Step Learning Model for the Diagnosis of Coronavirus Disease-19 Based on Chest X-ray Images with 3D Rotational Augmentation. *Appl. Sci.* **2022**, *12*, 8668. [CrossRef]
19. Ronneberger, O.; Fischer, P.; Brox, T. U-Net: Convolutional Networks for Biomedical Image Segmentation. In Proceedings of the Medical Image Computing and Computer-Assisted Intervention–MICCAI 2015: 18th International Conference, Munich, Germany, 5–9 October 2015; pp. 234–241.
20. Alsalamah, M.; Amin, S. Medical image inpainting with RBF interpolation technique. *Int. J. Adv. Comput. Sci. Appl.* **2016**, *7*. [CrossRef]
21. Guizard, N.; Nakamura, K.; Coupé, P.; Fonov, V.S.; Arnold, D.L.; Collins, D.L. Non-local means inpainting of MS lesions in longitudinal image processing. *Front. Neurosci.* **2015**, *9*, 456. [CrossRef]
22. Vlanek, P. Fuzzy image inpainting aimed to medical imagesl. In Proceedings of the International Conference in Central Europe on Computer Graphics, Visualization and Computer Vision, Prague, Czech Republic, 27 February 2018. [CrossRef]
23. Arnold, M.; Ghosh, A.; Ameling, S.; Lacey, G. Automatic segmentation and inpainting of specular highlights for endoscopic imaging. *EURASIP J. Image Video Process.* **2010**, *2010*, 1–12. [CrossRef]
24. Nazeri, K.; Ng, E.; Joseph, T.; Qureshi, F.; Ebrahimi, M. Edgeconnect: Structure guided image inpainting using edge prediction. In Proceedings of the IEEE/CVF International Conference on Computer Vision Workshops, Seoul, Republic of Korea, 27–28 October 2019; pp. 3265–3274.
25. Wang, Q.; Chen, Y.; Zhang, N.; Gu, Y. Medical image inpainting with edge and structure priors. *Measurement* **2021**, *185*, 110027. [CrossRef]
26. Wei, Y.; Gan, Z.; Li, W.; Lyu, S.; Chang, M.C.; Zhang, L.; Gao, J.; Zhang, P. MagGAN: High-Resolution Face Attribute Editing with Mask-Guided Generative Adversarial Network. In Proceedings of the Asian Conference on Computer Vision (ACCV), Kyoto, Japan, 30 November–4 December 2020.
27. Zeng, Y.; Fu, J.; Chao, H.; Guo, B. Learning Pyramid-Context Encoder Network for High-Quality Image Inpainting. In Proceedings of the IEEE/CVF Conference on Computer Vision and Pattern Recognition (CVPR), Long Beach, CA, USA, 15–20 June 2019.
28. Yi, Z.; Tang, Q.; Azizi, S.; Jang, D.; Xu, Z. Contextual Residual Aggregation for Ultra High-Resolution Image Inpainting. In Proceedings of the IEEE/CVF Conference on Computer Vision and Pattern Recognition (CVPR), Seattle, WA, USA, 13–19 June 2020.
29. Yan, Z.; Li, X.; Li, M.; Zuo, W.; Shan, S. Shift-net: Image inpainting via deep feature rearrangement. In Proceedings of the European Conference on Computer Vision (ECCV), Munich, Germany, 8–14 September 2018; pp. 1–17.
30. Wu, C.; Wang, Y.; Wang, F. Deep learning for ovarian tumor classification with ultrasound images. In Proceedings of the Pacific Rim Conference on Multimedia, Hefei, China, 21–22 September 2018; Proceedings, Part III; 2018; pp. 395–406.
31. Christiansen, F.; Epstein, E.; Smedberg, E.; Åkerlund, M.; Smith, K.; Epstein, E. Ultrasound image analysis using deep neural networks for discriminating between benign and malignant ovarian tumors: Comparison with expert subjective assessment. *Ultrasound Obstet. Gynecol.* **2021**, *57*, 155–163. [CrossRef]
32. Zhang, Z.; Han, Y. Detection of ovarian tumors in obstetric ultrasound imaging using logistic regression classifier with an advanced machine learning approach. *IEEE Access* **2020**, *8*, 44999–45008. [CrossRef]
33. Jin, J.; Zhu, H.; Zhang, J.; Ai, Y.; Zhang, J.; Teng, Y.; Xie, C.; Jin, X. Multiple U-Net-based automatic segmentations and radiomics feature stability on ultrasound images for patients with ovarian cancer. *Front. Oncol.* **2021**, *10*, 614201. [CrossRef] [PubMed]
34. He, K.; Zhang, X.; Ren, S.; Sun, J. Deep residual learning for image recognition. In Proceedings of the IEEE Conference on Computer Vision and Pattern Recognition, Ballari, India, 23–24 April 2016; pp. 770–778.
35. Chi, L.; Jiang, B.; Mu, Y. Fast fourier convolution. *Adv. Neural Inf. Process. Syst.* **2020**, *33*, 4479–4488.
36. Pathak, D.; Krahenbuhl, P.; Donahue, J.; Darrell, T.; Efros, A.A. Context encoders: Feature learning by inpainting. In Proceedings of the IEEE Conference on Computer Vision and Pattern Recognition, Las Vegas, NV, USA, 27–30 June 2016; pp. 2536–2544.
37. Yu, J.; Lin, Z.; Yang, J.; Shen, X.; Lu, X.; Huang, T.S. Generative image inpainting with contextual attention. In Proceedings of the IEEE Conference on Computer Vision and Pattern Recognition, Salt Lake City, UT, USA, 18–23 June 2018; pp. 5505–5514.
38. Iizuka, S.; Simo-Serra, E.; Ishikawa, H. Globally and locally consistent image completion. *ACM Trans. Graph. (ToG)* **2017**, *36*, 1–14. [CrossRef]
39. Liu, G.; Reda, F.A.; Shih, K.J.; Wang, T.C.; Tao, A.; Catanzaro, B. Image inpainting for irregular holes using partial convolutions. In Proceedings of the European Conference on Computer Vision (ECCV), Munich, Germany, 8–14 September 2018; pp. 85–100.
40. Yu, J.; Lin, Z.; Yang, J.; Shen, X.; Lu, X.; Huang, T.S. Free-form image inpainting with gated convolution. In Proceedings of the IEEE/CVF International Conference on Computer Vision, Seoul, Republic of Korea, 27 October–2 November 2019; pp. 4471–4480.
41. Johnson, J.; Alahi, A.; Fei-Fei, L. Perceptual losses for real-time style transfer and super-resolution. In Proceedings of the European Conference on Computer Vision, Amsterdam, The Netherlands, 8–16 October 2016; pp. 694–711.
42. Suvorov, R.; Logacheva, E.; Mashikhin, A.; Remizova, A.; Ashukha, A.; Silvestrov, A.; Kong, N.; Goka, H.; Park, K.; Lempitsky, V. Resolution-robust large mask inpainting with fourier convolutions. In Proceedings of the IEEE/CVF Winter Conference on Applications of Computer Vision, Waikoloa, HI, USA, 3–8 January 2022; pp. 2149–2159.
43. Wang, T.C.; Liu, M.Y.; Zhu, J.Y.; Tao, A.; Kautz, J.; Catanzaro, B. High-resolution image synthesis and semantic manipulation with conditional gans. In Proceedings of the IEEE Conference on Computer Vision and Pattern Recognition, Salt Lake City, UT, USA, 18–23 June 2018; pp. 8798–8807.

44. Wang, Z.; Bovik, A.C.; Sheikh, H.R.; Simoncelli, E.P. Image quality assessment: From error visibility to structural similarity. *IEEE Trans. Image Process.* **2004**, *13*, 600–612. [CrossRef]
45. Heusel, M.; Ramsauer, H.; Unterthiner, T.; Nessler, B.; Hochreiter, S. Gans trained by a two time-scale update rule converge to a local nash equilibrium. *Adv. Neural Inf. Process. Syst.* **2017**, *30*. [CrossRef]
46. Zhang, R.; Isola, P.; Efros, A.A.; Shechtman, E.; Wang, O. The unreasonable effectiveness of deep features as a perceptual metric. In Proceedings of the IEEE Conference on Computer Vision and Pattern Recognition, Salt Lake City, UT, USA, 18–23 June 2018; pp. 586–595.
47. Zhao, H.; Shi, J.; Qi, X.; Wang, X.; Jia, J. Pyramid Scene Parsing Network. In Proceedings of the IEEE Conference on Computer Vision and Pattern Recognition, Honolulu, HI, USA, 21–26 July 2017; pp. 2881–2890.

bioengineering

MDPI

Article

# Enhancement Technique Based on the Breast Density Level for Mammogram for Computer-Aided Diagnosis

Noor Fadzilah Razali [1], Iza Sazanita Isa [1,*], Siti Noraini Sulaiman [1,2], Noor Khairiah Abdul Karim [3,4], Muhammad Khusairi Osman [1] and Zainal Hisham Che Soh [1]

1 Centre for Electrical Engineering Studies, Universiti Teknologi MARA, Cawangan Pulau Pinang, Permatang Pauh Campus, Bukit Mertajam 13500, Pulau Pinang, Malaysia
2 Integrative Pharmacogenomics Institute (iPROMISE), Universiti Teknologi MARA Cawangan Selangor, Puncak Alam Campus, Puncak Alam 42300, Selangor, Malaysia
3 Department of Biomedical Imaging, Advanced Medical and Dental Institute, Universiti Sains Malaysia Bertam, Kepala Batas 13200, Pulau Pinang, Malaysia
4 Breast Cancer Translational Research Programme (BCTRP), Advanced Medical and Dental Institute, Universiti Sains Malaysia Bertam, Kepala Batas 13200, Pulau Pinang, Malaysia
* Correspondence: izasazanita@uitm.edu.my

**Abstract:** Mass detection in mammograms has a limited approach to the presence of a mass in overlapping denser fibroglandular breast regions. In addition, various breast density levels could decrease the learning system's ability to extract sufficient feature descriptors and may result in lower accuracy performance. Therefore, this study is proposing a textural-based image enhancement technique named Spatial-based Breast Density Enhancement for Mass Detection (SbBDEM) to boost textural features of the overlapped mass region based on the breast density level. This approach determines the optimal exposure threshold of the images' lower contrast limit and optimizes the parameters by selecting the best intensity factor guided by the best Blind/Reference-less Image Spatial Quality Evaluator (BRISQUE) scores separately for both dense and non-dense breast classes prior to training. Meanwhile, a modified You Only Look Once v3 (YOLOv3) architecture is employed for mass detection by specifically assigning an extra number of higher-valued anchor boxes to the shallower detection head using the enhanced image. The experimental results show that the use of SbBDEM prior to training mass detection promotes superior performance with an increase in mean Average Precision (mAP) of 17.24% improvement over the non-enhanced trained image for mass detection, mass segmentation of 94.41% accuracy, and 96% accuracy for benign and malignant mass classification. Enhancing the mammogram images based on breast density is proven to increase the overall system's performance and can aid in an improved clinical diagnosis process.

**Keywords:** breast density; CAD; image enhancement; breast cancer; deep learning; textural

**Citation:** Razali, N.F.; Isa, I.S.; Sulaiman, S.N.; Abdul Karim, N.K.; Osman, M.K.; Che Soh, Z.H. Enhancement Technique Based on the Breast Density Level for Mammogram for Computer-Aided Diagnosis. *Bioengineering* **2023**, *10*, 153. https://doi.org/10.3390/bioengineering10020153

Academic Editors: Paolo Zaffino and Maria Francesca Spadea

Received: 28 November 2022
Revised: 4 January 2023
Accepted: 16 January 2023
Published: 23 January 2023



## 1. Introduction

According to International Agency for Research on Cancer, an estimated 2.3 million new cases of breast cancer has overtaken lung cancer as the most prevalent cancer diagnosed, with cancer death rates significantly higher in transitioning nations [1]. Breast screening programs are a way to detect early signs of breast cancer and are dominated by utilizing digital mammography as the primary tool for cancer detection [2]. Additional modalities such as ultrasound are used in conjunction with mammography for denser breasts, whereas magnetic resonance imaging (MRI) is used for more progressive breast analysis for repeated and high-risk patients [3].

Breast density, as defined by the American College of Radiology (ACR), is used during clinical diagnosis that classifies the breast into four categories with increasing density: almost entirely fatty, scattered fibroglandular, heterogenous, and finally, extremely dense breast [4].

The heterogeneous dense breast as depicted in Figure 1A and the overlapped mass (red region) (in Figure 1B) on the dense region (blue region) is visually harder to distinguish compared to a non-dense breast that only contains mostly fatty (orange region) tissue. Diverse breast tissue structures cause mixed-intensity variations and limited visibility of breast features [5]. Due to this factor, the processed images may result in less acceptable breast tissue segmentation and inconsistent diagnosis by compromising the system's sensitivity and specificity to detect abnormalities [6,7]. Past studies concluded that mass detection decreased with increased density, due to the mass itself being similar to the surrounding dense tissue of the breast [8–10]. Additionally, image quality conditions also make it difficult to detect the lesion in dense breasts [11,12]. Specifying the edge of the mass from its surrounding dense tissue requires image processing that enhances the textural element of the image as one of the defining mass descriptors to assess a mammogram visually [13]. The textural analysis identifies distinctive descriptors in the form of a changing pattern or pixel intensity with various spatial arrangements. Its refinement aims to go beyond human-eye perception by defining semantic descriptors to extract quantitative radiological data [14].

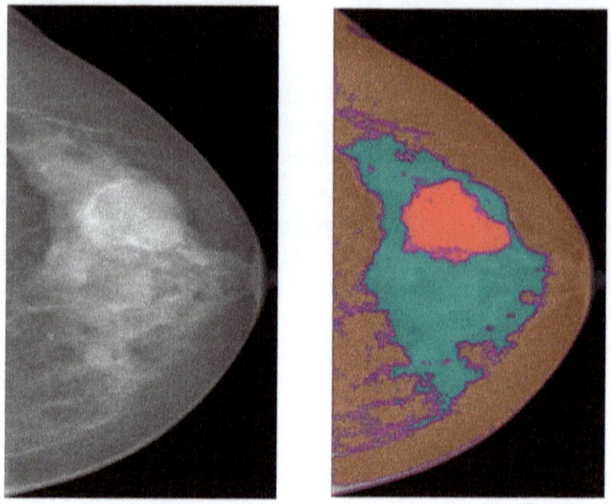

(A) (B)

**Figure 1.** (**A**) Original mammogram image example. (**B**) Mapped tissue region for the image on (**A**)—red: mass, green: dense tissue, orange: non-dense tissue.

To accommodate the analysis of mammographic mass, Computer-Aided Diagnosis (CAD) systems are introduced to breast cancer diagnosis stages, from improving the image quality [11,15], breast lesion detection, and segmentation [16], as well as benign or malignant classification [16–19]. Moreover, CAD implementation in mammography diagnostic could reduce the human rater's false-positive rate by 5.7% and false negative by 9.4%, as shown in a USA-based dataset [19], and an increase rate of 3% recall rate for a radiologist's mammogram analysis with CAD assistance for an expert radiologist [20]. CAD systems proved to aid radiologists in making a better diagnosis with the area under the curve (AUC) of 0.896 from 0.850 without affecting diagnosis timing [21]. Since deep-learning CAD systems performed best when trained using large datasets [22], it is harder to apply suitable image quality improvements individually on the images, leading to a need for special enhancement procedures and careful pre-processing for the images before they can be trained on a deep-learning architecture.

Most Convolutional Neural Network (CNN) applications for CAD systems have focused on direct mammogram images for detection and classification rather than the need for specific enhancement based on breast density level and the quality of the input images. This could unintentionally lead to reduced sensitivity for mass detection in dense mammograms, resulting from higher training weightage on non-dense breasts because of dataset class imbalance [23]. Enhancement techniques based on histogram manipulation, such as adaptive/histogram equalization (HE/AHE) and contrast-limited adaptive histogram equalization (CLAHE), have been extensively used to enhance the images before training. Nevertheless, the method's adaptability for different densities of the breast images and their effects needs to be paid attention. Several studies have included the analysis of the impact of breast density on the post-training level rather than countering its effect on the pre-processing level [10,18,24–26]. However, pre-processing analysis of the mass surrounded by dense tissue is essential to verify that the established CAD system is robust to dense breast images for accurate mass detection.

Based on this motivation, we proposed an enhancement technique that adapts non-dense and dense breast categories by subtly changing the non-dense region appearance within a mammogram image through textural refinement, mimicking the radiologist's manual contrast adjustment on individual images while maintaining the visual perceptual of the original image. The textural refinement on the mass edges boosts its feature vector representability during the convolutional process for detection and segmentation algorithm for better classification performance.

In summary, this work's contributions are focused on:

1. A breast density-based configuration is incorporated prior to the training detection algorithm.
2. An enhancement technique that enhances the textural appearance of the background and mass region by determining the threshold of the dense and non-dense region through a buffer region by manipulating the images' lower limit cap threshold value.

## 2. Past Literature

Image enhancement is required to optimize the image's overall quality in preparation for subsequent stages. Enhancements using histogram-based techniques have been proven to enhance mammogram images, such as through histogram equalization [26,27] and the widely used contrast-limited adaptive histogram equalization (CLAHE) [10,18,24,28,29]. Histogram-based image enhancement increases the contrast and dynamic range of the grayscale image by adjusting an image's contrast using its histogram and increases the image's contrast by dispersing the most common pixel intensity values by extending the image's intensity range [30]. Researchers also combined CLAHE with their proposed method to improve their performance. For instance, CLAHE was utilized in conjunction with unsharp masking filtering, with the effectiveness in demonstrating an enhancement for mass region segmentation [31]. In addition, breast cancer detection using a modified CLAHE method is used to sharpen the margins of the masses on three datasets [32]. Meanwhile, CLAHE, wavelet, and anisotropic diffusion combination were presented for mammography enhancement in [33] and obtained a sensitivity of 93% when tested on a limited number of abnormal and normal images from the mini-Mammographic Image Analysis Society (mini-MIAS) dataset. The introduction of multilevel Otsu's thresholding with wind-driven optimization for mass detection utilizing CLAHE enhancement on mini-MIAS and Digital Database for Screening Mammography (DDSM) mammogram datasets is conducted with 96.9% and 96.2% detection sensitivity [29].

Additionally, a different approach using top-hat transform-based mammography enhancement is established to increase the contrast between the suspicious area and normal breast tissues, increasing mass detection sensitivity using the proposed technique compared to unenhanced images [34]. Moreover, grayscale transformation applied by [35] helps reveal more information and increase contrast by selectively emphasizing or suppressing undesirable elements in the image, hence uniformizing the pixel distribution. Recently, a study to detect mass with its performance improved using contrast-based enhancement by employing a hyperbolic tangent function with an adjustable Tunicate swarm algorithm as optimization of the system via fitness function is demonstrated by [36] and shows improvement when compared to the CLAHE method. The use of another optimization through hybridized fast and robust fuzzy c-means clustering (FRFCM) and particle swarm optimization before mass detection was proposed on the mini-MIAS with 96.6% sensitivity [37]. A classification system for mammogram cancer by [38] using improved multi-fractal dimension features also included a pre-processing subsystem for denoising the mammogram following the cancer region segmentation.

These methods produced good final performance. However, these studies applied a straightforward object detection algorithm to analyze their method's effectiveness for the images to be trained in a full-scale CAD system. Moreover, the enhancement methods did not take the effect of variation of breast density into consideration, with some methods causing the final mass to be indistinguishable from the dense tissue [27,31], where the final output is in the form of classification of mass and non-mass only. This could raise the issue of losing crucial mass features if continued to the cancerous mass classification stage later. The studies were also not tested against any image quality metrics as an essential aspect of any image enhancement method proposal, by using metric performance such as applied by [36], which is not considered the best in the analysis of enhancement for breast density as it relies on the contrast and intensity of the images.

Existing state-of-the-art object identification techniques such as Faster Region-based CNN (R-CNN) [39], You Only Look Once (YOLO) versions [40,41], and Single Shot MultiBox Detector (SSD) [42] have been implemented in many vision studies for detection, following the image enhancement techniques. YOLO has been proven to be the most beneficial in terms of accurate and fast detection rate [43,44] compared to the other detection algorithms. For example, mass detection using the YOLO model was carried out as proved by Al-Antari et al. [45] and resulted in a detection accuracy of 98.96%. Similarly, [28] enhanced their approach by comparing feedforward CNN, ResNet-50, and InceptionResNet-V2 for classification before implementing the YOLO model for detection. Subsequently, this team [46] proposed a CAD system framework that classified breast masses into malignant and benign using Fully Connected Neural Networks (F-CNNs). This system framework first detected breast masses using the YOLO model with an overall accuracy of 99.7%. Meanwhile, [47] employed the YOLO fusion model for breast mass detection by fusing the best feature representation from single-class mass-based and calcification-based training models to a multiclass model that combined the feature maps. Their best performance observed was 98.1% for mass lesion accuracy detection. In [48], fusion YOLO was used for detection by introducing new classes of normal and architectural distortion abnormality on final prediction with mass detection accuracy at 93% $\pm$ 0.118.

Based on the discussions, although different strategies were implemented to boost mass detection performance, the study has severe limitations that have been conducted to adapt the breast density variance effect through enhancement techniques before training the system. A fully automated mass detection based on density through CAD is crucial, especially with its link with 2.2-fold more cancer risk in clinical profiling for denser breasts reported [49]. Studies conducted by [10,18,24,25] all pointed to a decrease in the model's performance when trained using denser breast images. One of the earliest studies of mammograms that includes adaptation to breast density developed their model using density-based spatial clustering of applications with noise (DB-SCAN), highlighting the breasts' internal structure before training [25]. Likewise, the same method was applied by [24] on a different dataset to improve the method proposed by [25], where the author introduces a two-stage false positive reduction process through bilateral breast analysis. Even though it has good results in preparing the models based on breast density, limitations include if only unilateral breast is available, and asymmetrical factors for both breasts might affect the performance.

## 3. Proposed Methodology

This section discusses the overall methodology for completing the framework's three main phases, as shown in Figure 2. Each phase is discussed further in the following subsections.

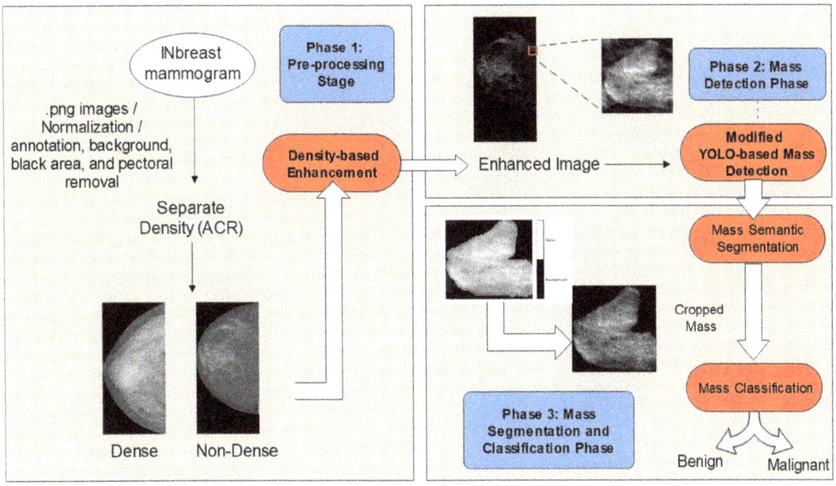


**Figure 2.** Overall Proposed Methodology for Breast Mammogram Mass Classification.

### 3.1. *Experimental Setting*

#### 3.1.1. Dataset: INbreast

The INbreast dataset has been widely used in previous studies [18,28,50,51] and was one of the first established datasets of full-field digital mammograms (FFDM) acquired in 2011 at Centro Hospitalar de S. Joo, Breast Centre, Porto [52]. A total of 410 images were extracted with 115 abnormal lesion cases ranging from mass, calcification, and architectural distortions, with both craniocaudal (CC) and mediolateral oblique (MLO) views. Subsequently, the extracted images were exclusively updated by the authors with permission, along with the annotated ground truth range of interest (ROI) of the segmented mass region. Note that 112 mass images were included for this study that ranges across four breast density classifications, further classified based on their mass types: benign and malignant. To avoid sampling bias, 80% of the images were randomly selected for training, with the remaining 20% used for testing and validation for all stages, and were independent of the breast mass types and density level. Finally, augmentation settings were set into degrees of rotation of 30° to 300°, horizontally flipped, and scaled to randomized 1.0 to 1.3 scale factor. Augmentation settings that alter the hue, contrast, brightness, and saturation were excluded to avoid unintentional intensity changes affecting the breast density.

#### 3.1.2. Experimental Setup

This study focuses on the effect of the proposed SbBDEM enhancement technique applied in the pre-processing to prepare the images for the subsequent stages. The performance was measured by comparing the performance with the system trained using original images and two established histogram-based enhancement techniques. The final classification stage used only the handcrafted learning features to reduce the overall computation, as the mass was already accurately detected and segmented from prior stages. To compare the breast density-wise performance, the initially randomized labeled image numbering was saved from the detection phase onto the following stages to make an unbiased comparison among the same test images. Additionally, a 5-fold cross-validation was performed on the classification stage to ensure the average of using all learning features to compare performance. These experiments were visualized and executed on a workstation equipped with CPU Intel(R) Core$^{TM}$ i7-10870H 2.3 GHz with single GPU graphic card NVIDIA GeForce RTX2060 6GB, 16 GB RAM, and trained and tested on MATLAB (Natick, MS, USA).

### 3.2. Stage 1: Proposed Image Pre-Processing for SbBDEM

#### 3.2.1. Image Preparation

The overall process for Stage 1 is illustrated in Figure 3. Standard morphological operations were applied to remove stray annotation marks to allow only the breast area to maximize the processing image area. To unify features between CC and MLO views, pectoral muscle was digitally removed from the MLO view images. To prepare the image to accommodate the needs of different breast densities, the images were segregated based on their supplied ACR density levels following the supplemented density scores to non-dense (1 = almost entirely fatty, 2 = scattered dense) and dense (3 = heterogeneous dense, 4 = extremely dense) categories.

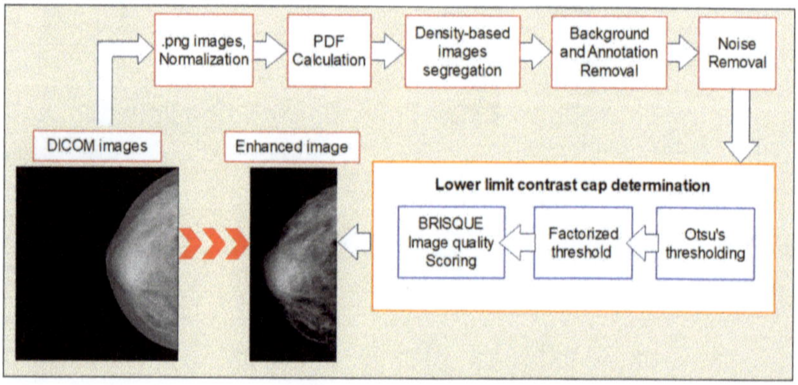


**Figure 3.** Stage 1: Proposed image SbBDEM technique as a pre-processing step.

#### 3.2.2. Lower Limit Contrast Cap Determination

As the next stage of the proposed framework includes mass detection process, it is essential to differentiate the mass from its background whether it is overlapped on the non-dense or dense background. To reduce the non-dense image information while enhancing features from the denser region (hence the mass), image modification was conducted by selecting the best lower-limit contrast of the image. The final output will be a breast image that have a less skin and non-dense region appearance and a pronounced textural definition of the dense region. This includes the mass region while keeping the textural features from the fibroglandular and vascular tissue of the lower-intensity fatty tissue in the background. To achieve this, the higher limit of contrast adjustment was set to the same as the original image.

#### 3.2.3. Factorized Otsu's Thresholding for Breast Density Group Segregation

Otsu's thresholding calculates the point value of intensity based on the image's intensity spread on a bimodal histogram and separates the image into its foreground and background [53]. Since the original mammogram was converted to a normalized grayscale image consisting of two main tissue types that are closely related to its intensity and contrast (higher intensity = dense region, lower intensity = non-dense region), the Otsu's value was definitive in determining the middle-intensity value that separates these tissue groups. Therefore, Otsu's method has been implemented in this study as a reference point for determining the lower limit contrast to be clipped from the input image. However, direct Otsu's threshold separates tissue that might belong to the other side of the histogram, such as the black background as a non-dense region and calcified vessels and the skin lining appearing white in the image as a dense region. To properly lessen this imbalance effect, the threshold value was interpolated on a scale of 1.0 to 1.9 for each non-dense and dense image group that has been separated in the previous step to subtly adapt the sudden change of region foreground to background image as a buffer intensity region. Subsequently, the training images were chosen based on their quality score, which is explained in the next stage.

#### 3.2.4. Blind/Reference-Less Image Spatial Quality Evaluator (BRISQUE)

When an image is altered, it is vital to assess it through an image quality assessment metric by referencing a gold-standard image for quality assessment in terms of its sharpness, contrast, etc., for comparison [54]. Common examples of tests where the referenced image must come from one of the images closely linked to the evaluated image include the mean-square error (MSE) and peak signal-to-noise (PSNR). However, when dealing with deep learning, possibly thousands of images are being trained, making it impossible to select only one for reference quality perspective. This is especially true if the dataset consists of multiple image acquisition techniques, which further vary the dataset's measurement range [55,56]. In this study, to separate the overlapped mass with its background, the non-dense region becomes darker, hence enhancing the mass's edge. This is expected to cause substantial image alteration, with mild changes on the mass and dense regions of the resulted image, causing noise to be increased in the final image. Hence, the MSE and PSNR scores are likely to produce unsatisfactory performance. Moreover, using quality assessments such as PSNR for reconstruction quality in determining the quality of an image used for a detection algorithm is unwarranted since a detection algorithm relies on its ability to separate a mass from its surroundings and, by extension, on the overall image, regardless of the final quality of the image used for training. Therefore, we chose the best Otsu's threshold factor with an image perceptual quality evaluator known as the Blind/Reference-less Image Spatial Quality Evaluator (BRISQUE) [57]. It performed as a spatial feature image assessment metric that is commonly known as opinion-aware and analyses images with similar distortion [57], similar to how visual perception is made. As image distortions affect the quality in term of its textural features (texture signifying the difference of pixel of dense region background and the overlapped mass), BRISQUE was chosen as the primary evaluation metrics in this study. The BRISQUE score guided in choosing the optimal quality factor that clearly defines the difference between non-dense and dense breast images without using any reference image. It provides a rating by generating matching differential mean opinion score (DMOS) values using a support vector machine (SVM) regression model trained on a spatial domain image database [57]. During the training of BRISQUE, the database contained both the clean and edited versions with different additive noise implementations such as Gaussian white noise and blur, compression artifacts, and Rayleigh fast fading channel simulation, serving as the distortion image version for comparison [57]. Besides that, BRISQUE uses scenic data from locally normalized luminosity coefficients to measure any loss of naturalness due to distortion, resulting in a holistic quality score compared to calculating user-defined quality, such as ringing or blurring, as what is being measured when using PSNR [55]. Recent studies of medical images such as mammogram [58–60], lung CT scans [15,58], kidney and brain MRIs [15] have moved towards reference-less image quality evaluators to evaluate their work with good results. In this study, the image group was ultimately selected as the input for mass detection in the subsequent step once the best image score of BRISQUE was obtained.

#### 3.2.5. Evaluation and Analysis of the Proposed Enhancement Technique

We measured the proposed SbBDEM enhancement quality and its direct application in the input of the detection stage based on both reference-less (BRISQUE) and referenced (MSE) measurements. BRISQUE was calculated based on the method proposed by [57], and MSE was given by Equation (1):

$$\text{Mean Squared Error, MSE} = \frac{1}{mn}\sum_{0}^{m}\sum_{0}^{n}||f(i,j)-g(i,j)||^2 \quad (1)$$

where $m$ and $n$ are the image's height and width, $i$ and $j$ are elements from the enhanced image, $f$, and referenced image, $g$, whereas additional textural features analysis was made on the images based on the Gray-Level Co-occurrence Matrix (GLCM) for comparison. The texture properties extracted from the produced matrix were four statistical feature descriptors defined as contrast, correlation, energy, and homogeneity as mathematically defined in Equations (2)–(5). For every element, $P$, it reflected the total number of occurrences of the pixel values of $i$ and $j$ respective to the number of gray levels where $\sigma$ and $\mu$ are the standard deviation and central moments derived in the form of means of variance and skewness.

$$\text{Contrast} = \sum_{i,j=0}^{levels-1} P_{i,j}(i-j)^2, \tag{2}$$

$$\text{Correlation} = \sum_{i,j=0}^{levels-1} P_{i,j}\left[\frac{(i-\mu_i)(j-\mu_j)}{\sqrt{(\sigma_i^2)(\sigma_j^2)}}\right], \tag{3}$$

$$\text{Energy} = \sqrt{\sum_{i,j=0}^{levels-1} P_{i,j}^2}, \tag{4}$$

$$\text{Homogeneity} = \sum_{i,j=0}^{levels-1} P_{i,j}|i-j|, \tag{5}$$

Additional analysis of the images' mean intensity was evaluated for comparison. The mean intensity is the normalized mean number of normalized pixel values in each RGB channel, divided by the total number of pixels in the image, $n$, given in Equation (6).

$$\text{Mean Intensity} = \frac{\sum_{n=0}^{n}(R+G+B)}{n}, \tag{6}$$

For pixel mapping evaluation, we assessed an example of True Positive (TP) and False Positive (FP) from a sample of mass edge from the enhanced testing image using the proposed SbBDEM technique. We assessed the probability of edge detection on the next-best performed on the BRISQUE and MSE scores. Note that mass edge detection's pixel analysis is emulated based on the first layer of modified YOLOv3 based on convolution process from Equation (7), zero padding, with a stride of two with maximum pooling downsampling to reveal the effect of pixel change made during enhancement that affects edge detection. On the other hand, diagonal edge analysis using kernel matrix $K$ = [110, 10-1, 0-1-1] was chosen with a window size of 3-by-3, slides on the image using the convolution process, where $I$ is the cropped mass image with $i, j$ element, $K$ represents the kernel with $x$, $y$ element, and $\eta_W$, $\eta_h$ and $\eta_C$ are the number of heights, widths, and channels of $I$, respectively. Consequently, the maximum pooling downsampled element was chosen to represent both suspected mass and background area. The edge pixel difference of Mass and Background edge detection is denoted as $\Delta$ in Equation (8). Higher $\Delta$ denotes the higher pixel difference between the neighboring pixel encapsulating the mass.

$$\text{Conv(I, K)}_{\text{x,y}} = \sum_{i=1}^{\eta_W}\sum_{j=1}^{\eta_W}\sum_{k=1}^{\eta_C} K_{i,j,k} I_{x+i-1,y+j-1,k} \tag{7}$$

$$\text{Edge pixel difference, } \Delta = \text{Max}_{\text{conv}}(\text{mass}) - \text{Max}_{\text{conv}}(\text{background}) \tag{8}$$

### *3.3. Stage 2: Mass Detection Using Modified YOLOv3*

#### 3.3.1. You Only Look Once (YOLO)

Object detection is a process of detecting a specifically trained object within an image. YOLO and its versions (v2, v3, and so on) implement a single forward-pass filter by splitting the original image into a grid of s-by-s size. Subsequently, a bounding box prediction will be made for each separated cell. The algorithm searches for the object's midpoint during training, where the specific cells containing the midpoint will be responsible for determining the target object's presence. The corresponding cells are linked to the cell with the midpoint, which is set up as the cell with the midpoints defined as the bounding box, which is made of four components [$x$, $y$, $w$, $h$]. Here, $x$ and $y$ are the top left-most coordinates of the bounding box with a value of 0 to 1.0, while $w$ and $h$ are the width and height of the box, respectively. Both $w$ and $h$ could be greater than 1.0 if the final detected box is wider than an entire s-by-s cell. In addition to the four components, each box has a probability value that indicates the presence of an object in the cell and the number of class predictions. Based on this prediction value, the trained network for each cell should be able to output a specific box coordinate that contains the highest probability value for the final detected output for class prediction.

#### 3.3.2. YOLOv3 Modification for Mass Detection

This study utilized the simplest form of YOLOv3 using SqueezeNet [61] as its base network and modified it to improve the overall detection result. Note that the SqueezeNet has only 1.2 million learnable parameters as opposed to the original DarkNet-53 [40] network, which has 41.6 million parameters. As a result, SqueezeNet-based YOLOv3 was chosen to lessen the burden of weightage parameter training. Among the benefits of using a simpler network architecture are more efficiently distributed training parameters, more use of spatial information, which leads to shorter training times, less bandwidth for future model updates, and the ability to be deployed with less memory configuration [62]. Aside from being lightweight, using predefined anchors and detection heads introduced in YOLOv3 architecture allows smaller objects to be detected [40]. Depending on the base network, the YOLOv3 could extract deep features to extract three-scale feature maps from the anchors used for the final bounding-box calculation to predict the best confidence score (CS). YOLOv3 has also been successfully implemented in recent mammogram studies [63,64], showing that its implementation is reliable with good results. A comparison of YOLOv3 and YOLOv4 conducted by [65] shows that even though YOLOv4 is an improvement, it shows no substantial difference between the two models, leading the author to infer that the performance of YOLO primarily depends on the features of the dataset and the representativity of the training images.

Figure 4 illustrates the modified SqueezeNet CNN architecture for the mass detection stage in this study. The input image size was set to 227-by-227, where the enhanced input training images were trained with whole mammogram images. The image went through a series of cascaded and parallel convolutions with concatenation along the nine repeated layers, reducing the information and computation by compacting feature maps as the network went deeper. Two detection heads were allocated when this architecture was modified for detection purposes in YOLOv3. The second detection head was double the size of the downsampled input (28-by-28) of the first detection head (14-by-14), causing smaller masses to be better detected. Since the mass size ranged from the aspect ratio of the breast size, with more than 50% of the training data containing mass with a size less than a sixth of the overall images, we have tried to resolve this problem by devising this architecture by modifying the input of the second detection head.

Hence, to improve the detection of small masses and overall detection performance, we proposed two strategies to solve this problem.

Strategy One: Residual feature mapping for the second detection head: Features from the shallower layer were included (depth concatenation four), containing higher spatial features from the skip connection, and were elementwise added with the semantic features from the deeper layer (depth concatenation nine), where the element-wise addition reduced feature degradation that occurred during downsampling which enhanced feature contrast and feature discrimination [51].

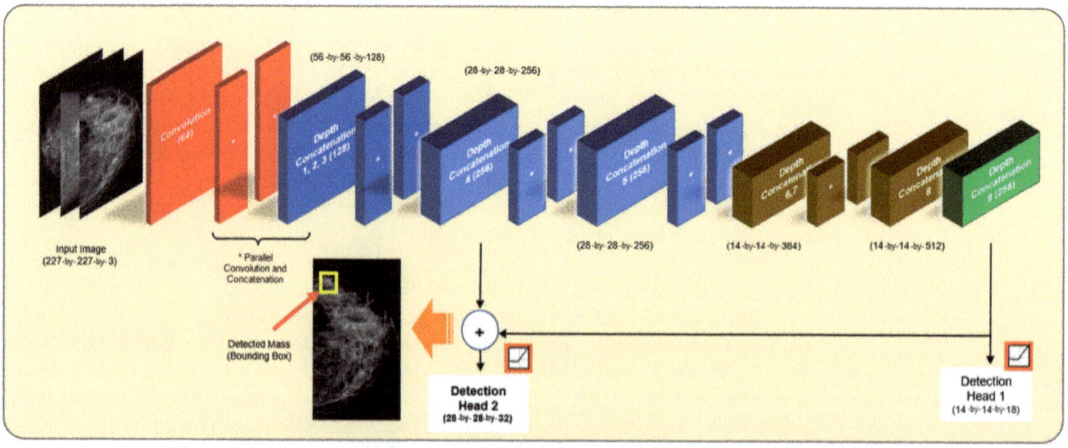


**Figure 4.** Modified SqueezeNet CNN architecture used for YOLOv3 training. The modified layer is in the Bold setting.

Strategy Two: An additional anchor box assigned to a smaller feature map: This anchor box was introduced to the lower scale of the anchor box number of the second detection head (ratio of 4:3 to first detection head). While simply increasing the number of anchor boxes increased the predefined mean intersection over union (IoU), this could only lead to lower performance due to overfitting the number of bounding boxes per image mapping [66]. However, assigning an extra anchor box only for the smaller feature map specifically will increase the bounding box refinement on the feature map allocated to features coming from Strategy One, which increases the possibility of detecting smaller mass sizes coming from the images' semantic information.

The image gave seven predictions with their confidence level scores on every single grid cell with the size of s-by-s. The network was trained on 80 epochs with 10 mini-batch sizes. The learnable parameters were updated through a loop of stochastic gradient descent momentum (sgdm) solver. The initial learning rate was set to 0.001, and a 0.5 confidence score (CS) threshold value was defined for determining the overall mean Average Precision (mAP) score for mass detection, with the largest CS bounding box score selected for final prediction. It is important to note that the hyper-parameter tuning values were chosen based on previous studies and this study's repeated trial processes [67].

### 3.3.3. Performance Evaluation of the Modified YOLOv3 Using Enhanced Images

In this study, mass detection performance was correlated with the image enhancement performance in the prior stage. Therefore, we assessed TP and FP, while the mAP was calculated from the area under the curve of recall and precision, following Equation (9):

$$\text{mean Average Precision, mAP} = \frac{1}{|classes|} \sum\nolimits_{c \in classes} \frac{|TP_c|}{|FP_c| + |TP_c|} \tag{9}$$

where $c$ is the number of classes. The mAP is the current metric used by computer vision researchers to evaluate the robustness of object identification models. It incorporates the trade-off between precision and recall, which optimizes the influence of both metrics, given that precision measures the prediction accuracy and recall measures the total number of predictions concerning the ground truth.

### 3.4. Stage 3: Mass Segmentation, Feature Extraction, and Classification

#### 3.4.1. Mass Segmentation and Evaluation

Following Stage 2, the final evaluation of the system's performance was based on its mass segmentation and classification. To fully separate the mass from its surrounding tissue, we utilized deep-learning-based semantic segmentation once the mass had been localized using the bounding box location obtained from the previous stage. Here, the highest CS was selected for more than one detection. Clearly, segmented mass is important in defining the area in which the features are extracted from the images when classifying the mass into benign or malignant in later stages. Therefore, the evaluation for segmentation performance from the Jaccard index, *J*, of the IoU score was calculated based on Equation (10):

$$\text{J(A, B) or Intersection over Union, IoU} = |A \bigcap B| / |A \bigcup B| \tag{10}$$

where A is the sample data being tested against sample data B (ground truth sample). A higher *J* or IoU score brings better similarity between the two sets. The accuracy of the segmentation was measured based on its testing performance on different input image settings, based on Equation (11), utilizing TP, FP, TN, and FN.

$$\text{Accuracy, Acc} = (TP + TN)/(TP + FN + TN + FP) \tag{11}$$

#### 3.4.2. Feature Extraction

In the final stage, the segmented mass was used to classify whether the mass is benign or malignant. Furthermore, handcrafted features were used to finally classify the mass into benign or malignant using a well-known machine learning technique. In this study, textural features were chosen as the main feature contributor. The segmented mass features were extracted based on three primarily used radiomics handcrafted features for mammography: textural feature (Gray-Level Co-occurrence Matrix (GLCM)), geometrical feature (mass circularity), and first-order statistics (mean intensity).

##### Feature Extraction: Gray-Level Co-Occurrence Matrix (GLCM)

The GLCM can highlight specific properties of the spatial distribution of the gray levels in the texture image. The proposed SbBDEM procedure was applied to increase the textural refinement of the dense and mass region in the earlier stage. Since both benign and malignant region segmented does not change in respect of illuminance when exposed to light, textural analysis is also essential in extracting important features to differentiate between two neighboring pixels [68]. The features were calculated based on Equations (2)–(5) as previously discussed in Section 3.2.5.

##### Feature Extraction: Circularity and Mean Intensity

A malignant breast mass varies in that its edges are uneven and likely to expand quicker, giving it a projecting look in a mammogram. In contrast, a benign mass differs because its geometric limits are more clearly defined, smooth, and consistently formed [26]. These are some of the features selected by radiologists when making visual clinical mammogram evaluations. As a result, one of the descriptors used in previous studies [25,26] is the mass's circularity characteristic, determined using Equation (12), that is implemented using the segmented region's area and perimeter.

$$\text{Circularity} = \frac{4(Area)(\pi)}{Perimeter^2} \tag{12}$$

Additionally, the inclusion of the supplementary characteristic of the mass's mean intensity is based on the notion that since malignant mass cells are more densely formed than benign mass, it may appear to have a greater overall image intensity. The features were calculated based on Equation (6).

#### 3.4.3. Mass Classification and Evaluation

All the features were trained with and without any feature selection or reduction method using a supervised weighted $k$-nearest neighbor ($k$-NN) algorithm [69,70]. To determine the proper $k$ for the training images, we ran the $k$-NN algorithm with different values of $k$ and chose the $k$ that minimizes errors while preserving the system's capability to make accurate predictions when given new testing data. To make an unbiased test performance of the features, 5-fold cross-validation was applied during training, with the final $k$-neighbors value set to 10, using Euclidean distance measurement, having inverse distance weighting for the multivariate interpolation of the data points applied.

The mass abnormality classification's performance was based on the testing accuracy as in Equation (11) and the area under the Receiver Operating Characteristic (ROC) curve. The ROC curve is a standard measuring the degree of separability of binary classification between the mass and its background on a plot of sensitivity (TP Rate) against the specificity (FN Rate), where the highest area under the ROC curve represents the model's ability to segregate the class better.

## 4. Results and Discussion

In this section, the results are discussed based on the stages of experimental procedures explained in the previous section. Comparison of the result of the proposed SbBDEM technique in the pre-processing stage is made based on the performance of the immediate stage of mass detection and is compared between original, adaptive histogram equalization (HE/AHE), contrast limited adaptive histogram equalization (CLAHE), and the proposed SbBDEM technique in this study on all mammogram images.

### 4.1. *Image Quality and Textural Elements*

The performance of the proposed image enhancement in the pre-processing stage before mass detection was investigated based on differently trained image input for the models. Figure 5 shows an example of mammogram and its respective histogram for comparison on the (A) original, (B) HE/AHE, (C) CLAHE, and (D) proposed SbBDEM techniques images. Comparison of histogram for the original in Figure 5A shows similar shape to the proposed SbBDEM in Figure 5D, however its pixel distribution has expanded and shifted to the left side of the histogram. This suggested that the proposed SbBDEM can retain the pixel distribution as similar as possible to the original image, but with the decrease of intensity resulted to increasing the pixel belonging to the non-dense region. More pixels of $<0.5$ are extrapolated causing non-dense area to be darkened, leaving the dense and mass area lighter for better edge difference for the network to learn.

Meanwhile, Table 1 shows the average scores for mean-square error (MSE), Blind/Reference-less Image Spatial Quality Evaluator (BRISQUE), image intensity, and GLCM statistical features comparison between the proposed SbBDEM against other enhancement techniques for all mammogram images. The BRISQUE score is improved from 43.5799 in the original image to 42.3841 and the lowest amongst others, suggesting that using the proposed SbBDEM produced an acceptable quality image in terms of better perceptual ability. Additionally, the average correlation feature for the proposed SbBDEM is the lowest at 0.9752. Since correlation measures how correlated a pixel is to its neighbor over the whole image, it is easy to conclude that neighboring pixels within the proposed SbBDEM image correlate the least with each other. This supports the better edge difference between the pixels within the image for better textural perception. Meanwhile, the energy property represents the estimated pixel attribute energy values that make up an image's energy properties [71,72]. The energy features combine to create an image weight model, which is a collection of weights reflecting the importance of the image pixels from the perspective of perception. The higher energy property in the proposed SbBDEM image suggests the overall pixel carrying more weight is expected to be represented during network training. Finally, the contrast and homogeneity properties show no reflection to the proposed SbBDEM technique as neither shows the least or the most out scores to form varying spatial pattern arrangements.

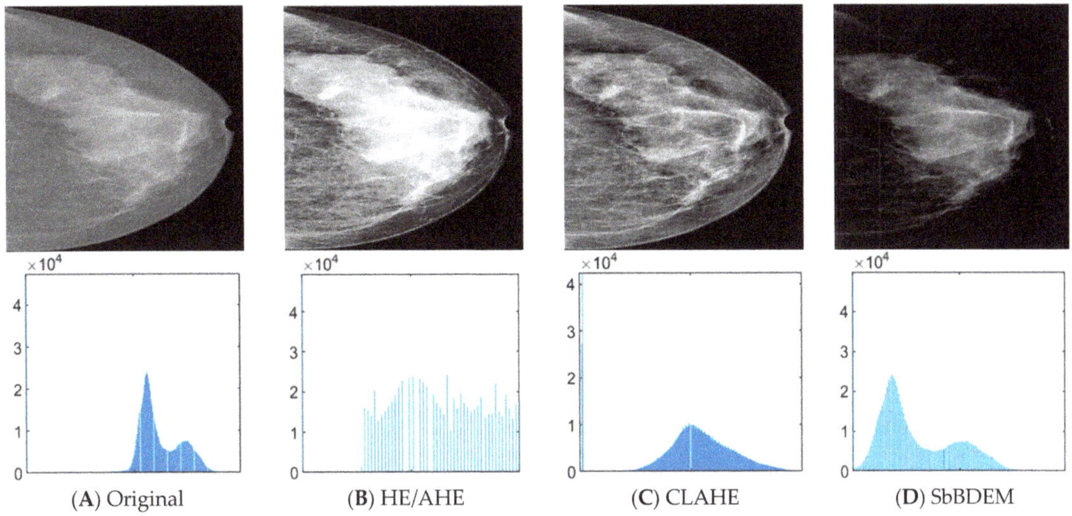


**Figure 5.** Sample images and histogram plots from columns of (**A**) Original, (**B**) HE/AHE, (**C**) CLAHE and (**D**) SbBDEM image enhancement techniques for comparison.

**Table 1.** Average Quality Tests and GLCM features on INbreast Images (N = 112) using Enhancement Techniques.

| No | Enhancement Techniques | MSE | BRISQUE | Mean Intensity | GLCM Textural Features | | | |
|---|---|---|---|---|---|---|---|---|
| | | | | | Contrast | Correlation | Energy | Homogeneity |
| 1 | Original | N/A | 43.5799 | 0.5914 | 0.0276 | 0.9957 | 0.3174 | 0.9876 |
| 2 | HE/AHE | 0.0214 | 42.4518 | 0.6584 | 0.0758 | 0.9901 | 0.2212 | 0.9640 |
| 3 | CLAHE | **0.0066** | 42.9427 | 0.3786 | 0.0856 | 0.9933 | 0.1709 | 0.9621 |
| 4 | SbBDEM | 0.1169 | **42.3841** | 0.2302 | 0.0399 | **0.9752** | **0.4339** | 0.9803 |

For breast mass analysis, the result from the CLAHE-enhanced image, the enhancement technique used in most past studies [10,18,24,45,51,68] is selected to be compared to the proposed SbBDEM method. Figure 6 illustrates sample images from the result of mass detection for both non-dense (Rows 1 and 2) and dense (Rows 3 and 4) images with the confidence score (CS) indicated in the yellow boxes obtained from the mass detection stage in this study. Here, the original image on the first column Figure 6A–E with the ground-truth labeled in red boxes is followed by its respective CLAHE-enhanced (second column) and the proposed SbBDEM technique (third column) images.

Visual evaluation of the images demonstrates increased and interpolated contrast stretching observed on the CLAHE-enhanced image in Figure 6F–J. Meanwhile, the proposed SbBDEM images produced darker overall contrast, as seen in Figure 6K–O, especially on the non-dense fatty tissue region, while preserving the mass and dense region intensity from the original image. Maintaining the pixel information of the mass is essential in feature extraction and convolution of the YOLOv3 algorithm, as this will also preserve the edge of the mass during enhancement.

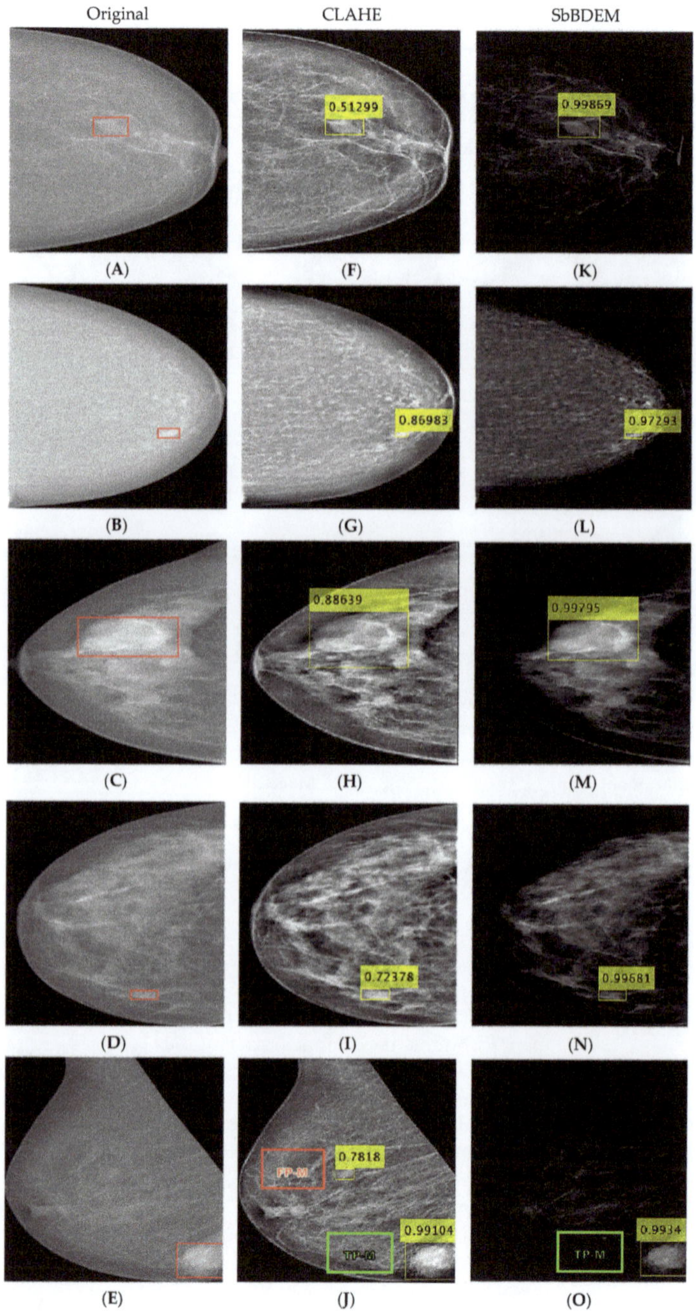


**Figure 6.** Result of Mass Detection for comparison. Rows 1 and 2: non-dense breasts. Rows 3 and 4: dense breasts. Row 5: Example of image with True Positive mass (**TP-M**) and False Positive mass (**FP-M**) detections. Yellow boxes indicate bounding boxes with a confidence score for mass detection. (**A–E**): Original images. (**F–J**): CLAHE-enhanced images. (**K–O**): proposed SbBDEM images.

Other than that, Row 5 of Figure 6E,J,O demonstrates an example of True-Positive Mass (TP) (TP-M) and False-Positive Mass (FP) (FP-M) detections during the mass detection stage. Further pixel analysis based on edge detection emulated by the network's convolutional process is extracted using an 8-by-8 grid window size on the edge of expected mass FP-M corresponding to Figure 7A,B, and mass TP-M in Figure 7C,D.

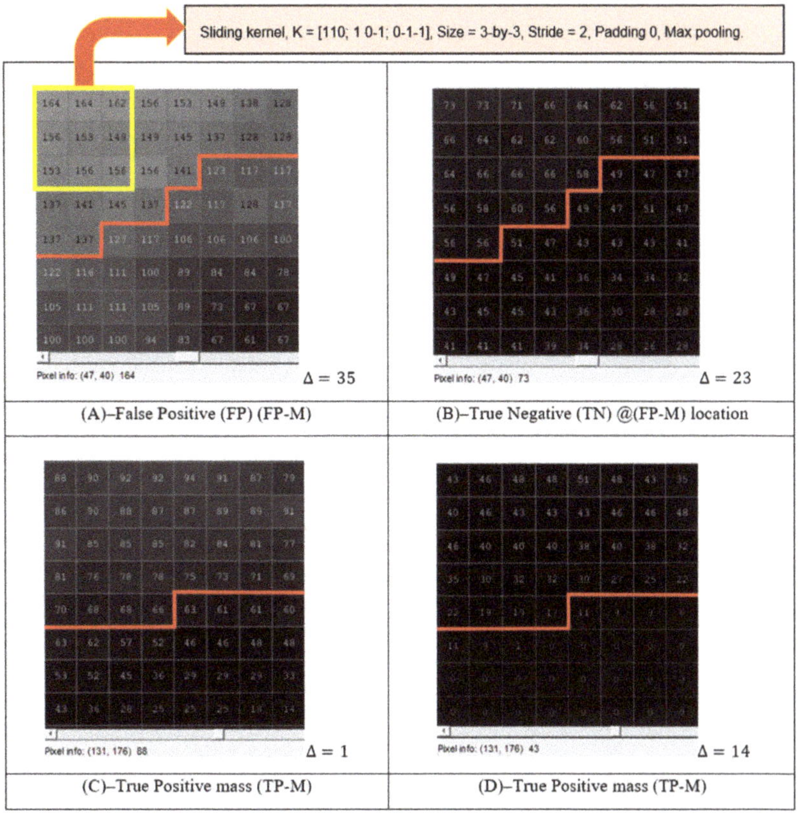


**Figure 7.** (**A**) FP detected mass edge on CLAHE image. (**B**) Corresponding location of TN mass location based on (**A**) on the proposed SbBDEM image result. (**C**) TP detected mass edge on the CLAHE image. (**D**) Corresponding TP location of detected mass location based on (**C**) on the proposed SbBDEM image result. The analysis is made from Figure 6E,J,O, where Δ is the pixel edge difference. The lighter region above the red lines indicates the mass region.

The mass edge analysis is based on the difference of maximum pixel Δ in the region where the region above the red line is the ground-truth-based mass, while the opposite is the background based on the convolution filtering process using kernel K = [110; 1 0-1; 0-1-1] and maximum pooling (Max pooling) downsampling. This revealed that the FP-M detected in Figure 7A on the CLAHE image has a higher probability of being detected based on its pixel region difference, Δ = 35 compared to Δ = 23 on the same pixel location on the proposed SbBDEM image in Figure 7B, as per the ground-truth in Figure 6E. Additionally, TP-M was detected on the CLAHE image and the proposed SbBDEM image. However, even though the proposed SbBDEM image is visually darker, the TP-M detected in Figure 7D for the proposed SbBDEM has a far higher mass edge detection difference at Δ = 14 compared to its counterpart in Figure 7C using CLAHE enhancement, having Δ = 1. This indicates that the new intensity value replacing the original pixel during the proposed SbBDEM process lowers FP detection on non-mass locations, as high-level spatial image features such as edge and coarse textures are extracted at the earliest learnable layer during YOLOv3 learning. At the same time, it increased the probability of detecting TP mass on the proposed SbBDEM image.

The mass detection performance of the overall image enhancement is made through the next stage. It is explained from the Recall-Precision Curves (RPC) in Figure 8 for models trained with the original, HE/AHE, CLAHE, and the proposed SbBDEM images. High recall and high precision are both represented as high areas under the RPC, where high precision is correlated with a low false-positive rate, and high recall is correlated with a low false-negative rate. Note that the proposed SbBDEM enhancement technique produced the highest mean Average Precision (mAP) as area under the RPC of 0.8125, followed by CLAHE images with mAP = 0.7496. In contrast, the HE/AHE images downgraded the performance from using the original images, with mAP at 0.5430 compared to 0.6842 for the original images. This result shows that refining the textural of the mass of the original pixel that further apart the difference between the mass and its neighboring non-dense or dense region background is important to preserve its edge without diminishing the mass itself. The result also justifies that improving the images based on breast density before extracting training features is essential to increase the final overall detection performance.

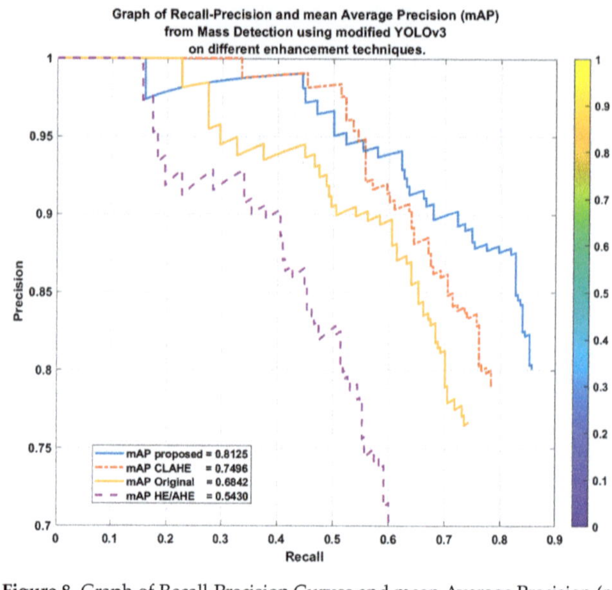


**Figure 8.** Graph of Recall-Precision Curves and mean Average Precision (mAP) from Mass Detection using modified YOLOv3 on different enhancement techniques.

Figure 9 presents a bar chart showing the comparison of performance between dense and non-dense breasts for mass detection on different image enhancement techniques. On average, the ability of the model to detect mass per image is shown on the overall performance showing the best mass detection when using the proposed SbBDEM images, followed by CLAHE, the original images, and finally, HE/AHE shows lesser performance compared to the original images. The lesser HE/AHE performance is in conformance with previous research [25] where HE/AHE might benefit in its application on RGB to HSV images in terms of gamma correction. Therefore, it is somewhat unsuitable in a grey-level image such as a mammogram, as it can only raise the contrast of the background noise while simultaneously reducing the amount of signal that can be utilized.

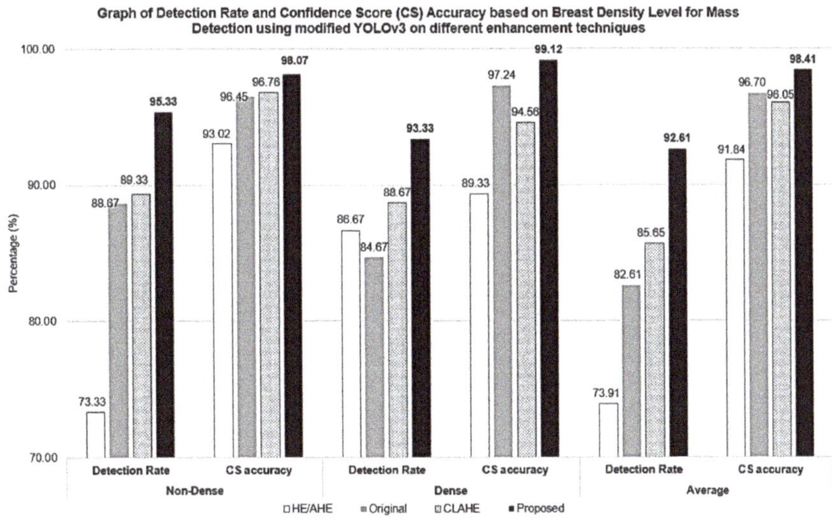


**Figure 9.** Graph of Detection Rate and Confidence Score (CS) Accuracy based on Breast Density Level for Mass Detection using modified YOLOv3 on different enhancement techniques.

As for CLAHE, although it improves mass rate detection by ±3%, the overall CS shows slightly lower performance than in the original image. Compared to other techniques, CLAHE operates on tiles rather than the overall image, in which the tiles are enhanced individually, resulting in a locally stretched contrast masking on the homogeneous areas that are limited to avoid amplifying any noise that might be present in the image [68]. This might contribute to the effect of introducing FP cases on the unrelated dense region within the image that was enhanced, giving a similar feature pattern to the mass. Meanwhile, an improvement of 10% from the original image for detection rate and a slight improvement of 2% for CS accuracy is observed when the proposed SbBDEM technique is applied for mass detection. This supports the reason that contributed to its higher performance is its ability to retain the mass and the denser region as it is while reducing the non-dense region pixel value in the background. In return, a prominent spatial feature defining a mass, such as its edge, is enhanced and contributed to the feature mapping extracted in the YOLO layers, resulting to better detection rate and CS accuracy.

On average, the detection rate of the proposed SbBDEM improved to 92.61% using the proposed SbBDEM technique, followed by CLAHE, original, and HE/AHE at 85.65%, 82.61%, and 73.91%, respectively. By standardizing all test images to only the detected images for all enhancement techniques, the CS accuracy, which measures the bounding box accuracy, is highest on average when the model is trained using the proposed SbBDEM with 98.41% accuracy. Nevertheless, CLAHE-enhanced images have a lower CS accuracy performance than the original image, which may be caused by additional FP detections where the overlapping bounding box may contribute to a wider range of overlapping intersections shared on the same image, resulting in a lower CS accuracy score.

On the other hand, non-dense breast exhibits better performance compared to dense breast, as supported by previous studies [10,18,25] on all enhancement techniques for both detection rate and CS. The highest CS accuracy using the proposed SbBDEM method is at 98.07%, showing a boost of 1.62% in performance from the original image for non-dense breast and increase of 9.79% of CS for dense breast. Even though the detection rate for dense breasts is slightly lower at 93.33% than non-dense breasts at 95.33%, the CS accuracy is observed to be slightly better at 99.12% in the dense breast than in non-dense breasts at 98.07%. Additionally, note that the dense breast detection rate improvement is the best, with an increase of 8.66% from the original image. The analysis of mass detection on the denser background proves that by using the proposed SbBDEM technique, the overlapped mass detection could be improved.

### 4.2. Analysis of Modified YOLOv3 Performance

In this study, a modified convolutional neural network (CNN) for YOLOv3 is developed to evaluate the input images. Furthermore, the modification is utilized to detect the mass's location in the mammograms by improving its ability to receive spatial features enhanced from the proposed SbBDEM technique. Table 2 presents the result of mAP performance for mass detection on the original and other enhancement image input settings with and without YOLOv3 modification for comparison.

**Table 2.** mAP performance for mass detection before and after YOLOv3 modification using different image enhancement techniques.

| Image Condition | Mean Average Precision (mAP) Using YOLOv3 (%) | |
|---|---|---|
| | Without Modification | With Modification |
| Original | 64.01 | 68.42 |
| CLAHE | 67.92 | 74.96 |
| HE/AHE | 57.40 | 54.35 |
| Proposed | 78.33 | **81.25** |

The result displays a pattern of increasing detection performance for all image input settings on the modified YOLOv3 model, except the HE/AHE enhancement input image. The highest mAP of 81.25% is observed using the proposed SbBDEM on the modified model, with an increase in performance of 17.25% compared to using the original image on the non-modified YOLOv3 model. In this study, the modification is crafted to focus on the use of spatial features retained from the proposed SbBDEM training images. Its textural features have been improved based on the result observed from using the proposed SbBDEM technique discussed previously in Table 1. This textural refinement is further taken advantage of as an essential higher-level spatial feature extracted during training by adding the features from the earlier YOLOv3 layer to the second detection head specifically used to detect a smaller object from its initial development setting [40]. Moreover, the extra larger anchor box value that is assigned to these features gives extra weightage and encapsulates the detected mass region through the overlapping of bounding box tiled across the image, further improving the detection performance, resulting in better intersection over union (IoU) placement, given the multi-sizes of the mass on the input images [49].

### 4.3. Performance of Mass Segmentation and Classification

After localizing the position of the mass on the image, the mass region is segmented for the ease of feature extraction for classification in this study. Table 3 compares segmentation results by applying the proposed SbBDEM against the original HE/AHE and CLAHE enhancement techniques. A slight improvement in segmentation accuracy can be observed when using the proposed SbBDEM technique by achieving a mean accuracy of 0.9437 from 0.9431 from the original image. Since the mass is well contained within the bounding-box, less overlapping of mass and dense background issue needs to be resolved using the proposed SbBDEM technique. Nevertheless, the proposed SbBDEM technique also produces the highest accuracy along with IoU for both classes of mass and its background.

**Table 3.** Result of semantic segmentation for mass using different image input settings.

| No | Image Input | Mean Accuracy | Mean IoU | IoU | |
|---|---|---|---|---|---|
| | | | | Mass | Background |
| 1 | Original | 0.9438 | **0.8921** | 0.8873 | 0.8970 |
| 2 | HE/AHE | 0.9385 | 0.8830 | 0.8775 | 0.8885 |
| 3 | CLAHE | 0.9423 | 0.8891 | 0.8844 | 0.8938 |
| 4 | SbBDEM | **0.9441** | 0.8917 | **0.8878** | **0.8984** |

Meanwhile, we employed handcrafted features from the segmented mass region with and without using the principal component analysis (PCA) feature reduction method for benign and malignant classification. Comparison is also made using the chi-square-based feature selection method by removing features having a chi-square score of less than 1.0 as correlated features during training. The result shows the highest testing accuracy for benign vs. malignant mass of 96.0% is achieved on the training time at 0.670 s.

Additionally, a comparison of mass detection results of the past studies and similar methods are listed in Table 4, with and without breast density consideration before or after analysis performance, as well as the computational cost for each algorithm's deployment. In this study, the main objective is to validate the performance of object detection utilizing the simplest CNN of SqueezeNet for a modified YOLOv3 using a differently enhanced input image, specifically to improve the performance for the detection of mass in dense breast mammograms. Similar works addressed the problem of mass detection while disregarding the probable issue of class training imbalance caused by higher non-dense images in the training images that could contribute to lower Computer-Aided Diagnosis (CAD) establishment in clinical settings. In contrast, our study specifically brings the breast density into the focus of the learned parameter of the training images to adapt the class imbalance and improve the image before it can be trained to conduct mass detection, consequently bringing a good mass abnormality classifier. Nonetheless, limited studies have used metrics to compare their performance between non-dense and dense images before and after implementing their proposed work, making it difficult to make a suitable state-of-the-art analysis.

**Table 4.** Comparison of CAD for mammogram mass detection previous works.

| No | Authors | Enhancement Technique | Dense | Non-dense | mAP @0.5 Threshold | Overall Detection Acc (%) | Classification Acc (%) | Segmentation Acc (%) | Detection Time per Test Image |
|---|---|---|---|---|---|---|---|---|---|
| 1 | [10] | CLAHE | ROC = 0.902 | ROC = 0.984 | - | - | - | - | - |
| 2 | [24] | CLAHE | Acc = 91.00% | Acc = 94.80% | - | - | - | - | - |
| 3 | [25] | HE/AHE | Acc = 84.08% | Acc = 88.69% | - | - | - | - | - |
| 4 | [18] | CLAHE | - | - | - | - | 99.91 | - | - |
| 5 | [45] | CLAHE | - | - | - | 98.96 | 95.64 | 92.97 | 12.3 s |
| 6 | [28] | HE/AHE | - | - | - | 97.27 | 95.32 | - | 71 fps |
| 7 | [66] | - | - | - | 0.9420 [1]<br>0.8460 [2] | 89.50 | - | - | 0.009 s |
| 8 | This Study | Proposed-SbBDEM | Acc = 93.33% | Acc = 95.33% | 0.8125 | 92.61 | 96.00 | 94.41 | 1.78 s |

[1] Benign, [2] Malignant, Acc = Accuracy, ROC = Area under ROC curve, fps = frame per second.

Although direct comparison is essentially incomparable between these works, both detection accuracy rate and testing time indicate that we achieved a better overall performance, which plays a significant role in showing that the proposed SbBDEM technique indeed increases the density-based performance. Our method outperformed works by [24,25] in terms of accuracy for non-dense and dense images. However, their work uses different datasets for a fair comparison. To the best of our knowledge, no study has been conducted using specifically the INbreast dataset with the metrics included for density-based mass detection. Meanwhile, work by [18] achieves 99.91% accuracy for benign and malignant classification compared to our method at 96% accuracy based on different breast densities. However, since the study's augmentation process brings almost 7000 images from, originally, 112 images in INbreast, their work may cause unreliable results if the same technique is applied to a newer dataset. In contrast, the work of [45] exceeded our detection results for the same dataset. Nevertheless, it required more testing time than our approach due to the simpler training architecture employed. Additionally, since most of the studies listed applied CLAHE in their pre-processing stage, given that our enhancement method improves the detection model by mAP of 13.33% for CLAHE compared to the proposed SbBDEM technique as discussed in the result section, it is also expected to increase these studies detection stage if our pre-processing method is applied beforehand. Indeed, low accuracy limitations could be overcome by applying a more complex algorithm with more sophisticated hardware for training, which is expected to further improve the currently proposed SbBDEM technique for mass detection.

## 5. Conclusions

This work presents an image enhancement method according to the breast density level for Computer-Aided Diagnosis (CAD) stages for mammogram image analysis. Based on the result, the proposed SbBDEM technique could increase the performance for all stages of mass detection, segmentation, and classification for mammogram images. An improvement is observed when the proposed SbBDEM method is compared to the original image and the most widely used enhancement technique, i.e., contrast-limited adaptive histogram equalization (CLAHE) and histogram equalization (HE). The adjustment of the lower limit cap acts as a threshold value to separate the dense and mass to non-dense regions. This helps refine the textural information as a feature that represents both regions and through textural feature extraction in the classification stage, boosting the accuracy to 96% for the 5-fold cross-validation of benign vs. malignant classification experiment. The result also presents an improvement of mass detection with mean Average Precision (mAP) = 0.6401 to mAP = 0.8125, with mass detection in non-dense and dense accuracy of 93.33% and 95.33%, respectively. We achieved an increase of 98.41% confidence scores (CS) as opposed to 91.84% in the original image and a slight improvement of 0.03% in the mass segmentation using the proposed SbBDEM technique.

Meanwhile, in its original documentation, You Only Look Once v3 (YOLOv3) specializes in detecting smaller objects with the implementation of the second detection head. We further utilize this by modifying the second detection head into receiving the textural features that were already enhanced in the pre-processing stage through our proposed SbBDEM technique by adding these features to the deeper learning layer that contains more semantic information of the same image to improve the feature discrimination.

Our proposed method is limited by the unavailability of standardized image quality metrics that can determine the best image for all training images based on textural elements while considering the need for thousands of images for deep-learning purposes. While a high-quality image might be good for measuring accuracy, it is unnecessarily true to measure its textural aspect. Although statistical information for textural analysis is available, more suitable metrics can be investigated for more reliable metrics that relate image quality and texture. Additionally, with a running GPU capability of only 6 GB, the study is limited by the unavailability of a more sophisticated computing facility to employ higher-functioned YOLO, such as versions 4, 5, 6, and 7 without affecting the performance by reducing the mini-batches. However, the implementation of YOLOv3 in this study is sufficient as a way to demonstrate the effectiveness of density-based enhancement on the dataset before training and was modified based on its simplicity, which only runs on 5 MB or 1.2 million learnable parameters. Future studies could be explored by using other breast mammogram datasets with validation from a trained radiologist to enable CAD implementation in the medical field. Finally, the result obtained was comparable to the state-of-the-art performance from other methods discussed and can work as a base model for future updates by employing a more complex model on another dataset as well.

**Author Contributions:** Conceptualization, N.F.R. and I.S.I.; Formal analysis, N.F.R., N.K.A.K. and M.K.O.; Funding acquisition, N.F.R. and Z.H.C.S.; Investigation, N.F.R., I.S.I. and S.N.S.; Methodology, N.F.R., I.S.I., S.N.S. and M.K.O.; Project administration, I.S.I. and Z.H.C.S.; Resources, Z.H.C.S.; Software, I.S.I. and S.N.S.; Supervision, I.S.I.; Validation, N.K.A.K.; Visualization, N.F.R.; Writing—original draft, N.F.R.; Writing—review and editing, N.F.R. and I.S.I. All authors have read and agreed to the published version of the manuscript.

**Funding:** This research was funded by the Fundamental Research Grant Scheme (FRGS) under the Ministry of Higher Education Malaysia, grant number FRGS/1/2022/SKK06/UITM/02/3.

**Institutional Review Board Statement:** Not applicable.

**Informed Consent Statement:** Not applicable.

**Data Availability Statement:** Updated dataset of INbreast is acquired through personal communication via email with the original author for latest updates.

**Acknowledgments:** The authors would like to acknowledge the Breast Research Group, INESC Porto, Portugal, for providing the updated INbreast dataset and its accompanying ground truth for this study. The authors would also like to express their gratitude to members of the Advanced Control System and Computing Research Group (ACSCRG), Advanced Rehabilitation Engineering in Diagnostic and Monitoring Research Group (AREDiM), Integrative Pharmacogenomics Institute (iPROMISE), and Centre for Electrical Engineering Studies, Universiti Teknologi MARA (UiTM), Cawangan Pulau Pinang for their assistance during the duration of this study. Finally, the authors are grateful to UiTM, Cawangan Pulau Pinang, for their immense administrative support.

**Conflicts of Interest:** The authors declare no conflict of interest.

## References

1. Sung, H.; Ferlay, J.; Siegel, R.L.; Laversanne, M.; Soerjomataram, I.; Jemal, A.; Bray, F. Global Cancer Statistics 2020: GLOBOCAN Estimates of Incidence and Mortality Worldwide for 36 Cancers in 185 Countries. *CA Cancer J. Clin.* **2021**, *71*, 209–249. [CrossRef] [PubMed]
2. Iranmakani, S.; Mortezazadeh, T.; Sajadian, F.; Ghaziani, M.; Ghafari, A.; Khezrloo, D.; Musa, A. A Review of Various Modalities in Breast Imaging: Technical Aspects and Clinical Outcomes. *Egypt. J. Radiol. Nucl. Med.* **2020**, *51*, 1–22. [CrossRef]
3. Radhakrishna, S.; Agarwal, S.; Parikh, P.M.; Kaur, K.; Panwar, S.; Sharma, S.; Dey, A.; Saxena, K.K.; Chandra, M.; Sud, S. Role of Magnetic Resonance Imaging in Breast Cancer Management. *South Asian J. Cancer* **2018**, *7*, 69–71. [CrossRef]
4. Ciritsis, A.; Rossi, C.; Vittoria De Martini, I.; Eberhard, M.; Marcon, M.; Becker, A.S.; Berger, N.; Boss, A. Determination of Mammographic Breast Density Using a Deep Convolutional Neural Network. *Br. J. Radiol.* **2019**, *92*, 20180691. [CrossRef] [PubMed]
5. Advani, S.M.; Zhu, W.; Demb, J.; Sprague, B.L.; Onega, T.; Henderson, L.M.; Buist, D.S.M.; Zhang, D.; Schousboe, J.T.; Walter, L.C.; et al. Association of Breast Density With Breast Cancer Risk Among Women Aged 65 Years or Older by Age Group and Body Mass Index. *JAMA Netw. Open* **2021**, *4*, e2122810. [CrossRef] [PubMed]
6. Kim, Y.J.; Kim, K.G. Detection and Weak Segmentation of Masses in Gray-Scale Breast Mammogram Images Using Deep Learning. *Yonsei Med. J.* **2022**, *63*, S63–S73. [CrossRef]
7. Kim, Y.J.; Lee, E.H.; Jun, J.K.; Shin, D.-R.; Park, Y.M.; Kim, H.-W.; Kim, Y.; Kim, K.W.; Lim, H.S.; Park, J.S.; et al. Analysis of Participant Factors That Affect the Diagnostic Performance of Screening Mammography: A Report of the Alliance for Breast Cancer Screening in Korea. *Korean J. Radiol.* **2017**, *18*, 624–631. [CrossRef]
8. Li, S.; Wei, J.; Chan, H.-P.; Helvie, M.; Roubidoux, M.; Lu, Y.; Zhou, C.; Hadjiiski, L.; Samala, R. Computer-Aided Assessment of Breast Density: Comparison of Supervised Deep Learning and Feature Based Statistical Learning. *Phys. Med. Biol.* **2017**, *63*, 025005. [CrossRef]
9. Boyd, N.F.; Guo, H.; Martin, L.J.; Sun, L.; Stone, J.; Fishell, E.; Jong, R.A.; Hislop, G.; Chiarelli, A.; Minkin, S.; et al. Mammographic Density and the Risk and Detection of Breast Cancer. *N. Engl. J. Med.* **2007**, *356*, 227–236. [CrossRef]
10. Suh, Y.J.; Jung, J.; Cho, B.-J. Automated Breast Cancer Detection in Digital Mammograms of Various Densities via Deep Learning. *J. Pers. Med.* **2020**, *10*, 211. [CrossRef]
11. Boita, J.; Van Engen, R.E.; Mackenzie, A.; Tingberg, A.; Bosmans, H.; Bolejko, A.; Zackrisson, S.; Wallis, M.G.; Ikeda, D.M.; Van Ongeval, C.; et al. How Does Image Quality Affect Radiologists' Perceived Ability for Image Interpretation and Lesion Detection in Digital Mammography? *Eur. Radiol.* **2021**, *31*, 5335–5343. [CrossRef]
12. Warren, L.M.; Given-Wilson, R.M.; Wallis, M.G.; Cooke, J.; Halling-Brown, M.D.; Mackenzie, A.; Chakraborty, D.P.; Bosmans, H.; Dance, D.R.; Young, K.C. The Effect of Image Processing on the Detection of Cancers in Digital Mammography. *AJR Am. J. Roentgenol.* **2014**, *203*, 387–393. [CrossRef]
13. Corrias, G.; Micheletti, G.; Barberini, L.; Suri, J.; Saba, L. Texture Analysis Imaging "What a Clinical Radiologist Needs to Know". *Eur. J. Radiol.* **2021**, *146*, 110055. [CrossRef]

14. Malik, J.; Belongie, S.; Leung, T.; Shi, J. Contour and Texture Analysis for Image Segmentation. *Int. J. Comput. Vis.* **2001**, *43*, 7–27. [CrossRef]
15. Jalab, H.A.; Ibrahim, R.W.; Hasan, A.M.; Karim, F.K.; Al-Shamasneh, A.R.; Baleanu, D. A New Medical Image Enhancement Algorithm Based on Fractional Calculus. *Comput. Mater. Contin.* **2021**, *68*, 1467–1483. [CrossRef]
16. Al-Antari, M.A.; Al-Masni, M.A.; Kim, T.-S. Deep Learning Computer-Aided Diagnosis for Breast Lesion in Digital Mammogram. *Adv. Exp. Med. Biol.* **2020**, *1213*, 59–72. [CrossRef]
17. Malebary, S.J.; Hashmi, A. Automated Breast Mass Classification System Using Deep Learning and Ensemble Learning in Digital Mammogram. *IEEE Access* **2021**, *9*, 55312–55328. [CrossRef]
18. Huang, M.-L.; Lin, T.-Y. Considering Breast Density for the Classification of Benign and Malignant Mammograms. *Biomed. Signal Process Control* **2021**, *67*, 102564. [CrossRef]
19. McKinney, S.M.; Sieniek, M.; Godbole, V.; Godwin, J.; Antropova, N.; Ashrafian, H.; Back, T.; Chesus, M.; Corrado, G.S.; Darzi, A.; et al. International Evaluation of an AI System for Breast Cancer Screening. *Nature* **2020**, *577*, 89–94. [CrossRef]
20. Bargalló, X.; Santamaría, G.; Del Amo, M.; Arguis, P.; Ríos, J.; Grau, J.; Burrel, M.; Cores, E.; Velasco, M. Single Reading with Computer-Aided Detection Performed by Selected Radiologists in a Breast Cancer Screening Program. *Eur. J. Radiol.* **2014**, *83*, 2019–2023. [CrossRef]
21. He, Z.; Li, Y.; Zeng, W.; Xu, W.; Liu, J.; Ma, X.; Wei, J.; Zeng, H.; Xu, Z.; Wang, S.; et al. Can a Computer-Aided Mass Diagnosis Model Based on Perceptive Features Learned From Quantitative Mammography Radiology Reports Improve Junior Radiologists' Diagnosis Performance? An Observer Study. *Front Oncol* **2021**, *11*, 1–12. [CrossRef] [PubMed]
22. Oza, P.; Sharma, P.; Patel, S.; Adedoyin, F.; Bruno, A. Image Augmentation Techniques for Mammogram Analysis. *J. Imaging* **2022**, *8*, 141. [CrossRef]
23. Dabass, J.; Dabass, M. Segmentation of Noisy Mammograms Using Hybrid Techniques. In *Advances in Communication and Computational Technology. Lecture Notes in Electrical Engineering*; Springer: Singapore, 2021; Volume 668, pp. 1371–1382, ISBN 978-981-15-5340-0.
24. Bandeira Diniz, J.O.; Bandeira Diniz, P.H.; Azevedo Valente, T.L.; Corrêa Silva, A.; De Paiva, A.C.; Gattass, M. Detection of Mass Regions in Mammograms by Bilateral Analysis Adapted to Breast Density Using Similarity Indexes and Convolutional Neural Networks. *Comput. Methods Programs Biomed.* **2018**, *156*, 191–207. [CrossRef] [PubMed]
25. Sampaio, W.B.; Diniz, E.M.; Silva, A.C.; De Paiva, A.C.; Gattass, M. Detection of Masses in Mammogram Images Using CNN, Geostatistic Functions and SVM. *Comput. Biol. Med.* **2011**, *41*, 653–664. [CrossRef] [PubMed]
26. Shrivastava, N.; Bharti, J. Breast Tumor Detection and Classification Based on Density. *Multimed. Tools Appl.* **2020**, *79*, 26467–26487. [CrossRef]
27. Singh, N.; Suraparaju, V. Breast Cancer Segmentation Using Global Thresholding and Region Merging. *Int. J. Comput. Sci. Eng.* **2018**, *6*, 292–297. [CrossRef]
28. Al-Antari, M.A.; Han, S.-M.; Kim, T.-S. Evaluation of Deep Learning Detection and Classification towards Computer-Aided Diagnosis of Breast Lesions in Digital X-Ray Mammograms. *Comput. Methods Programs Biomed.* **2020**, *196*, 105584. [CrossRef]
29. Laishram, R.; Rabidas, R. WDO Optimized Detection for Mammographic Masses and Its Diagnosis: A Unified CAD System. *Appl. Soft Comput.* **2021**, *110*, 107620. [CrossRef]
30. Ghosh, S.; Biswas, B.; Ghosh, A. Development of Intuitionistic Fuzzy Special Embedded Convolutional Neural Network for Mammography Enhancement. *Comput. Intell.* **2020**, *37*, 47–69. [CrossRef]
31. Al-Najdawi, N.; Biltawi, M.; Tedmori, S. Mammogram Image Visual Enhancement, Mass Segmentation and Classification. *Appl. Soft Comput.* **2015**, *35*, 175–185. [CrossRef]
32. Maqsood, S.; Damaševičius, R.; Maskeliūnas, R. TTCNN: A Breast Cancer Detection and Classification towards Computer-Aided Diagnosis Using Digital Mammography in Early Stages. *Appl. Sci.* **2022**, *12*, 3273. [CrossRef]
33. Kurt, B.; Nabiyev, V.V.; Turhan, K. Comparison of Enhancement Methods for Mammograms with Performance Measures. *Stud. Health Technol. Inform.* **2014**, *205*, 486–490.
34. Singh, L.; Alam, A. An Efficient Hybrid Methodology for an Early Detection of Breast Cancer in Digital Mammograms. *J. Ambient Intell. Humaniz. Comput.* **2022**, 1–24. [CrossRef]
35. Deng, J.; Ma, Y.; Deng-Ao, L.; Zhao, J.; Liu, Y.; Zhang, H. Classification of Breast Density Categories Based on SE-Attention Neural Networks. *Comput. Methods Programs Biomed.* **2020**, *193*, 105489. [CrossRef]
36. Laishram, R.; Rabidas, R. Optimized Hyperbolic Tangent Function-Based Contrast-Enhanced Mammograms for Breast Mass Detection. *Expert Syst. Appl.* **2023**, *213*, 118994. [CrossRef]
37. Laishram, R.; Rabidas, R. Detection of Mammographic Masses Using FRFCM Optimized by PSO. In Proceedings of the 2020 13th International Congress on Image and Signal Processing, BioMedical Engineering and Informatics (CISP-BMEI), Chengdu, China, 17 October 2020; pp. 327–332.
38. Zebari, D.A.; Ibrahim, D.A.; Zeebaree, D.Q.; Mohammed, M.A.; Haron, H.; Zebari, N.A.; Damaševičius, R.; Maskeliūnas, R. Breast Cancer Detection Using Mammogram Images with Improved Multi-Fractal Dimension Approach and Feature Fusion. *Appl. Sci.* **2021**, *11*, 12122. [CrossRef]
39. Ren, S.; He, K.; Girshick, R.; Sun, J. Faster R-CNN: Towards Real-Time Object Detection with Region Proposal Networks. *IEEE Trans. Pattern Anal. Mach. Intell.* **2015**, *39*, 1137–1149. [CrossRef] [PubMed]
40. Redmon, J.; Farhadi, A. YOLOv3: An Incremental Improvement. *arXiv* **2018**, arXiv:1804.02767.

41. Redmon, J.; Divvala, S.; Girshick, R.; Farhadi, A. You Only Look Once: Unified, Real-Time Object Detection. In Proceedings of the IEEE Conference on Computer Vision and Pattern Recognition, Las Vegas, NV, USA, 8 June 2016; pp. 779–788.
42. Liu, W.; Anguelov, D.; Erhan, D.; Szegedy, C.; Reed, S.; Fu, C.-Y.; Berg, A.C. SSD: Single Shot MultiBox Detector. In Proceedings of the Computer Vision—ECCV 2016, Amsterdam, The Netherlands, 11–14 October 2016; pp. 21–37.
43. Tan, L.; Huangfu, T.; Wu, L.; Chen, W. Comparison of RetinaNet, SSD, and YOLO v3 for Real-Time Pill Identification. *BMC Med. Inform. Decis. Mak.* **2021**, *21*, 324–334. [CrossRef] [PubMed]
44. Du, J. Understanding of Object Detection Based on CNN Family and YOLO. *J. Phys. Conf. Ser.* **2018**, *1004*, 012029. [CrossRef]
45. Al-Antari, M.A.; Al-Masni, M.A.; Choi, M.-T.; Han, S.-M.; Kim, T.-S. A Fully Integrated Computer-Aided Diagnosis System for Digital X-Ray Mammograms via Deep Learning Detection, Segmentation, and Classification. *Int. J. Med. Inform.* **2018**, *117*, 44–54. [CrossRef]
46. Al-Masni, M.; Al-Antari, M.A.; Park, J.; Gi, G.; Kim, T.-Y.; Rivera, P.; Valarezo Añazco, E.; Choi, M.-T.; Han, S.-M.; Kim, T.-S. Simultaneous Detection and Classification of Breast Masses in Digital Mammograms via a Deep Learning YOLO-Based CAD System. *Comput. Methods Programs Biomed.* **2018**, *157*, 85–94. [CrossRef]
47. Baccouche, A.; Garcia-Zapirain, B.; Olea, C.C.; Elmaghraby, A.S. Breast Lesions Detection and Classification via YOLO-Based Fusion Models. *Comput. Mater. Contin.* **2021**, *69*, 1407–1425. [CrossRef]
48. Baccouche, A.; Garcia-Zapirain, B.; Zheng, Y.; Elmaghraby, A.S. Early Detection and Classification of Abnormality in Prior Mammograms Using Image-to-Image Translation and YOLO Techniques. *Comput. Methods Programs Biomed.* **2022**, 106884. [CrossRef]
49. Lee, A.; Mavaddat, N.; Wilcox, A.N.; Cunningham, A.P.; Carver, T.; Hartley, S.; Babb de Villiers, C.; Izquierdo, A.; Simard, J.; Schmidt, M.K.; et al. BOADICEA: A Comprehensive Breast Cancer Risk Prediction Modelincorporating Genetic and Nongenetic Risk Factors. *Genet. Med.* **2019**, *21*, 1708–1718. [CrossRef]
50. Adedigba, A.P.; Adeshina, S.A.; Aibinu, A.M. Performance Evaluation of Deep Learning Models on Mammogram Classification Using Small Dataset. *Bioengineering* **2022**, *9*, 161. [CrossRef]
51. Abdelhafiz, D.; Nabavi, S.; Ammar, R.; Yang, C.; Bi, J. Residual Deep Learning System for Mass Segmentation and Classification in Mammography. In Proceedings of the 10th ACM International Conference on Bioinformatics, Computational Biology and Health Informatics, New York, NY, USA, 4 September 2019; pp. 475–484.
52. Moreira, I.C.; Amaral, I.; Domingues, I.; Cardoso, A.; Cardoso, M.J.; Cardoso, J.S. INbreast: Toward a Full-Field Digital Mammographic Database. *Acad. Radiol.* **2012**, *19*, 236–248. [CrossRef]
53. Otsu, N. A Threshold Selection Method from Gray-Level Histograms. *IEEE Trans. Syst. Man Cybern.* **1979**, *9*, 62–66. [CrossRef]
54. Rajevenceltha, J.; Gaidhane, V.H. An Efficient Approach for No-Reference Image Quality Assessment Based on Statistical Texture and Structural Features. *Eng. Sci. Technol. Int. J.* **2022**, *30*, 101039. [CrossRef]
55. Huynh-Thu, Q.; Ghanbari, M. The Accuracy of PSNR in Predicting Video Quality for Different Video Scenes and Frame Rates. *Telecommun. Syst.* **2012**, *49*, 35–48. [CrossRef]
56. Huynh-Thu, Q.; Ghanbari, M. Scope of Validity of PSNR in Image/Video Quality Assessment. *Electron. Lett.* **2008**, *44*, 800–801. [CrossRef]
57. Mittal, A.; Moorthy, A.K.; Bovik, A.C. No-Reference Image Quality Assessment in the Spatial Domain. *IEEE Trans. Image Process.* **2012**, *21*, 4695–4708. [CrossRef] [PubMed]
58. Ibrahim, R.W.; Jalab, H.A.; Karim, F.K.; Alabdulkreem, E.; Ayub, M.N. A Medical Image Enhancement Based on Generalized Class of Fractional Partial Differential Equations. *Quant Imaging Med. Surg.* **2022**, *12*, 172–183. [CrossRef] [PubMed]
59. Tumakov, D.; Kayumov, Z.; Zhumaniezov, A.; Chikrin, D.; Galimyanov, D. Elimination of Defects in Mammograms Caused by a Malfunction of the Device Matrix. *J. Imaging* **2022**, *8*, 128. [CrossRef] [PubMed]
60. Oyelade, O.N.; Ezugwu, A.E.; Almutairi, M.S.; Saha, A.K.; Abualigah, L.; Chiroma, H. A Generative Adversarial Network for Synthetization of Regions of Interest Based on Digital Mammograms. *Sci. Rep.* **2022**, *12*, 6166. [CrossRef]
61. Iandola, F.; Han, S.; Moskewicz, M.; Ashraf, K.; Dally, W.; Keutzer, K. SqueezeNet: AlexNet-Level Accuracy with 50x Fewer Parameters and <0.5 MB Model Size. *arXiv* **2016**, arXiv:1602.07360. [CrossRef]
62. Bressem, K.K.; Adams, L.C.; Erxleben, C.; Hamm, B.; Niehues, S.M.; Vahldiek, J.L. Comparing Different Deep Learning Architectures for Classification of Chest Radiographs. *Sci. Rep.* **2020**, *10*, 13590. [CrossRef]
63. Zhao, J.; Chen, T.; Cai, B. A Computer-Aided Diagnostic System for Mammograms Based on YOLOv3. *Multimed. Tools Appl.* **2021**, *81*, 19257–19281. [CrossRef]
64. Zhang, L.; Li, Y.; Chen, H.; Wu, W.; Chen, K.; Wang, S. Anchor-Free YOLOv3 for Mass Detection in Mammogram. *Expert Syst. Appl.* **2022**, *191*, 116273. [CrossRef]
65. Ammar, A.; Koubaa, A.; Ahmed, M.; Saad, A.; Benjdira, B. Vehicle Detection from Aerial Images Using Deep Learning: A Comparative Study. *Electronics* **2021**, *10*, 820. [CrossRef]
66. Aly, G.H.; Marey, M.; El-Sayed, S.A.; Tolba, M.F. YOLO Based Breast Masses Detection and Classification in Full-Field Digital Mammograms. *Comput. Methods Programs Biomed.* **2021**, *200*, 105823. [CrossRef]
67. Saranyaraj, D.; Manikandan, M.; Maheswari, S. A Deep Convolutional Neural Network for the Early Detection of Breast Carcinoma with Respect to Hyper- Parameter Tuning. *Multimed. Tools Appl.* **2020**, *79*, 11013–11038. [CrossRef]
68. Hazarika, M.; Mahanta, L.B. A New Breast Border Extraction and Contrast Enhancement Technique with Digital Mammogram Images for Improved Detection of Breast Cancer. *Asian Pac. J. Cancer Prev.* **2018**, *19*, 2141–2148. [CrossRef]

69. Htay, T.; Maung, S. Early Stage Breast Cancer Detection System Using GLCM Feature Extraction and K-Nearest Neighbor (k-NN) on Mammography Image. In Proceedings of the 2018 18th International Symposium on Communications and Information Technologies (ISCIT), Bangkok, Thailand, 26–29 September 2018; pp. 171–175.
70. Murtaza, G.; Shuib, L.; Wahab, A.W.A.; Mujtaba, G.; Raza, G. Ensembled Deep Convolution Neural Network-Based Breast Cancer Classification with Misclassification Reduction Algorithms. *Multimed Tools Appl.* **2020**, *79*, 18447–18479. [CrossRef]
71. Lyasheva, M.M.; Lyasheva, S.A.; Shleymovich, M.P. Image Texture Model Based on Energy Features. *J. Phys. Conf. Ser.* **2021**, *1902*, 012120. [CrossRef]
72. Gizatullin, Z.M.; Lyasheva, S.A.; Morozov, O.G.; Shleymovich, M.P. A Method of Contour Detection Based on an Image Weight Model. *Comput. Opt.* **2020**, *44*, 393–400. [CrossRef]

*bioengineering* MDPI

*Article*

# 2D/3D Non-Rigid Image Registration via Two Orthogonal X-ray Projection Images for Lung Tumor Tracking

**Guoya Dong [1,2,3,†], Jingjing Dai [1,2,3,4,†], Na Li [5], Chulong Zhang [4], Wenfeng He [4], Lin Liu [4], Yinping Chan [4], Yunhui Li [4], Yaoqin Xie [4] and Xiaokun Liang [4,*]**

1 School of Health Sciences and Biomedical Engineering, Hebei University of Technology, Tianjin 300130, China
2 Hebei Key Laboratory of Bioelectromagnetics and Neural Engineering, Tianjin 300130, China
3 Tianjin Key Laboratory of Bioelectromagnetic Technology and Intelligent Health, Tianjin 300130, China
4 Shenzhen Institute of Advanced Technology, Chinese Academy of Sciences, Shenzhen 518055, China
5 Department of Biomedical Engineering, Guangdong Medical University, Dongguan 523808, China
* Correspondence: xk.liang@siat.ac.cn
† These authors contributed equally to this work.

**Abstract:** Two-dimensional (2D)/three-dimensional (3D) registration is critical in clinical applications. However, existing methods suffer from long alignment times and high doses. In this paper, a non-rigid 2D/3D registration method based on deep learning with orthogonal angle projections is proposed. The application can quickly achieve alignment using only two orthogonal angle projections. We tested the method with lungs (with and without tumors) and phantom data. The results show that the Dice and normalized cross-correlations are greater than 0.97 and 0.92, respectively, and the registration time is less than 1.2 seconds. In addition, the proposed model showed the ability to track lung tumors, highlighting the clinical potential of the proposed method.

**Keywords:** 2D/3D registration; orthogonal X-ray; deep learning

**Citation:** Dong, G.; Dai, J.; Li, N.; Zhang, C.; He, W.; Liu, L.; Chan, Y.; Li, Y.; Xie, Y.; Liang, X. 2D/3D Non-Rigid Image Registration via Two Orthogonal X-ray Projection Images for Lung Tumor Tracking. *Bioengineering* **2023**, *10*, 144. https://doi.org/10.3390/bioengineering10020144

Academic Editors: Paolo Zaffino and Maria Francesca Spadea

Received: 26 December 2022
Revised: 10 January 2023
Accepted: 16 January 2023
Published: 21 January 2023



## 1. Introduction

Medical imaging has helped a lot with diagnosing and treating diseases as modern medical technology has grown quickly. Image registration is crucial in medical image processing because it helps predict, diagnose, and treat diseases. For the images to be registered, three-dimensional (3D) medical images with rich anatomical and structural information are an inevitable choice for clinical problems. Unfortunately, 3D images have a higher radiation dose and a slower imaging speed, which inconveniences real-time clinical problems, such as image-guided radiotherapy and interventional surgery. On the other hand, two-dimensional (2D) images lack some spatial structure information, while the imaging speed is very fast. Therefore, in recent years, 2D/3D image registration with faster speed and simple imaging equipment has attracted much attention. The types of 2D images are usually X-ray [1–4], fluoroscopic [5], digital subtraction angiography (DSA) [6,7], or ultrasound [8], whereas 3D images are chosen from computed tomography (CT) [1–4] or magnetic resonance imaging(MRI) [8].

2D/3D registration methods can be divided into traditional and deep learning-based image registration. In traditional image registration, 2D/3D alignment usually translates into the problem of solving for the maximum similarity between digitally reconstructed radiographs (DRR) and X-ray images. Similarity metrics are usually based on intensity-based mutual information [9–11], normalized cross-correlation (NCC) [12] and Pearson correlation coefficients [13], or gradient-based similarity metrics [14]. To minimize the dimensionality of the transformation parameters, regression models that rely on a priori information are usually built using B-spline [15] or principal component analysis (PCA) [16–19]. However, organ motion and deformation can cause errors in regression models, which rely too much on prior information. By incorporating finite element information into the regression model, Zhang et al. [18,19] obtained more realistic and effective deformation parameters. However, adding finite element information makes the model-driven method of finding the optimal solution iteratively more inefficient. Therefore, this process is a constraint for developing real-time 2D/3D registration and tumor tracking algorithms.

With the development of artificial intelligence and deep learning, learning-based methods replace the tedious iterative optimization process with predicted values in the testing process, greatly improving computing efficiency. Zhang [20] proposed an unsupervised 2D-3D deformable registration network that addresses 2D/3D registration based on finite angles. Li et al. [4] proposed an unsupervised multiscale encode decode framework to achieve non-rigid 2D/3D registration based on a single 2D lateral brain image and 3D CBCT image. Ketcha et al. [21] used multi-stage rigid registration based on convolutional neural networks (CNN) to obtain a deformable spine model. Finally, Zhang et al. [22] achieved a deformable registration of the skull surface. Unfortunately, the above learning-based approach evaluates the similarity between DRR and X-ray, a 2D/3D registration reduced dimension to 2D/2D registration. Therefore, it is inevitable that spatial information will be lost to some extent. In addition, even with Graphic Processing Unit (GPU) support, forward projection, backward projection, and DRR generation involved in the above methods are computationally expensive. Then, the researchers completed end-to-end 2D/3D registration by integrating the forward/inverse projection spatial transformation layer into a neural network [3,23]. Frysch et al. [2] used Grangeat's relation instead of expensive forward/inverse projection to complete the 2D/3D registration method based on a single projection of arbitrary angle, which greatly accelerated the computational speed. However, this is a rigid transformation which is difficult to apply to elastic organs. Likewise, deep learning researchers have attempted to use statistical deformation models to build deep learning-based regression models. Using a priori information to build patient-specific deformation spaces, convolutional neural networks are used to accomplish regression on PCA coefficients [1,24,25] or B-spline parameter coefficients [26,27] to achieve patient-specific registration networks. Tian et al. [28] obtained the predicted deformation field based on the regression coefficients. However, this deformation space, which is completely based on a priori information, may lead to mistakes in the clinical application stage. In addition, some researchers [29,30] also accomplished 2D/3D image registration by extracting feature points. With the maturity of point cloud technology, many researchers have also built point-to-plane alignment models by extracting global point clouds to complete 2D/3D alignment models, but the anomaly removal for 2D/3D alignment models presents a challenge [31–33]. Graphical neural networks are also used for 2D/3D registration in low-contrast conditions [34]. Shao et al. [35] tracked liver tumors by adding finite element modeling. Still, the introduction of finite elements also brought some trouble to the registration time.

Therefore, we developed a deep learning-based method for non-rigid 2D/3D image registration of the same subject. Compared with traditional algorithms based on iterative optimization, this approach significantly improves the registration speed. Compared with the downscaled optimization of DRR and X-ray similarity, we optimized the similarity of 3D/3D images, which can effectively moderate the loss of spatial information. Additionally, only two projections based on orthogonal angles were chosen for 2D images to reduce the irradiation dose further. The proposed method is used to study the process of changes in the elastic organ as respiratory motion proceeds. More significantly, we also investigated the change in tumor position with respiratory motion, which can be used to achieve tracking of tumors based on orthogonal angular projections during radiotherapy.

The contributions of our work are summarized as follows:

1. We propose a 2D/3D elastic alignment framework based on deep learning, which can be applied to achieve organ shape tracking at lower doses using only two orthogonal angles of X-rays.

2. Our framework is expected to be used for tumor tracking with tumor localization accuracy up to 0.97 and registration time within 1.2 s, which may be a potential solution for image-guided surgery and radiotherapy.

The organizational structure of this article is as follows. Section 2 describes the experimental method. Section 3 describes the experiment setups. Section 4 shows the Result. Section 5 is the discussion and Section 6 concludes the paper and the references.

## 2. Methods

### *2.1. Overview of the Proposed Method*

The framework of this method is shown in Figure 1. We design a non-rigid 2D/3D registration framework based on deep learning of orthogonal angle projection. Since it is a deep learning-based model, a large amount of data is needed to participate in training. The real paired 2D/3D medical images at the same time are very scarce, so the first task that needs to be done is data augmentation. We chose 4D CT of the lungs as the experimental subject. The expiratory end was used as a moving image $M_{CT}$ and hybrid data augmentation [36,37] was used to obtain a large number of CT $F_{CT}$ representing each respiratory phase of the lung (this procedure will be described in Section 3.1). Then, the ray casting method obtains a pair of 2D DRRs of $F_{CT}$ with orthogonal angles. After that, the orthogonal DRR and the moving image $M_{CT}$ are input into the 2D/3D registration network. The network outputs a 3D deformation field $\phi_p$. Then, the moving image $M_{CT}$ is transformed by the spatial transformation layer [38] to obtain the corresponding predicted CT image. The maximum similarity between the predicted CT image and the ground truth $F_{CT}$ is calculated. Through continuous iterative optimization, we can complete the model training. In the inference phase, only the X-ray projections or DRRs and the moving image must be input to the trained network to get corresponding 3D images.

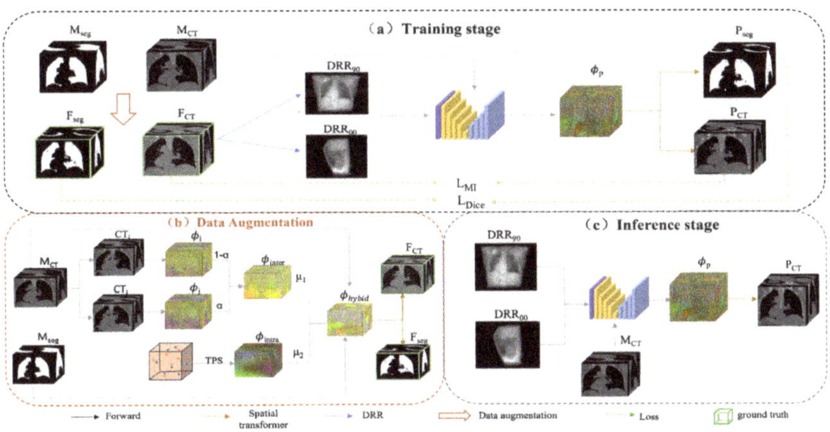


**Figure 1.** Overview of the proposed method. (**a**) Flowchart of the training phase of the method. First, a large number of CT $F_{CT}$ and segmentation $F_{seg}$ representing each phase are obtained by performing hybrid data augmentation of the moving image $M_{CT}$ and the corresponding segmentation image $M_{seg}$. Then, the $F_{CT}$ images are projected to obtain the 2D $DRR_{90}$ and $DRR_{00}$. After that, they are fed into the registration network with the moving image $M_{CT}$ to obtain the predicted deformation field $\phi_p$. Finally, the moving image $M_{CT}$ and the moving segmentation map $M_{seg}$ are transformed to obtain the corresponding predicted images, $P_{CT}$ and $P_{seg}$. (**b**) The process of hybrid data augmentation. The deformation field $\phi_{inter}$ is first obtained by inter-phase registration using traditional image registration. The small deformation $\phi_{intra}$ is simulated by TPS interpolation. The hybrid deformation field $\phi_{hybrid}$ is obtained by summing with random weights for data augmentation. (**c**) Inference stage. The 2D projection and moving images are directly input to the trained network to get the prediction $\phi_p$, and then the registration can be completed by transformation.

### 2.2. 2D/3D Registration Network

Figure 2 shows the registration network. For 2D/3D image registration, the first thing to consider is the consistency of spatial dimension. As a result, we use the extracted feature up-dimensional approach to transform the 2D/3D registration problem into the 3D/3D registration problem. We used the residual network to get the 2D features. The most important step is identity mapping, stopping the gradient from going away, and helping train the network. Thus, when two DRRs with orthogonal angles are input to the network, they are first concatenated in the channel layer as the input of the residual network and then passed through the convolution layer, the max pooling layer, and two output channels with 64 and 128 residual blocks in turn. The channel layer is the third dimension to form a 3D feature map, which is input to the feature extraction network together with the moving image.

We selected the 3D Attention-U-net [37,39] (3D Attu) as the feature extraction network in this study. It can be called the 3D/3D matching network. The network 3D Attu adds an attention gate mechanism to the original U-net, which can automatically distinguish the target shape and scale, and learn more useful information. It also employs encoding and decoding mechanisms and skips connection mechanisms. It effectively blends high- and low-level semantic information while widening the perceptual domain. It has been used in many medical image processing tasks with excellent results. As a result, in this model, we feed the moving image $M_{CT}$ and the 3D feature map into the 3D Attu. The output is the predicted deformation field.

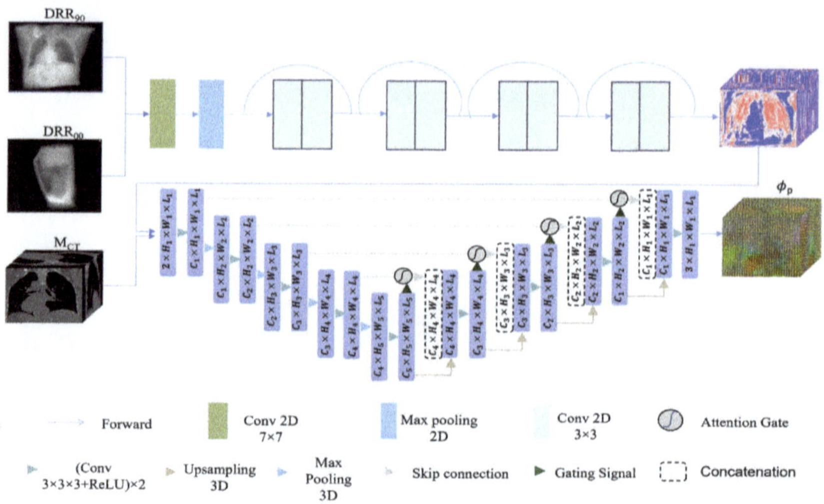


**Figure 2.** 2D/3D registration network. First, 2D DRRs at orthogonal angles are processed by residual blocks to obtain 3D feature maps. Then, the feature maps and moving images are fed into a 3D Attu-based encode–decode network. The final output of this network is the predicted 3D deformation field.

### 2.3. Loss Function

The mutual information (MI) between the ground truth $F_{CT}$ and the predicted 3D CT $P_{CT}$ obtained by the registration network constitute the loss function $L_{MI}(F_{CT}, P_{CT})$. The other part of the loss function is $L_{Dice}(F_{seg}, P_{seg})$, obtained by computing the Dice between the corresponding segmented images, which allows the model to focus more on the lung region. Lastly is the regularized smoothing constraint $L_{Reg}(\phi_p)$ for the deformation field.

$$L_{Dice}(F_{seg}, P_{seg}) = \sum_{i=0}^{n} \frac{1}{n} \frac{2\left|F_{seg}^{i} \cap P_{seg}^{i}\right|}{\left|F_{seg}^{i} + P_{seg}^{i}\right|} \quad (1)$$

$$L = \lambda_1 L_{Dice}(F_{seg}, P_{seg}) + \lambda_2 L_{MI}(F_{CT}, P_{CT}) + \lambda_3 L_{Reg}(\phi_p) \quad (2)$$

where $n$ denotes the number of categories in the image, $i$ denotes the $i$-th category of the image. $\varphi$ denotes all elements in the entire deformation field. $\lambda_1$, $\lambda_2$, $\lambda_3$ denote the weights of $L_{Dice}$, $L_{MI}$, $L_{Reg}$ respectively, which were chosen as 0.5, 0.5, and 0.1 in this experiment.

## 3. Experiment Setups

### 3.1. Data and Augmentation

We conducted experiments on three different types of lung data, TCIA [40–43] patient with a tumor, Dirlab [44] lung CT without tumor, and CIRS phantom. ITK-SNAP is used for automatic segmentation to obtain labels. In the TCIA patient data, we selected one of the patients for the experiment. In Dirlab, we selected the first five sets of data for the experiment. In the CIRS phantom, we simulated the lung tumor with a water sphere. In the experiment, we resampled the 3D CT image to 128 $*$ 128 $*$ 128 with a voxel spacing of 1 mm $*$ 1 mm $*$ 1 mm. Since our experiment is a 2D/3D registration, paired 2D projections and 3D medical images of the same moment are rare. It is unethical to expose the human body to additional radiation doses, so the first task is data augmentation. However, for 2D/3D registration of the treatment phase (e.g., radiotherapy, surgical navigation), it is obvious that the focus is more on the specific person. Therefore, we chose a hybrid data augmentation approach to train a deep learning-based 2D/3D registration model for a specific human body.

In the hybrid data augmentation shown in Figure 1b, we first selected the end-expiratory phase of 4D CT as the moving image $M_{CT}$ and the remaining phases as the fixed image $CT_{1,\ldots,i,j,9}$. Then, we used the conventional intensity-based image registration method to obtain nine deformation fields in the order of $\phi_{1,\ldots,i,j,9}$. The deformation fields used for data augmentation were arbitrarily selected from two of the nine deformation fields and superimposed with random weights to obtain many inter-phase deformations. The lung may also change during respiratory motion. Therefore, we use thin plate spline (TPS) interpolation to simulate small changes in specific phases. The number of control points N was randomly chosen between 20 and 60. The movement distance of control points was chosen between 0 mm and 20 mm to obtain many phase-specific random deformations. To obtain more morphologically diverse images, we combined inter-phase and intra-phase specific deformation with random weights to obtain many hybrid deformation fields. Spatial warping of the moving and segmented images was performed to obtain CT and segmented images representing each respiratory phase of the lung.

### 3.2. DRR Image Generation

The orthogonal angle X-ray projection system in this experiment is shown in Figure 3. Two-point light sources at orthogonal angles emit rays through the object and project them on two detectors perpendicular to the central axis. We assume that the initial intensity of $I_0$ at the light source, $\mu$ is the internal attenuation coefficient of the object to the rays, $I$ is the thickness of the ray through the object, and $I_n$ is the intensity of the ray after passing through the object. The formula $I_n = I_0 e^{-\int \mu(l)dl}$ arises. After the projection of one ray is finished, the attenuation coefficient obtained by accumulating the whole path and then converting it to CT value is the X-ray image. In this experiment, like most researchers, DRR images with the same imaging principle are used instead of X-ray. Virtual X-rays were used to pass through the CT images, and after attenuation, they were projected onto the imaging plane to reconstruct the DRR images. The 3D CT images representing each respiratory phase after data augmentation are projected using this method to obtain the DRR images at the corresponding moment. This technique has been widely used for 2D/3D registration methods.

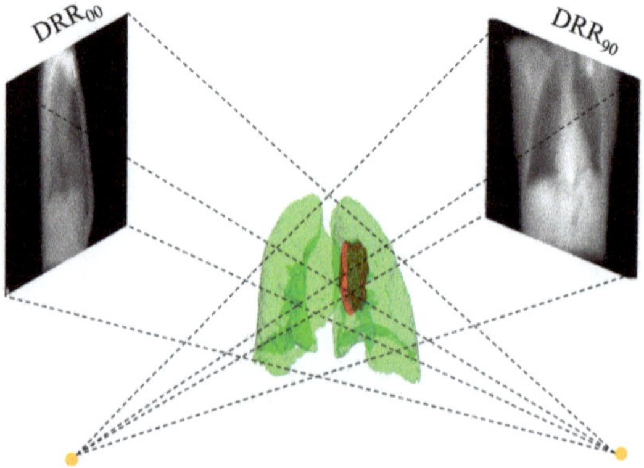


**Figure 3.** Schematic diagram of DRR image generation.

*3.3. Experiment Detail*

We used hybrid data augmentation to obtain 6000 samples from the three types of experimental data. Of these, 5400 were used as the training set, 300 as the validation set, and 300 as the test set. Our experiment was implemented using the deep learning framework Pytorch 1.10 on a NVIDIA A6000 GPU with 48 G of memory, and an AMD Ryzen 7 3700X 8-core processor with 128 GB of internal memory. The learning rate is set to $10^{-4}$. For all datasets, the batch size was set to 8 and the optimization algorithm is Adam.

*3.4. Experiment Evaluation*

In order to verify that our model can achieve 2D/3D registration by two orthogonal angular projections, we selected the end of expiration as the moving image and aligned it toward the remaining phases. We evaluated the three-lung data using NCC, MI, 95% Hausdorff surface distance, and Dice. In addition, to explore the tracking of lung tumors that can be achieved by our model, we compared between predicted and ground truth values for the dataset with tumors and quantitatively evaluated using Dice and the tumor center of mass.

## 4. Result

*4.1. Registration from the Expiratory End to Each Phase*

Here, we demonstrate the registration results of each phase from the end of expiration to the end of inspiration for the TCIA, Dirlab, and phantom. For the qualitative assessment, Figure 4a shows the results of our selected experiments on patients with tumors on TCIA, Figure 4b shows a randomly selected set of experiments from Dirlab, and Figure 4c shows the effect of registration of the phantom data. The odd rows are the unaligned ones, and the even rows are the aligned results. Based on the results, both TCIA patients with tumors and without tumors in Dirlab, as well as the phantom model with water balloons that simulate tumors, can achieve registration from the end of expiration to the rest of the stages.

For the quantitative analysis, we used Dice of the segmentation map, 95% Hausdorff surface distance, NCC, and MI of grayscale images to evaluate our scheme separately. The results are shown in Table 1. It can be seen that good registration results are obtained for all three types of data, not only on the grayscale images, but also on the lung of interest. The Dice values of all three data types are above 0.97, the Hausdorff surface distances are below 2 mm, NCC are above 0.92 and MI are above 0.90. Compared with the real human lung, the NCC and MI of the phantom data are relatively small because the lung of the phantom itself does not change. Only the internal water sphere changes, which is more rigidly transformed relative to the real patient, so the NCC and MI are relatively small at higher Dice. However, the total accuracies are still above 0.92 and 0.90. Therefore, quantitative and qualitative results show that the proposed method can achieve non-rigid 2D/3D registration for a specific subject by two orthogonal angular projections.

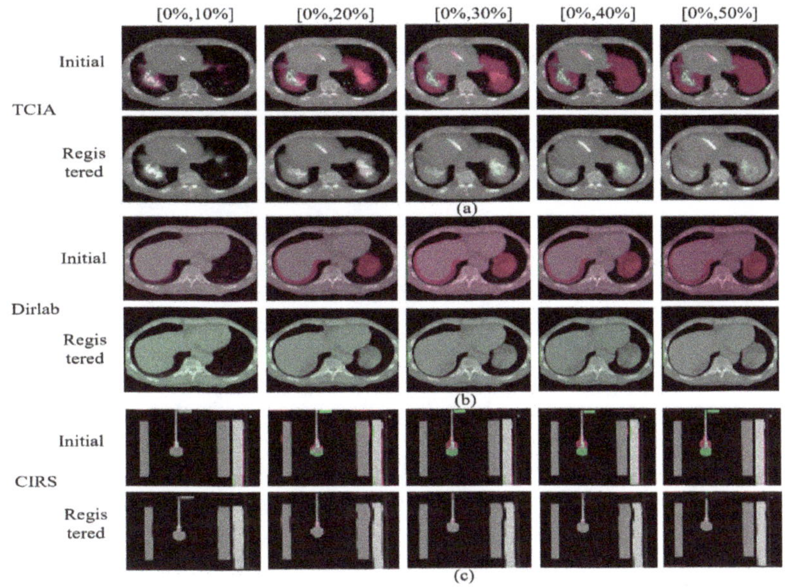


**Figure 4.** Registration from the exhalation end to the other stages. (**a**) shows the results of our registration on TCIA, (**b**) a randomly selected set of experiments from Dirlab, and (**c**) the registration results of the phantom data. The odd-numbered rows are the unregistered contrast images, and the even-numbered rows are the registered contrast images.

**Table 1.** The accuracy of subjects' registration from the expiratory end to each phase.

| | | Dice | Hauf (95%) | NCC | MI |
|---|---|---|---|---|---|
| TCIA | [0%,10%] | 0.9814 | 1.1210 | 0.9846 | 0.9796 |
| | [0%,20%] | 0.9791 | 1.3811 | 0.9846 | 0.9760 |
| | [0%,30%] | 0.9795 | 1.3811 | 0.9847 | 0.9540 |
| | [0%,40%] | 0.9785 | 1.3811 | 0.9846 | 0.9578 |
| | [0%,50%] | 0.9806 | 1.4196 | 0.9848 | 0.9650 |
| Dirlab | [0%,10%] | 0.9857 | 1.8839 | 0.9762 | 0.9590 |
| | [0%,20%] | 0.9857 | 1.8620 | 0.9723 | 0.9535 |
| | [0%,30%] | 0.9853 | 1.8280 | 0.9691 | 0.9511 |
| | [0%,40%] | 0.9853 | 1.8290 | 0.9680 | 0.9424 |
| | [0%,50%] | 0.9854 | 1.8290 | 0.9753 | 0.9608 |
| CIRS | [0%,10%] | 0.9862 | 0.9043 | 0.9338 | 0.9135 |
| | [0%,20%] | 0.9907 | 1.6713 | 0.9291 | 0.9023 |
| | [0%,30%] | 0.9888 | 2.0000 | 0.9360 | 0.9064 |
| | [0%,40%] | 0.9885 | 1.9087 | 0.9348 | 0.9065 |
| | [0%,50%] | 0.9894 | 1.9087 | 0.9349 | 0.9178 |

### 4.2. Tumor Location

Both TCIA patient and the phantom contained tumors. The accuracy of tumor localization was evaluated qualitatively and quantitatively.

Figure 5 shows the qualitative evaluation of the 3D tumor with two types of data, where (a) is a 3D visualization image of the patient's overall lung and tumor and (b) is of the phantom data. Table 2 presents the quantitative results, where we evaluated the tumor center mass and Dice. The tumor center of mass deviation is within 0.15 mm for the real patient. The phantom tumor center of mass is less than 0.05 mm. The Dice of both are above 0.88. It can be seen that the proposed method can achieve registration both for the whole lung and for the tumor. In addition, the fact that local tumors are well aligned suggests that our model could be useful for clinical applications such as tracking tumors.

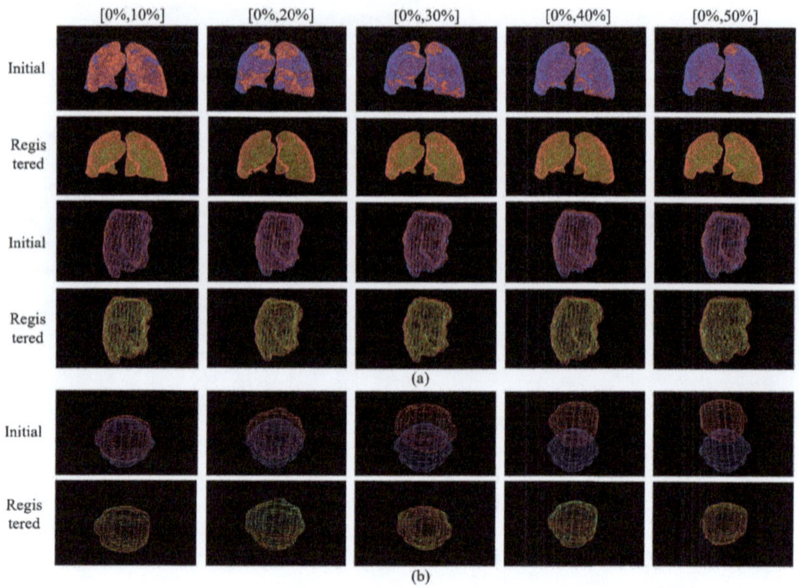


**Figure 5.** Tumor registration results from the exhalation end to other stages. Where (**a**) is the 3D presentation of the results before and after a real patient's lung and tumor registration, and (**b**) is of the phantom data of the 3D results of the tumor display. The odd rows are the unregistered images, and the even rows are the post-registered images. The red image indicates the ground truth. Blue is the moving image, and green is the predicted result obtained by the model.

**Table 2.** The accuracy of tumor location from the expiratory end to each phase.

| | | Center Mass (mm) | | | | Dice |
|---|---|---|---|---|---|---|
| | | **X(LR)** | **Y(AP)** | **Z(LR)** | **Center** | **Tomor** |
| TCIA | [0%,10%] | 0.0003 | 0.0473 | 0.0844 | 0.0968 | 0.9440 |
| | [0%,20%] | 0.0117 | 0.0133 | 0.0260 | 0.0315 | 0.9434 |
| | [0%,30%] | 0.0031 | 0.0023 | 0.0473 | 0.0022 | 0.9023 |
| | [0%,40%] | 0.0032 | 0.0078 | 0.0339 | 0.0350 | 0.9080 |
| | [0%,50%] | 0.0227 | 0.0510 | 0.1251 | 0.1370 | 0.8984 |
| CIRS | [0%,10%] | 0.0224 | 0.0011 | 0.0270 | 0.0351 | 0.9717 |
| | [0%,20%] | 0.0061 | 0.0025 | 0.0081 | 0.0104 | 0.9764 |
| | [0%,30%] | 0.0086 | 0.0018 | 0.0158 | 0.0181 | 0.9609 |
| | [0%,40%] | 0.0174 | 0.0177 | 0.0084 | 0.0262 | 0.9702 |
| | [0%,50%] | 0.0022 | 0.0043 | 0.0477 | 0.0479 | 0.8826 |

## 5. Discussion

### *5.1. Traditional Registration in Data Augmentation*

We used traditional intensity-based image registration for data augmentation to complete the registration between phases. Here we present the two most distorted parts of the three data, the end of expiration and the end of inspiration, for evaluation. The experimental results are shown in Figure 6.

Figure 6 shows the results from three directions before and after the registration. The odd columns are the unregistered images. The even columns are the results after the traditional registration method. Conventional image registration can be accomplished for real patients and models from exhalation to the end of inspiration, ensuring that our augmentation data encompasses all respiratory phases of the lung. In addition, the registration covers the larger deformations at both ends.

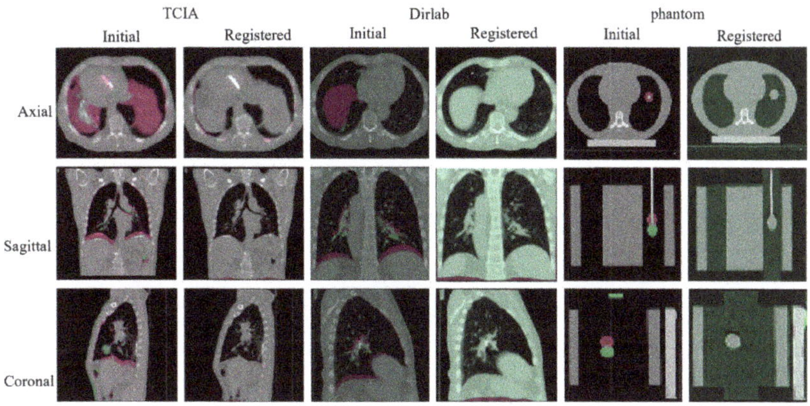


**Figure 6.** The traditional registration method results from the end of exhalation to the end of inhalation. The first two columns are the registration results on the patient, the middle two are the registration results on the normal human lung, and the last two are the registration results on the phantom. The odd columns are the unregistered images, and the even columns are the results of the registered images.

### *5.2. Landmark Error*

For the Dirlab data, the landmark points and the deformation field for performing data augmentation are known. Thus, the landmark points of the image generated after data augmentation are also known and used as our ground truth. The mean target registration error (mTRE) is evaluated with our model-predicted images.

The data obtained from the evaluation are shown in Table 3. The corresponding box plot is shown in Figure 7, in which green indicates the data before registration, yellow is the data after registration of the proposed method, and purple represents the data after 3D/3D registration using Demons [45]. The results show that the proposed model can achieve effective 2D/3D registration, but the accuracy is lower than the existing advanced 3D/3D registration models because the experimental data are only two 2D X-rays with orthogonal angles. Although the proposed method transforms 2D/3D registration into a 3D/3D registration problem, some image details are indeed lost compared with the 3D images, resulting in the loss of information on tiny details, such as capillaries, leading to a lower accuracy of landmark error based on detailed information. However, in the 2D/3D registration mission, more attention is paid to the global overall changes in the lung and the tumor location. The proposed method greatly reduces the irradiation dose and improves the registration speed.

**Table 3.** Mean target registration error of landmarks in Dirlabs.

| (mm) | Initial | Proposed (2D/3D) | Demons (3D/3D) |
|---|---|---|---|
| Dirlab1 | 3.9776 (1.8616) | 2.0065 (0.6748) | 1.6297 (0.3196) |
| Dirlab2 | 6.3989 (2.1719) | 4.0079 (1.0077) | 3.3807 (0.5089) |
| Dirlab3 | 6.2138 (1.7843) | 2.9219 (0.7237) | 2.1556 (0.2991) |
| Dirlab4 | 7.6437 (2.3978) | 4.0682 (0.9898) | 3.2525 (0.3918) |
| Dirlab5 | 6.6075 (2.1448) | 2.6253 (0.7719) | 1.6408 (0.4824) |

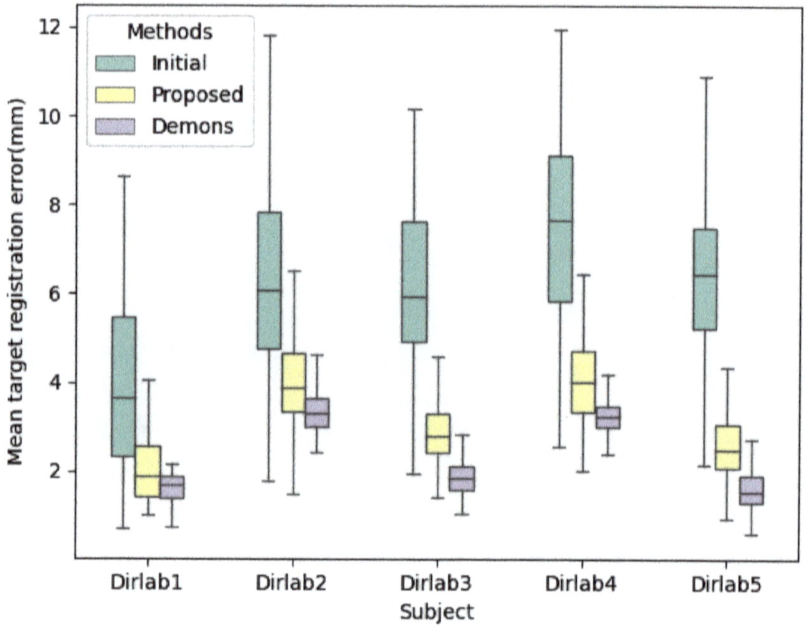


**Figure 7.** Box plot of landmark points in Dirlab; green indicates the data before registration, yellow is the data after registration of the proposed method, and purple represents the data after 3D/3D registration using Demons.

In addition, our method can complete 2D/3D registration in 1.2 s. In contrast, other data-driven 2D/3D registration models, such as [20], may take a few seconds. On the other hand, traditional image registration methods may take tens of minutes or even hours. Our method also only needs two different angles of X-rays, which greatly reduces the amount of radiation and makes the hardware in the clinic easier to use. Of course, our method also has some limitations. First of all, since real medical images do not exist at the same moment of paired orthogonal angles of X-ray and corresponding 3D CT, we use 2D DRR. Although DRR and real X-ray use the same imaging way, it is undeniable that there are some grayscale and noise differences between the two. However, it can be corrected by using existing methods, such as histogram matching [24], network of GAN [25], etc., which is not the main focus of our study. We will also make some improvements to the program to speed up the processing speed for radiotherapy or interventional procedures that require more real-time, etc. In addition, since there are few non-rigid 2D/3D registration articles, the code is not open source. We have yet to choose a suitable comparison experiment, and we will continue to look for it in the future.

## 6. Conclusions

This study proposes a deep learning-based 2D/3D registration method using two orthogonal angular X-ray projection images. The proposed algorithm has been verified on lung data with and without tumor and phantom data, and obtained high registration accuracy, where Dice and NCC are greater than 0.97 and 0.92. In addition, we evaluated the accuracy on the data containing tumor, and the tumor center-of-mass error was within 0.15 mm, which indicates the promising use of our model for tumor tracking. The registration time is within 1.2 s, and this is promising for clinical applications, such as radiotherapy or surgical navigation, to track the shape of organs in real time. Moreover, we only need to use two orthogonal angles of X-rays to achieve 2D/3D deformable image registration, which can greatly reduce the extra dose during treatment and simplify the hardware system required.

**Author Contributions:** Conceptualization, G.D., J.D., Y.X. and X.L.; methodology, J.D. and X.L.; software, J.D.; validation, N.L.; investigation, N.L., H.W. and L.X.; data curation, G.D. and J.D.; writing—original draft preparation, G.D. and J.D.; writing—review and editing, C.Z., W.H., L.L., Y.C., L.L. Y.L. and X.L.; visualization, L.L. and X.L.; supervision, X.L. All authors have read and agreed to the published version of the manuscript.

**Funding:** This work is partly supported by grants from the National Natural Science Foundation of China (U20A201795, U21A20480, 82202954, 61871374, 62001464, 11905286), Young S&T Talent Training Program of Guangdong Provincial Association for S&T, China (SKXRC202224), and the Chinese Academy of Sciences Special Research Assistant Grant Program.

**Institutional Review Board Statement:** Not applicable.

**Informed Consent Statement:** Not applicable.

**Data Availability Statement:** The dataset images used for this study is publicly available on TCIA at https://wiki.cancerimagingarchive.net/pages/viewpage.action?pageId=21267414 (accessed on 27 October 2022) and Dirlab at https://med.emory.edu/departments/radiation-oncology/research-laboratories/deformable-image-registration/downloads-and-reference-data/4dct.html (accessed on 27 October 2022).

**Acknowledgments:** We thank The Cancer Imaging Archive (TCIA) and Emory University School of Medicine (4D CT) for public access and for sharing their datasets.

**Conflicts of Interest:** The authors declare no conflict of interest.

## Abbreviations

The following abbreviations are used in this manuscript:

| | |
|---|---|
| 2D | Two-dimensional |
| 3D | Three-dimensional |
| DSA | Digital Subtraction Angiography |
| CT | Computed Tomography |
| MRI | Magnetic Resonance Imaging |
| DRR | Digitally Reconstructed Radiographs |
| NCC | Normalized cross-correlation |
| PCA | Principal Component Analysis |
| CNN | Convolutional Neural Network |
| GPU | Graphic Processing Unit(GPU) |
| 3D Attu | 3D Attention-U-net |
| MI | Mutual Information |
| TPS | Thin plate spline |
| mTRE | Mean Target Registration Error |

## References


1. Foote, M.D.; Zimmerman, B.E.; Sawant, A.; Joshi, S.C. Real-time 2D-3D deformable registration with deep learning and application to lung radiotherapy targeting. In Proceedings of the International Conference on Information Processing in Medical Imaging, Hong Kong, 2–7 June 2019 ; Springer: Berlin/Heidelberg, Germany, 2019; pp. 265–276.
2. Frysch, R.; Pfeiffer, T.; Rose, G. A novel approach to 2D/3D registration of X-ray images using Grangeat's relation. *Med. Image Anal.* **2021**, *67*, 101815. [CrossRef]
3. Van Houtte, J.; Audenaert, E.; Zheng, G.; Sijbers, J. Deep learning-based 2D/3D registration of an atlas to biplanar X-ray images. *Int. J. Comput. Assist. Radiol. Surg.* **2022**, *17*, 1333–1342. [CrossRef]
4. Li, P.; Pei, Y.; Guo, Y.; Ma, G.; Xu, T.; Zha, H. Non-rigid 2D-3D registration using convolutional autoencoders. In Proceedings of the 2020 IEEE 17th International Symposium on Biomedical Imaging (ISBI), Iowa City, IA, USA, 3–7 April 2020; IEEE: New York, NY, USA, 2020; pp. 700–704.
5. Wang, C.; Xie, S.; Li, K.; Wang, C.; Liu, X.; Zhao, L.; Tsai, T.Y. Multi-View Point-Based Registration for Native Knee Kinematics Measurement with Feature Transfer Learning. *Engineering* **2021**, *7*, 881–888. [CrossRef]
6. Guan, S.; Wang, T.; Sun, K.; Meng, C. Transfer learning for nonrigid 2d/3d cardiovascular images registration. *IEEE J. Biomed. Health Inform.* **2020**, *25*, 3300–3309. [CrossRef]
7. Miao, S.; Wang, Z.J.; Liao, R. A CNN regression approach for real-time 2D/3D registration. *IEEE Trans. Med. Imaging* **2016**, *35*, 1352–1363. [CrossRef]
8. Markova, V.; Ronchetti, M.; Wein, W.; Zettinig, O.; Prevost, R. Global Multi-modal 2D/3D Registration via Local Descriptors Learning. *arXiv* **2022**, arXiv:2205.03439.
9. Zheng, G. Effective incorporating spatial information in a mutual information based 3D–2D registration of a CT volume to X-ray images. *Comput. Med. Imaging Graph.* **2010**, *34*, 553–562. [CrossRef]
10. Zollei, L.; Grimson, E.; Norbash, A.; Wells, W. 2D-3D rigid registration of X-ray fluoroscopy and CT images using mutual information and sparsely sampled histogram estimators. In Proceedings of the 2001 IEEE Computer Society Conference on Computer Vision and Pattern Recognition, CVPR 2001, Kauai, HI, USA, 8–14 December 2021; IEEE: New York, NY, USA, 2001; Volume 2, p. II.
11. Gendrin, C.; Furtado, H.; Weber, C.; Bloch, C.; Figl, M.; Pawiro, S.A.; Bergmann, H.; Stock, M.; Fichtinger, G.; Georg, D.; et al. Monitoring tumor motion by real time 2D/3D registration during radiotherapy. *Radiother. Oncol.* **2012**, *102*, 274–280. [CrossRef]
12. Gao, C.; Grupp, R.B.; Unberath, M.; Taylor, R.H.; Armand, M. Fiducial-free 2D/3D registration of the proximal femur for robot-assisted femoroplasty. *IEEE Trans. Med. Robot. Bionics* **2020**, *11315*, 350–355.
13. Munbodh, R.; Knisely, J.P.; Jaffray, D.A.; Moseley, D.J. 2D–3D registration for cranial radiation therapy using a 3D kV CBCT and a single limited field-of-view 2D kV radiograph. *Med. Phys.* **2018**, *45*, 1794–1810. [CrossRef]
14. De Silva, T.; Uneri, A.; Ketcha, M.; Reaungamornrat, S.; Kleinszig, G.; Vogt, S.; Aygun, N.; Lo, S.; Wolinsky, J.; Siewerdsen, J. 3D–2D image registration for target localization in spine surgery: Investigation of similarity metrics providing robustness to content mismatch. *Phys. Med. Biol.* **2016**, *61*, 3009. [CrossRef]
15. Yu, W.; Tannast, M.; Zheng, G. Non-rigid free-form 2D–3D registration using a B-spline-based statistical deformation model. *Pattern Recognit.* **2017**, *63*, 689–699. [CrossRef]
16. Li, R.; Jia, X.; Lewis, J.H.; Gu, X.; Folkerts, M.; Men, C.; Jiang, S.B. Real-time volumetric image reconstruction and 3D tumor localization based on a single x-ray projection image for lung cancer radiotherapy. *Med. Phys.* **2010**, *37*, 2822–2826. [CrossRef]
17. Li, R.; Lewis, J.H.; Jia, X.; Gu, X.; Folkerts, M.; Men, C.; Song, W.Y.; Jiang, S.B. 3D tumor localization through real-time volumetric x-ray imaging for lung cancer radiotherapy. *Med. Phys.* **2011**, *38*, 2783–2794. [CrossRef]
18. Zhang, Y.; Tehrani, J.N.; Wang, J. A biomechanical modeling guided CBCT estimation technique. *IEEE Trans. Med. Imaging* **2016**, *36*, 641–652. [CrossRef]
19. Zhang, Y.; Folkert, M.R.; Li, B.; Huang, X.; Meyer, J.J.; Chiu, T.; Lee, P.; Tehrani, J.N.; Cai, J.; Parsons, D. 4D liver tumor localization using cone-beam projections and a biomechanical model. *Radiother. Oncol.* **2019**, *133*, 183–192. [CrossRef]
20. Zhang, Y. An unsupervised 2D–3D deformable registration network (2D3D-RegNet) for cone-beam CT estimation. *Phys. Med. Biol.* **2021**, *66*, 074001. [CrossRef]
21. Ketcha, M.; De Silva, T.; Uneri, A.; Jacobson, M.; Goerres, J.; Kleinszig, G.; Vogt, S.; Wolinsky, J.; Siewerdsen, J. Multi-stage 3D–2D registration for correction of anatomical deformation in image-guided spine surgery. *Phys. Med. Biol.* **2017**, *62*, 4604. [CrossRef]
22. Zhang, Y.; Qin, H.; Li, P.; Pei, Y.; Guo, Y.; Xu, T.; Zha, H. Deformable registration of lateral cephalogram and cone-beam computed tomography image. *Med. Phys.* **2021**, *48*, 6901–6915. [CrossRef]
23. Gao, C.; Liu, X.; Gu, W.; Killeen, B.; Armand, M.; Taylor, R.; Unberath, M. Generalizing spatial transformers to projective geometry with applications to 2D/3D registration. In Proceedings of the International Conference on Medical Image Computing and Computer-Assisted Intervention, Lima, Peru, 4–8 October 2020; Springer: Berlin/Heidelberg, Germany, 2020; pp. 329–339.
24. Wei, R.; Liu, B.; Zhou, F.; Bai, X.; Fu, D.; Liang, B.; Wu, Q. A patient-independent CT intensity matching method using conditional generative adversarial networks (cGAN) for single x-ray projection-based tumor localization. *Phys. Med. Biol.* **2020**, *65*, 145009. [CrossRef]
25. Wei, R.; Zhou, F.; Liu, B.; Bai, X.; Fu, D.; Liang, B.; Wu, Q. Real-time tumor localization with single X-ray projection at arbitrary gantry angles using a convolutional neural network (CNN). *Phys. Med. Biol.* **2020**, *65*, 065012. [CrossRef]

26. Van Houtte, J.; Gao, X.; Sijbers, J.; Zheng, G. 2D/3D registration with a statistical deformation model prior using deep learning. In Proceedings of the 2021 IEEE EMBS international conference on biomedical and health informatics (BHI), Athens, Greece, 27–30 July 2021; IEEE: New York, NY, USA, 2021, pp. 1–4.
27. Pei, Y.; Zhang, Y.; Qin, H.; Ma, G.; Guo, Y.; Xu, T.; Zha, H. Non-rigid craniofacial 2D-3D registration using CNN-based regression. In *Deep Learning in Medical Image Analysis and Multimodal Learning for Clinical Decision Support*; Springer: Berlin/Heidelberg, Germany, 2017; pp. 117–125.
28. Tian, L.; Lee, Y.Z.; San José Estépar, R.; Niethammer, M. LiftReg: Limited Angle 2D/3D Deformable Registration. In Proceedings of the Medical Image Computing and Computer Assisted Intervention—MICCAI 2022, Singapore, 18–22 September 2022; Wang, L., Dou, Q., Fletcher, P.T., Speidel, S., Li, S., Eds.; Springer Nature: Cham, Switzerland, 2022; pp. 207–216.
29. Liao, H.; Lin, W.A.; Zhang, J.; Zhang, J.; Luo, J.; Zhou, S.K. Multiview 2D/3D rigid registration via a point-of-interest network for tracking and triangulation. In Proceedings of the IEEE/CVF Conference on Computer Vision and Pattern Recognition, Long Beach, CA, USA, 15–20 June 2019; pp. 12638–12647.
30. Markova, V.; Ronchetti, M.; Wein, W.; Zettinig, O.; Prevost, R. Global Multi-modal 2D/3D Registration via Local Descriptors Learning. In Proceedings of the Medical Image Computing and Computer Assisted Intervention—MICCAI 2022, Singapore, 18–22 September 2022; Wang, L., Dou, Q., Fletcher, P.T., Speidel, S., Li, S., Eds.; Springer Nature: Cham, Switzerland, 2022; pp. 269–279.
31. Schaffert, R.; Wang, J.; Fischer, P.; Borsdorf, A.; Maier, A. Learning an attention model for robust 2-D/3-D registration using point-to-plane correspondences. *IEEE Trans. Med. Imaging* **2020**, *39*, 3159–3174. [CrossRef]
32. Schaffert, R.; Wang, J.; Fischer, P.; Borsdorf, A.; Maier, A. Metric-driven learning of correspondence weighting for 2-D/3-D image registration. In Proceedings of the German Conference on Pattern Recognition, Stuttgart, Germany, 9–12 October 2019; Springer: Berlin/Heidelberg, Germany, 2019, pp. 140–152.
33. Schaffert, R.; Wang, J.; Fischer, P.; Maier, A.; Borsdorf, A. Robust multi-view 2-d/3-d registration using point-to-plane correspondence model. *IEEE Trans. Med. Imaging* **2019**, *39*, 161–174. [CrossRef]
34. Nakao, M.; Nakamura, M.; Matsuda, T. Image-to-Graph Convolutional Network for 2D/3D Deformable Model Registration of Low-Contrast Organs. *IEEE Trans. Med. Imaging* **2022**, *41*, 3747–3761. [CrossRef]
35. Shao, H.C.; Wang, J.; Bai, T.; Chun, J.; Park, J.C.; Jiang, S.; Zhang, Y. Real-time liver tumor localization via a single x-ray projection using deep graph neural network-assisted biomechanical modeling. *Phys. Med. Biol.* **2022**, *67*, 115009. [CrossRef]
36. Liang, X.; Zhao, W.; Hristov, D.H.; Buyyounouski, M.K.; Hancock, S.L.; Bagshaw, H.; Zhang, Q.; Xie, Y.; Xing, L. A deep learning framework for prostate localization in cone beam CT-guided radiotherapy. *Med. Phys.* **2020**, *47*, 4233–4240. [CrossRef]
37. Liang, X.; Li, N.; Zhang, Z.; Xiong, J.; Zhou, S.; Xie, Y. Incorporating the hybrid deformable model for improving the performance of abdominal CT segmentation via multi-scale feature fusion network. *Med. Image Anal.* **2021**, *73*, 102156. [CrossRef]
38. Jaderberg, M.; Simonyan, K.; Zisserman, A.; Kavukcouglu, K. Spatial transformer networks. *Adv. Neural Inf. Process. Syst.* **2015**, *28*, 2017–2025.
39. Schlemper, J.; Oktay, O.; Schaap, M.; Heinrich, M.; Kainz, B.; Glocker, B.; Rueckert, D. Attention gated networks: Learning to leverage salient regions in medical images. *Med. Image Anal.* **2019**, *53*, 197–207. [CrossRef]
40. Hugo, G.D.; Weiss, E.; Sleeman, W.C.; Balik, S.; Keall, P.J.; Lu, J.; Williamson, J.F. Data from 4d lung imaging of nsclc patients. *Med. Phys.* **2017**, *44*, 762–771. [CrossRef]
41. Balik, S.; Weiss, E.; Jan, N.; Roman, N.; Sleeman, W.C.; Fatyga, M.; Christensen, G.E.; Zhang, C.; Murphy, M.J.; Lu, J.; et al. Evaluation of 4-dimensional computed tomography to 4-dimensional cone-beam computed tomography deformable image registration for lung cancer adaptive radiation therapy. *Int. J. Radiat. Oncol. Biol. Phys.* **2013**, *86*, 372–379. [CrossRef] [PubMed]
42. Clark, K.; Vendt, B.; Smith, K.; Freymann, J.; Kirby, J.; Koppel, P.; Moore, S.; Phillips, S.; Maffitt, D.; Pringle, M.; et al. The Cancer Imaging Archive (TCIA): Maintaining and operating a public information repository. *J. Digit. Imaging* **2013**, *26*, 1045–1057. [CrossRef]
43. Roman, N.O.; Shepherd, W.; Mukhopadhyay, N.; Hugo, G.D.; Weiss, E. Interfractional positional variability of fiducial markers and primary tumors in locally advanced non-small-cell lung cancer during audiovisual biofeedback radiotherapy. *Int. J. Radiat. Oncol. Biol. Phys.* **2012**, *83*, 1566–1572. [CrossRef]
44. Castillo, R.; Castillo, E.; Guerra, R.; Johnson, V.E.; McPhail, T.; Garg, A.K.; Guerrero, T. A framework for evaluation of deformable image registration spatial accuracy using large landmark point sets. *Phys. Med. Biol.* **2009**, *54*, 1849. [CrossRef] [PubMed]
45. Vercauteren, T.; Pennec, X.; Perchant, A.; Ayache, N. Diffeomorphic demons: Efficient non-parametric image registration. *NeuroImage* **2009**, *45*, S61–S72. [CrossRef]

bioengineering

MDPI

Article

# Deep Feature Engineering in Colposcopy Image Recognition: A Comparative Study

**Shefa Tawalbeh [1], Hiam Alquran [1,*] and Mohammed Alsalatie [2]**

1 Department of Biomedical Systems and Informatics Engineering, Yarmouk University, Irbid 21163, Jordan
2 The Institute of Biomedical Technology, King Hussein Medical Center, Royal Jordanian Medical Service, Amman 11855, Jordan
* Correspondence: heyam.q@yu.edu.jo

**Abstract:** Feature fusion techniques have been proposed and tested for many medical applications to improve diagnostic and classification problems. Specifically, cervical cancer classification can be improved by using such techniques. Feature fusion combines information from different datasets into a single dataset. This dataset contains superior discriminant power that can improve classification accuracy. In this paper, we conduct comparisons among six selected feature fusion techniques to provide the best possible classification accuracy of cervical cancer. The considered techniques are canonical correlation analysis, discriminant correlation analysis, least absolute shrinkage and selection operator, independent component analysis, principal component analysis, and concatenation. We generate ten feature datasets that come from the transfer learning of the most popular pre-trained deep learning models: Alex net, Resnet 18, Resnet 50, Resnet 10, Mobilenet, Shufflenet, Xception, Nasnet, Darknet 19, and VGG Net 16. The main contribution of this paper is to combine these models and then apply them to the six feature fusion techniques to discriminate various classes of cervical cancer. The obtained results are then fed into a support vector machine model to classify four cervical cancer classes (i.e., Negative, HISL, LSIL, and SCC). It has been found that the considered six techniques demand relatively comparable computational complexity when they are run on the same machine. However, the canonical correlation analysis has provided the best performance in classification accuracy among the six considered techniques, at 99.7%. The second-best methods were the independent component analysis, least absolute shrinkage and the selection operator, which were found to have a 98.3% accuracy. On the other hand, the worst-performing technique was the principal component analysis technique, which offered 90% accuracy. Our developed approach of analysis can be applied to other medical diagnosis classification problems, which may demand the reduction of feature dimensions as well as a further enhancement of classification performance.

**Keywords:** cervical cancer; feature fusion; feature selection; deep learning structures; support vector machine; disease discrimination accuracy; performance comparisons

**Citation:** Tawalbeh, S.; Alquran, H.; Alsalatie, M. Deep Feature Engineering in Colposcopy Image Recognition: A Comparative Study. *Bioengineering* **2023**, *10*, 105. https://doi.org/10.3390/bioengineering10010105

Academic Editor: Larbi Boubchir

Received: 6 December 2022
Revised: 27 December 2022
Accepted: 9 January 2023
Published: 12 January 2023



## 1. Introduction

In 2020, 604,000 new cases of cervical cancer were estimated, and 342,000 deaths were reported; 90% of the new cases and deaths were reported in middle- and low-income countries [1]. These cases were due to the lack of health awareness as well as the limited access to screening methodologies. According to the World Health Organization (WHO), appropriate screening reduces morbidity and mortality among women [2]. In this regard, a pap smear is the most common early screening and diagnostic tool for cervical cancer. Hundreds of sub-pap smear images are examined under a microscope by a cytopathologist. This makes such manual analysis a subjective, error-prone, and time-consuming process.

Computer-aided design (CAD) tools can play an important role in overcoming the inconsistency, inaccuracy, and time-consuming problems of manual analysis. In the last few decades, automated methods have been developed and then approved by the food and drug administration (FDA) to diagnose and classify cervical cancer [3–6].

The recent advances in computing and the large growing data repository have supported efficient machine learning (ML) and deep learning (DL) algorithms to aid medical decisions. In recent years, pap smear images have been efficiently processed by adequate machine learning algorithms for cervical cancer classification [7–12]. One of the first steps in building such models is to identify the features that best describe the input data.

In this paper, we mainly focus on providing comprehensive testing results for the estimation accuracy of various data fusion techniques when they are applied to cervical cancer classification. It is noted that data fusion can occur at different levels, such as the feature level, matching score level, or the decision level [13]. The main aim of feature fusion is to combine information from two or more feature sets into a single dataset that has more discriminant power than each feature vector. Accordingly, in this paper, we are interested in utilizing this discriminant power in separating classes more efficiently. We are conducting a comparative analysis to test the effectiveness of selected feature fusion techniques in enhancing the accuracy of cervical cancer classification. These techniques are applied on the feature level, which reduces the dimensionality of the feature datasets while enhancing the accuracy of classification.

The following literature review highlights recent studies that show the effectiveness of data fusion techniques for cervical cancer detection. However, due to the limited number of studies that use feature-level fusion for cervical cancer classification, which is the main purpose of this paper, the literature review is followed by other related studies that use feature fusion on other medical images for diagnostic and classification purposes.

## 2. Related Work

In this section, we have selected the most recent studies that use feature engineering specifically on cervical cancer classification. In each article, the authors used a different fusion technique and showed how this improved the classification accuracy. Alquran et al. [14], proposed a computer-aided diagnosis of cervical cancer classification based on feature fusion between the well-known Shuffle Net DL structure and a novel Cervical Net structure. The novel Cervical Net structure was proposed by Alquran. The authors used a principal component analysis (PCA) and canonical correlation analysis (CCA) as the feature reduction and fusion techniques. The resultant features were fed into different ML classifiers. The best accuracy of 99.1% was obtained using a support vector machine (SVM) to classify between five classes of pap smear images. On the other hand, Liu et al. [15] proposed a framework to classify cervical cancer cell classification based on DL. Specifically, they extracted local and global features using a convolutional neural network (CNN) module and a visual transformer module, respectively, from cervical cancer cell images. Then these features were fused using a multilayer perceptron module. The framework proposed by Liu et al. obtained an accuracy of classification of 91.72% by combining two datasets (CRIC and SIPaKMeD datasets) for an 11-class classification problem.

Rahman et al. [16] proposed a method for enhancing computer-aided diagnosis of cervical pap smear images using a hybrid deep feature fusion (HDFF) method. This method was tested on the SIPaKMeD dataset and performance was compared with multiple DL models alongside the late fusion method. The late fusion, sometimes called decision-level fusion, leverages predictions from multiple models to make a final decision. In their paper using the SIPaKMeD dataset, they obtained a classification accuracy of 99.85%, 99.38%, and 99.14%, for a 2-class, 3-class, and 5-class classification. They also tested their model on the Herlev dataset and achieved an accuracy of 98.32% for a 2-class and 90.32% for a 7-class classification. Moreover, Hussain et al. [17] proposed a computer-assisted screening system based on DL. The paper explored six deep learning structures, namely Alexnet, Vggnet (vgg-16 and vgg-19), Resnet (resnet-50 and resnet-101), and Googlenet architectures, for a four-class diagnosis of cervical cancer lesions. The authors fused the best three DL models yielding the best accuracy for class classification. The output of each deep learning structure mentioned above was evaluated based on performance, then the best three models (Resnet-50, Resnet-101, and Googlenet) were combined (fused) to generate their ensemble classifier. Their results showed that the proposed classifier achieved the highest area under curve (AUC) = 97% between two positive and negative classes.

The above articles applied some sort of data fusion method to enhance the decision accuracy from cervical cancer pap smear images. However, not all the above studies used feature-level fusion. Rahman et al. and Hussain et al. used decision-level fusion. Alquran et al. used CCA to fuse features from two datasets, and finally, Lui et al. used a multilayer perceptron model. Due to the limited number of studies that use feature fusion for cervical cancer classification, we listed other studies that highlight the effectiveness of using feature fusion and reduction analysis to improve other medical image classification problems. In the below references, we have selected articles that used feature analysis CCA, discriminant correlation analysis (DCA), least absolute shrinkage and selection operator (LASSO), independent component analysis (ICA), PCA, and others. Most of the feature fusion techniques mentioned in the below articles were selected in our comparative study.

Zhang et al. [18] studied four different feature fusion and reduction techniques between two independent feature sets, namely, LungTrans features and PrRadomics features. In their paper, the authors proposed a method for feature fusion named the 'risk score based' feature fusion method. Their paper showed that the proposed risk score-based feature fusion method improves the prognosis performance for predicting the survival of pancreatic ductal adenocarcinoma patients, yielding an increase of 40% of AUC compared with AUC without fusion. The feature fusion and reduction techniques used were PCA, LASSO, Boruto, Univariant Cox proportional-hazards CPH, and the proposed risk score-based technique. The latest was performed by feeding each feature set to two different random forest classification models, and the resulting most significant features were fed into another random forest-based prognosis model. In summary, Zhang et al. compared five different feature fusion techniques on two feature datasets (lungtrans features and PrRadomics) to improve the prognosis of PDAC. Moreover, Fan et al. [19] integrated dynamic contrast-enhanced magnetic resonance imaging and T2-weighted imaging radiomic features by a CCA. The paper aimed to provide related complementary information between the fused feature datasets to improve breast cancer prediction. After fusing the two datasets, they used SVM-based recursive feature elimination (SVM-RFE) to identify the optimal features for prediction. They noticed an enhancement in the AUC after using fused features. Moreover, they reported that using CCA was more beneficial than using concatenation-based feature fusion or classifier fusion methods. Another method for feature-level fusion is the DCA, which was proposed by Haghighat et al. [20] where they introduce DCA as an effective feature fusion method to enhance class separation. They tested DCA on multiple biometric datasets showing the effectiveness of this approach. Using DCA combines the information from more than one feature dataset into a single dataset that has more discriminant power. This was applied to different medical diagnostic applications, for example, Wang et al. [21] extracted features from four datasets for COVID-19 CCT images using a novel feature learning algorithm. Then, they proposed a selection algorithm to select the best two models. Finally, they used the DCA to fuse the two features from the two models. The final determined model was named CCSHNET. Their proposed CCSHNET model based on fusing features using DCA showed high-performance measures when compared to other COVID-19 detection methods.

In this paper, we focus on the existing feature engineering techniques. The utilization of pre-trained DL structures to extract features from whole-slice pap smear images is a promising idea, alongside exploiting feature fusion and reduction techniques to obtain the highest level of confidential computer-aided diagnosis system for colposcopy images. To our knowledge, this is the first paper that employs ten deep-learning models to extract representative descriptors, which can be utilized for the recognition of pap smear image diseases via feature engineering algorithms. The novelty in our approach is using existing feature-level fusion to extract the most representative features from ten DL models to enhance classification accuracy.

## 3. Materials and Methods

The method that is proposed in this paper is illustrated in Figure 1.

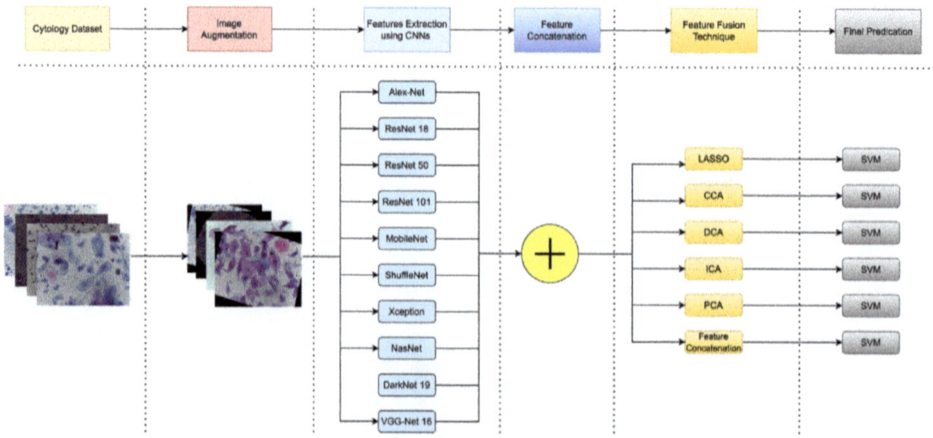


**Figure 1.** The proposed method. Showing all sequential steps of the proposed methodology in this paper.

The methodology followed in this paper consists of six steps. Step one: collect the cytology dataset that consists of 1000 samples for 4 different cervical cancer classes. Step two: perform image augmentation. Step three: extract features using CNN from ten deep learning structures (4 features for each DL structure total of 40 features). Step four: concatenate all the features from the ten DL structures to be fed into the feature fusion step. Step five: apply different feature fusion techniques to fuse or select features. Step six: feed the features into an SVM to measure classification performance. The details of each step are described in detail in the following section. Cytology dataset acquisition and augmentation are described in Sections 3.1 and 3.2. Extracting features using deep learning structures are described in Section 3.3. The theoretical background of the six selected fusion techniques is described in Sections 3.4 and 3.5. Finally, the SVM method is described in Section 3.6.

### 3.1. Image Acquisition

One of the cervical screening tests is liquid-based cytology (LBC). A total of 963 LBC images are separated into four sets to reflect the four classes, namely, NILM, LSIL, HSIL, and SCC, that make up the whole repository. It includes cervical cancer-related precancerous and cancerous lesions that meet the Bethesda System requirements (TBS). A total of 460 patients visited the obstetrics and gynecology (O&G) department of the public hospital with varied gynecological issues and were examined using the ICC50 HD microscope to take the images at a magnification of 40×. The pathology department's professionals then examined and categorized the images [22].

### 3.2. New Image Augmentation

Data augmentation is a strategy used to expand the amount of data by adding slightly changed copies of either existing data or freshly created synthetic data from existing data. It serves as a regularizer and helps minimize overfitting. This paper used rotation images at random angles in the range of [−45, 45] degrees, image resizing with random scale factors between [0.2, 1], and translation in both directions X and Y are [−3, 3], to accomplish image augmentation for the abnormal cases [23]. Table 1 describes the number of images before and after augmentation.

**Table 1.** The number of images before and after augmentation for abnormal cells. After augmentation the number of images becomes equal.

| Abnormal Cells | Before Augmentation | After Augmentation |
|---|---|---|
| 1. Low-grade squamous intraepithelial lesion (LSIL) | 113 | 250 |
| 2. High-grade squamous intraepithelial lesion (HSIL) | 163 | 250 |
| 3. Squamous cell carcinoma (SCC) | 74 | 250 |

### 3.3. Deep Learning Features

Several pre-trained deep learning models are employed to extract the most representative features from the last fully connected model in each one. The selected deep-learning structures were trained on the ImageNet database to distinguish between 1000 classes from nature. Transfer learning techniques were used to make these structures compatible with the designed problem statement, which focused on classifying four types of whole-slice cervical cells. The transfer learning appeared by augmenting the input size of the image to be appropriated with the input layer of each one and removing the last fully connected layer to make it four neurons for four classes. The represented features for each model are extracted from the last fully connected layer. Each one provides four distinguished features for four classes. The networks that are utilized for feature extractions are AlexNet, ResNet18,50, and 101, Mobile Net, Shuffle Net, Xception Net, Nasnet, Dark-19, and VGG16.

#### 3.3.1. AlexNet

AlexNet is one of the most popular convolutional networks. It was first introduced in 2012 for ImageNet recognition of 1000 nature classes. AlexNet architecture consists of five convolutional layers, three max-pooling layers, two normalization layers, and two fully connected layers with a softmax layer beside input and output layers. Each convolutional layer is composed of convolutional filters, which are responsible for extracting the graphical features, and a nonlinear activation function named ReLU. Max pooling is in charge of the down sampling of activated extracted features. The image input size should be $227 \times 227 \times 3$ to accommodate the parameters of the following layers [24].

#### 3.3.2. ResNets

Residual neural networks (18, 50, and 101) are pre-trained convolutional neural networks. They are distinguished by their residual block property. This feature solves the problems of vanishing or exploding gradients due to deep learning. ResNets allow the formation of a skip connection, which enables the activation of a layer to further layers by skipping some layers in between. That is the architecture of the residual block. ResNets consist of stacking such blocks. Several versions of ResNet have existed that depend mainly on the number of connected layers, such as ResNet 18, ResNet50, and ResNet101. The input size of these networks is $224 \times 224 \times 3$ [25].

#### 3.3.3. Mobile Net

Mobile Net is a pre-trained convolutional neural network. It was designed for mobile and computer vision applications. One of the most prominent properties is depth-wise separable convolution, which reduces the number of parameters that contain problems in the existing convolutional layers in the existing networks. That depends mainly on depth-wise convolution, which is named channel-wise spatial convolution, followed by pointwise convolution, with a kernel size of $1 \times 1$ that combines the resultant features from the depth-wise convolution. On the other hand, it reduces the dimension of generated feature map. Their advantages are low latency and a low number of parameters [26,27].

#### 3.3.4. Shuffle Net

Shuffle Net is one of the most efficient networks that is designed for mobile applications. To maintain a high level of accuracy, Shuffle Net performs point-wise group convolution and channel convolution. These distinguished properties make Shuffle Net more accurate, while reducing the complex time computation. It consists of a stacking of shuffle netblocks, each one consisting of two grouped convolutional layers, channel shuffle layer, in addition to depth-wise convolutional layers. The process within one block considers depth-wise convolutional and point-wise convolution as well. The output from each block passes to the ReLU layer for mapping purposes. The designed input layer is compatible with image size 224 × 224 × 3 [28].

#### 3.3.5. Xception Net

The insight behind the 3D convolutional layer is the capability to allow the filter to learn within the 2D spatial domain alongside the depth via channel dimension. Therefore, the output is obtained by the correlation between the spatial and the channel convolutions. The idea behind the inception blocks makes the process easy and forward by using several explicit series of operations ended by cross-channel correlation and spatial correlations. The process operation starts with cross-channel correlation to reduce the dimension via 1 × 1 convolution that maps the input data into 3 or 4 spaces that are lower dimensional than the original input space. After that, the process proceeds via regular 3 × 3 or 5 × 5 convolutions.

The new version of the inception module is called the "extreme". The Xception module performs the channel convolution and obtains a spatial convolution for each channel separately. The Xception architecture consists of 36 convolutional layers forming the feature extraction base of the network. Moreover, the Xception structure is formed as linear stacking of inception modules [29].

#### 3.3.6. NasNet

Neural search architecture (NAS) networks stand for NASNET. It is a predefined architecture that is trained over an ImageNet database of over 1000 categories from nature. It consists of a series of cells. These cells are the normal and reduction cell, where the normal cell is responsible for constructing the feature map via convolutional filters, and the reduction cell oversees the reduction of the size of the feature map in terms of width and height by factor two. Moreover, the structure of NASNET ended by the softmax layer yields the probability for the last classification layer [30].

#### 3.3.7. Dark-19 Net

Darknet is one of the most known deep learning structures that is used to detect objects from images in the available dataset. Dark Net-19 consists of 19 layers, which yields to its name. The Darknet has various applications in object detection, alongside counting as the most known algorithm in YOLO, which stands for you only look once [31].

#### 3.3.8. VGG-16 Net

VGG stands for visual geometry group convolutional network, which is trained on the ImageNet database. VGG16 consists of 16 layers: thirteen are convolutional layers, and the rest are fully connected layers. The input layer is compatible in design with image size 224 × 224 × 3. The VGG network has a small perspective field where the convolutional filter size is 3 × 3, which influences capturing more details in the image in both left-right and up-down directions. Moreover, the convolution of 1 × 1 acts as a linear transformation for the input data. This network utilizes transfer learning techniques to extract the most significant features for four pap smear image classes [25].

### 3.4. Feature Fusion

Feature extraction is the genesis of the recognition between various classes in machine learning algorithms. However, the leverage of most representative features may appear in the performance of the designed classifier. Therefore, looking for the most influential attributes is a crucial challenge in computer-aided diagnosis systems. This paper compares techniques in engineering features to classify whole-slice images with highly confidential results. Employing deep learning semantic descriptors alongside one of the most known feature processing methods is a hot topic presented in this paper. This paper applies two types of fusion algorithms: CCA and DCA.

#### 3.4.1. Canonical Correlation Analysis

CCA is one state-of-art statistical analysis of multivariate data that measures the linear relationship between two datasets. It is one of the most commonly used methods in data fusion. CCA focuses on maximizing the correlation between the variables of the two datasets and ignores the relationship between the variables within the same datasets [32].

CCA is defined as two sets of basis feature vectors, where x and y, the correlation of the features between these bases, are mutually maximized.

These two datasets x and y can be written as linear combinations of their internal features:

$$\mathrm{x} = \mathrm{x}^{\mathrm{T}}\hat{\mathrm{w}}_{\mathrm{x}}$$
$$\mathrm{y} = \mathrm{y}^{\mathrm{T}}\hat{\mathrm{w}}_{\mathrm{y}}$$

To maximize the above two functions, the corresponding function should be maximized

$$\rho = \frac{\mathrm{E[xy]}}{\sqrt{\mathrm{E[x^2]\,E[y^2]}}}$$

The maximum values of ρ in respect to the weights of subsets x and y are called canonical variates.

#### 3.4.2. Discriminant Correlation Analysis

Feature fusion aims to find the highly correlated features between two separate datasets. In DCA, the class is considered a membership of correlation analysis that enhances the fusion process. DCA needs low computational complexity, which leads to minimizing time in real-world applications. Moreover, it reduces the number of features that best describes the original ones [21]. The corresponding equations illustrate the process of DCA. The training features are:

$$E = \{(x_1, y_1), (x_2, y_2), \ldots, (x_n, y_n)\}\text{ , where } y_i = \{1, \ldots, k\}$$

where $k$ is the number of classes, $x$, and $y$ are features and their corresponding class.

The first step is calculating the mean of each class separately:

$$\overline{x_i} = \frac{1}{m_i}\sum_{j=1}^{m_i} x_j^i,$$

where $m_i$ is the number of samples in each class. Then, evaluate the overall mean of the training set by:

$$\overline{x} = \frac{1}{n}\sum_{v=1}^{k} m_v\overline{x_v}$$

The covariance matrix is calculated by the following equation:

$$sigma = \varphi^T\varphi,$$

where $\varphi = \sqrt{m_1}(\overline{x_1} - \overline{x}), \ldots, \sqrt{m_k}(\overline{x_k} - \overline{x})$.

The singular value matrices (SVD):

$$sigma = U\Lambda U^T,$$

where $\Lambda$ is the diagonal of eigenvalues $\lambda_i$, $U$ donates to eigenvectors. The eigenvectors are ordered in ascending form with their corresponding eigenvectors. The transformation matrix is given by:

$$R = \varphi U_t \Lambda_{t\times t}^{-1/2}$$

All previous steps are performed on each set separately. Then, data points are transformed as:

$$Z_1 = R_1^T X_1$$
$$Z_2 = R_2^T X_2$$

After that, the covariance matrix of transformed features between two sets:

$$S_b = Z_1 Z_2^T$$

Then, SVD is calculated for $S_b$:

$$S_b = V\Sigma V^T$$

Then, the transformation matrix is given by:

$$T = V\Sigma^{-\frac{1}{2}}$$

Then, the data are generated by the following equation:

$$X_1' = T^T Z_1$$
$$X_2' = T^T Z_2$$

Finally, the output features are generated according to the equations:

$$X' = X_1' + X_2'$$

### 3.5. Feature Selection

Feature selection is one of the prominent topics in machine learning, processing, and data analysis. The mean goal of attributes selection maintains that the best representative attributes have high variance, which reduces the dimensionality of the feature maps and reduces the time computation and complexity. Various feature selection techniques are proposed in the literature. In this paper LASSO, PCA, and ICA are used. Below is a description of each method.

#### 3.5.1. Least Absolute Shrinkage and Selection Operator

LASSO is one type of penalized logistic regression, where a penalty is imposed on the logistic model for having too many variables. This leads to shrinking the coefficients of the least contributive variables to zero. Specifically, LASSO forces the less contributive variables to become exactly zero. For LASSO regression, a constant lambda should be specified to adjust the amount of the coefficient shrinkage. The best lambda can be defined as the lambda that minimizes the cross-validation mean square error rate. The mean squared error (MSE) measures how close a regression line is to a set of data points. In our method, we have chosen the one standard deviation lambda λ1se to select the final model [33].

#### 3.5.2. Principal Component Analysis

The PCA is well known as an unsupervised learning algorithm used to obtain the most significant features using dimensionality reduction. First, the dataset is standardized using the Z score a

$z_i = \frac{x_i - \mu_c}{\sigma_c}$, where $x_i$ is the feature value for each sample, $\mu_c$ the mean of each feature column, and $\sigma_c$ is the standard deviation for each column as well.

Then, the covariance matrix is built for all standardized features, where the diagonal represents the variance of each feature, and the off-diagonal describes the covariance between two features in the whole dataset. Then, calculate the eigenvector and eigenvalues that represent the 95% variation for the constructing covariance matrix. Finally, the eigenvalues are ascended from the highest to lowest principal components. The projection is calculated to find the original significant features from the original dataset [34,35].

3.5.3. Independent Component Analysis

ICA is a statistical technique that reveals hidden factors (sources) from sets of random variables, or signals [36], and these sources are maximally independent. ICA has been used in unsupervised learning classification problems. Many studies have shown the utility of ICA to extract independent features from the original feature dataset to reduce the feature space and thus, improve classification accuracy [37–39]. Mathematically speaking, assuming that x(t) = x1(t); x2(t); ... ; xn(t) are the set of observed variables that are a combination of the original and mutually independent sources (original features), source s(t) = s1(t); s2(t); ... ; sn(t), the relation can be expressed by x(t) = As(t), where A is called the mixing matrix. In other words, the equation can be written as y = Wx, where W is the demixing matrix W = $A^{-1}$, and y = y1; y2; ... ; yn, are the independent components. The demixing matrix and the independent variables can be found from mixed observations using one of the ICA algorithms such as fastICA [40], which was used in this paper. Furthermore, the set of extracted components (y = y1; y2; ... ; yn) are non-Gaussian and maximally independent. One way to measure this is using the kurtosis [41] measure, which was adopted in this paper to rank the extracted independent components.

### *3.6. Support Vector Machine Classifier*

SVM classifier is a well-known supervised machine learning algorithm, which was developed in 1963 by Vladimir N. Vapnik. SVM selects the extreme training points from different classes to specify the boundary region between various labels, which is called the margin region. If the training points are linearly separable, then the discrimination between them is an easy task. If it is not linearly separable, then the SVM has a distinguished property to represent this feature into higher space using the kernel trick to be linearly separable in higher space. These kernels are radial basis functions, polynomial-Gaussian, and many forms of kernels.

## 4. Results & Discussion

The whole-slice images are passed independently to ten pre-trained deep learning structures. Each pre-trained CNN is modified using transfer learning so that the last fully connected layers become compatible with four classes. Four features were extracted from each CNN. The generated feature map consists of 40 features from 1000 samples. Each class consists of 250 samples; 250 slices of the normal class; 250 samples for HISL; samples for 250 LSIL; and samples for 250 SCC.

The generated maps are passed to different feature selection and fusion methods. The resultant feature map is divided into a 70% training set for the SVM classifier and a 30% to test the generated SVM model. The corresponding results describe the performance of the SVM in discriminating four colposcopy whole-slice images using feature fusion and selection techniques.

### *4.1. CCA*

The whole mapped features were passed to the CCA, which resulted in the six most correlated attributes. These were then split into 70% as a training set and 30% as a test set for the SVM classifier. The resultant confusion matrix shown in Figure 2a shows the performance of the trained model. The HSIL samples are classified correctly with a sensitivity of 100% and a precision of 98.7%. Moreover, the LSIL achieves 100% positive predictive value (PPV) and 100% recall. The same prominent results are obtained in the normal class, with a true positive rate of 100% and precision of 100%. For the lowest sensitivity obtained in the SCC, the PPV is 100%. Finally, the overall accuracy achieved is 99.7. Figure 3 illustrates the receiver operating characteristics (ROC), which defines the area under the curve (AUC) for each feature selection technique. The AUC represents the relation between the false positive rate (specificity) on the x-axis and the true positive rate (sensitivity or recall) on the y-axis for each class. As is clear from Figure 3a, the AUC for all classes in the case of the CCA is one.

**Figure 2.** Six confusion matrices of the SVM model when considering six feature fusion techniques. **(a)** Using CCA, **(b)** DCA, **(c)** LASSO, **(d)** concatenation, **(e)** PCA, and **(f)** ICA. The matrices show the performance of the SVM after using different fusion techniques.

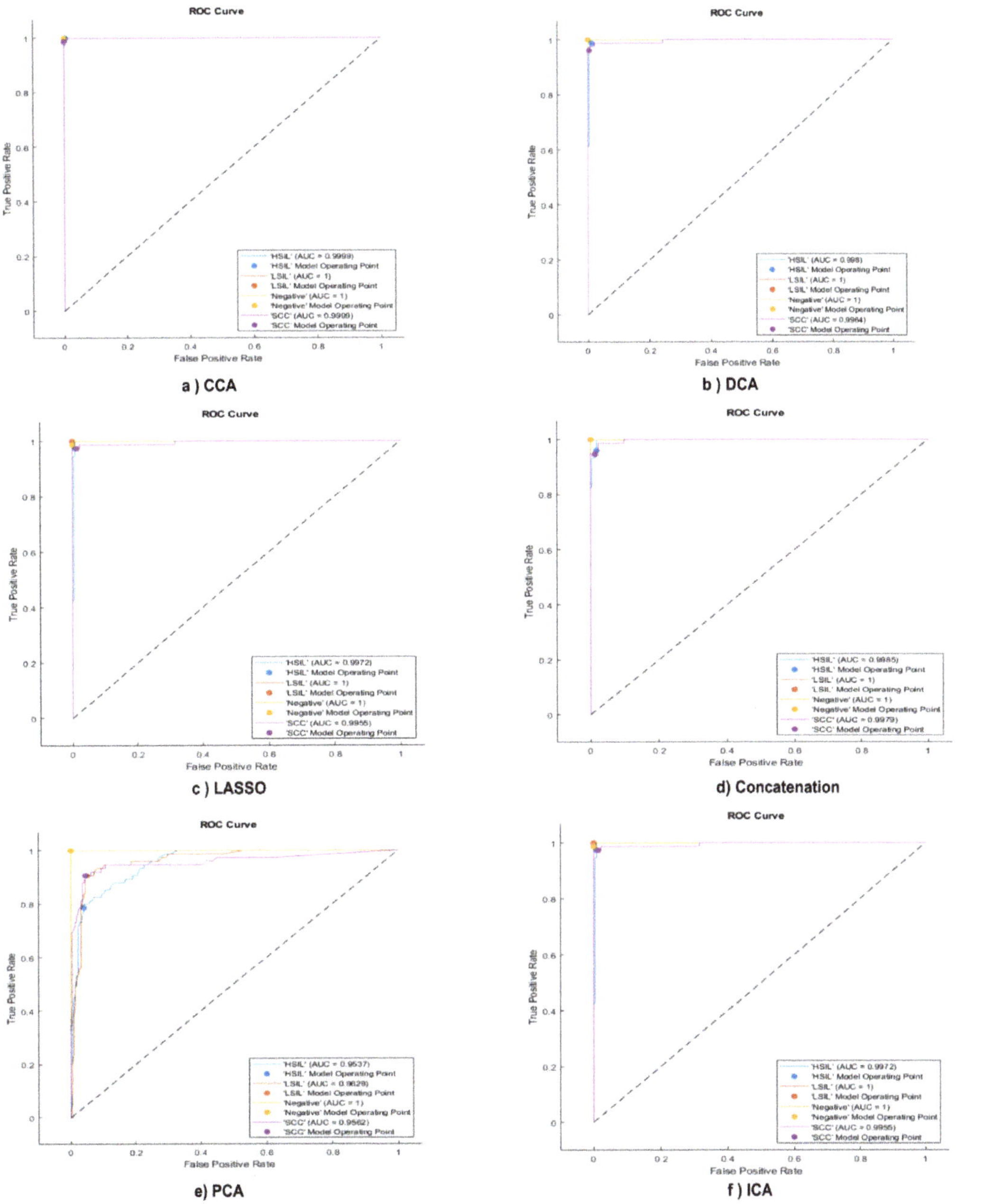


**Figure 3.** Six receiver operating characteristic curves of the SVM model when considering six feature fusion techniques. (**a**) Using CCA, (**b**) DCA, (**c**) LASSO, (**d**) concatenation, (**e**) PCA, and (**f**) ICA. The figure shows the performance of the SVM via AUC after using different fusion techniques.

### 4.2. DCA

The same procedure was performed for the discriminant correlated analysis. Forty DL-labeled features were passed to the DCA. The performance of the trained SVM model reached 98.7% for sensitivity to the HISL category, with a low PPV of 96.1%. However, the prominent results appear in both the LSIL and normal classes, where recall and precision reach 100%. The behavior of the designed classifier in the SCC samples is similar to the HSIL, with the lowest sensitivity of no more than 96%, and a precision of 98.6%. The overall accuracy of the SVM using the DCA feature fusion method is 98.7%. The confusion matrix is shown in Figure 2b. On the other hand, the performance of the combination between the DCA and SVM is represented in Figure 3b. Almost all classes have the highest level of AUC.

### 4.3. LASSO

The feature set was passed to the LASSO algorithm to select the most representative features. Figure 4 shows the cross-validated mean square error (MSE) for the LASSO model. Each red dot represents a lambda (λ) value with confidence intervals for the error rate. Two vertical lines are drawn between the lambda that achieves the lowest MSE (λmin) and the lambda that indicates the highest value of MSE within one standard deviation of the minimum lambda (λ1se). The numbers at the top of the plot represent the number of features of the model at a given lambda value.

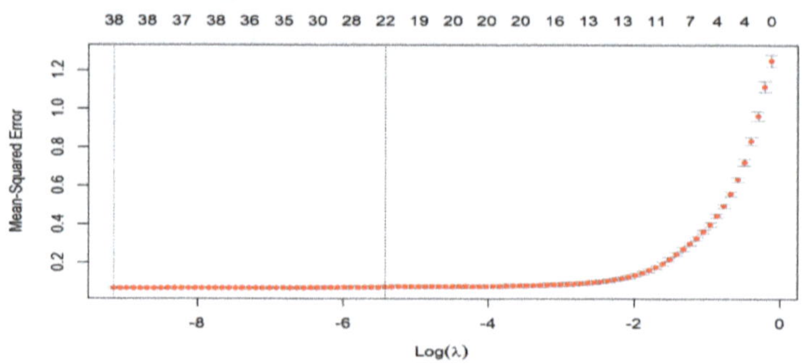


**Figure 4.** MSE of LASSO model. Showing how the number of features selected is affected by the MSE value.

In our methodology, we have selected λ1se = 0.004 to be fed into the LASSO model, which resulted in the extraction of 19 features from a total of 40. Therefore, the selected features that passed to the SVM were 19. The corresponding confusion matrix is shown in Figure 2c, which clarifies the performance of the SVM model using the 19 selected features by the LASSO algorithm. The SVM correctly distinguishes LSIL, with higher sensitivity and precision reaching 100%. However, the lowest true positive rate in the HISL class and its PPV do not exceed 97.3%. The performance of the normal class is 98.7% and 100% for recall and precision, respectively. Furthermore, the SCC has the lowest precision of 96.1% and a moderate value sensitivity of 97.3%. Moreover, the AUC for all classes is almost equal to one. This shows the effectiveness of the proposed method.

### 4.4. Feature Concatenation

The feature concatenation is performed by unionizing all features into a single dataset. All deep learning features are concatenated to obtain 40 attributes, which are split into 70% for SVM training and the rest to evaluate the classifier. The corresponding confusion matrix shown in Figure 2d illustrates the outputs of the test data using the fused 40 features. It is clear from the confusion matrix of the fused 40 features, that 72 cases of HSIL are classified correctly among the 75 cases, with recall reaching 96% and precision reaching 94.7%. For the

LSIL 75 samples, they are classified correctly with a sensitivity and precision of 100%. The same applies for the normal classes, where the performance is 100% in both the TPR and PPV. The worst behavior appeared in the SCC category, with the lowest sensitivity reaching 94.7% and a precision of 95.9%. The overall accuracy is 97.7%, and the misclassification rate is 2.3%. Furthermore, Figure 3d describes the AUC for each class, which is nearly one for all categories.

### 4.5. PCA

The principal component analysis is employed to select the most significant features that represent the four classes. Depending on a 95% variance among features, the most independent features are selected. As shown in the corresponding Figure 5, three principal components describe most of the variability in the data. However, the rest features have low significance in class representation.

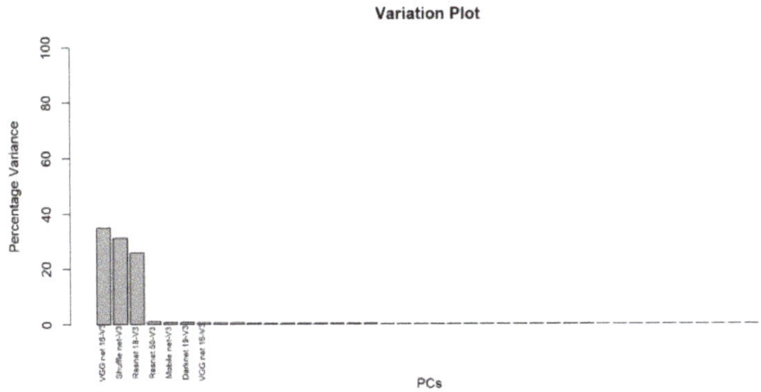


**Figure 5.** Percentage variance of each feature according to PCA. The first three features contribute the most to the variability of the data.

Figure 6 shows the relationship between two principal components. The scatter representation visually shows how these two principals are capable of discriminating between classes. The clustering describes the classification capability of these two PCAs, where the red cluster indicates HSIL, the dark green cluster indicates normal, the cyan color represents LSIL, and the purple grouping distinguishes the SCC. The three significant features are exploited to train polynomial SVM. The corresponding confusion matrix in Figure 2e shows the performance of the classifier using the three independent features. The capability of the SVM to discriminate HSIL is low in terms of sensitivity and precision. On the other hand, recall and precision are low for LSIL. The normal class is the best distinguished, with a sensitivity and PPV of 100%. The precision of the SCC class is lower, at 87.2%, whereas the sensitivity is a moderate value that does not exceed 90%. The overall accuracy of the designed SVM using the most significant features is 90%, and the misclassification rate reaches 10%, which is too high. Moreover, Figure 3e illustrates the AUC for each class, which is the lowest in the SCC class with 0.95, and the highest in the normal class where the AUC is one.

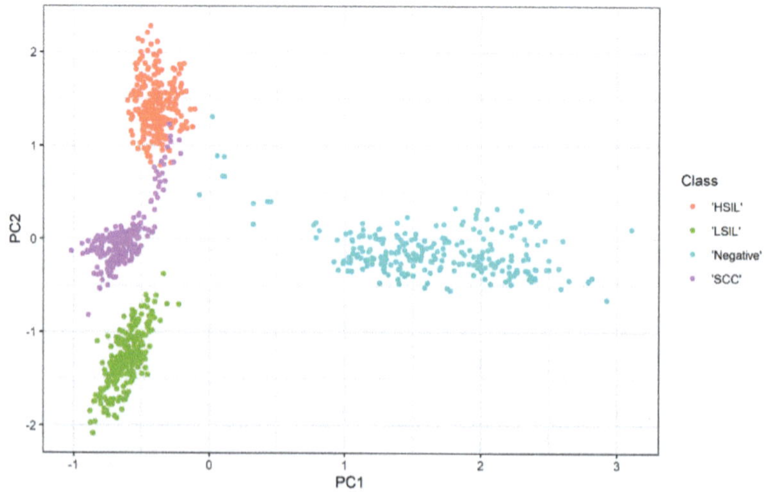


**Figure 6.** Scatter plot showing the first two principal components and how they visually discriminate between the four classes. The figure shows the effectiveness of separation between classes after selecting the most representative components using PCA.

### 4.6. ICA

Forty features are passed to the independent component analysis algorithm to achieve the best independent and representative features among all. The best six features are a candidate. Figure 7 illustrates the scatter representation between the best two independent components. The grouping of the scattered points indicates the capability of the ICA to select the best representative features. According to Figure 6, the red group represents the HSIL class, the dark green cluster illustrates the normal (negative) class, the cyan bunch shows the LSIL class, and the purple color describes the SCC category. The best six independent features are passed to the third polynomial SVM. The corresponding confusion matrix shown in Figure 2f illustrates the output of the test phase. The best results were obtained in the LSIL class, with the sensitivity and precision reaching 100%. However, the lowest recall values in both the HSIL and the SCC classes were 97.3%. Furthermore, the lowest precision value in the SCC was 96.1%. On the other hand, the precision value of the LSIL was 100%. The overall accuracy using the ICA and SVM is 98.3% for all four classes, with a misclassification rate of 1.7%. Finally, Figure 3f shows the AUC for all the classes that are almost equal to one.

Figure 8 shows the comparison between the features engineering algorithm and its impact on the accuracy of the SVM classifier in discriminating whole-slice cervical images. The same data are shown in tabular form in Table 2. As illustrated in Figure 7, the highest accuracy achieved was by the CCA feature fusion, with a maximum accuracy reaching 99.7%. However, the performance of the other algorithms is almost the same with slight differences, apart from PCA, which exhibits the lowest accuracy value. These results show the influence of various feature processing algorithms on obtaining accurate computer-aided diagnosis systems.

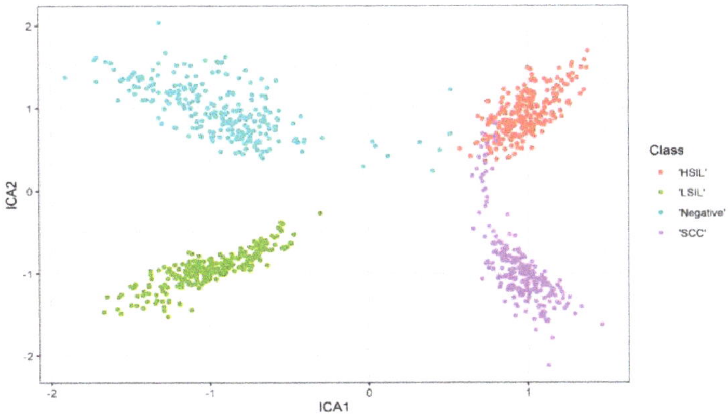


**Figure 7.** Scatter plot showing the first two independent components and how they discriminant between the four classes. The figure shows the effectiveness of separation between classes after selecting the most representative components using ICA.

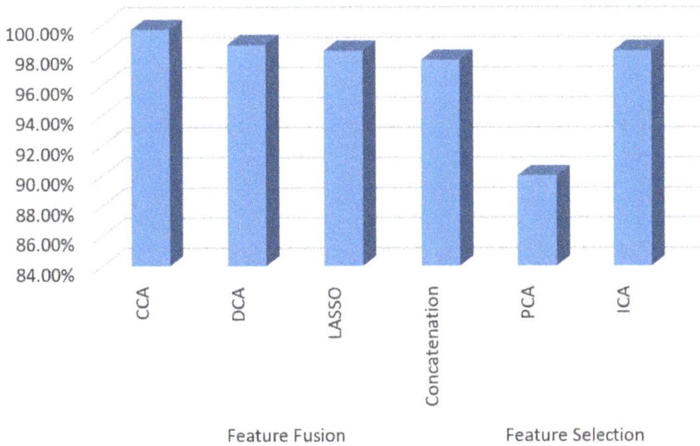


**Figure 8.** Bar plot showing the accuracy of the SVM when using different feature fusion techniques.

**Table 2.** Comparison between various scenarios. Showing the SVM accuracy for the six different feature analysis techniques.

| Data Fusion Method | Number of Features Selected or Fused from the Original 40 | SVM Accuracy |
|---|---|---|
| Concatenation | 40 | 97.7% |
| LASSO | 19 | 98.3 |
| DCA | 6 | 97.3% |
| CCA | 6 | 99.7% |
| PCA | 3 | 90% |
| ICA | 3 | 98.3% |

Table 3 shows the study comparison for the most recent studies that used data fusion on cervical cancer images. All mentioned studies showed the effectiveness of data fusion in improving the classification accuracy of cervical cancer. Comparing the previous studies that focused on cervical cancer diagnosis, the proposed approach in this paper achieves the highest accuracy with automated features. This paper deals with whole-slice cervical images, ignoring the overlapping and non-overlapping issues for cells. On top of that, all the previous studies focused on the diagnostics of single cells, whereas this paper deals with the whole-slice image, which is more practical for physicians and medical fields. Due to the limited work on feature-level fusion in cervical cancer, other studies with different medical diagnostic problems were shown in Table 4. These studies were selected based on the feature fusion technique used. All of the studies in Table 4 used data fusion analysis on the feature level. All the studies showed an improvement in classification accuracy when using feature-level fusion or selection. In our paper, we have adopted some of these existing methods. The studies listed in Tables 3 and 4 have different perspectives on dealing with data fusion. They could be grouped into two perspectives: The first perspective is the data level that is being fused (feature level, matching score, or decision-level fusion) listed in Table 3. The second perspective is on the method used for fusion, the approaches mainly used either feature reduction techniques (such as PCA, ICA, and LASSO), or feature fusion techniques (such as CCA and DCA) listed in Table 4. These approaches have been used to fuse different types of data to enhance diagnostic decisions.

**Table 3.** Comparison with literature study used on cervical cancer images. The mentioned study focuses on using a fusion technique for the cervical cancer classification problem.

| Study Author (Year) | Feature Fusion Method | Number of Fused Datasets | Best Accuracy |
|---|---|---|---|
| Alquran et al. (2022) | CCA | Two datasets from Shuffle Net and novel Cervical Net | 99.1 (four-class classifications) |
| Liu et al. (2022) | Multilayer perceptron module | Two (CRIC and SIPaKMeD) | 91.7 (eleven-class classifications) |
| Rahman et al. (21) | Late fusion | SIPaKMeD dataset | 99.14 % (five-class classification) |
| Hussain et al. (2020) | Ensemble classifier based on selecting the best three DL models | Six datasets from (Alexnet, Vgg-16, Vgg-19, Resnet-50, Resnet-101, and Googlenet) | 97% (two classes) |
| This paper | LASSO, CCA, DCA, PCA, and ICA | Ten datasets from (Alex Net, Resnet 18, 50, and 10, Mobilenet, Shufflenet, Xception, Nasnet, Darknet 19, and VGG Net 16) | 99.7% (four classes) |

**Table 4.** Most recent studies show the effectiveness of data feature fusion in improving classification accuracy on other medical diagnostic problems. Most of the mentioned feature fusion methods were selected as a part of our comparative study.

| Study Author (Year) | Classification Problem | Feature Fusion Method |
|---|---|---|
| Fan et al. (2019) | Breast cancer prediction | CCA |
| Zhang et al. (2021) | Pancreatic ductal adenocarcinoma prediction | PCA, LASSO, Boruto, and proposed feature fusion method by Zhang et al. |
| Wnag et al. (2021) | COVID-19 classification | DCA |
| Haghighat et al. (2016) | Multimodal biometric recognition | DCA |
| This paper | Cervix cancer images four classes | LASSO, CCA, DCA, PCA, and ICA |

### 4.7. Computational Complexity

As explained above, extracting the features using DL models has demanded substantial time, which took hours of computation. Thereafter, feature fusion and SVM analyses have required seconds of computational time for each of the considered techniques. Therefore, the considered six techniques have demanded relatively comparable computational complexity when they are run on the same machine.

### 4.8. Future Work and Real-Life Applications

To the best of our knowledge, this paper presents a unique approach of using ten pre-trained DL models with the most common feature selection techniques to diagnose whole-slice cervical images. The relatively high level of accuracy obtained herein can act as a background to building robust and reliable computer-aided detection and diagnosis systems for assessing colposcopy images. These findings can help reduce the mortality rate and enhance the chances of survival among women. Further enhancement on the proposed approach of analysis can be implemented in future works to expand the extracted features and to provide more robust results for medical diagnosis under different deep learning models.

## 5. Conclusions

This paper has focused on employing feature fusion techniques to enhance the classification accuracy of cervical cancer. It involved the generation of a new, uncorrelated dataset of features while faithfully conveying the output information. Using the new dataset of features, we have been able to reduce the dimension of feature space without degrading the performance of disease classification. This paper constructed a comparative analysis of the existing feature fusion techniques to extract the best representative features from ten independent datasets. These datasets came from ten pre-trained DL models, which were trained on a huge ImageNet database. Our approach to this analysis involved applying six sequential steps. The first step consisted of collecting a cytology dataset that contained 1000 samples for four different cervical cancer classes. The second step performed image augmentation, which was then followed by extracting features using CNN from ten DL models (4 features for each DL model for a total of 40 features). The next step concatenated all features from the ten DL models to be fed into the feature fusion step. Step five applied six different feature fusion techniques to extract features. Finally, the extracted features were input into an SVM to test the classification performance. The approach of this analysis revealed the highest accuracy of 99.7% using CCA fusion. The key benefit was reducing the number of features introduced to SVM and obtaining state-of-the-art accuracy. Therefore, the use of data fusion at the feature level, which was proposed in this paper, can indeed enhance classification accuracy for colposcopy images. The presented approach herein can be used as a guideline for other CAD medical applications to aid diagnostic decisions.

**Author Contributions:** Conceptualization, S.T. and H.A.; methodology, S.T. and H.A.; software, S.T. and M.A.; validation, S.T. and H.A.; formal analysis, S.T. and M.A.; investigation, S.T. and H.A.; data curation, S.T., H.A. and M.A.; writing—original draft preparation, S.T.; writing—review and editing, H.A.; visualization, S.T. and H.A.; supervision, H.A.; project administration, H.A.; funding acquisition, H.A. All authors have read and agreed to the published version of the manuscript.

**Funding:** This research received no external funding.

**Institutional Review Board Statement:** Not applicable.

**Informed Consent Statement:** Not applicable.

**Data Availability Statement:** The dataset in this study was obtained from the Liquid-based cytology pap smear images for multi-class diagnosis of cervical cancer, which was published in Mendeley on 15 November 2019. It consists of 963 images that were collected from 460 patients and that were diagnosed into four classes. This dataset has been publicly available online since 2019. It is available on the corresponding website: https://data.mendeley.com/datasets/zddtpgzv63/3 (accessed on 15 March 2022).

**Conflicts of Interest:** The authors declare no conflict of interest.

## References

1. Sung, H.; Ferlay, J.; Siegel, R.L.; Laversanne, M.; Soerjomataram, I.; Jemal, A.; Bray, F. Global Cancer Statistics 2020: GLOBOCAN Estimates of Incidence and Mortality Worldwide for 36 Cancers in 185 Countries. *CA Cancer J. Clin.* **2021**, *71*, 209–249. [CrossRef] [PubMed]
2. WHO. *WHO Guidelines for Screening and Treatment of Precancerous Lesions for Cervical Cancer Prevention*; WHO: Geneva, Switzerland, 2013.
3. Erhardt, R.; Reinhardt, E.R.; Schlipf, W.; Bloss, W.H. FAZYTAN: A system for fast automated cell segmentation, cell image analysis and feature extraction based on TV-image pickup and parallel processing. *Anal. Quant. Cytol.* **1980**, *2*, 25–40.
4. Tanaka, N.; Ikeda, H.; Ueno, T.; Mukawa, A.; Watanabe, S.; Okamoto, K.; Hosoi, S.; Tsunekawa, S. Automated cytologic screening system (CYBEST model 4): An integrated image cytometry system. *Appl. Opt.* **1987**, *26*, 3301–3307. [CrossRef] [PubMed]
5. Chivukula, M.; Saad, R.S.; Elishaev, E.; White, S.; Mauser, N.; Dabbs, D.J. Introduction of the Thin Prep Imaging System™ (TIS): Experience in a high volume academic practice. *Cytojournal* **2007**, *4*, 6. [CrossRef] [PubMed]
6. Zahniser, D.J.; Oud, P.S.; Raaijmakers, M.C.T.; Vooys, G.P.; Van de Walle, R.T. Field test results using the BioPEPR cervical smear prescreening system. *Cytometry* **1980**, *1*, 200–203. [CrossRef]
7. Sharma, M.; Singh, S.K.; Agrawal, P.; Madaan, V. Classification of Clinical Dataset of Cervical Cancer using KNN. *Indian J. Sci. Technol.* **2016**, *9*, 1–5. [CrossRef]
8. Kumar, R.; Srivastava, R.; Srivastava, S.K. Detection and Classification of Cancer from Microscopic Biopsy Images Using Clinically Significant and Biologically Interpretable Features. *J. Med. Eng.* **2015**, *2015*, 457906. [CrossRef]
9. Chankong, T.; Theera-Umpon, N.; Auephanwiriyakul, S. Automatic cervical cell segmentation and classification in Pap smears. *Comput. Methods Programs Biomed.* **2014**, *113*, 539–556. [CrossRef]
10. Talukdar, J. Fuzzy Clustering Based Image Segmentation of Pap smear Images of Cervical Cancer Cell Using FCM Algorithm. *Markers* **2013**, *3*, 460–462.
11. Ampazis, N.; Dounias, G.; Jantzen, J. Pap-Smear Classification Using Efficient Second Order Neural Network Training Algorithms. In *Hellenic Conference on Artificial Intelligence*; Springer: Berlin, Heidelberg, 2004; pp. 230–245.
12. Sreedevi, M.T.; Usha, B.S.; Sandya, S. Pap smear Image based Detection of Cervical Cancer. *Int. J. Comput. Appl.* **2012**, *45*, 35–40.
13. Lip, C.C.; Ramli, D.A. Comparative Study on Feature, Score and Decision Level Fusion Schemes for Robust Multibiometric Systems. In *Frontiers in Computer Education*; Sambath, S., Zhu, E., Eds.; Springer: Berlin Heidelberg, 2012; pp. 941–948.
14. Alquran, H.; Alsalatie, M.; Mustafa, W.A.; Abdi, R.M.A.; Ismail, A.R. Cervical Net: A Novel Cervical Cancer Classification Using Feature Fusion. *Bioengineering* **2022**, *9*, 578. [CrossRef]
15. Liu, W.; Li, C.; Xu, N.; Jiang, T.; Rahaman, M.; Sun, H.; Wu, X.; Hu, W.; Chen, H.; Sun, C.; et al. CVM-Cervix: A Hybrid Cervical Pap-Smear Image Classification Framework Using CNN, Visual Transformer and Multilayer Perceptron. *Pattern Recognit.* **2022**, *130*, 108829. [CrossRef]
16. Rahaman, M.; Li, C.; Yao, Y.; Kulwa, F.; Wu, X.; Li, X.; Wang, Q. DeepCervix: A deep learning-based framework for the classification of cervical cells using hybrid deep feature fusion techniques. *Comput. Biol. Med.* **2021**, *136*, 104649. [CrossRef] [PubMed]
17. Hussain, E.; Mahanta, L.B.; Das, C.R.; Talukdar, R.K. A comprehensive study on the multi-class cervical cancer diagnostic prediction on pap smear images using a fusion-based decision from ensemble deep convolutional neural network. *Tissue Cell* **2020**, *65*, 101347. [CrossRef] [PubMed]
18. Zhang, Y.; Lobo-Mueller, E.M.; Karanicolas, P.; Gallinger, S.; Haider, M.A.; Khalvati, F. Improving prognostic performance in resectable pancreatic ductal adenocarcinoma using radiomics and deep learning features fusion in CT images. *Sci. Rep.* **2021**, *11*, 1378. [CrossRef]
19. Fan, M.; Liu, Z.; Xie, S.; Xu, M.; Wang, S.; Gao, X.; Li, L. Integration of dynamic contrast-enhanced magnetic resonance imaging and T2-weighted imaging radiomic features by a canonical correlation analysis-based feature fusion method to predict histological grade in ductal breast carcinoma. *Phys. Med. Biol.* **2019**, *64*, 215001. [CrossRef] [PubMed]
20. Haghighat, M.; Abdel-Mottaleb, M.; Alhalabi, W. Discriminant Correlation Analysis: Real-Time Feature Level Fusion for Multimodal Biometric Recognition. *IEEE Trans. Inf. Forensics Secur.* **2016**, *11*, 1984–1996. [CrossRef]
21. Wang, S.-H.; Nayak, D.R.; Guttery, D.S.; Zhang, X.; Zhang, Y.-D. COVID-19 classification by CCSHNet with deep fusion using transfer learning and discriminant correlation analysis. *Inf. Fusion* **2020**, *68*, 131–148. [CrossRef] [PubMed]
22. Hussain, E.; Mahanta, L.B.; Borah, H.; Das, C.R. Liquid based-cytology Pap smear dataset for automated multi-class diagnosis of pre-cancerous and cervical cancer lesions. *Data Brief* **2020**, *30*, 105589. [CrossRef]
23. Alsalatie, M.; Alquran, H.; Mustafa, W.A.; Yacob, Y.M.; Alayed, A.A. Analysis of Cytology Pap Smear Images Based on Ensemble Deep Learning Approach. *Diagnostics* **2022**, *12*, 2756. [CrossRef]
24. Alom, Z.; Taha, T.M.; Yakopcic, C.; Westberg, S.; Sidike, P.; Nasrin, S.; Van Esesn, B.; Awwal, A.S.; Asari, V.K. The history began from alexnet: A comprehensive survey on deep learning approaches. *arXiv* **2018**, arXiv:1803.01164.

25. He, K.; Zhang, X.; Ren, S.; Sun, J. Deep residual learning for image recognition. In Proceedings of the IEEE Conference on Computer Vision and Pattern Recognition, Las Vegas, NV, USA, 27–30 June 2016; pp. 770–778.
26. Howard, A.G.; Zhu, M.; Chen, B.; Kalenichenko, D.; Wang, W.; Weyand, T.; Andreetto, M.; Adam, H. Mobilenets: Efficient convolutional neural networks for mobile vision applications. *arXiv* **2017**, arXiv:1704.04861.
27. Alqudah, A.M.; Qazan, S.; Al-Ebbini, L.; Alquran, H.; Qasmieh, I.A. ECG heartbeat arrhythmias classification: A comparison study between different types of spectrum representation and convolutional neural networks architectures. *J. Ambient Intell. Humaniz. Comput.* **2022**, *13*, 4877–4907. [CrossRef]
28. Zhang, X.; Zhou, X.; Lin, M.; Sun, J. ShuffleNet: An Extremely Efficient Convolutional Neural Network for Mobile Devices. *arXiv* **2017**, arXiv:1707.01083v2.
29. Chollet, F. Xception: Deep learning with depthwise separable convolutions. In Proceedings of the IEEE Conference on Computer Vision and Pattern Recognition, Honolulu, HI, USA, 21–26 July 2017; pp. 1251–1258.
30. Qin, X.; Wang, Z. Nasnet: A neuron attention stage-by-stage net for single image deraining. *arXiv* **2019**, arXiv:1912.03151.
31. Al Shehri, W.; Almalki, J.; Mehmood, R.; Alsaif, K.; Alshahrani, S.M.; Jannah, N.; Alangari, S. A Novel COVID-19 Detection Technique Using Deep Learning Based Approaches. *Sustainability* **2022**, *14*, 12222. [CrossRef]
32. Hardoon, D.R.; Szedmak, S.; Shawe-Taylor, J. Canonical Correlation Analysis: An Overview with Application to Learning Methods. *Neural Comput.* **2004**, *16*, 2639–2664. [CrossRef]
33. Zou, H. The adaptive lasso and its oracle properties. *J. Am. Stat. Assoc.* **2006**, *101*, 1418–1429. [CrossRef]
34. Abdi, H.; Williams, L.J. Principal component analysis. *Wiley Int. Rev. Comput. Stat.* **2010**, *2*, 433–459. [CrossRef]
35. Alquran, H.; Abu-Qasmieh, I.; Khresat, S.; Younes, A.B.; Almomani, S. Weight estimation for anesthetic administration using singular value decomposition and template matching for supine subject of different obesity levels. *Health Technol.* **2018**, *8*, 265–269. [CrossRef]
36. Bell, A.J.; Sejnowski, T.J. The "independent components" of natural scenes are edge filters. *Vis. Res.* **1997**, *37*, 3327–3338. [CrossRef] [PubMed]
37. Fan, L.; Poh, K.-L.; Zhou, P. A sequential feature extraction approach for naïve bayes classification of microarray data. *Expert Syst. Appl.* **2009**, *36*, 9919–9923. [CrossRef]
38. Fan, L.; Poh, K.-L.; Zhou, P. Partition-conditional ICA for Bayesian classification of microarray data. *Expert Syst. Appl.* **2010**, *37*, 8188–8192. [CrossRef]
39. Korde, K.S.; Paikrao, P.; Jadhav, N. Analysis of eeg signals and biomedical changes due to meditation on brain by using ica for feature extraction. In Proceedings of the 2018 Second International Conference on Intelligent Computing and Control Systems (ICICCS), Madurai, India, 14–15 June 2018; pp. 1479–1484.
40. Oja, E.; Yuan, Z. The FastICA Algorithm Revisited: Convergence Analysis. *IEEE Trans. Neural Networks* **2006**, *17*, 1370–1381. [CrossRef]
41. DeCarlo, L.T. On the meaning and use of kurtosis. *Psychol. Method.* **1997**, *2*, 292. [CrossRef]

*bioengineering*

Article

# Artificial Hummingbird Algorithm with Transfer-Learning-Based Mitotic Nuclei Classification on Histopathologic Breast Cancer Images

Areej A. Malibari [1], Marwa Obayya [2], Abdulbaset Gaddah [3], Amal S. Mehanna [4], Manar Ahmed Hamza [5,*], Mohamed Ibrahim Alsaid [5], Ishfaq Yaseen [5] and Amgad Atta Abdelmageed [5]

1 Department of Industrial and Systems Engineering, College of Engineering, Princess Nourah bint Abdulrahman University, Riyadh 11671, Saudi Arabia
2 Department of Biomedical Engineering, College of Engineering, Princess Nourah bint Abdulrahman University, Riyadh 11671, Saudi Arabia
3 Department of Computer Sciences, College of Computing and Information System, Umm Al-Qura University, Mekkah 24211, Saudi Arabia
4 Department of Digital Media, Faculty of Computers and Information Technology, Future University in Egypt, New Cairo 11845, Egypt
5 Department of Computer and Self Development, Preparatory Year Deanship, Prince Sattam bin Abdulaziz University, Al-Kharj 16278, Saudi Arabia
* Correspondence: ma.hamza@psau.edu.sa

**Abstract:** Recently, artificial intelligence (AI) is an extremely revolutionized domain of medical image processing. Specifically, image segmentation is a task that generally aids in such an improvement. This boost performs great developments in the conversion of AI approaches in the research lab to real medical applications, particularly for computer-aided diagnosis (CAD) and image-guided operation. Mitotic nuclei estimates in breast cancer instances have a prognostic impact on diagnosis of cancer aggressiveness and grading methods. The automated analysis of mitotic nuclei is difficult due to its high similarity with nonmitotic nuclei and heteromorphic form. This study designs an artificial hummingbird algorithm with transfer-learning-based mitotic nuclei classification (AHBATL-MNC) on histopathologic breast cancer images. The goal of the AHBATL-MNC technique lies in the identification of mitotic and nonmitotic nuclei on histopathology images (HIs). For HI segmentation process, the PSPNet model is utilized to identify the candidate mitotic patches. Next, the residual network (ResNet) model is employed as feature extractor, and extreme gradient boosting (XGBoost) model is applied as a classifier. To enhance the classification performance, the parameter tuning of the XGBoost model takes place by making use of the AHBA approach. The simulation values of the AHBATL-MNC system are tested on medical imaging datasets and the outcomes are investigated in distinct measures. The simulation values demonstrate the enhanced outcomes of the AHBATL-MNC method compared to other current approaches.

**Keywords:** breast cancer; mitotic nuclei classification; histopathology images; artificial hummingbird algorithm; medical imaging

**Citation:** Malibari, A.A.; Obayya, M.; Gaddah, A.; Mehanna, A.S.; Hamza, M.A.; Ibrahim Alsaid, M.; Yaseen, I.; Abdelmageed, A.A. Artificial Hummingbird Algorithm with Transfer-Learning-Based Mitotic Nuclei Classification on Histopathologic Breast Cancer Images. *Bioengineering* **2023**, *10*, 87. https://doi.org/10.3390/bioengineering10010087

Academic Editors: Paolo Zaffino and Maria Francesca Spadea

Received: 16 October 2022
Revised: 9 December 2022
Accepted: 13 December 2022
Published: 9 January 2023



## 1. Introduction

Mitosis can be defined as a process of cell cycle where a replicated chromosome is split into dual new nuclei that produce genetically identical cells which retain the chromosome number. This method can be split into four phases: telophase, prophase, metaphase, and anaphase. It culminates into two daughter nuclei that are genetically identical [1]. Then, the cell might perform division by cytokinesis to produce dual daughter cells. Producing more than three daughter cells rather than two normal cells is a mitotic fault that might tempt mutations or apoptosis, initiating specific kinds of cancer [2]. In the tissue samples, haematoxylin and eosin (H&E)-stained slides lead to histopathology images where mitosis rate is a significant parameter to determine the tumor aggressiveness, especially breast tumor, and recognition of a typical way of mitosis is utilized as a prognostic and diagnostic marker. Breast tumor is the main factor that leads to higher mortality amongst women and is a frequently diagnosed tumor amongst females; if diagnosed at earlier stages, tit can be the most curable form of tumor [3]. Breast tumor, where survival rate is under 40% in lower-income nations, is the primary tumor type in females globally that costs a great number of lives per annum. As stated by the National Tumor Institution, up to 20% of each breast tumor fails to be found by X-ray mammography (using ionizing radiation) [4]. Mitosis count assists in tumor diagnosis and provides an assessment of tumor aggressiveness that assists in tumor grading. The high number of mitotic cells in a region represents fast-growing or higher-grade tumor.

The visual detection of mitotic nuclei through pathologists is a time-intensive and subjective job with poor reproducibility because of many difficulties. Mitotic nuclei are hyperchromatic objects having different morphological sizes and shapes [5]. Furthermore, the occurrence of mitotic nuclei differs according to tumor stage and tumor grade. In aggressive tumors, generally, mitotic nuclei are nondifferentiable and appear in smaller sizes with higher frequency. The accurate detection of mitotic nuclei depends on the experience and knowledge of the pathologist [6]. Object-level interobserver analysis exposes pathologist disagreement on individual objects. The limitation of manual workflows generates the necessity to automate the count of mitotic nuclei to enhance the decision of the pathologist [7]. For the development of the detection of mitotic nuclei in histopathology images, thus far, various methods have been introduced based on segmentation, classification, and detection methods [8]. The current approaches frequently exploit data balancing methods, namely, rotation, translation, and mirror imaging-oriented techniques for augmenting mitotic examples. Likewise, various researchers implemented a two-step recognition technique to reduce class imbalance and enhance precision [9]. With regard to the complicated nature of mitoses, several research workers used the method of ensemble learning, while few approaches simultaneously trained two deep learning (DL) models to make the concluding decision.

This study designs an artificial hummingbird algorithm with transfer-learning-based mitotic nuclei classification (AHBATL-MNC) on histopathologic breast cancer images. The goal of the AHBATL-MNC technique lies in identification of mitotic and nonmitotic nuclei on histopathology images (HIs). For HI segmentation process, the PSPNet model is utilized to identify the candidate mitotic patches. Next, the residual network (ResNet) model is employed as feature extractor, and the extreme gradient boosting (XGBoost) model is applied as a classifier. To enhance the classification performance, the parameter tuning of the XGBoost model takes place, utilizing the AHBA algorithm. The simulation values of the AHBATL-MNC approach are tested on a medical imaging dataset and the results are investigated in distinct measures.

## 2. Related Works

Shwetha and Dharmanna [10] modeled a new technique for automatic identification and detection by DL model. In this presented technique, the work can be split into five phases. In the initial phase, histopathological images are preprocessed to boost the contrast of the nonmitotic and mitotic cells through image adjustment method. In the next phase, using Otsu segmentation method, the background and foreground are divided. In [11], the author devised a new structure called SmallMitosis for identifying mitotic cells that are very small in size undergoing mitosis out of the H&E-stained breast histological images. SmallMitosis structure has a deep multiscale (MS-RCNN) detector and an atrous fully convolution-oriented segmentation (A-FCN) method. In the A-FCN technique, the atrous convolution concept aids in predict bounding box annotations and mitosis masks of very-small-sized mitotic cells.

Sohail et al. [12] devised an innovative deep convolutional neural network (DCNN)-related heterogeneous ensemble method, "DHE-Mit-Classifier", for examining mitotic nuclei in breast histopathological imageries. Sebai et al. [13] proposed an accurate and robust algorithm for detecting mitoses automatically from histology breast cancer slides by making use of the multitask DL structure for instance segmentation mask region-based convolutional neural network (RCNN) and object detection. Lei et al. [14] devised an accurate and fast approach to automatically identify mitosis from histopathology images. This presented algorithm is capable of detecting the mitotic candidates automatically from histological units for mitosis screening. In particular, this technique uses DCNN for extracting high-level features of mitosis to find mitotic applicants. After that, the author employed spatial attention elements to re-encode mitotic features that enabled the method to very effectively study features.

Das and Dutta [15] introduced an innovative technique for mitotic cell recognition in breast histology images, exploiting wavelet decomposed image patches and DCNN. In this method, Haar wavelet is used to formulate a DCNN technique for automatic recognition of mitotic cells. The decomposition step reduces convolutional period for mitotic cell recognition related to the usage of raw image patches in traditional DCNN approaches. Beevi et al. [16] explored the feasibility of transfer learning (TL) for mitosis recognition. A pretrained convolutional neural network (CNN) was shown by merging RF method with the initial FC layers for deriving discriminant features from nuclei patches and to accurately prognosticate class labels of cell nuclei. The altered CNN precisely categorizes the identified cell nuclei with limited trained datasets. This structure would establish maximum classifier accuracy by prudently preprocessing the extracted features and fine-tuning the pretrained methods.

## 3. The Proposed Mitotic Nuclei Classification Model

In this study, we develop a new AHBATL-MNC technique for effective identification of mitotic and nonmitotic nuclei on HIs. The presented AHBATL-MNC technique encompasses a series of processes, namely, PSPNet segmentation, ResNet feature extraction, XGBoost classification, and AHBA parameter tuning. Figure 1 defines the overall work flow of the AHBATL-MNC system.

**Figure 1.** Overall working process of AHBATL-MNC system.

### 3.1. Segmentation Process

In the AHBATL-MNC technique, the PSPNet model is utilized for segmentation process. PSPNet is the renowned network architecture for semantic segmentations [17]. The PSPNet was initially introduced for scene parsing. To aggregate multiscale contextual datasets, one pyramid pooling network (PPM) was introduced in PSPNet. At first, max pooling is enforced to generate a feature map using three pyramid scales that can be attained by Equation (1), wherein FDS and $\lambda$, correspondingly, signify input and downsampling method through max pooling, and stride of max pooling layer can also be attained using Equation (2):

$$F_j = DS(F, \lambda_j.)j = 1,2,3 \quad (1)$$

$$\frac{w - \lambda_j}{\lambda_j} + 1 = 0_j \Rightarrow \lambda_j = \frac{w}{o_j} \quad (2)$$

whereas $w$ and 0 signify input and output size of feature maps.

After applying convolution method to these multiscale feature maps, bilinear interpolation can be performed to resize feature maps, whereas $W_j^T$ and $b_j$, correspondingly, denote the weight and bias of $j$-$th$ $1 \times 1$ convolutional layer, and $BI(.)$ denotes the bilinear interpolation.

$$O_j = BI\left(W_j^T \otimes F_j + b_j\right)j = 1,2,3 \quad (3)$$

Likewise, the feature maps having the new input and pyramid scale were concatenated, and $1 \times 1$ convolution was implemented to reduce channel number of output, whereas $W_j^T$ and $b_j$ demonstrate weight and bias of the $1 \times 1$ convolution layer.

$$C = W_{rd}^T(concat(F, O_1, O_2, O_3)) + b_{rd} \quad (4)$$

Dissimilar to the original PPM, feature maps having four pyramid scales, which include 1, 2, 3, and 6, are constructed by the new PPM, whereas feature maps having three pyramid scales, including 1, 2, and 6, are constructed by max pooling.

Furthermore, the $1 \times 1$ convolution layer is interconnected with the concatenation layer for dimensionality reduction.

Based on the UNet structure, a multilevel PSPNet is introduced as the decoder. The 1, 2, and 3 attention gates are enforced to correspondingly generate initial convolutional layer and the attention maps of third and fifth identity blocks. In addition, to incorporate multilevel features, the attention gate and the output of PPM are concatenated densely with the following equation:

$$Y_j = concat(US(C_j, 3)M\_output_j)j = 1,2,3 \quad (5)$$

### 3.2. Feature Extraction Process

In this study, the ResNet model was employed as feature extractor. We adapted the CNN, ResNet50, to characterize the image, and the deep network has 50 layers [18]. The depth of network was crucial for neural network (NN), but a deep network can be tough to train. The ResNet50 infrastructure facilitates the network training and permits it to be deeper which leads to enhanced efficacy in diverse tasks. ResNet50 is deeper than simple counterparts, but parameter count of these networks is smaller. A DCNN resulted in a series of breakthroughs for image classification. Many nontrivial visual detection techniques have benefitted from deep methods. Once the network depth rises, performance of the network degrades quickly (saturated) and rapidly increases. Meanwhile, deep networks have large representation power. It can be possible for ResNet50 to accomplish a deep model that is not worse than lesser deep networks. It is implemented by adding numerous identity layers, viz., levels that skip signal without further amendment. ResNet50 deep level has to predict variations amongst the main function and outcome of the previous layer.

The method considers the image and generates the caption, encrypted as a series of $1-K$ codewords.

$$y = \{y_1, y_2, \cdots, y_c\}, y_j \in R^K \tag{6}$$

From the expression, $K$ indicates the dictionary size and $c$ represents caption length. The extractor will produce $L$-vectors, each having $D$-dimensional representation of the image.

The hyperparameter tuning of the ResNet model is performed by the Adamax optimizer [19]. It is an amended form of the Adam optimizer where the distributed variance is projected ∞. In addition, the maximized weight can be determined as follows:

$$w_t^i = w_{t-1}^i - \frac{\eta}{v_t + \epsilon} \times \hat{m}_t \tag{7}$$

whereas

$$\hat{m}_t = \frac{m_t}{1 - \beta_1^t} \tag{8}$$

$$v_t = max(\beta_2 \times v_{t-1}, |G_t|) \tag{9}$$

$$m_t = \beta_1 m_{t-1} + (1 - \beta_1) G \tag{10}$$

$$G = \nabla_w C(w_t) \tag{11}$$

From the expression, η denotes the rate of learning, $w_t$ refers to the weights at steps $t$, $C(.)$ signifies the cost function, and $\nabla_w C(w_t)$ suggests the gradient of weight variable $w_t$ $x$ and equal label $y$. $\beta_i$ is employed to select the data needed for older upgrades, when $\beta_i \in [0,1]$. $m_t$ and $v_t$ are the first and second moments as explained in Algorithm 1.

**Algorithm 1:** Pseudocode of Adamax

$\eta$: Rate of Learning
$\beta_1, \beta_2 \in [0,1)$: Exponential decomposed values to moment candidate
$C(w)$: Cost function with $w$ variable
$w_0$: parameter vector
$m_0 \leftarrow 0$
$u_0 \leftarrow 0$
$i \leftarrow 0$ (Implement time step)
while $w$ does not converge do

$$i \leftarrow i + 1$$
$$m_i \leftarrow \beta_1 \times m_{i-1} + (1 - \beta_1) \times \frac{\partial C}{\partial w}(w_i)$$
$$u_i \leftarrow \max\left(\beta_2 \times u_{i-1}, \left|\frac{\partial C}{\partial w}(w_i)\right|\right)$$

$w_{i+1} \leftarrow w_i - \left(\eta / \left(1 - \beta_1^i\right)\right) \times m_i / u_i$
end while
show $w_i$ (end parameter)

### 3.3. Optimal Classification Process

Finally, the XGBoost model is exploited for classification purposes. XGBoost is used to classify the regression tree model that comes from the gradient lifting decision tree (DT) [20]. The presented algorithm is used for the pedestrian detection classifier. Firstly, it learns a tree from a sample to attain the initial assessment outcome $Y1$, and then learns with $y$ based on the variance between the predictive and the real labels in the prior step. Likewise, the model error can be reduced successfully. Equations (4)–(8) provide the assessment flow of XGBoost training. The subsequent formula is to evaluate the target of $n-th$ tree models. The primary behavior determines a regularization term that could decrease overfitting to enhance the generalization ability. Taylor's expansion has first and second derivatives and constant terms.

Among them, the objective function of every round is evaluated as follows, and $f_t$ can be selected for minimizing the main function, viz., the error between actual outcome and the predictive outcome is decreased after adding $f_t$. Here, $l$ represents the error function and $\Omega$ denotes a regularization term, the error function tries to fit the training dataset, and the regularization term encourages a simple method. The randomness of the outcomes of the finite data fitting is very small, which is not easy for overfitting, making the prediction of the concluding model more stable.

$$Obj^{(t)} = \sum_{i=1}^{n} l\left(y_i, \hat{y}_i^{(t-1)} + f_t(x_i)\right) + \Omega(f_i) + constant \tag{12}$$

Once the error function $l$ is not a square error, the first three terms of the Taylor equation are utilized for approximating original objective function.

$$Obj^{(t)} = \sum_{i=1}^{n} [l(y_i, \hat{y}^{(t-1)}) + g_i f_t(x_t) + \frac{1}{2} h_l f_t^2(x_i)] + \Omega(f_t) + constant \tag{13}$$

where $g_i$ and $h_i$ refer to the initial and second derivatives of the error function.

$$g_i = \partial_{\hat{y}_i^{(t-1)}} l(y_i, \hat{y}^{(t-1)}) \tag{14}$$

$$h_i = \partial^2_{\hat{y}_i^{(t-1)}} l(y_i, \hat{y}_i^{(t-1)}) \tag{15}$$

Next, we eliminate the constant terms, such as the variance between real value and predicted value of the previous round.

$$Obj^{(t)} = \sum_{i=1}^{n} [g_l f_t(x_i) + \frac{1}{2} h_i f_t^2(x_i)] + \Omega(f_t) \tag{16}$$

According to the realization of XGBoost, the model initially ranks the eigenvalue, since the tree model should define the better segmentation points and later store them in numbers of blocks. This architecture is reutilized in later iterations, which significantly decreases the computation difficulty. Furthermore, the data gain of every feature should be evaluated in the procedure of node splitting, hence the computation of data gain is parallelized through the data structure.

For improving the classification performance, the parameter tuning of the XGBoost model is performed by the AHBA technique. AHBA is a population-related metaheuristic approach that primarily simulates three foraging behaviors of hummingbirds (HB): migratory, guided, and territorial foraging [21]. In the foraging process, the three flight skills include axial, diagonal, and modeled-omnidirectional flights. Simultaneously, an access table simulating HB remarkable memory capability is created for guiding HB to carry out global optimization. The three flying skills are described in the following: the flight skill simulation is expanded to d-D space with axial flight and can be given in Equation (17):

$$D^{(i)} = \begin{cases} 1 & if\ i = randi([1, d]) i = 1, \cdots, d \\ 0 & else \end{cases} \tag{17}$$

Diagonal flight can be determined by Equation (18):

$$D^{(i)} = \begin{cases} 1, if\ i = p(j) P = randperm(k), k \in [2, \lceil r_1(d-2) \rceil + 1] \\ 0, else. \end{cases} \tag{18}$$

Omnidirectional flight is defined below:

$$D^{(i)} = 1 i = 1, \cdots, d \tag{19}$$

In Equation (19), $randi([1, d])$ creates a random number from 1 to $d$, $randperm(k)$ generates a random permutation of integer from 1 to $k$, and $r_1$ indicates a random integer that ranges from zero to one. First, the AHA initializes a visiting table and a set of random solutions. In all the iterations, territorial or guided foraging can be carried out 50% of the time. Hummingbirds move toward the food sources using guided foraging, viz., depending on a visiting table and nectar filling rate. Territorial foraging enables HBs to find new food sources as candidates and easily move toward the neighboring region within their territory. Migration foraging can be performed in each of two iterations. Until the stopping condition is met, each operation and calculation are interactively performed. At last, the food source with the maximum rate of nectar refilling is returned as near-global optimal.

(1) A population of $n$ HBs is initialized at random to $n$ food source in the following:

$$\chi_j = Low + r \times (Up - Low) i = 1, \cdots \cdots n \tag{20}$$

In Equation (20), Low and $Up$ indicate the lower and upper limitations for $d$-dimension problems, correspondingly; $r$ refers to a random integer within the range of zero and one; $x_i$ signifies the location of the $i$-$th$ food sources.

$$VT_{i,j} = \begin{cases} 0 & if i \neq j \\ null & i = j \end{cases} \tag{21}$$

where $i = j$, $VT_{i,j} = null$ shows that an HB takes food from a certain food source; $i \neq j$, $VT_{i,j} = 0$ denotes that the $j$-$th$ food sources were visited by $i$-$th$ HB in the present iteration.

(2) Guided foraging: With the abovementioned flight abilities, an HB could access its targeted food sources to attain candidate food source, hence the following mathematical expression simulates candidate food source and guiding foraging behaviors:

$$v_i(t+1) = x_{i,tar}(t) + a \times D \times (x_i(t) - x_{i,tar}(t)) \tag{22}$$

$$a \sim N(0,1) \tag{23}$$

From the expression, $x_i(t)$ and $x_{i,tar}(t)$ are the position of $i$-$th$ hummingbird food and target source at $t$ time; $a$ is distributed uniformly, with standard deviation of 1 and mean = 0.

The location updating of $i$-$th$ food sources is given below:

$$x_i(t+1) = \begin{cases} \chi_{i(t)} & f(\chi_i(t)) \leq f(v_j(t+1)) \\ v_i(t+1) & f(x_i(t)) > f(v_i(t+1)) \end{cases} \tag{24}$$

In Equation (24), $f(\cdot)$ denotes function fitness value. Equation (24) represents that if the nectar refilling rate of candidate food sources is superior to the present one, the HB will abandon the existing food source and stay at a candidate one for feeding.

(3) Territorial foraging: After attaining targeted food sources where nectar was eaten, an HB seeks innovative food sources. Thus, an HB could move towards a neighboring region within its own territory whereby a novel food source is found that is the best candidate solution. The mathematical expression to stimulate local search of an HB for territorial foraging strategy and candidate food source is shown below:

$$v_i(t+1) = x_i(t) + b \times D \times x_i(t) \tag{25}$$

$$b \sim N(0,1) \tag{26}$$

Now, $b$ is distributed uniformly, with a standard deviation of 1 and mean = 0.

(4) Once food becomes frequently scarce in a territory visited by an HB, the bird frequently migrates to more distant food sources for foraging.

## 4. Results and Discussion

The proposed model is simulated using Python 3.6.5 tool on PC i5-8600k, GeForce 1050Ti 4 GB, 16 GB RAM, 250 GB SSD, and 1 TB HDD. The parameter settings are given as follows: learning rate: 0.01, dropout: 0.5, batch size: 5, epoch count: 50, and activation: ReLU. The experimental validation of the AHBATL-MNC method on mitosis cell classification is tested using a dataset [22] that has 150 images and two classes, as represented in Table 1. Figure 2 depicts some sample images of mitosis and nonmitosis.

**Table 1.** Dataset details.

| Class | No. of Images |
|---|---|
| Mitosis | 75 |
| Nonmitosis | 75 |
| Total Number of Images | 150 |

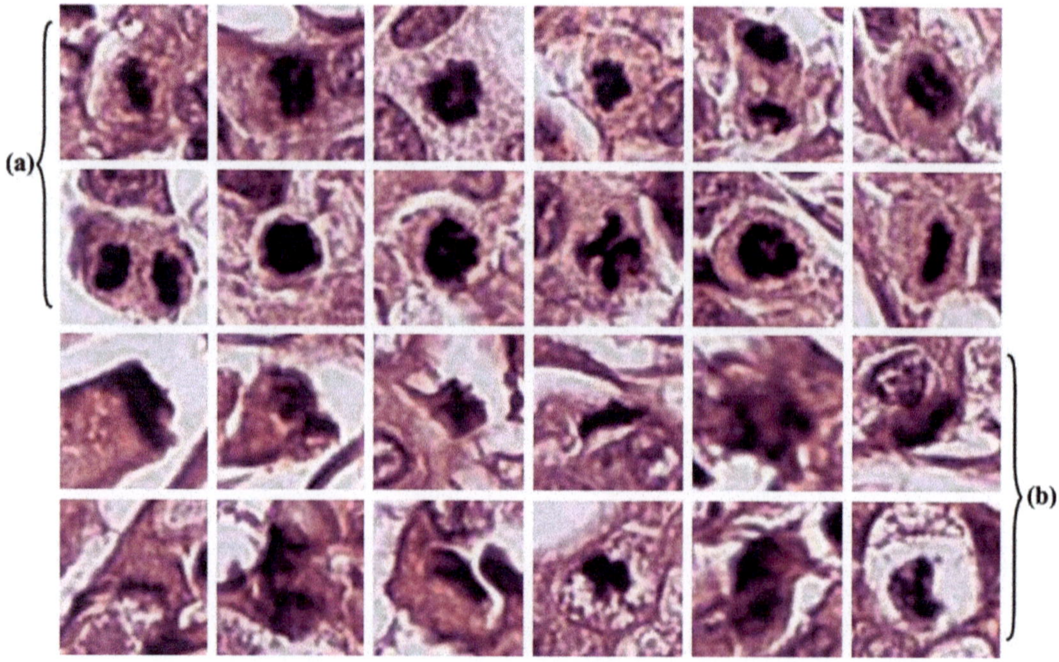


**Figure 2.** Sample images of (**a**) mitosis; (**b**) nonmitosis.

The binary classification outcomes of the AHBATL-MNC method on the applied dataset are portrayed in the form of confusion matrix in Figure 3. On 60% of the training (TR) database, the AHBATL-MNC model detected 39 samples into mitosis class and 42 samples into nonmitosis class. Meanwhile, on 40% of the testing (TS) database, the AHBATL-MNC method detected 29 samples into mitosis class and 29 samples into nonmitosis class. Eventually, on 70% of the TR database, the AHBATL-MNC system detected 42 samples into mitosis class and 55 samples into nonmitosis class. Finally, on 30% of the TS database, the AHBATL-MNC algorithm detected 22 samples into mitosis class and 19 samples into nonmitosis class.

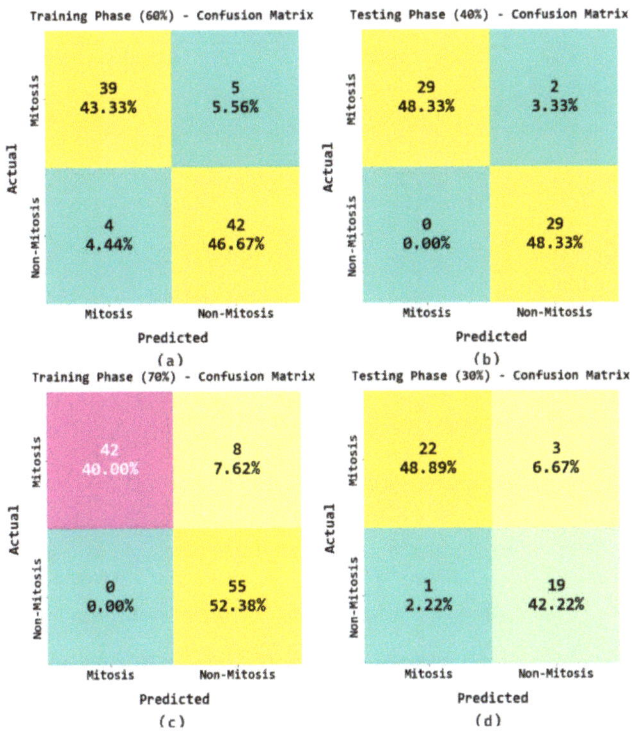


**Figure 3.** Confusion matrices of AHBATL-MNC system. (**a**,**b**) TR and TS databases of 60:40; (**c**,**d**) TR and TS databases of 70:30.

In Table 2, overall mitosis classification results of the AHBATL-MNC model under 60% of TR and 40% of TS databases are given. Figure 4 exhibits the detailed classifier outcome of the AHBATL-MNC model on 60% of the TR database. The outcomes depict that the AHBATL-MNC model properly classified mitosis and nonmitosis class images. It is noted that the AHBATL-MNC model attained average $accu_{bal}$ of 89.97%, $prec_n$ of 90.03%, $reca_l$ of 89.93%, $F_{score}$ of 89.99%, MCC of 80%, and $G_{measure}$ of 89.99%.

**Table 2.** Mitosis classification outcome of AHBATL-MNC approach under 60:40 of TR/TS databases.

| Class | Accuracy$_{bal}$ | Precision | Recall | F-Score | MCC | G-Measure |
|---|---|---|---|---|---|---|
| Training Phase (60%) | | | | | | |
| Mitosis | 88.64 | 90.70 | 88.64 | 89.66 | 80.00 | 89.66 |
| Nonmitosis | 91.30 | 89.36 | 91.30 | 90.32 | 80.00 | 90.33 |
| Average | 89.97 | 90.03 | 89.97 | 89.99 | 80.00 | 89.99 |
| Testing Phase (40%) | | | | | | |
| Mitosis | 93.55 | 100.00 | 93.55 | 96.67 | 93.55 | 96.72 |
| Nonmitosis | 100.00 | 93.55 | 100.00 | 96.67 | 93.55 | 96.72 |
| Average | 96.77 | 96.77 | 96.77 | 96.67 | 93.55 | 96.72 |

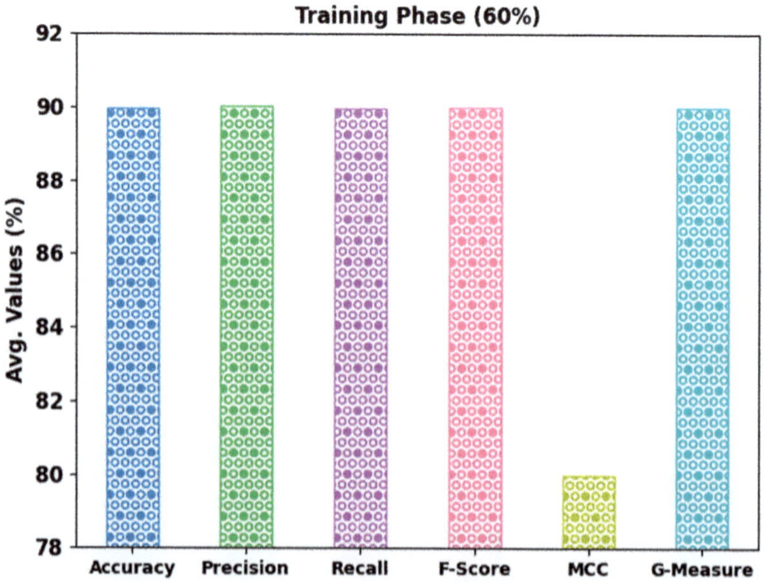


**Figure 4.** Average analysis of AHBATL-MNC approach under 60% of TR database.

Figure 5 reveals a comprehensive classifier outcome of the AHBATL-MNC system on 40% of the TS database. The outcomes show that the AHBATL-MNC approach properly classified the mitosis and nonmitosis class images. It can be seen that the AHBATL-MNC method reached average $accu_{bal}$ of 96.77%, $prec_n$ of 96.77%, $reca_l$ of 96.77%, $F_{score}$ of 96.67%, MCC of 93.55%, and $G_{measure}$ of 96.72%.

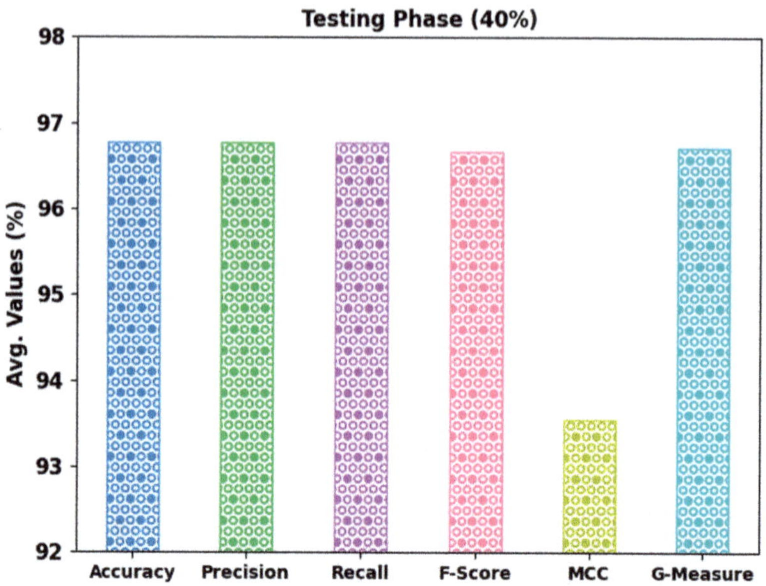


**Figure 5.** Average analysis of AHBATL-MNC approach under 40% of TS database.

In Table 3, the overall mitosis classification outcome of the AHBATL-MNC algorithm under 70% of the TR and 30% of the TS databases is given. Figure 6 demonstrates the detailed classifier outcome of the AHBATL-MNC method on 70% of the TR database. The outcomes represent that the AHBATL-MNC system properly classified the mitosis and nonmitosis class images. It is clear that the AHBATL-MNC methodology obtained average $accu_{bal}$ of 92%, $prec_n$ of 93.65%, $reca_l$ of 92%, $F_{score}$ of 92.26%, MCC of 85.63%, and $G_{measure}$ of 92.54%.

**Table 3.** Mitosis classification outcome of AHBATL-MNC approach under 60:40 of TR/TS databases.

| Class | Accuracy$_{bal}$ | Precision | Recall | F-Score | MCC | G-Measure |
|---|---|---|---|---|---|---|
| | | | Training Phase (70%) | | | |
| Mitosis | 84.00 | 100.00 | 84.00 | 91.30 | 85.63 | 91.65 |
| Nonmitosis | 100.00 | 87.30 | 100.00 | 93.22 | 85.63 | 93.44 |
| Average | 92.00 | 93.65 | 92.00 | 92.26 | 85.63 | 92.54 |
| | | | Testing Phase (30%) | | | |
| Mitosis | 88.00 | 95.65 | 88.00 | 91.67 | 82.51 | 91.75 |
| Nonmitosis | 95.00 | 86.36 | 95.00 | 90.48 | 82.51 | 90.58 |
| Average | 91.50 | 91.01 | 91.50 | 91.07 | 82.51 | 91.16 |

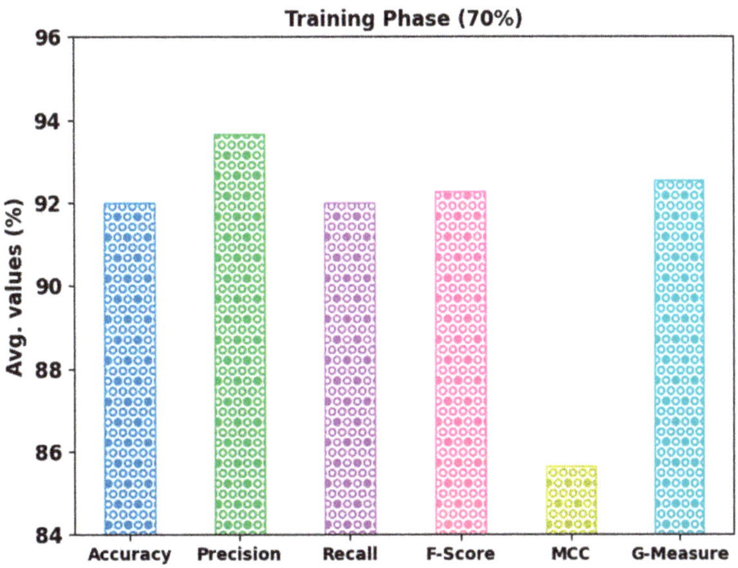


**Figure 6.** Average analysis of AHBATL-MNC approach under 70% of TR database.

Figure 7 shows a brief classifier outcome of the AHBATL-MNC approach on 30% of the TS database. The outcome demonstrates that the AHBATL-MNC algorithm properly classified the mitosis and nonmitosis class images. It can be stated that the AHBATL-MNC algorithm accomplished average $accu_{bal}$ of 91.50%, $prec_n$ of 91.01%, $reca_l$ of 91.50%, $F_{score}$ of 91.07%, MCC of 82.51%, and $G_{measure}$ of 91.16%.

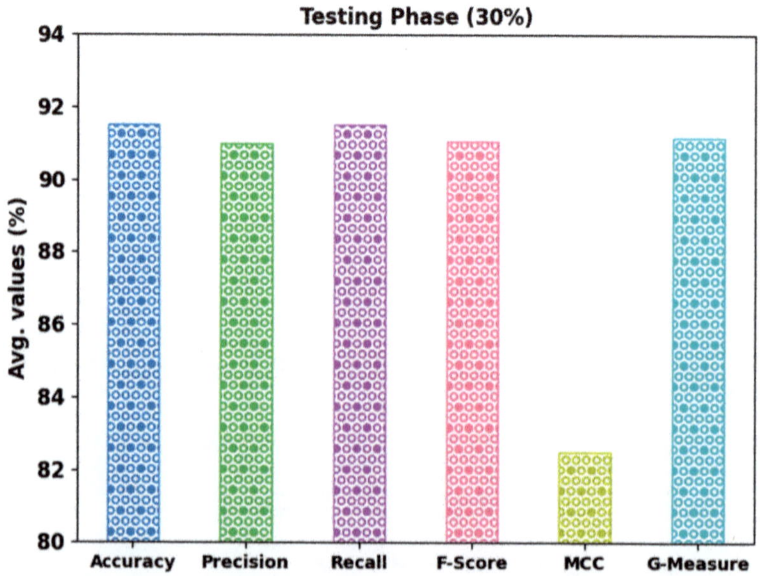


**Figure 7.** Average analysis of AHBATL-MNC approach under 30% of TS database.

The training accuracy (TACC) and validation accuracy (VACC) of the AHBATL-MNC system are inspected on breast cancer performance in Figure 8. The figure reveals that the AHBATL-MNC approach shows improved performance with improved values of TACC and VACC. It is noticeable that the AHBATL-MNC system gained higher TACC outcomes.

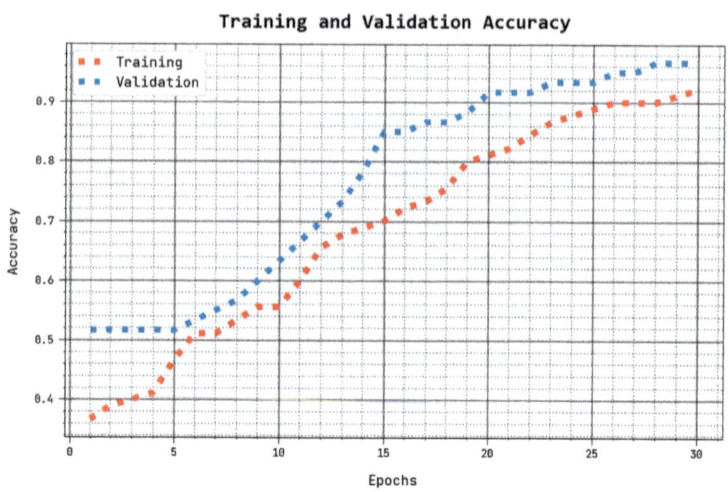


**Figure 8.** TACC and VACC analysis of AHBATL-MNC approach.

The training loss (TLS) and validation loss (VLS) of the AHBATL-MNC methodology are tested on breast cancer performance in Figure 9. The figure points out that the AHBATL-MNC algorithm revealed better performance with lower values of TLS and VLS. It is observable that the AHBATL-MNC methodology resulted in minimal VLS outcomes.

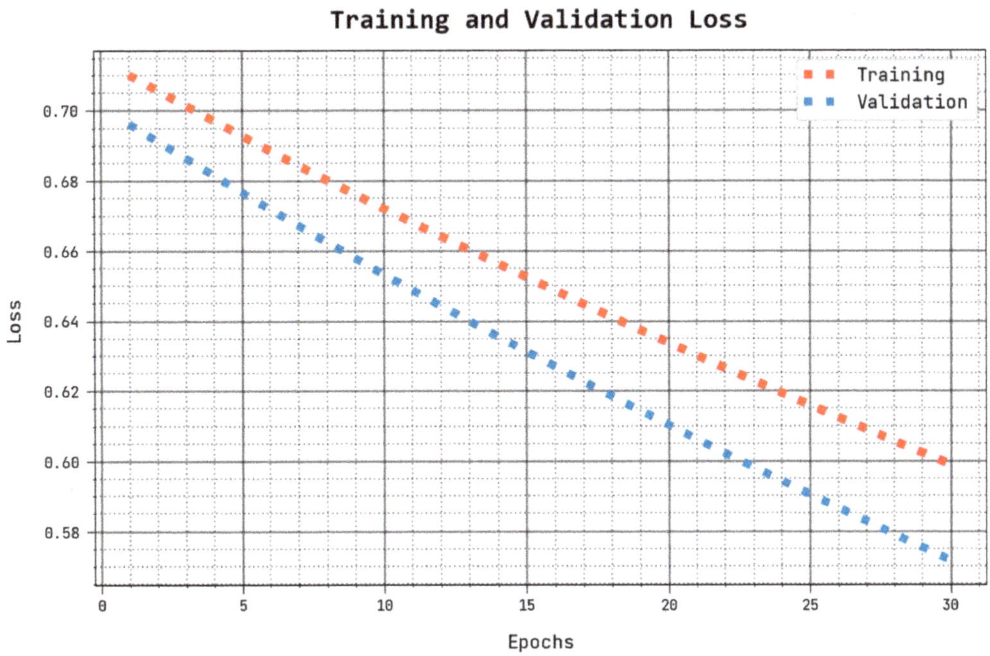


**Figure 9.** TLS and VLS analysis of AHBATL-MNC approach.

Table 4 reports an overall comparative inspection of the AHBATL-MNC method with recent approaches [13]. Figure 10 offers a comparative inspection of the AHBATL-MNC method in terms of $accu_y$ and $F_{score}$. The outcomes indicate that the AHBATL-MNC method achieved improved performance. For instance, based on $accu_y$, the AHBATL-MNC model obtained higher $accu_y$ of 96.77%. In contrast, the DHE-Mit, DenseNet-201, and ResNet-18 models attained lower $accu_y$ of 85.23%, 83.96%, and 82.01%, respectively. Eventually, with respect to $F_{score}$, the AHBATL-MNC approach gained maximal $F_{score}$ of 96.67%. In contrast, the DHE-Mit, DenseNet-201, and ResNet-18 systems obtained decreased $F_{score}$ of 77.33%, 76.38%, and 74.05%, correspondingly.

**Table 4.** Comparative analysis of AHBATL-MNC system with other approaches.

| Methods | $accu_y$ | $prec_n$ | $reca_l$ | $F_{score}$ |
|---|---|---|---|---|
| AHBATL-MNC | 96.77 | 96.77 | 96.77 | 96.67 |
| DHE-Mit model | 85.23 | 84.45 | 75.26 | 77.33 |
| DenseNet-201 model | 83.96 | 83.20 | 73.85 | 76.38 |
| ResNet-18 model | 82.01 | 81.26 | 71.73 | 74.05 |
| Inception-V3 model | 78.54 | 77.51 | 68.18 | 70.64 |
| ResNext-50 model | 77.48 | 76.20 | 66.73 | 69.49 |
| ResNet-101 model | 76.03 | 74.83 | 65.89 | 68.65 |
| VGG-16 model | 74.72 | 73.93 | 65.00 | 67.66 |

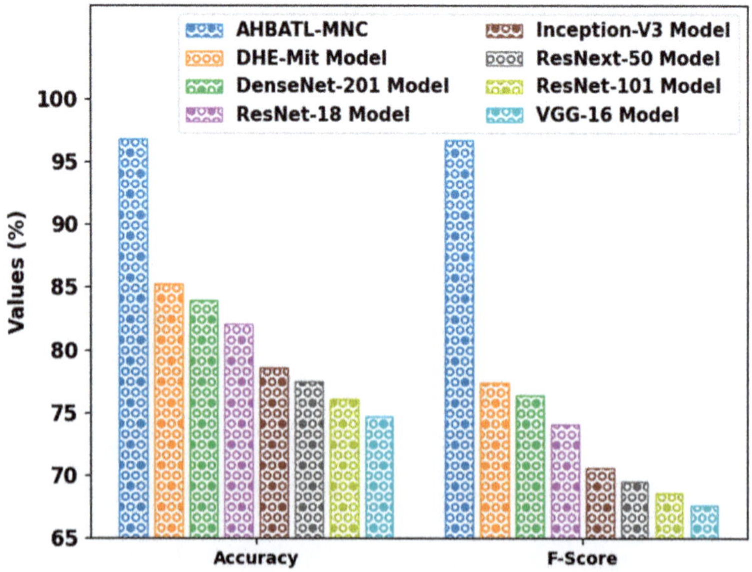


**Figure 10.** $Accu_y$ and $F_{score}$ analysis of AHBATL-MNC system compared with other approaches.

Figure 11 provides a comparative examination of the AHBATL-MNC approach with respect to $prec_n$ and $reca_l$. The outcomes state that the AHBATL-MNC approach gained enhanced performance. For example, in terms of $prec_n$, the AHBATL-MNC model obtained higher $prec_n$ of 96.77%. In contrast, the DHE-Mit, DenseNet-201, and ResNet-18 models attained lower $prec_n$ of 84.45%, 83.20%, and 81.26%, correspondingly. Finally, with respect to $reca_l$, the AHBATL-MNC model gained enhanced $reca_l$ of 96.77%. In contrast, the DHE-Mit, DenseNet-201, and ResNet-18 methods accomplished lower $reca_l$ of 75.26%, 73.85%, and 71.73%, respectively.

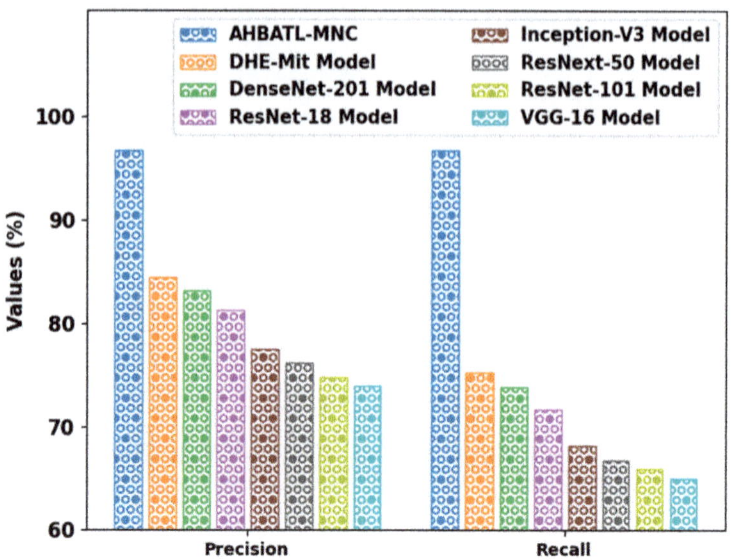


**Figure 11.** $Prec_n$ and $Reca_l$ analysis of AHBATL-MNC system compared with other approaches.

Table 5 offers a detailed computation time (CT) examination of the proposed model with existing models. The experimental results indicate that the proposed model shows better performance with minimum CT of 12.34 s. On the contrary, the existing models attained increased CT values compared to the AHBATL-MNC model. These results confirm the improvement of the AHBATL-MNC model over other models. The proposed model accomplished superior performance to other existing techniques due to the hyperparameter selection of ResNet using Adamax optimizer and AHBA for XGBoost classifier.

**Table 5.** Comparative CT analysis of AHBATL-MNC system with other approaches.

| Methods | Computational Time (s) |
|---|---|
| AHBATL-MNC | 12.34 |
| DHE-Mit model | 25.17 |
| DenseNet-201 model | 42.58 |
| ResNet-18 model | 41.03 |
| Inception-V3 model | 59.67 |
| ResNext-50 model | 39.36 |
| ResNet-101 model | 44.60 |
| VGG-16 model | 56.14 |

## 5. Conclusions

In this study, we developed a new AHBATL-MNC technique for effective identification of mitotic and nonmitotic nuclei on His. Primarily, in the AHBATL-MNC technique, the PSPNet model is utilized for segmentation process, which identifies the candidate mitotic patches. Followed by this, the ResNet model is employed as feature extractor, and the XGBoost model is applied as a classifier. For improving the classification performance, the parameter tuning of the XGBoost model was performed by the AHBA technique. The performance evaluation of the AHBATL-MNC technique was tested on medical imaging datasets and the outcomes were examined in distinct measures. The simulation values validated the improved outcomes of the AHBATL-MNC algorithm over other recent approaches. In future, the performance of the AHBATL-MNC method can be improved by the use of ensemble learning methodologies. In addition, the proposed model needs to be tested on large-scale databases and can be extended to detect other kinds of cancer.

**Author Contributions:** Conceptualization, A.A.M. and M.O.; methodology, A.G.; software, I.Y.; validation, M.A.H., A.A.M., A.G. and M.O.; formal analysis, I.Y.; investigation, A.S.M.; resources, A.S.M.; data curation, M.I.A.; writing—original draft preparation, M.A.H., A.A.M., A.G. and M.O.; writing—review and editing, I.Y., A.A.A. and M.I.A.; visualization, A.A.A.; supervision, A.A.M.; project administration, M.A.H.; funding acquisition, A.A.M. and A.G. All authors have read and agreed to the published version of the manuscript.

**Funding:** Princess Nourah bint Abdulrahman University Researchers Supporting Project number (PNURSP2022R151), Princess Nourah bint Abdulrahman University, Riyadh, Saudi Arabia. The authors would like to thank the Deanship of Scientific Research at Umm Al-Qura University for supporting this work by Grant Code: (22UQU4320484DSR10). This study is supported via funding from Prince Sattam bin Abdulaziz University project number (PSAU/2023/R/1444).

**Institutional Review Board Statement:** Not applicable.

**Informed Consent Statement:** Not applicable.

**Data Availability Statement:** Data sharing not applicable to this article as no datasets were generated during the current study.

**Conflicts of Interest:** The authors declare that they have no conflict of interest. The manuscript was written through contributions of all authors. All authors have given approval to the final version of the manuscript.

## References

1. Sigirci, I.O.; Albayrak, A.; Bilgin, G. Detection of mitotic cells in breast cancer histopathological images using deep versus handcrafted features. *Multimed. Tools Appl.* **2022**, *81*, 13179–13202. [CrossRef]
2. Maroof, N.; Khan, A.; Qureshi, S.A.; ul Rehman, A.; Khalil, R.K.; Shim, S.O. Mitosis detection in breast cancer histopathology images using hybrid feature space. *Photodiagnosis Photodyn. Ther.* **2020**, *31*, 101885. [CrossRef] [PubMed]
3. Li, C.; Wang, X.; Liu, W.; Latecki, L.J.; Wang, B.; Huang, J. Weakly supervised mitosis detection in breast histopathology images using concentric loss. *Med. Image Anal.* **2019**, *53*, 165–178. [CrossRef] [PubMed]
4. Lakshmanan, B.; Priyadharsini, S.; Selvakumar, B. Computer assisted mitotic figure detection in histopathology images based on DenseNetPCA framework. *Mater. Today Proc.* **2022**, *62*, 4936–4939. [CrossRef]
5. Sohail, A.; Mukhtar, M.A.; Khan, A.; Zafar, M.M.; Zameer, A.; Khan, S. Deep Object Detection based Mitosis Analysis in Breast Cancer Histopathological Images. *arXiv* **2020**, arXiv:2003.08803.
6. Samah, A.A.; Fauzi, M.F.A.; Khor, S.Y.; Lee, J.T.H.; Teoh, K.H.; Looi, L.M.; Mansor, S. Mitotic cells detection in H&E-stained breast carcinoma images. *Int. J. Biomed. Eng. Technol.* **2022**, *40*, 54–69.
7. Mahmood, T.; Arsalan, M.; Owais, M.; Lee, M.B.; Park, K.R. Artificial intelligence-based mitosis detection in breast cancer histopathology images using faster R-CNN and deep CNNs. *J. Clin. Med.* **2020**, *9*, 749. [CrossRef] [PubMed]
8. Cai, D.; Sun, X.; Zhou, N.; Han, X.; Yao, J. Efficient mitosis detection in breast cancer histology images by RCNN. In Proceedings of the 2019 IEEE 16th International Symposium on Biomedical Imaging (ISBI 2019), Venice, Italy, 8–11 April 2019; pp. 919–922.
9. Razavi, S.; Khameneh, F.D.; Nouri, H.; Androutsos, D.; Done, S.J.; Khademi, A. MiNuGAN: Dual Segmentation of Mitoses and Nuclei Using Conditional GANs on Multi-center Breast H&E Images. *J. Pathol. Inform.* **2022**, *13*, 100002.
10. Shwetha, S.V.; Dharmanna, L. An automatic recognition, identification and classification of mitotic cells for the diagnosis of breast cancer stages. *Int. J. Image Graph. Sign. Process.* **2021**, *13*, 1–11.
11. Kausar, T.; Wang, M.; Ashraf, M.A.; Kausar, A. SmallMitosis: Small size mitotic cells detection in breast histopathology images. *IEEE Access* **2020**, *9*, 905–922. [CrossRef]
12. Sohail, A.; Khan, A.; Nisar, H.; Tabassum, S.; Zameer, A. Mitotic nuclei analysis in breast cancer histopathology images using deep ensemble classifier. *Med. Image Anal.* **2021**, *72*, 102121. [CrossRef]
13. Sebai, M.; Wang, X.; Wang, T. MaskMitosis: A deep learning framework for fully supervised, weakly supervised, and unsupervised mitosis detection in histopathology images. *Med. Biol. Eng. Comput.* **2020**, *58*, 1603–1623. [CrossRef]
14. Lei, H.; Liu, S.; Elazab, A.; Gong, X.; Lei, B. Attention-guided multi-branch convolutional neural network for mitosis detection from histopathological images. *IEEE J. Biomed. Health Inform.* **2020**, *25*, 358–370. [CrossRef] [PubMed]
15. Das, D.K.; Dutta, P.K. Efficient automated detection of mitotic cells from breast histological images using deep convolution neutral network with wavelet decomposed patches. *Comput. Biol. Med.* **2019**, *104*, 29–42. [CrossRef]
16. Beevi, K.S.; Nair, M.S.; Bindu, G.R. Automatic mitosis detection in breast histopathology images using convolutional neural network based deep transfer learning. *Biocybern. Biomed. Eng.* **2019**, *39*, 214–223. [CrossRef]
17. Yan, L.; Liu, D.; Xiang, Q.; Luo, Y.; Wang, T.; Wu, D.; Chen, H.; Zhang, Y.; Li, Q. PSP net-based automatic segmentation network model for prostate magnetic resonance imaging. *Comput. Methods Programs Biomed.* **2021**, *207*, 106211. [CrossRef]
18. Chu, Y.; Yue, X.; Yu, L.; Sergei, M.; Wang, Z. Automatic image captioning based on ResNet50 and LSTM with soft attention. *Wirel. Commun. Mob. Comput.* **2020**, *2020*, 8909458. [CrossRef]
19. Xiao, B.; Liu, Y.; Xiao, B. Accurate state-of-charge estimation approach for lithium-ion batteries by gated recurrent unit with ensemble optimizer. *IEEE Access* **2019**, *7*, 54192–54202. [CrossRef]
20. Jiang, Y.; Tong, G.; Yin, H.; Xiong, N. A pedestrian detection method based on genetic algorithm for optimize XGBoost training parameters. *IEEE Access* **2019**, *7*, 118310–118321. [CrossRef]
21. Wang, L.; Zhang, L.; Zhao, W.; Liu, X. Parameter Identification of a Governing System in a Pumped Storage Unit Based on an Improved Artificial Hummingbird Algorithm. *Energies* **2022**, *15*, 6966. [CrossRef]
22. Veta, M.; Pluim, J.P.W.; Stathonikos, N.; van Diest, P.J.; Beca, F.; Beck, A. Tumor Proliferation Assessment Challenge 2016, MICCAI Grand Challenge [WWW Document]. 2016. Available online: http://tupac.tue-image.nl/ (accessed on 12 July 2022).

bioengineering

MDPI

*Article*

# Accurate Image Reconstruction in Dual-Energy CT with Limited-Angular-Range Data Using a Two-Step Method

**Buxin Chen [1], Zheng Zhang [1], Dan Xia [1], Emil Y. Sidky [1], Taly Gilat-Schmidt [2] and Xiaochuan Pan [1,3,*]**

1 Department of Radiology, The University of Chicago, Chicago, IL 60637, USA
2 Department of Biomedical Engineering, Marquette University and Medical College of Wisconsin, Milwaukee, WI 53201, USA
3 Department of Radiation and Cellular Oncology, The University of Chicago, Chicago, IL 60637, USA
* Correspondence: xpan@uchicago.edu

**Abstract:** Dual-energy CT (DECT) with scans over limited-angular ranges (LARs) may allow reductions in scan time and radiation dose and avoidance of possible collision between the moving parts of a scanner and the imaged object. The beam-hardening (BH) and LAR effects are two sources of image artifacts in DECT with LAR data. In this work, we investigate a two-step method to correct for both BH and LAR artifacts in order to yield accurate image reconstruction in DECT with LAR data. From low- and high-kVp LAR data in DECT, we first use a data-domain decomposition (DDD) algorithm to obtain LAR basis data with the non-linear BH effect corrected for. We then develop and tailor a directional-total-variation (DTV) algorithm to reconstruct from the LAR basis data obtained basis images with the LAR effect compensated for. Finally, using the basis images reconstructed, we create virtual monochromatic images (VMIs), and estimate physical quantities such as iodine concentrations and effective atomic numbers within the object imaged. We conduct numerical studies using two digital phantoms of different complexity levels and types of structures. LAR data of low- and high-kVp are generated from the phantoms over both single-arc (SA) and two-orthogonal-arc (TOA) LARs ranging from 14° to 180°. Visual inspection and quantitative assessment of VMIs obtained reveal that the two-step method proposed can yield VMIs in which both BH and LAR artifacts are reduced, and estimation accuracy of physical quantities is improved. In addition, concerning SA and TOA scans with the same total LAR, the latter is shown to yield more accurate images and physical quantity estimations than the former. We investigate a two-step method that combines the DDD and DTV algorithms to correct for both BH and LAR artifacts in image reconstruction, yielding accurate VMIs and estimations of physical quantities, from low- and high-kVp LAR data in DECT. The results and knowledge acquired in the work on accurate image reconstruction in LAR DECT may give rise to further understanding and insights into the practical design of LAR scan configurations and reconstruction procedures for DECT applications.

**Keywords:** dual-energy CT; two-step method; limited-angular range; directional total variation

**Citation:** Chen, B.; Zhang, Z.; Xia, D.; Sidky, E.Y.; Gilat-Schmidt, T.; Pan, X. Accurate Image Reconstruction in Dual-Energy CT with Limited-Angular-Range Data Using a Two-Step Method. *Bioengineering* **2022**, *9*, 775. https://doi.org/10.3390/bioengineering9120775

Academic Editors: Paolo Zaffino and Maria Francesca Spadea

Received: 25 October 2022
Accepted: 29 November 2022
Published: 6 December 2022



## 1. Introduction

Dual-energy computed tomography (DECT) has found applications in clinical and industrial settings. In current DECT, one generally acquires data of low- and high-kVp X-ray spectra over a full-angular range (FAR) of $2\pi$, or over at least a short-scan angular range [1–4]. Interest remains in the development of DECT imaging over limited-angular ranges (LARs) that are considerably less than the FAR of $2\pi$ (or than the short-scan angular range,) because such LAR scans may bear implications for radiation dose reduction, scan time minimization, and collision avoidance between the scanner and the imaged object. Inspired by the directional-total-variation (DTV) work on image reconstruction from LAR data in conventional single-energy CT (SECT) [5], we have investigated image reconstruction previously from LAR data in DECT [6,7] by focusing on the correction only for LAR artifacts and using DTV constraints in the reconstruction of kVp images followed by an image-domain decomposition. Other methods have also been developed for DECT with LAR data, but the angular ranges are generally not smaller than 90° [8,9].

In this work, we propose a two-step method to reconstruct quantitatively accurate images in DECT from LAR data by correcting for both BH and LAR artifacts, thus improving the quantitative accuracy of images reconstructed and physical quantities estimated. In the method, from LAR data of low- and high-kVp, a data-domain decomposition (DDD) algorithm [10] is used first for obtaining LAR basis data in which the BH artifacts are compensated for; and a DTV algorithm [5] is then developed and tailored to reconstruct basis images from LAR basis data obtained. The reconstructed basis images can be combined to form virtual monochromatic images (VMIs), i.e., the X-ray linear attenuation coefficients, for visual inspection, and can be used also for estimating physical quantities such as iodine-contrast concentrations and effective atomic numbers within the imaged object [11–14]. We hypothesize that images and physical quantities with both BH and LAR artifacts corrected for in LAR DECT are quantitatively comparable with those obtained in FAR DECT. Therefore, in this work the results obtained for LAR DECT are compared with those obtained from FAR data in DECT in which BH artifacts are corrected for by using the DDD algorithm.

Numerical studies are conducted with a chest phantom [15] and a suitcase phantom [6] containing distinct anatomies and structures of potential relevance in medical and security applications [11,16–19]. Low- and high-kVp data are collected with single-arc (SA) or two-orthogonal-arc (TOA) scans of LAR [6], ranging from 14° to 180°. Using the DDD and DTV algorithms, we estimate basis data and then reconstruct basis images, followed by the formation of VMIs at energies of interest from the basis images reconstructed. In addition to visual inspection and quantitative analysis of VMIs obtained, we also estimate iodine-contrast concentrations in chest images and effective atomic numbers in suitcase images from data of different LARs. Furthermore, we investigate image reconstructions from data acquired with SA and TOA scans of possible implications for potential non-diagnostic imaging applications involving, e.g., C-arm DECT, in which workflow or safety concerns may limit the scan angular range. The two-step method and the study design in the work can also be applied to investigations concerning image reconstruction in DECT and multi-spectral CT using techniques with sandwiched detectors [2], sequential scans [20], or advanced photon-counting detectors [21,22]. DECT with fast-kVp-switching X-ray tubes can also collect approximately overlapping rays [4].

## 2. Materials and Methods

### 2.1. Scans of Limited-Angular Ranges

In this work, we consider single-arc (SA) or two-orthogonal-arc (TOA) scans in a fan-beam DECT, as shown schematically in Figure 1a,b. The SA scan includes a pair of completely overlapping arcs of LAR $\alpha_\top$, whereas the TOA scan includes two pairs of completely overlapping arcs of LARs $\alpha_1$ and $\alpha_2$. For each pair of the completely overlapping arcs in the SA and TOA scans, low- and high-kVp data are collected over one of the paired arcs. In this work, we assume that the $x$- or $y$-axis intersects with the middle point of each pair of the completely overlapping arcs, and that the tangential directions at the middle points of the two pairs of completely overlapping arcs in the TOA scan are orthogonal with each other in Figure 1b. We use $\alpha_\top$ to denote the LAR of an SA scan and investigate image reconstruction from data collected over SAs of LARs $\alpha_\top = 14°, 20°, 30°, 45°, 60°, 90°, 120°$, $150°$, and $180°$. For an SA of LAR $\alpha_\top$, we also consider a TOA scan with two arcs of equal LARs satisfying $\alpha_1 = \alpha_2 = 0.5\alpha_\top$. (The work can readily be generalized to a TOA scan with two arcs of different LARs [23]).

Dual-energy data are generated from a chest phantom and a suitcase phantom in Figure 2 with two different fan-beam geometries used in the numerical study: for the chest phantom, the source-to-rotation distance (SRD) and source-to-detector distance (SDD) are 100 cm and 150 cm, with a linear detector of 70 cm comprising 896 bins, whereas for the suitcase phantom, the SRD and SDD are 100 cm and 150 cm, with a linear detector of 32 cm including 512 bins. The imaged objects are assumed to be completely within the field-of-view of the scan configurations, resulting in no truncation. In the studies involving both phantoms, the angular interval is fixed at 0.25° between two adjacent views. Dual-energy data are also collected over two full rotations, or the FAR of 360°, and images reconstructed from FAR data may be used as references in the work.

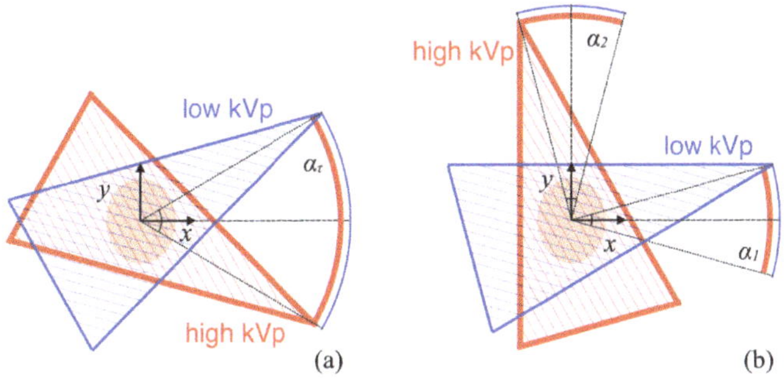


**Figure 1.** Schematics of SA (**a**) and TOA (**b**) scans of LARs in fan-beam DECT. The SA scan includes a pair of completely overlapping arcs of LAR $\alpha_T$, and the $x$-axis intersects with the middle point of the two arcs, whereas the TOA scan includes two pairs of completely overlapping arcs of LARs $\alpha_1$ and $\alpha_2$, and the $x$- and $y$-axis intersect with the middle points of the two pairs of arcs. For each pair of the completely overlapping arcs in the SA and TOA scans, low- and high-kVp data are collected over one of the paired arcs. In this work, we consider $\alpha_1 = \alpha_2 = 0.5\alpha_T$.

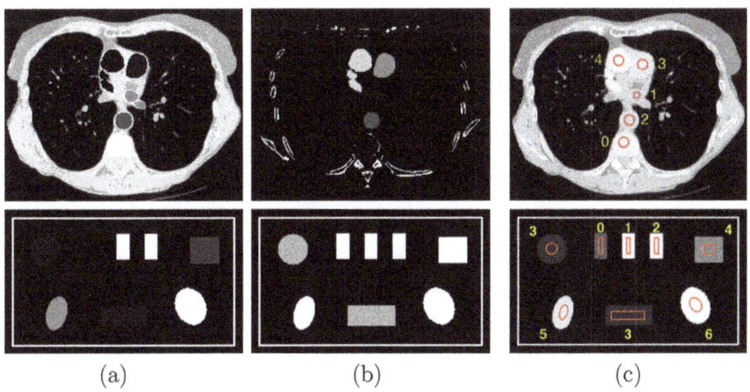


**Figure 2.** Row 1: (**a**) water and (**b**) iodine basis images and (**c**) VMI at 100 keV of the chest phantom; and row 2: (**a**) photoelectric effect (PE) and (**b**) Compton scattering (KN) basis images and (**c**) VMI at 40 keV of the suitcase phantom. Display windows for the chest phantom are [0, 1.2] for the two basis images and [0, 0.22] cm$^{-1}$ for the VMI, while those for the suitcase phantom are [0, 0.22] and [0.1, 0.65] cm$^{-1}$, respectively.

### 2.2. Imaging Model

In DECT, data are collected at ray $j$ with two distinct spectra, referred to as low- and high-kVp spectra, and an imaging model can be expressed as [8]

$$\begin{aligned} g_j^L &= -\ln \sum_m^M q_{jm}^L \exp\left(-\sum_i^I a_{ji} f_{mi}\right), \\ g_j^H &= -\ln \sum_m^M q_{jm}^H \exp\left(-\sum_i^I a_{ji} f_{mi}\right), \end{aligned} \tag{1}$$

where $g_j^L$ and $g_j^H$ denote model data of the low- and high-kVp scans; $q_{jm}^L$ and $q_{jm}^H$ the low- and high-kVp spectra after normalization (including possibly filtered tube spectra and detector response) at energy bin $m$; $a_{ji}$ the contribution of image pixel $i$ to data of ray $j$; and $f_{mi}$ the image value at pixel $i$ within energy bin $m$ of the monochromatic image, i.e., the linear attenuation coefficient.

In the absence of the basis-decomposition error, $f_{mi}$ can be written as the combination of two basis images, i.e.,

$$f_{mi} = \mu_{0m} b_{0i} + \mu_{1m} b_{1i}, \tag{2}$$

where $b_{ki}$ denotes basis image $k$ at pixel $i$ and $\mu_{km}$ the linear attenuation coefficient at energy bin $m$ for basis material $k$ ($k = 0$ or 1). Image $\mathbf{f}_m$, with $f_{mi}$ as its elements, obtained with Equation (2) is referred to also as the virtual monochromatic image (VMI).

In the work, assuming $\mu_{km}$, $q_{jm}^L$, and $q_{jm}^H$ are known, the two-step method is proposed for accurately reconstructing basis images $b_{ki}$, or, equivalently, VMI $f_{mi}$, from data collected over an SA or TOA of LARs in fan-beam DECT.

### 2.3. Numerical Phantoms Studied

We consider in the work two phantoms, i.e., the chest phantom [15] and suitcase phantom [6] shown in Figure 2, motivated by their possible implications in medical and security imaging, two distinct DECT imaging applications, and their distinctly different anatomic structures for evaluating algorithm performance. The chest phantom contains four regions of interest (ROIs) 1–4 with iodine-contrast agents at concentrations of 5 mg/mL, 10 mg/mL, 15 mg/mL, and 20 mg/mL, respectively, and other ROIs with mixed materials, such as muscle, lung tissue, and bone; whereas the suitcase phantom includes three ROIs 0–2 of single-element calibration materials, i.e., carbon, aluminum, and calcium, and four more ROIs 3–6 of mixed materials, corresponding to water, ANFO (Ammonium Nitrate and Fuel Oil [11]), teflon, and PVC, respectively.

In DECT, basis images may be selected according to the task considered. For the chest phantom, to estimate iodine concentrations, we select material-based basis images of water and iodine concentration of 20 mg/mL, with the corresponding $\mu_{km}$'s obtained from the NIST database [24]. For the suitcase phantom, in order to estimate effective atomic numbers, we select interaction-based basis images of the photoelectric effect (PE) and Compton scattering (KN) with $\mu_{km}$'s that are $1/E^3$, where $E$ denotes X-ray energy, and obtained with the Klein–Nishina formula [1], respectively. The basis images and VMIs of the chest and suitcase phantoms are formed on image arrays of $200 \times 256$ and $150 \times 256$ square pixels of size 0.7 mm, as displayed in rows 1 and 2, respectively, in Figure 2.

### 2.4. Image Reconstruction Approach

In an attempt to compensate for the BH effect inherent in $g_j^L$ and $g_j^H$, we rewrite Equation (1) as

$$\begin{aligned} g_j^L &= -\ln \sum_m^M q_{jm}^L \exp(-\mu_{0m} l_{0j} - \mu_{1m} l_{1j}), \\ g_j^H &= -\ln \sum_m^M q_{jm}^H \exp(-\mu_{0m} l_{0j} - \mu_{1m} l_{1j}), \end{aligned} \tag{3}$$

where $l_{kj} = \sum_i^I a_{ji} b_{ki}$, $k = 0$ or 1, denotes the sinogram of basis image $k$, also referred to as basis data, which is independent of energy $m$. Therefore, applying the DDD algorithm [10] to Equation (3), we can obtain basis sinograms $l_{kj}$ from knowledge of $g_j^L$ and $g_j^H$ for each ray $j$. It has been shown empirically [25] that the DDD algorithm can recover accurately basis sinograms from $g_j^L$ and $g_j^H$. Using existing algorithms such as the FBP algorithm, one can reconstruct readily accurate basis images from full knowledge of basis sinograms $l_{kj}$ in a FAR or short scan. In the work, because knowledge of $l_{kj}$ can be available only over a

SA or TOA of LARs, the FBP algorithm yield basis images of significant artifacts. We thus develop and tailor the DTV algorithm to reconstruct basis images with minimized LAR artifacts from knowledge of $l_{kj}$'s available only over a SA or TOA of LARs.

Using vectors $\mathbf{b}_k$ and $\mathbf{L}_k$ ($k = 0$ or 1) of sizes $I$ and $J$, respectively, to denote basis images and their sinograms with elements $b_{ki}$ and $l_{kj}$ in concatenated forms, we formulate the reconstruction problem of basis images from their sinograms as a convex optimization problem

$$\begin{aligned} \mathbf{b}_k^{\star} &= \underset{\mathbf{b}_k}{\text{argmin}} \ \frac{1}{2} \parallel \mathbf{L}_k - \mathcal{A}\,\mathbf{b}_k \parallel_2^2 \\ \text{s.t.}\ &||\mathcal{D}_x\mathbf{b}_k||_1 \le t_{kx},\ ||\mathcal{D}_y\mathbf{b}_k||_1 \le t_{ky},\ \text{and}\ b_{ki} \ge 0, \end{aligned} \tag{4}$$

where matrix $\mathcal{A}$ of size $J \times I$ denotes the discrete fan-beam X-ray transform with element $a_{ji}$; $\|\cdot\|_2$ the $\ell_2$-norm of a vector; and $||\mathcal{D}_x\mathbf{b}_k||_1$ and $||\mathcal{D}_y\mathbf{b}_k||_1$ are the image's directional total variations (DTVs) [5] of the basis image $\mathbf{b}_k$ along the $x$- and $y$-axis, respectively.

The DTV algorithm used to reconstruct basis images from knowledge of the basis sinograms in DECT through solving Equation (4) shares the same general structure as that of the algorithm for image reconstruction from LAR data in conventional SECT [5]. The pseudo-code is thus summarized in Appendix A for clarity.

### 2.5. Visual Inspection and Quantitative Analysis of Images

As VMIs are of visualization interest in DECT, we first obtain VMIs at energy levels of interest from basis images reconstructed by using Equation (2) and then visually inspect LAR artifacts in the VMIs. Additionally, two quantitative metrics, normalized root-mean-square error (nRMSE) and Pearson correlation coefficient (PCC) [5,26,27] are calculated. Metric nRMSE evaluates quantitative difference, while metric PCC assesses visual correlation, between a VMI obtained from LAR data and a reference image obtained from FAR data. In particular, higher PCCs suggest better visual correlation between the VMI and its reference image. The VMI and its reference are identical when PCC $\rightarrow$ 1 and nRMSE $\rightarrow$ 0.

In the chest phantom study, we seek to estimate iodine-contrast concentration within ROIs 1–4 shown in the basis images in row 1 of Figure 2. Using the estimated basis image of 20-mg/mL iodine-contrast agent, we estimate the concentration of iodine-contrast agent within ROIs 1–4 with a linear fitting [6]. Constants in the linear relationship are determined by using pixel values within iodine-contrast ROIs 1–4 in the reference image of the chest phantom obtained from the FAR data by use of the two-step method, and fitting into the corresponding known concentrations. In the work, the calibrated slope and intercept of the linear fitting were computed as 19.3 mg/mL and −0.0074 mg/mL. In general, the linear fitting, as compared to the default setting of 20 and 0 as slope and intercept, yields more accurate estimation of the iodine concentration, because the mean pixel values within ROI 0 in the 20-mg/mL iodine basis image could be non-zero. This occurs as a result of the incomplete basis set in the material decomposition model by using 2 materials. On the other hand, in the study involving the suitcase phantom, we seek to estimate the effective atomic number of materials [11]. As the basis images are estimated as PE and KN components, their ratios are used in an affine transform with the effective atomic number in the log-log domain [6]. The effective atomic numbers are then computed for ROIs 3–6 of the suitcase phantom, as shown in row 2 of Figure 2. Constants in the affine transformation are determined by using the pixel values within single-element ROIs 0–2 in the reference image of the suitcase phantom obtained from the FAR data by use of the two-step method, and fitting into the corresponding known atomic numbers.

## 3. Results

### 3.1. Numerical Study Design and Data Generation

In our numerical studies with noiseless and noisy LAR data, the TASMIC model [28] was used for generating filtered tube spectra of given low- and high-kVps. Taking into account the detector's energy-integrating response, we then obtain $q^L_{jm}$ and $q^H_{jm}$ by multiplying the filtered tube spectra with corresponding X-ray energies $E$. For both phantoms, the low- and high-kVp spectra are set at 80 and 140 kVp, with a 5-mm Al filter used in both.

For each of the chest or suitcase phantom in an SA or TOA scan described in Section 2.1 above, basis sinograms $l_{kj}$ are first generated from basis images shown in Figure 2, and noiseless low- and high-kVp data $g^L_j$ and $g^H_j$ can be generated subsequently by use of Equation (3) with $l_{kj}$, and knowledge of $\mu_{km}$, $q^L_{jm}$, and $q^H_{jm}$ determined. The aims of the noiseless data study are (1) to verify that the two-step method, including the DDD and DTV algorithms, can recover numerically accurate basis images and VMIs first from FAR-scan data and (2) to use the two-step method verified to explore empirically its performance upper bound , i.e., the performance in the best case scenario without any inconsistencies, such as noise and decomposition error, in the data, as a function of LARs for yielding accurate reconstruction of VMIs and physical quantity estimation in DECT with LAR scans.

**Table 1.** NEQs per detector bin in air scans of either the low- or high-kVp scans for the chest and suitcase phantoms with LARs ranging from 14° to 180°, as well as with the FAR of 360°.

| LAR | 14° | 20° | 30° | 45° | 60° |
|---|---|---|---|---|---|
| NEQ | $9.64 \times 10^5$ | $6.75 \times 10^5$ | $4.5 \times 10^5$ | $3 \times 10^5$ | $2.25 \times 10^5$ |
| LAR | 90° | 120° | 150° | 180° | 360° |
| NEQ | $1.5 \times 10^5$ | $1.125 \times 10^5$ | $9 \times 10^4$ | $7.5 \times 10^4$ | $3.75 \times 10^4$ |

Using noiseless data as the means of the Poisson noise model, we obtain noisy data containing Poisson noise. For both chest and suitcase phantoms, Table 1 shows the noise-equivalent quanta's (NEQs) of each detector bin for the SA or TOA scans studied, which are determined such that the means in SA or TOA scans studied have a fixed total number of quanta of $\sim 6.9 \times 10^9$ in an air scan, amounting to 75% of that in a FAR scan with 360 projection views, 512 detector bins, and $\sim 5 \times 10^4$ NEQs per detector bin [23]. The purpose of the noisy data study is to yield some preliminary insights into the reconstruction robustness of the two-step method. Clearly, its reconstruction accuracy depends not only on the LAR extent but also on the characteristics and level of data noise. No additional data or image processing is applied in the study with noisy data, although such processing may improve the quality of VMI visualization and physical quantity estimation.

Constraint parameters $t_{kx}$ and $t_{ky}$ have an impact on image reconstruction by defining the feasible solution set of Equation (4). In the study below with consistent noiseless data, the DTV values of the phantom basis images in Figure 2 are selected as the values of parameters $t_{kx}$ and $t_{ky}$, in order to form the tightest feasible solution set that still contain the desired solution (i.e., the truth basis images in this case). In the study with noisy data, the values of parameters $t_{kx}$ and $t_{ky}$ are selected in terms of visual evaluation of reconstructed VMIs with minimum artifacts [5,6]. In general, parameter selection is accomplished through surveying the parameter space within relevant ranges and optimizing a well-defined image-quality metric, e.g., image visualization for artifact reduction or quantitative estimation of iodine-contrast concentration, for studies with inconsistent data, including those with real data where the truth images are not available. In our experience, $t_{kx}$ and $t_{ky}$ selected in the noisy data studies are generally smaller than those in the corresponding noiseless data studies. In the work, the $t_{kx}$ values selected are in general smaller than $t_{ky}$ in the SA scans, so as to suppress horizontal streaks along the $y$-axis, while both $t_{kx}$ and $t_{ky}$ selected in the TOA scans are slightly larger than those in the SA scans, as the improved conditioning of the system matrix leads to fewer artifacts overall. Basis images are also reconstructed from $\mathbf{L}_k$ estimated by using the FBP algorithm with a Hanning kernel and a cutoff frequency at 0.5, which are then combined into the VMIs with Equation (2). The FBP algorithm is used only for demonstrating the LAR artifacts associated with the phantoms and data conditions in the work.

### 3.2. Image Reconstruction of the Chest Phantom

#### 3.2.1. Verification Study with the Chest Phantom

A study is performed to first verify that (1) the DDD algorithm can invert the non-linear model in Equation (3) and numerically accurately recover the basis sinogram $l_{kj}$ from noiseless low- and high-kVp data of the chest phantom acquired with the FAR scan of 360°; and (2) the DTV algorithm developed and tailored can reconstruct numerically accurate basis images and VMIs from noiseless basis sinogram. In Figure 3a, we display the VMI at 100 keV [29], along with a zoomed-in view, and show in Figure 3b their differences from the truth counterparts in Figure 2c (top row). The result confirms that the two-step method can yield accurate reconstructions from noiseless FAR data. In an attempt to demonstrate possible LAR artifacts associated with the phantom, we apply both the DTV and FBP algorithms to reconstructing images from noiseless basis sinogram acquired with a SA scan of LAR $\alpha_\top = 30°$, and display them in Figure 3c,d, respectively. It can be observed that the two-step method can significantly reduce the LAR artifacts observed in the FBP image.

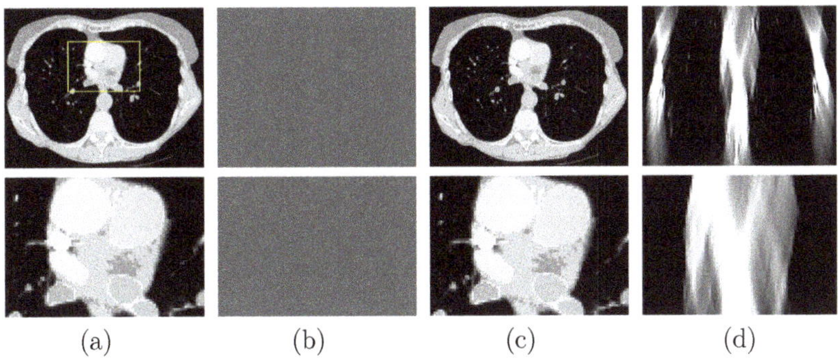

**Figure 3.** Row 1: (**a**) VMI of the chest phantom at 100 keV obtained with the two-step method from FAR data, (**b**) difference between the VMI in (**a**) and its truth in Figure 2c, and VMIs at 100 keV obtained with (**c**) the two-step method and (**d**) the FBP algorithm from noiseless data acquired over a SA of LAR 30°; Row 2: zoomed-in views of their corresponding images in row 1. The zoomed-in area is enclosed by the rectangular box depicted in the VMI in (**a**). Display windows [0, 0.22] $cm^{-1}$ for columns (**a**,**c**,**d**), and $[-10^{-4}, 10^{-4}]$ $cm^{-1}$ for column (**b**).

#### 3.2.2. Image Reconstruction from Noiseless Data Acquired with SA and TOA Scans of LARs

We subsequently apply the two-step method verified to reconstructing basis images of water and iodine from noiseless data of the chest phantom collected in SA or TOA scans of $\alpha_\top = 20°, 30°, 45°, 60°, 90°, 120°$, 150°, and 180°. In Figure 4, we display VMIs at 100 keV, along with their zoomed-in views within the ROI, reconstructed for the SA (rows 1&2 and 5&6) and TOA (rows 3&4 and 7&8) scans. It can be observed that the two-step method yields visually comparable images for these scans of LARs, revealing quantitatively possible performance upper bounds of the two-step method in accurate image reconstruction, i.e., numerically accurately inverting Equation (3), for SA and TOA scans of LARs studied in the work.

From VMIs in Figure 4, we compute nRMSEs and PCCs, which are displayed in row 1 of Figure 5. Using the method described in Section 2.5, we also estimate iodine concentrations in ROIs 1–4 indicated in the top row of Figure 2c, and plot them as functions of LARs in row 1 of Figure 6. These results reveal that, from the chest phantom noiseless data collected over the range of LARs as low as 20°, the two-step method can yield VMIs visually and quantitatively close to the reference VMIs from FAR data of 360° in terms of PCC and estimated iodine concentrations. Regarding metric nRMSE, it increases as LAR decreases, mainly due to the increasing null spaces present in the system matrices of the LAR scans, while TOA scans can lower nRMSE by an order of magnitude especially for small LARs as compared to SAs of the same LAR.

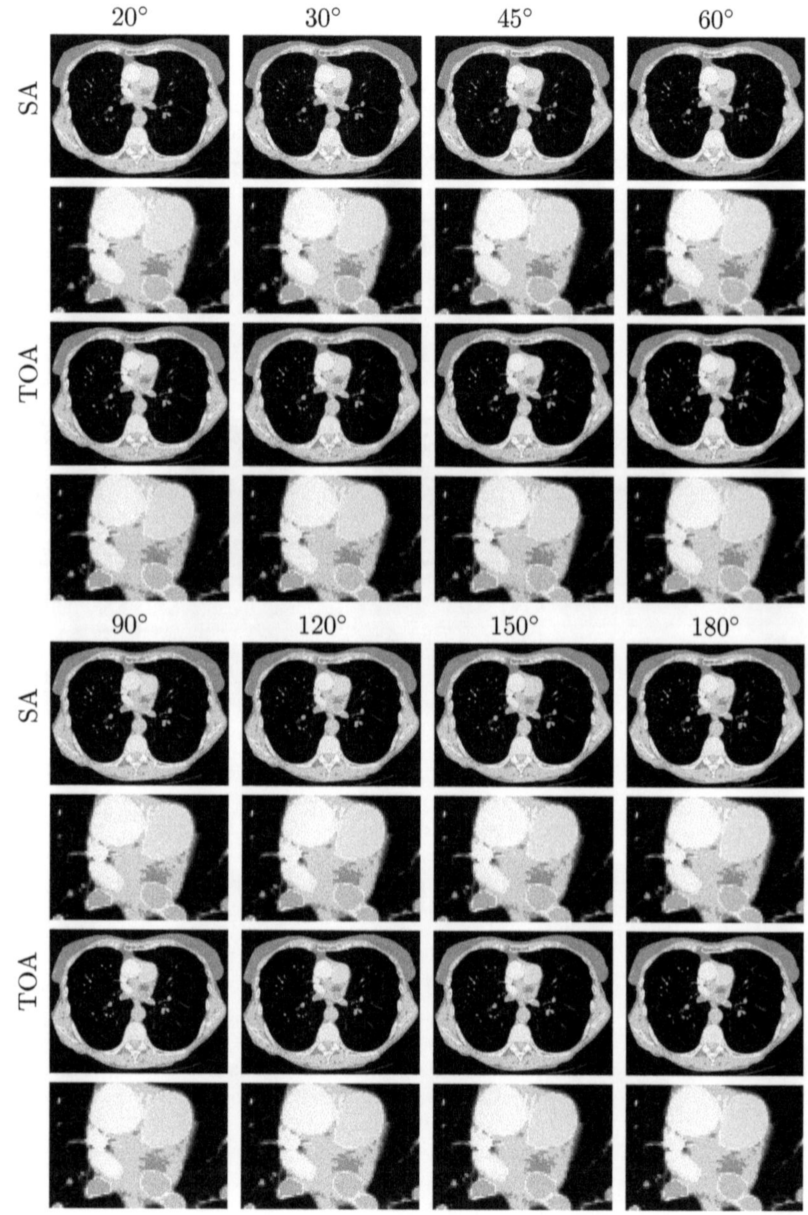


**Figure 4.** VMIs (rows 1, 3, 5, and 7), along with their respective zoomed-in views (rows 2, 4, 6, and 8), of the chest phantom at 100 keV obtained from noiseless data over SAs (rows 1&2 and 5&6) and TOAs (rows 3&4 and 7&8) of LAR 20°, 30°, 45°, 60°, 90°, 120°, 150°, and 180°, respectively, by use of the two-step method. Display window: [0, 0.22] $cm^{-1}$.

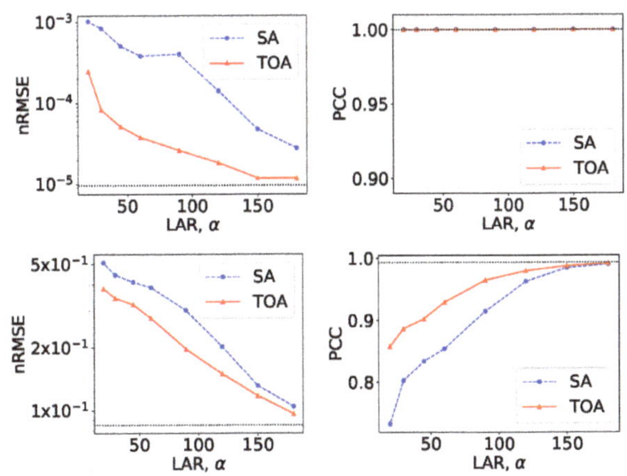


**Figure 5.** Metrics nRMSE and PCC, computed over VMIs of the chest phantom from noiseless data in Figure 4 (row 1) and those from noisy data in Figure 7 (row 2) as functions of LARs $\alpha_{\mathcal{T}}$ for SA (blue, dashed) and TOA (red, solid) scans. The horizontal lines (black, dotted) indicate the reference values from FAR data of 360°.

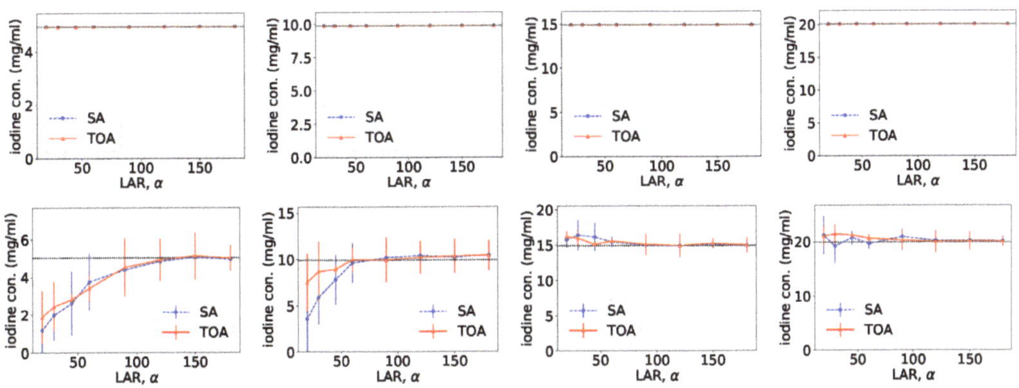


**Figure 6.** Iodine concentrations, along with their respective error bars, in ROIs 1–4 (from left to right) within the chest phantom, as functions of LARs $\alpha_{\mathcal{T}}$ for SA (blue, dashed) and TOA (red, solid) scans, estimated from basis images reconstructed from noiseless (row 1) and noisy (row 2) data by use of the two-step method.

### 3.2.3. Image Reconstruction from Noisy Data Acquired with SA and TOA Scans of LARs

We repeat the study by applying the DTV algorithm to noisy data of the chest phantom for SA and TOA scans considered in the noiseless study above. In Figure 7, we display VMIs at 100 keV, along with their zoomed-in views. For the noise levels considered, it can be observed that (1) LAR artifacts can be amplified by noise, (2) LAR artifacts are reduced substantially in VMIs for SA and TOA scans of LARs $\alpha_{\mathcal{T}} \geq 120°$, and (3) TOA scans can more effectively suppress LAR artifacts than SA scans for the chest phantom and noise level studied in the work. Such observations may provide insights into the design of practical procedures for image reconstruction from LAR data that contain additional inconsistencies. We note that no processing is applied to the data or images reconstructed in our study.

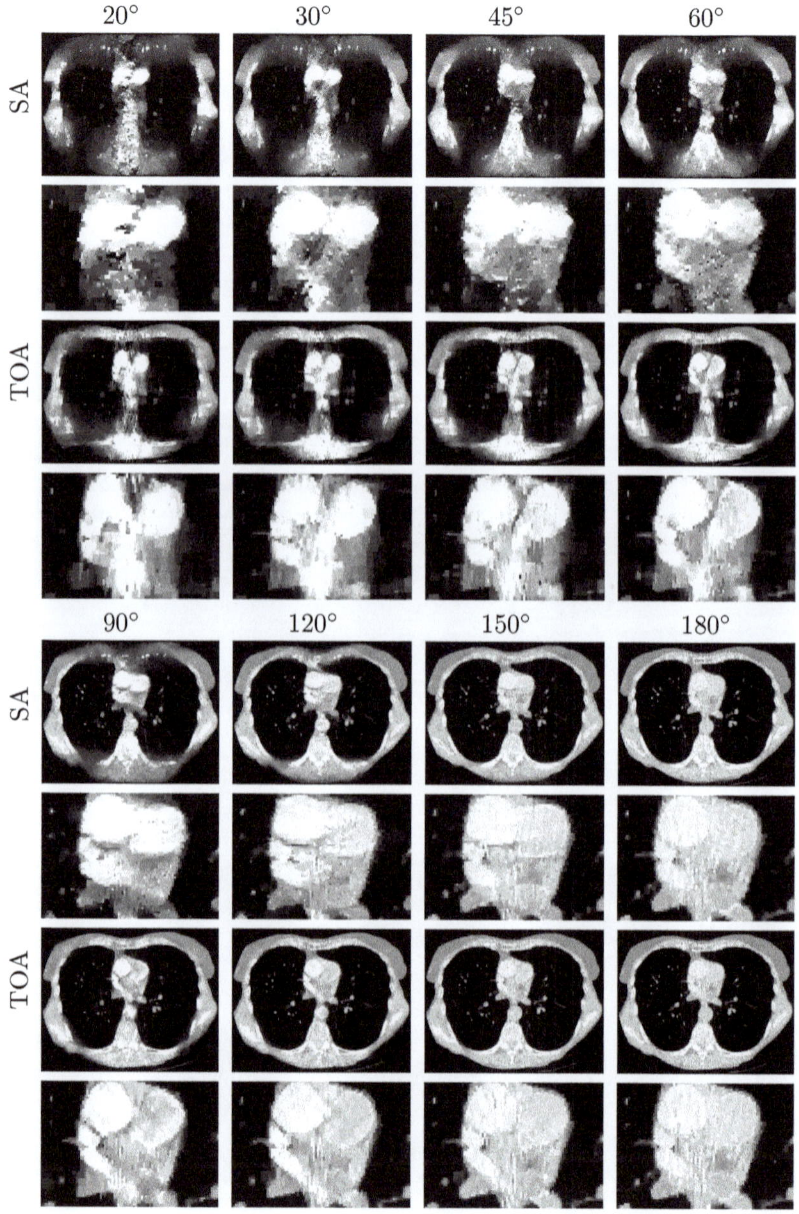


**Figure 7.** VMIs (rows 1, 3, 5, and 7), along with their respective zoomed-in views (rows 2, 4, 6, and 8), of the chest phantom at 100 keV obtained from noisy data over SAs (rows 1&2 and 5&6) and TOAs (rows 3&4 and 7&8) of LAR 20°, 30°, 45°, 60°, 90°, 120°, 150°, and 180°, respectively, by use of the two-step method. Display window: [0, 0.22] $cm^{-1}$.

Similar to the noiseless-data study, we compute nRMSEs and PCCs from VMIs in Figure 7, and plot them as functions of LARs in row 2 of Figure 5. We also estimate iodine concentrations and plot them as functions of LAR in row 2 of Figure 6. In the noisy-data study, error bars, i.e., standard deviations, are calculated over the chest-phantom ROIs indicated in Figure 2c, and they are plotted in row 2 of Figure 6. The horizontal lines (black, dotted) indicate the reference values from FAR data of 360°. Quantitative results of PCC appear consistent with the visual inspection, suggesting that VMI images for $\alpha_T \geq 120°$ in SA and $\alpha_T \geq 90°$ in TOA scans visually resemble the reference VMI obtained from noisy FAR data, and the degree of resemblance drops understandably as LAR decreases. The estimation accuracy of iodine concentration for $\alpha_T \geq 90°$ remains comparable to those obtained from the reference images reconstructed from noisy FAR data.

### 3.3. *Image Reconstruction of the Suitcase Phantom*

Next, we repeat the studies in Section 3.2 with the suitcase phantom. We show in Figure 8a the VMI and its zoomed-in view reconstructed from FAR data and in Figure 8b their differences from the truth counterparts in Figure 2c (bottom row). The result again confirms the reconstruction accuracy of the two-step method using the suitcase phantom, which is of different complexity and structure to the chest phantom. To reveal the LAR artifacts associated with the suitcase phantom, we apply the DTV and FBP algorithms to reconstruct images from noiseless basis sinogram over an SA of $\alpha_T = 30°$ and display them in Figure 8c,d, respectively. It can be observed that the LAR artifacts in the FBP image are almost eliminated in the image reconstructed by use of the two-step method.

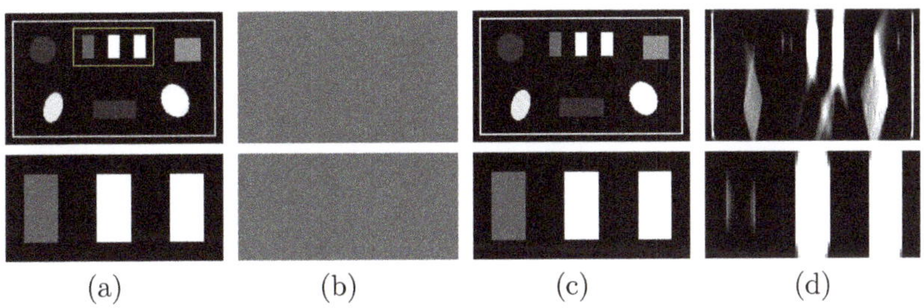


**Figure 8.** Row 1: (**a**) VMI of the suitcase phantom at 40 keV obtained with the two-step method from FAR data, (**b**) difference between the VMI in (**a**) and its truth in Figure 2c, VMIs at 40 keV obtained with (**c**) the two-step method and (**d**) the FBP algorithm from data acquired over a SA of LAR 30°; and row 2: zoomed-in views of their corresponding images in row 1. The zoomed-in area is enclosed by the rectangular box depicted in the VMI in (**a**). Display windows [0.1, 0.65] $cm^{-1}$ for columns (**a**,**c**,**d**), and $[-10^{-4}, 10^{-4}]$ $cm^{-1}$ for column (**b**).

#### 3.3.1. Image Reconstruction from Noiseless Data Acquired with SA and TOA Scans of LARs

Next, we apply the algorithm verified to reconstructing basis images of PE and KN from noiseless data of the suitcase phantom collected in SA or TOA scans of $\alpha_T = 14°, 20°, 30°, 60°, 90°, 120°$, $150°$, and $180°$. The lowest LAR studied for the suitcase phantom, 14°, is smaller than that for the chest phantom, 20°. In Figure 9, we display the VMIs at 40 keV, along with their zoomed-in views, reconstructed from data collected with SA (rows 1&2 and 5&6) and TOA (rows 3&4 and 7&8) scans. It can be observed that the two-step method yields almost visually identical images for these LARs, revealing possible performance upper bounds of the method in numerically accurately inverting Equation (3) for scans with SA and TOA of LARs.

From VMIs in Figure 9, we compute nRMSEs and PCCs, and display them in row 1 of Figure 10. Using the method described in Section 2.5, we also estimate effective atomic numbers in ROIs 3–6 indicated in bottom row of Figure 2c, and plot them as functions of LARs in row 1 of Figure 11. These results reveal that, from the suitcase phantom noiseless data collected over the range of LARs as low as 14°, the two-step method can yield VMIs visually and quantitatively close to the reference VMIs from FAR data of 360° in terms of PCC and estimated effective atomic numbers. With regard to metric nRMSE, it increases as LAR decreases, largely due to the increasing null spaces in the system matrices of the LAR scans, while TOA scans can lower nRMSE by an order of magnitude especially for small LARs as compared to SAs of the same LAR.

3.3.2. Image Reconstruction from Noisy Data Acquired with SA and TOA Scans of LARs

We apply the two-step method to reconstructing images from noisy data of the suitcase phantom collected over the same LARs in SA and TOA. In Figure 12, we display the VMIs at 40 keV, along with their zoomed-in views. For the suitcase phantom, LAR artifacts are substantially reduced in VMIs from data collected over $\alpha_\mathcal{T} \geq 90^\circ$ in SA and $\alpha_\mathcal{T} \geq 60^\circ$ in TOA scans. Similar to the chest phantom results, TOA configurations can more effectively suppress LAR artifacts than SA ones, especially recovering the distorted edges around the circular and elliptical disks, for the suitcase phantom under noise level studied in the work.

**Figure 9.** VMIs (rows 1, 3, 5, and 7), along with their respective zoomed-in views (rows 2, 4, 6, and 8), of the suitcase phantom at 40 keV obtained from noiseless data acquired over SAs (rows 1&2 and 5&6) and TOAs (rows 3&4 and 7&8) of LAR 14°, 20°, 30°, 60°, 90°, 120°, 150°, and 180°, respectively, by use of the two-step method. Display window: [0.1, 0.65] cm$^{-1}$.

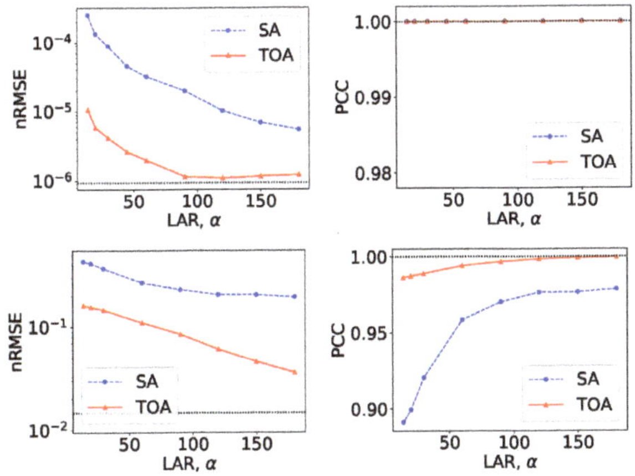


**Figure 10.** Metrics nRMSE and PCC, computed over VMIs of the suitcase phantom from noiseless data in Figure 9 (row 1) and those from noisy data in Figure 12 (row 2) as functions of LARs $\alpha_\mathcal{T}$ for SA (blue, dashed) and TOA (red, solid) scans. The horizontal lines (black, dotted) indicate the reference values from FAR data of 360°.

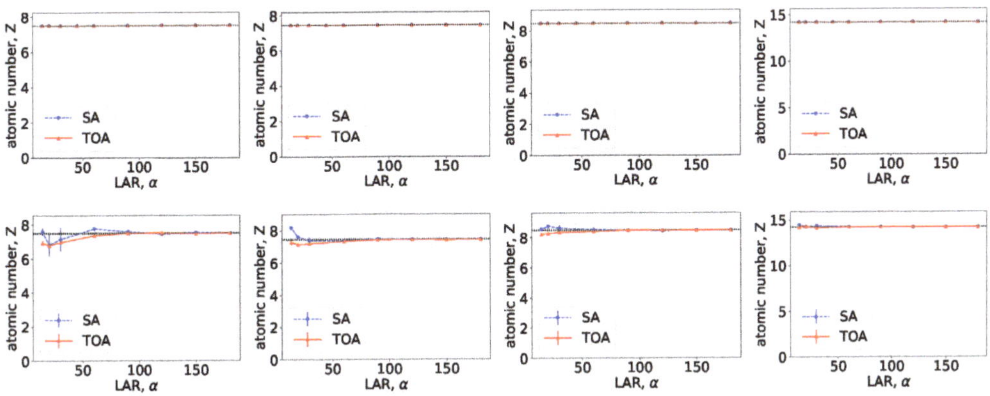


(a) (b) (c) (d)

**Figure 11.** Effective atomic numbers of (**a**) water, (**b**) ANFO, (**c**) Teflon, and (**d**) PVC, along their respective error bars, within the suitcase phantom estimated as functions of LAR $\alpha_\mathcal{T}$ for SA (blue, dashed) and TOA (red, solid) scans, computed from basis images reconstructed from noiseless (row 1) and noisy (row 2) data by use of the two-step method.

We compute nRMSEs and PCCs from VMIs in Figure 12, and plot them as functions of LARs in row 2 of Figure 10. We also estimate effective atomic numbers and plot them as functions of LAR in row 2 of Figure 11. In the noisy-data study, error bars, i.e., standard deviations, are calculated over the suitcase-phantom ROIs indicated in Figure 2c, and they are plotted in row 2 of Figure 11. The quantitative results suggest that VMI images visually resemble the reference VMI from FAR data for noisy LAR data collected over $\alpha_\mathcal{T} \geq 60°$ in SA and $\alpha_\mathcal{T} \geq 14°$ in TOA scans, and the resemblance decreases understandably as LAR decreases and that the estimation accuracy from noisy LAR data collected over $\alpha_\mathcal{T} \geq 60°$ is comparable to those from the FAR data.

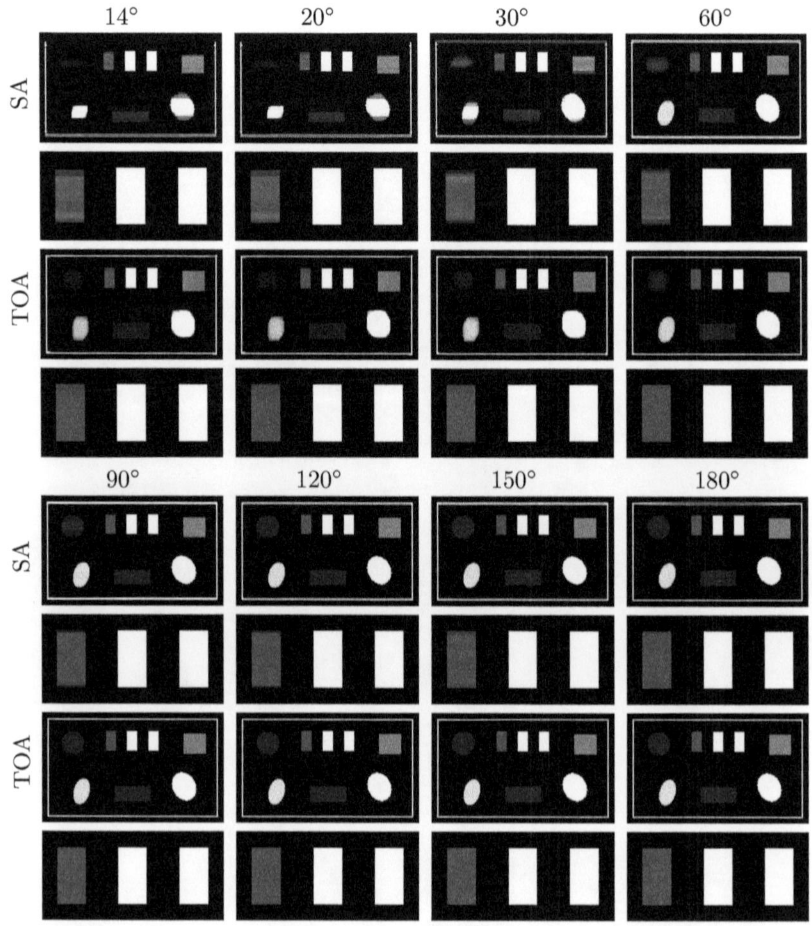


**Figure 12.** VMIs (rows 1, 3, 5, and 7), along with their respective zoomed-in views (rows 2, 4, 6, and 8), of the suitcase phantom at 40 keV obtained from noisy data acquired over SAs (rows 1&2 and 5&6) and TOAs (rows 3&4 and 7&8) of LAR 14°, 20°, 30°, 60°, 90°, 120°, 150°, and 180°, respectively, by use of the two-step method. Display window: [0.1, 0.65] $cm^{-1}$.

## 4. Discussion

In this work, we have investigated and developed a two-step method for image reconstruction from low- and high-kVp data collected with SA and TOA scans of LARs in DECT. The method combines the DDD and DTV algorithms to effectively compensate for both BH and LAR artifacts, yielding accurate VMIs and physical-quantity estimation. For the study conditions such as phantoms and noise levels considered, visual inspection of VMIs at energies of interest indicates that the method can yield from noiseless LAR data VMIs that are visually comparable to the reference VMI from FAR data, and from noisy LAR data VMIs with reduced BH and LAR artifacts; and quantitative observations can be made that the accurate estimation of physical quantities such as iodine concentrations and effective atomic numbers can be obtained for noiseless data of LAR as low as 20° and for noisy data of LAR as low as 60°. For the SA and TOA scans of the same total angular range studied, the latter appear to yield more accurate images and estimations of physical quantities than the former, due to the improved conditioning of the system matrix.

We used two distinct phantoms, i.e., chest and suitcase phantoms, of varying complexity levels and structures of different application interest. The chest phantom contains lung tissue, airways, and blood vessels within the pulmonary anatomy, while the suitcase phantom contains various materials of interest in baggage screening. Results of the numerical study indicate that the effectiveness of the two-step method, like any other algorithm, is understandably dependent on the anatomic complexity of an object imaged with varying contrast and spatial resolution. Results from the suitcase phantom are less impacted, in terms of image artifacts and quantitative accuracy of the estimated physical quantities, by the decreasing LAR than the chest phantom, possibly due to its structure and the noise levels in the data. We have studied additional phantoms of different anatomies, and corroborative observations can be made.

In the work, we have investigated the DTV algorithm for numerically accurately solving the optimization program in Equation (4) with DTV constraints. Additionally, we have conducted noisy data studies to provide some preliminary insights into the stability of the two-step method in the presence of data inconsistencies. While a fixed total number of quanta is used for Poisson noise simulation, the visualization of VMIs and estimation accuracy of physical quantities obtained can be dependent on the noise levels and characteristics of different applications. In addition, other sources of inconsistency, such as metal, scatter, imperfect spectra, low- and high-kVp X-ray mismatch, and decomposition error, may also impact the reconstruction quality and estimation accuracy. Blooming artifacts usually stem from highly attenuating materials present in the patient, such as metal implants and calcification plaques. While it is important to investigate the effectiveness of the two-step method in studies containing these physical effects, such an investigation nevertheless is beyond the scope of this work, and the proposed method may be used as the basis for future investigative efforts that focus on correcting other physical factors in DECT with LAR data.

The studies and results in this work may provide insights into the possible development for practical approaches to reducing radiation dose and scanning time and to avoiding collision between the moving gantry of the scanner and the imaged object in clinical and industrial applications. One limitation of the proposed two-step method is the requirement of completely overlapping arcs of low- and high-kVp scans, imposed by the data-domain decomposition step. This can be avoided by performing the image-domain decomposition in a two-step method [30]; however, linear data models are usually assumed and the non-linear BH effect is not explicitly corrected for, which may impact the quantitative accuracy of the reconstruction. On the other hand, one-step methods [25] may accommodate LAR scanning configurations with partially or non-overlapping arcs of low- and high-kVp scans, while using the non-linear data model and correcting for the BH effect. Therefore, future investigations will include studies on one-step methods for DECT reconstruction with LAR data. It is worthy of a separate, comprehensive investigation, since existing studies on one-step methods focus largely on full- or short-angular-range scans and leverage image constraints, such as TV, not specifically designed for LAR data [15,25].

## 5. Conclusions

In this work, we investigated and developed a two-step method to reconstruct images accurately from low- and high-kVp LAR data by correcting for both BH and LAR effects in DECT. Numerical studies conducted reveal that the two-step method can yield VMIs with reduced BH and LAR artifacts, and estimation of physical quantities with improved accuracy, and that for SA and TOA scans with identical total LARs, the latter generally yields more accurate image reconstruction and physical-quantity estimation than the former. Results and knowledge acquired in the work on accurate image reconstruction in LAR DECT may give rise to further understanding and insights into the practical design of LAR scan configurations and reconstruction procedures for DECT applications. Future works will investigate the impact of additional inconsistencies and the one-step method for accommodating non-overlapping scans in DECT with LAR data.

**Author Contributions:** Conceptualization, B.C. and X.P.; methodology, B.C. and Z.Z.; software, B.C. and Z.Z.; validation, B.C., Z.Z. and D.X.; formal analysis, B.C.; investigation, B.C.; resources, X.P.; data curation, B.C. and T.G.-S.; writing—original draft preparation, B.C.; writing—review and editing, B.C., Z.Z., D.X., E.Y.S., T.G.-S. and X.P.; visualization, B.C.; supervision, X.P.; project administration, B.C.; funding acquisition, T.G.-S., E.Y.S. and X.P. All authors have read and agreed to the published version of the manuscript.

**Funding:** This work was funded in part by NIH R01 Grant Nos. EB026282 and EB023968, and R21 Grant No. CA263660-01A1. The contents of this article are solely the responsibility of the authors and do not necessarily represent the official views of the National Institutes of Health.

**Institutional Review Board Statement:** Not applicable.

**Informed Consent Statement:** Not applicable.

**Data Availability Statement:** The data presented in this study are available on request from the corresponding author.

**Conflicts of Interest:** The authors declare no conflict of interest. The funders had no role in the design of the study; in the collection, analyses, or interpretation of data; in the writing of the manuscript; or in the decision to publish the results.

## Abbreviations

The following abbreviations are used in this manuscript:

| | |
|---|---|
| DECT | Dual-energy computed tomography |
| LAR | Limited angular range |
| BH | Beam hardening |
| DDD | Data-domain decomposition |
| DTV | Directional total variation |
| VMI | Virtual monochromatic image |
| SA | Single-arc |
| TOA | Two-orthogonal-arc |
| FAR | Full angular range |
| nRMSE | Normalized root-mean-square error |
| PCC | Pearson correlation coefficient |
| PE | Photoelectric |
| KN | Klein–Nishina |
| NEQ | Noise-equivalent quanta |

## Appendix A. Pseudo-Code of the DTV Algorithm

**Algorithm A1** Pseudo-code of the DTV algorithm for solving Equation (4)

INPUT: $\mathbf{L}_k, t_{kx}, t_{ky}, \mathcal{A}, \rho$

$L \leftarrow ||\mathcal{K}||_2, \tau \leftarrow \rho/L, \sigma \leftarrow 1/(\rho L), \nu_1 \leftarrow ||\mathcal{A}||_2/||\mathcal{D}_x||_2, \nu_2 \leftarrow ||\mathcal{A}||_2/||\mathcal{D}_y||_2, \mu \leftarrow ||\mathcal{A}||_2/||\mathcal{I}||_2$

$n \leftarrow 0$

INITIALIZE: $\mathbf{b}^{(0)}$, $\mathbf{w}^{(0)}$, $\mathbf{p}^{(0)}$, $\mathbf{q}^{(0)}$, and $\mathbf{t}^{(0)}$ to zero

$\bar{\mathbf{b}}^{(0)} \leftarrow \mathbf{b}^{(0)}$

**repeat**

$\mathbf{w}^{(n+1)} = (\mathbf{w}^{(n)} + \sigma(\mathcal{A}\bar{\mathbf{b}}^{(n)} - \mathbf{L}))/(1+\sigma)$

$\mathbf{p}'^{(n)} = \mathbf{p}^{(n)} + \sigma\nu_1\mathcal{D}_x\bar{\mathbf{b}}^{(n)}$

$\mathbf{q}'^{(n)} = \mathbf{q}^{(n)} + \sigma\nu_2\mathcal{D}_y\bar{\mathbf{b}}^{(n)}$

$\mathbf{p}^{(n+1)} = \mathbf{p}'^{(n)} - \sigma\frac{\mathbf{p}'^{(n)}}{|\mathbf{p}'^{(n)}|}\ell_1\text{ball}_{\nu_1 t_{kx}}(\frac{|\mathbf{p}'^{(n)}|}{\sigma})$

$\mathbf{q}^{(n+1)} = \mathbf{q}'^{(n)} - \sigma\frac{\mathbf{q}'^{(n)}}{|\mathbf{q}'^{(n)}|}\ell_1\text{ball}_{\nu_2 t_{ky}}(\frac{|\mathbf{q}'^{(n)}|}{\sigma})$

$\mathbf{t}^{(n+1)} = \text{neg}(\mathbf{t}^{(n)} + \sigma\mu\bar{\mathbf{b}}^{(n)})$

$\mathbf{b}^{(n+1)} = \mathbf{b}^{(n)} - \tau(\mathcal{A}^\top\mathbf{w}^{(n+1)} + \nu_1\mathcal{D}_x^\top\mathbf{p}^{(n+1)} + \nu_2\mathcal{D}_y^\top\mathbf{q}^{(n+1)} + \mu\mathbf{t}^{(n+1)})$

$\bar{\mathbf{b}}^{(n+1)} = 2\mathbf{b}^{(n+1)} - \mathbf{b}^{(n)}$

$n \leftarrow n+1$

**until** the convergence conditions are satisfied

OUTPUT: $\mathbf{b}^{(n)}$ as the estimate of $\mathbf{b}_k$

In the pseudo-code of the derived algorithm instance, the definitions of the auxiliary variables, including matrices $\mathcal{K}$ and $\mathcal{I}$ and vectors $\mathbf{w}^{(n)}$, $\mathbf{p}'^{(n)}$, $\mathbf{q}'^{(n)}$, $\mathbf{p}^{(n)}$, $\mathbf{q}^{(n)}$, and $\mathbf{t}^{(n)}$, and operators, including $||\cdot||_2$, $\text{neg}(\cdot)$, $\ell_1\text{ball}_\beta(\cdot)$, and $|\mathbf{q}'^{(n)}|$, are intentionally kept consistent with those used in Ref. [6]. In each reconstruction from a set of basis data in an SA and TOA scan, the DTV algorithm reconstructs the basis images through solving Equation (4) until the convergence conditions described in Ref. [5] are satisfied numerically.

## References


1. Alvarez, R.E.; Macovski, A. Energy-selective reconstructions in X-ray computerised tomography. *Phys. Med. Biol.* **1976**, *21*, 733–744. [CrossRef] [PubMed]
2. Carmi, R.; Naveh, G.; Altman, A. Material separation with dual-layer CT. In Proceedings of the IEEE Nuclear Science Symposium Conference Record, 2005, Fajardo, PR, USA, 23–29 October 2005; Volume 4, p. 3.
3. Flohr, T.G.; McCollough, C.H.; Bruder, H.; Petersilka, M.; Gruber, K.; Süβ, C.; Grasruck, M.; Stierstorfer, K.; Krauss, B.; Raupach, R.; et al. First performance evaluation of a dual-source CT (DSCT) system. *Eur. Radiol.* **2006**, *16*, 256–268. [CrossRef]
4. Xu, D.; Langan, D.A.; Wu, X.; Pack, J.D.; Benson, T.M.; Tkaczky, J.E.; Schmitz, A.M. Dual energy CT via fast kVp switching spectrum estimation. In Proceedings of the SPIE Medical Imaging 2009: Physics of Medical Imaging, Lake Buena Vista, FL, USA, 7–12 February 2009; Volume 7258, p. 72583T.
5. Zhang, Z.; Chen, B.; Xia, D.; Sidky, E.Y.; Pan, X. Directional-TV algorithm for image reconstruction from limited-angular-range data. *Med. Image Anal.* **2021**, *70*, 102030. [CrossRef] [PubMed]
6. Chen, B.; Zhang, Z.; Xia, D.; Sidky, E.Y.; Pan, X. Dual-energy CT imaging with limited-angular-range data. *Phys. Med. Biol.* **2021**, *66*, 185020. [CrossRef]
7. Chen, B.; Zhang, Z.; Xia, D.; Sidky, E.Y.; Pan, X. Dual-energy CT imaging over non-overlapping, orthogonal arcs of limited-angular ranges. *J. X-ray Sci. Technol.* **2021**, *29*, 975–985. [CrossRef] [PubMed]
8. Chen, B.; Zhang, Z.; Sidky, E.Y.; Xia, D.; Pan, X. Image reconstruction and scan configurations enabled by optimization-based algorithms in multispectral CT. *Phys. Med. Biol.* **2017**, *62*, 8763. [CrossRef]
9. Sheng, W.; Zhao, X.; Li, M. A sequential regularization based image reconstruction method for limited-angle spectral CT. *Phys. Med. Biol.* **2020**, *65*, 235038. [CrossRef]
10. Zou, Y.; Silver, M.D. Analysis of fast kV-switching in dual energy CT using a pre-reconstruction decomposition technique. In Proceedings of the SPIE Medical Imaging 2008: Physics of Medical Imaging, San Diego, CA, USA, 17–19 February 2008; Volume 6913, p. 691313.
11. Ying, Z.; Naidu, R.; Crawford, C.R. Dual energy computed tomography for explosive detection. *J. X-ray Sci. Technol.* **2006**, *14*, 235–256.
12. Goodsitt, M.M.; Christodoulou, E.G.; Larson, S.C. Accuracies of the synthesized monochromatic CT numbers and effective atomic numbers obtained with a rapid kVp switching dual energy CT scanner. *Med. Phys.* **2011**, *38*, 2222–2232. [CrossRef]
13. Chandarana, H.; Megibow, A.J.; Cohen, B.A.; Srinivasan, R.; Kim, D.; Leidecker, C.; Macari, M. Iodine quantification with dual-energy CT: Phantom study and preliminary experience with renal masses. *Am. J. Roentgenol.* **2011**, *196*, W693–W700. [CrossRef]
14. Faby, S.; Kuchenbecker, S.; Sawall, S.; Simons, D.; Schlemmer, H.P.; Lell, M.; Kachelrieß, M. Performance of today's dual energy CT and future multi energy CT in virtual non-contrast imaging and in iodine quantification: A simulation study. *Med. Phys.* **2015**, *42*, 4349–4366. [CrossRef] [PubMed]
15. Barber, R.F.; Sidky, E.Y.; Schmidt, T.G.; Pan, X. An algorithm for constrained one-step inversion of spectral CT data. *Phys. Med. Biol.* **2016**, *61*, 3784–3818. [CrossRef] [PubMed]
16. Iwano, S.; Ito, R.; Umakoshi, H.; Ito, S.; Naganawa, S. Evaluation of lung cancer by enhanced dual-energy CT: Association between three-dimensional iodine concentration and tumour differentiation. *Br. J. Radiol.* **2015**, *88*, 20150224. [CrossRef]
17. Koonce, J.D.; Vliegenthart, R.; Schoepf, U.J.; Schmidt, B.; Wahlquist, A.E.; Nietert, P.J.; Bastarrika, G.; Flohr, T.G.; Meinel, F.G. Accuracy of dual-energy computed tomography for the measurement of iodine concentration using cardiac CT protocols: Validation in a phantom model. *Eur. Radiol.* **2014**, *24*, 512–518. [CrossRef]
18. Pelgrim, G.J.; van Hamersvelt, R.W.; Willemink, M.J.; Schmidt, B.T.; Flohr, T.; Schilham, A.; Milles, J.; Oudkerk, M.; Leiner, T.; Vliegenthart, R. Accuracy of iodine quantification using dual energy CT in latest generation dual source and dual layer CT. *Eur. Radiol.* **2017**, *27*, 3904–3912. [CrossRef]
19. Mouton, A.; Breckon, T.P. A review of automated image understanding within 3D baggage computed tomography security screening. *J. X-ray Sci. Technol.* **2015**, *23*, 531–555. [CrossRef] [PubMed]
20. McCollough, C.H.; Leng, S.; Yu, L.; Fletcher, J.G. Dual- and multi-energy CT: Principles, technical approaches, and clinical applications. *Radiology* **2015**, *276*, 637–653. [CrossRef]
21. Taguchi, K.; Iwanczyk, J.S. Vision 20/20: Single photon counting x-ray detectors in medical imaging. *Med. Phys.* **2013**, *40*, 100901. [CrossRef]
22. Danielsson, M.; Persson, M.; Sjölin, M. Photon-counting x-ray detectors for CT. *Phys. Med. Biol.* **2021**, *66*, 03TR01. [CrossRef] [PubMed]

23. Zhang, Z.; Chen, B.; Xia, D.; Sidky, E.Y.; Pan, X. Image reconstruction from data over two orthogonal arcs of limited-angular ranges. *Med. Phys.* **2022**, *49*, 1468–1480. [CrossRef]
24. Hubbell, J.; Seltzer, S. Tables of X-ray Mass Attenuation Coefficients and Mass Energy-Absorption Coefficients (Version 1.4). Available online: http://physics.nist.gov/xaamdi (accessed on 12 March 2016).
25. Chen, B.; Zhang, Z.; Xia, D.; Sidky, E.Y.; Pan, X. Non-convex primal-dual algorithm for image reconstruction in spectral CT. *Comput. Med. Imaging Graph.* **2021**, *87*, 101821. [CrossRef] [PubMed]
26. Pearson, K. Notes on Regression and Inheritance in the Case of Two Parents. *Proc. R. Soc. Lond.* **1895**, *58*, 240–242.
27. Bian, J.; Siewerdsen, J.H.; Han, X.; Sidky, E.Y.; Prince, J.L.; Pelizzari, C.A.; Pan, X. Evaluation of sparse-view reconstruction from flat-panel-detector cone-beam CT. *Phys. Med. Biol.* **2010**, *55*, 6575–6599. [CrossRef] [PubMed]
28. Hernandez, A.M.; Boone, J.M. Tungsten anode spectral model using interpolating cubic splines: Unfiltered X-ray spectra from 20 kV to 640 kV. *Med. Phys.* **2014**, *41*, 042101. [CrossRef] [PubMed]
29. Delesalle, M.A.; Pontana, F.; Duhamel, A.; Faivre, J.B.; Flohr, T.; Tacelli, N.; Remy, J.; Remy-Jardin, M. Spectral optimization of chest CT angiography with reduced iodine load: Experience in 80 patients evaluated with dual-source, dual-energy CT. *Radiology* **2013**, *267*, 256–266. [CrossRef]
30. Maass, C.; Baer, M.; Kachelriess, M. Image-based dual energy CT using optimized precorrection functions: A practical new approach of material decomposition in image domain. *Med. Phys.* **2009**, *36*, 3818–3829. [CrossRef]

bioengineering

Article

# A High-Accuracy Detection System: Based on Transfer Learning for Apical Lesions on Periapical Radiograph

Yueh Chuo [1], Wen-Ming Lin [1,†], Tsung-Yi Chen [2,†], Mei-Ling Chan [1,3,*,†], Yu-Sung Chang [2], Yan-Ru Lin [2], Yuan-Jin Lin [4], Yu-Han Shao [2], Chiung-An Chen [5,*], Shih-Lun Chen [2,*] and Patricia Angela R. Abu [6]

1 Department of General Dentistry, Chang Gung Memorial Hospital, Taoyuan City 33305, Taiwan
2 Department of Electronic Engineering, Chung Yuan Christian University, Taoyuan City 32023, Taiwan
3 School of Physical Educational College, Jiaying University, Meizhou City 514000, China
4 Department of Electrical Engineering and Computer Science, Chung Yuan Christian University, Chungli City 32023, Taiwan
5 Department of Electrical Engineering, Ming Chi University of Technology, New Taipei City 243303, Taiwan
6 Department of Information Systems and Computer Science, Ateneo de Manila University, Quezon City 1108, Philippines
* Correspondence: lynn321630@gmail.com (M.-L.C.); joannechen@mail.mcut.edu.tw (C.-A.C.); chrischen@cycu.edu.tw (S.-L.C.)
† These authors contributed equally to this work.

**Abstract:** Apical Lesions, one of the most common oral diseases, can be effectively detected in daily dental examinations by a periapical radiograph (PA). In the current popular endodontic treatment, most dentists spend a lot of time manually marking the lesion area. In order to reduce the burden on dentists, this paper proposes a convolutional neural network (CNN)-based regional analysis model for spical lesions for periapical radiographs. In this study, the database was provided by dentists with more than three years of practical experience, meeting the criteria for clinical practical application. The contributions of this work are (1) an advanced adaptive threshold preprocessing technique for image segmentation, which can achieve an accuracy rate of more than 96%; (2) a better and more intuitive apical lesions symptom enhancement technique; and (3) a model for apical lesions detection with an accuracy as high as 96.21%. Compared with existing state-of-the-art technology, the proposed model has improved the accuracy by more than 5%. The proposed model has successfully improved the automatic diagnosis of apical lesions. With the help of automation, dentists can focus more on technical and medical diagnoses, such as treatment, tooth cleaning, or medical communication. This proposal has been certified by the Institutional Review Board (IRB) with the certification number 202002030B0.

**Keywords:** PA; CNN; tooth disease recognition; image segmentation; image preprocessing

**Citation:** Chuo, Y.; Lin, W.-M.; Chen, T.-Y.; Chan, M.-L.; Chang, Y.-S.; Lin, Y.-R.; Lin, Y.-J.; Shao, Y.-H.; Chen, C.-A.; Chen, S.-L.; et al. A High-Accuracy Detection System: Based on Transfer Learning for Apical Lesions on Periapical Radiograph. *Bioengineering* **2022**, *9*, 777. https://doi.org/10.3390/bioengineering9120777

Academic Editors: Paolo Zaffino and Maria Francesca Spadea

Received: 3 October 2022
Accepted: 23 November 2022
Published: 6 December 2022



## 1. Introduction

X-rays have been used in medical images since 1896, and they also help doctors determine whether a patient is healthy. X-rays are widely used in dental treatments [1], such as periapical radiographs (PA), bitewing radiographs (BW), and panoramic radiographs (PANO). The PA film is important in routine dental X-ray examinations because it requires lower radiation dose exposure and could identify periapical pathology efficiently. Periapical radiographs are commonly the result of trauma, caries, or tooth wear. These conditions will bring out root canal infection if the dental treatment has no intervention, and pulp necrosis may occur [2,3]. Numerous studies have confirmed that this oral problem requires prompt and thorough treatment. Otherwise, it may lead to tooth loss and repeated inflammation [4–6]. Detection of the peri-apical lesion and opportune endodontic procedure intervention can treat root canal infection caused by these problems. The PA film can capture local teeth, which can effectively and quickly enable dental professionals to find local details of tooth lesions [7,8] and undertake the treatment [9]. Despite advances in treatment and access to extensive care, the prevalence of apical lesions remains high [10]. Dentists use PA to find diseased areas, but it is very time-consuming to find lesions that do not contrast well with normal areas. In addition, the dentist may become tired after long working hours, thereby ignoring subtle differences. In a day, a dentist must review hundreds of X-rays of a patient and diagnose and document the patient. However, the current method for detecting periapical noise still needs to be judged by the dentist. Dentists may inevitably make mistakes as the length and number of consultations increase.

With the continuous advancement of technology, artificial intelligence has shined in many industries, such as vehicle image recognition [11], smart city [12], product recommendation systems [13], and emotion recognition [14]. The development of medical imaging is also changing with each passing day. More and more cases have shown that the combination of artificial intelligence and medical imaging has good results, such as breast cancer (BC) [15], arrhythmia [16], and lung function prediction [17]. There are also related studies in dentistry, for example, adding machine learning to the color recognition of dentures [18] or adding deep learning to the detection of tooth decay [19]. Artificial intelligence was used to improve the apical lesion detection accuracy rate of the dental Panoramic Radiograph Identification System to about 75.53% in [20], while the accuracy rate of CT image identification is about 82% [21]. For clinical use, there is still a lot of room for improvement. Therefore, in order to reduce the workload of dentists and provide more objective data, this study proposes an automatic detection system for periapical radiographs using CNN transfer learning. The purpose is to improve the symptom enhancement technology in the existing technology and to analyze and compare the final research results. In [22], a system combining PA and CNN was proposed. The authors improved the classification accuracy of batch normalization by adjusting the parameters such as the layers of convolutional blocks in the CNN. The accuracy rate obtained was as high as 92.75%. This article will use this as the main reference. However, its image enhancement improvements are not outstanding. Therefore, in this proposal, the focus is on improving image segmentation and image enhancement. In order to increase the reference value of the data obtained from the Alexnet classification model, three different CNN models are constructed, namely Googlenet, Resnet50, and Resnet101 which are trained, tested, and verified through the same database. During this period, the parameters of each model were kept consistent by the control variable method which was convenient for subsequent comparison. The research in this paper also utilizes CNN technology for dentists to diagnose symptoms and ultimately provide patients with more effective and better adjuvant treatment. The innovations of this method are as follows:

1. In the image cropping preprocessing part, this study adds the adaptive threshold and angle rotation technology. Compared with the existing methods, this method significantly improves the image clarity and accuracy of a single tooth image.
2. This study proposes an advanced image enhancement technique for apical lesions. It adds raw grayscale images and Gaussian high-pass filtered images to highlight the possible lesion areas and changes the color of the possible lesion area to green. Experiments show that the accuracy of the model is improved by more than 10% which proves that the proposed method is intuitive and effective.
3. The innovation of this work is to realize the classification of various diseases. It can simultaneously judge a variety of different types of dental diseases (such as apical lesions, fillings, etc.), and the obtained final accuracy of the model proposed in this paper is as high as 93%. AlexNet even improves the accuracy up to 96.21% which is 4% higher than the state-of-the-art in [23].

The presentation structure of this proposal is as follows: Section 2 introduces the materials and methods for apical lesion detection on periapical radiographs based on transfer learning. Section 3 mainly describes and analyzes the evaluation method of the model and the experimental results. The results of the study are discussed in Section 4. Finally, the conclusions and future prospects are given in Section 5. The purpose of this paper is to predict root apical lesions located at the base of a tooth by means of a convolutional neural network (CNN).

## 2. Materials and Methods

This study is divided into three parts: image cropping, image preprocessing and CNN training. The image cropping part extracts a single tooth which helps model training more efficiently. In addition, through a series of image preprocessing techniques, possible lesion areas can be highlighted, resulting in more accurate detection. The output of these image preprocessing steps is saved in the CNN database. The clinical images used in this research were collected by attending physicians with more than three years of experience in hospital dentistry. All clinical images utilized in this research had been approved by the Institutional Review Board Statement (permission number 202002030B0). For enhancement, the most challenging problem is that after image segmentation, there is too much noise in the original apical slice and the resected part of the lesion area. Therefore, the angle of cutting or image noise reduction became the challenge of this project. This proposal uses the same Gaussian high-pass filter as that of [23] to achieve the best noise reduction result. The flow chart of this study is shown in Figure 1.

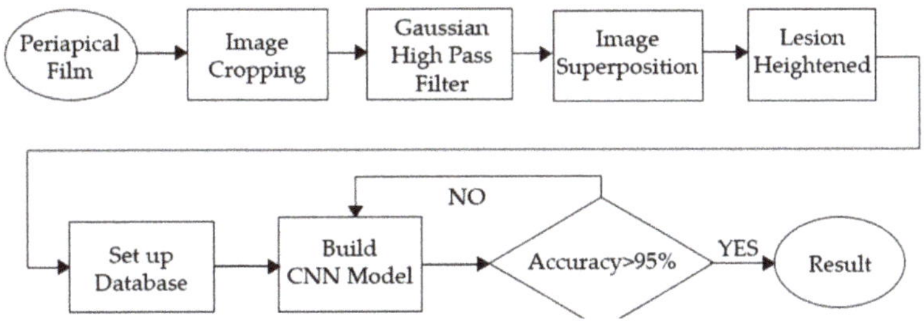


**Figure 1.** The flow chart of this research.

### 2.1. *Image Segmentation and Retouching*

In order to build a high-precision model and conform to the judgment of dentists, this research uses a single tooth image to build a clinical image database. However, since the original image is a PA composed of about three to four separate vertical teeth, segmentation of individual teeth must be performed on the original image. In [24], the vertical cutting method of dividing the image is a very good idea. However, its target image is a BW film; there would be some flaws in the process of cutting the PA film. Therefore, the segmentation method that this proposal focuses on is improved on the basis of [24] to make the segmentation more accurate. The next step is to retouch the segmented photo by adding a technique to block non-target areas on the segmented image.

#### 2.1.1. Vertical Cutting

The core concept of vertical cutting is to calculate the sum of the horizontal pixels of the image and find the point with the smallest sum making that point the segmentation point. Before performing the calculation, the image is first converted from an RGB image to a grayscale image for easier calculation. In this research, the method of iterative thresholding in [23] is improved and adaptive thresholding is used for transformation. The feature of adaptive threshold processing is that each image behaves differently from RGB to grayscale which means that each image has its own most suitable adaptive threshold. In this proposal, the most appropriate adaptive threshold is selected for each image and image transformation is performed. Figure 2 is the image result applying the conversion method using [23], and Figure 2b is the image result of the improved adaptive threshold. From a data point of view, this method can effectively improve the accuracy of image segmentation.

(a) (b)

**Figure 2.** The result of the images. (**a**) image binarization, (**b**) adaptive threshold processing.

This proposal summarizes the range of the sum of pixels where the appropriate threshold is located by calculating the sum of the pixels of each grayscale image. However, the calculation cannot be performed immediately after converting the image because not every interdental gap is vertical. Therefore, rotating the image is a necessary step before calculation. Additionally, image rotation is performed by rotating the image 12 degrees clockwise and 12 degrees counterclockwise [24]. After rotating the image, the 24 positions and value of the minimum pixel sum for each angle can be computed. The 24 values are then compared with the smallest value being the most suitable position and angle for segmentation. Since the number of PA image teeth in the database is at most four teeth, a maximum of four cutting lines are required. This means that the above steps need to be repeated four times. However, the number of teeth in the PA images is not always four and some images have only three. In this case, the redundant dividing line needs to be removed. Therefore, this study designed two methods to address this problem. The first method is based on the relative position of the fourth cutting line to the other cutting lines. Assuming that the distance between the fourth cutting line and the other cutting lines is less than the width of one tooth, the fourth cutting line should be removed. The second method is that if the value of the fourth row is greater than the average, it means that the fourth row is very likely not on the tooth. Thus, the fourth row can be removed by this feature. Figure 3a shows the result of finding all the dividing lines. Figure 3b is the result of removing the redundant dividing lines. When the cutting line is inclined, the image segmented by the cutting line is not rectangular. However, for the subsequent training of CNN when normalized, the target in this step is a rectangular one. Therefore, this proposal adds a green vertical line to the far right and a blue vertical line to the far left of each cutting line. Moreover, it segments the first segmented image with the leftmost position of the original image and the first green vertical line as the boundary. The second segmented image is bounded by the first blue vertical line and the second green vertical line. The third and fourth cut images are similarly divided.

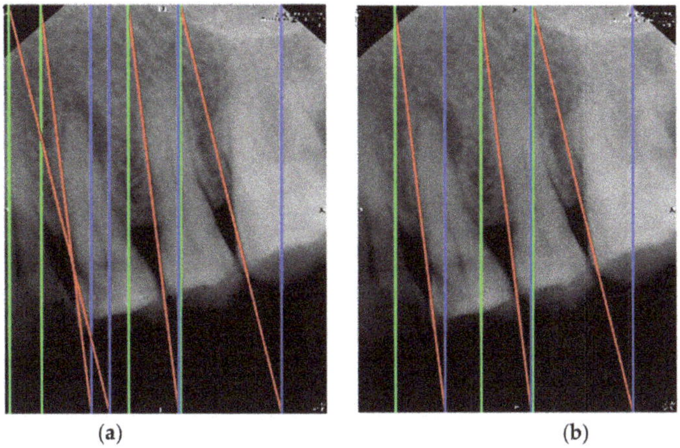

**Figure 3.** The result of cutting lines. (**a**) all cutting lines, (**b**) removed unnecessary cutting lines.

#### 2.1.2. Image Masks

The inclination of the cutting line is designed to match the inclination of the teeth. However, most of the cutting lines have oblique angles which means that most segmented images will contain a small fraction of adjacent teeth as shown in Figure 4a. This leads to disturbances in the accuracy of the CNN model. In view of this, this study retouches the segmented images according to the cutting line (red line in Figure 3). It sets the mask template according to the clipping line and the mask template will be superimposed with the original clipping image. This can effectively mask the non-target area as shown in Figure 4b. The modified image will be the final output and result. The retouched image will optimally preserve the desired feature areas.

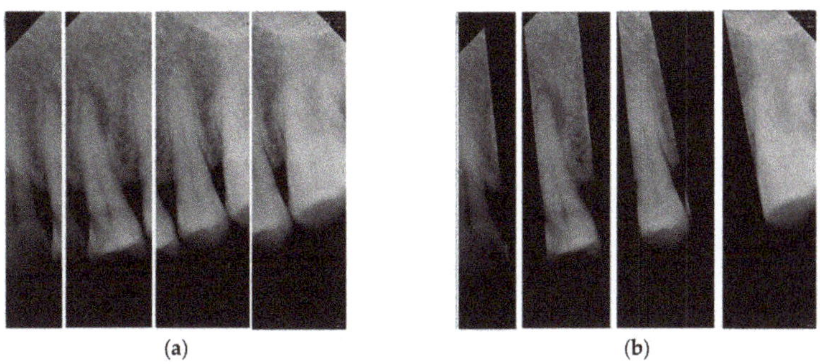

**Figure 4.** The results of the masking image. (**a**) original segmented, (**b**) retouched segmented.

### *2.2. Enhancing Lesion*

In the collected original images, the root apical lesions will be affected by factors such as shooting angle, dose, and operator, thus, the lesions are sometimes inconspicuous. In this regard, this research proposes an advanced and intuitive enhancement method that can highlight the lesions. The first work is converting the RGB image to a grayscale image using a conversion formula. It then uses the Gaussian high-pass filter to filter out the noise. In order to make the lesion more obvious, the result of the Gaussian high-pass filter is used to superimpose it back to a grayscale image. Finally, the simple enhancement technique is used to change the color of the possible lesion area to green. In this way, clinical images for efficient training of CNN models can be obtained.

### 2.2.1. Grayscale Image

The original apical section is an RGB image. However, this is not very friendly to the subsequent image processing step. To make image processing easier and accelerate the subsequent CNN training, the first step in lesion enhancement is to convert the image from an RGB three-channel image to a grayscale single-channel image. This step generates all the points needed for the subsequent steps described by the $x$- and $y$-axes of the grayscale image and at the same time achieves a more efficient process.

### 2.2.2. Gaussian High Pass Filter

The biggest challenge in judging symptoms is the noisy points in the image. Therefore, attribute filters that reduce these noisy points are crucial. In the existing technology, there are many different filters. How to choose the most suitable filter is the key. Gaussian filters are used in two different ways. The Gaussian low-pass filter on the other hand is used to reduce certain noise points while the Gaussian high-pass filter is used to enhance dark areas. For the purpose of preprocessing to make possible apical lesions as evident as possible, the Gaussian high-pass filter is clearly the most suitable filter. This filter is able to pass high-frequency pixels and block low-frequency terms. Marginal and possibly apical lesion areas belong to high frequencies in the frequency domain. Hence, these pixels will remain in the resulting image as shown in Figure 5. The Gaussian high-pass filter [25] can be represented by Equation (1).

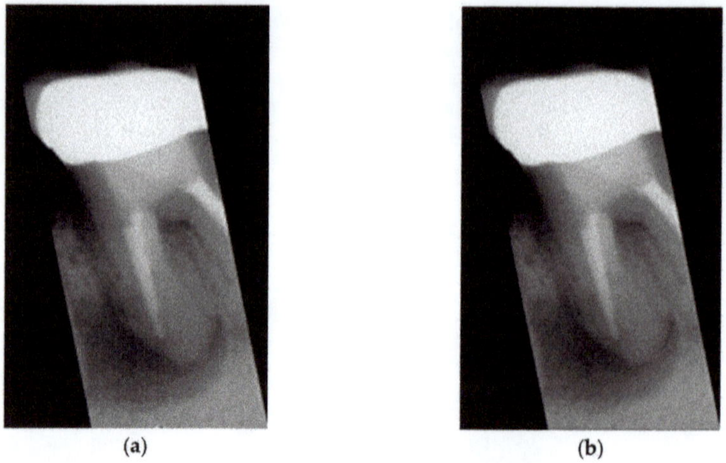

**Figure 5.** The results of the Gaussian high-pass filter. (**a**) original apical lesion image, (**b**) image after Gaussian filtering.

$$H(u,\ v) = 1 - e^{-D^2(u,v)/2D_0^2} \quad (1)$$

Although the image lesions after applying Gaussian high-pass filtering are obvious, the results are not as expected after model training. Therefore, other methods must be explored to enhance the lesions. In [23], in order to obtain better edges, the image preprocessing is performed by taking the grayscale image array minus the Gaussian high-pass filtered films array. The reason is that the filtered image will preserve the noisy areas and the result may not be significantly different from the input image. Therefore, by subtracting the filtered film from the input array, a clear tooth outline image can be obtained. Based on the above method, this study attempts to improve it by adding a film array to the grayscale image array after passing through a Gaussian high-pass filter to increase the likelihood of the lesion area. After applying a Gaussian high-pass filter, the filtered image retains the high-frequency pixels thus including the noise points, edges and possible lesion pixels.

After superimposing this filtered image, the contrast between the light and dark areas of the grayscale original image and the filtered image becomes evident as shown in Figure 6.

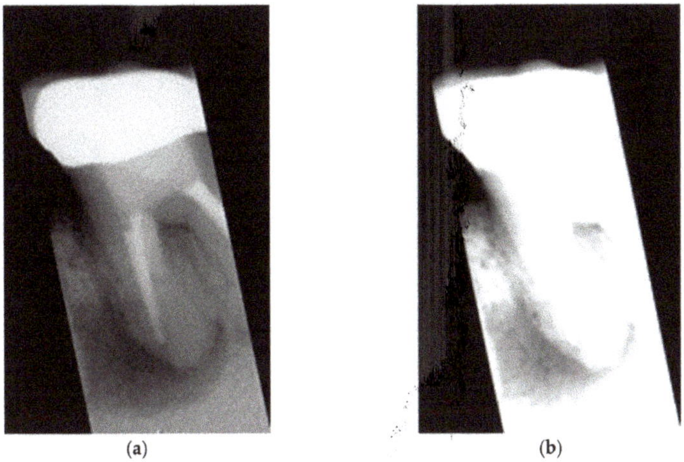

**Figure 6.** The image after adding the result of the Gaussian high-pass filter. (**a**) before, (**b**) after.

2.2.3. Lesion Heightened

The last step of feature enhancement is to adjust the color of the suspected lesion area where the color of the root tip is darkened. The biggest challenge for this step is how to lock the dark area of the suspected lesion. The filtered film makes it easier to find dark areas than the original image because bright areas have much larger pixel values than dark areas. However, directly using the threshold to adjust the whole block will cause other non-lesion areas in the image to be adjusted at the same time. Hence, this proposal calculates the pixel average and selects the threshold range. It can be simply reserved for possible lesions and excluded for bright areas as shown in Figure 7a. Moreover, this research used the method of calculating the pixel value difference between each pixel and its upper and lower pixels to ensure that possible lesion areas were determined. If the pixel value difference between the pixel and the pixel above or below is higher than the standard pixel value, it means that the pixel may be a point in the lesion area or just a noise point. At this time, a standard pixel value is selected. Figure 7b shows the experimental results. After using the above method of calculating the mean and calculating the difference in pixels, an AND operator is performed on the processed film to keep the points where the result is true after the AND operator. MATLAB programming provides a function to find several large regions in an image which this project uses to get the three regions that are larger than the others. Figure 7c presents the superimposed image. Finally, comparing these regions with their position in the film and their distance from the center pixel of the film, the regions that meet the above conditions will be changed into a green color and overlap with the original image. The final result is shown in Figure 7d.

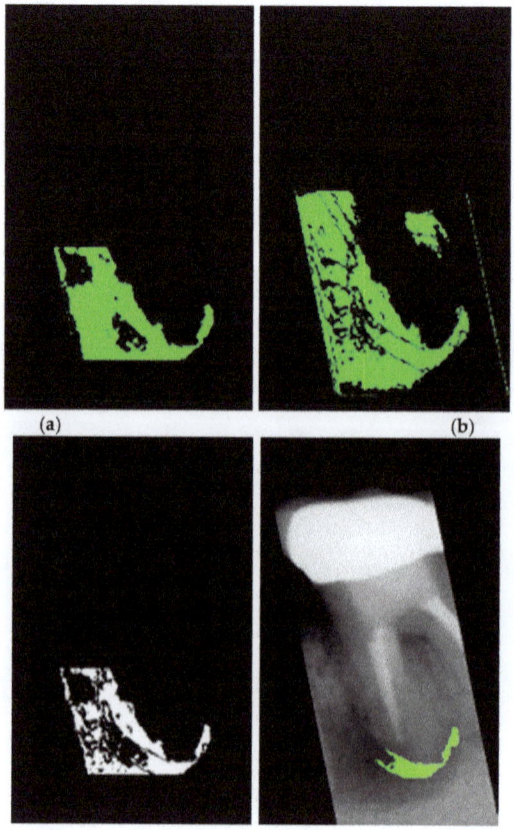


**Figure 7.** The image in different computing methods. (**a**) computed average image, (**b**) calculated pixel value difference for every pixel to the upper pixels and to the lower pixels, (**c**) superimposed image after going through the previous methods, (**d**) the final preprocessed image.

### *2.3. Image Identification*

In order to obtain a more scientific and reliable experimental model, the project first divides the original tooth images into a training set and a validation set according to the ratio of 4:1, as shown in Table 1. According to the transfer learning theory, the separated tooth images are cut and preprocessed according to symptoms and are then classified into the database. After that, the number of diseased teeth in the training set was expanded using horizontal and vertical mirror flips to increase the number of datasets and make it consistent with the number of normal teeth in the training set as listed in Table 2. The expanded dataset is only used to train various classification network models of CNN. It would not be used in the validation set data.

**Table 1.** Data Classification of the periapical image after preprocessing.

| The Number of Periapical Images after Classification | | | |
|---|---|---|---|
| | **Training Set** | **Validation Set** | **Total** |
| Normal | 332 | 83 | 415 |
| Lesion | 330 (Expanded) | 15 | 345 |

**Table 2.** Data distribution of the original periapical image from clinical.

| | The Number of Original Periapical Images | | |
|---|---|---|---|
| | Normal | Lesion | Total |
| Quantity | 415 | 75 | 490 |

2.3.1. CNN Model

In terms of deep learning, this proposal uses the tools in Matlab that support transfer learning for software development. The software environments and the hardware environments used in the proposal are listed in Table 3. To speed up the training efficiency of the CNN model, this study uses AMD R7-5800H CPU, Nvidia GeForce RTX 3070 GPU and DDR4 3200 16GB DRAM in terms of hardware performance. The architecture of each layer of the model takes AlexNet as the example of this research, as shown in Table 4. In the input stage of the model, the real training set and test set are put into the ratio of 4:1. The CNN model is trained through the classified dataset. Then, the purpose of adding a test set is to check whether the training effect deviates from the subsequent validation accuracy, thus making the experimental results more rigorous.

**Table 3.** The hardware and software platform.

| Hardware Platform | Version |
|---|---|
| CPU | AMD R7-5800H |
| GPU | GeForce RTX 3070 |
| DRAM | DDR4 3200 16GB |
| Software platform | Version |
| MATLAB | R2021a |
| Deep Network designer | 14.2 |

**Table 4.** The input and output of AlexNet model.

| | Type | Activations |
|---|---|---|
| 1 | Image Input | 227 × 227 × 3 |
| 2 | Convolution | 55 × 55 × 96 |
| 3 | ReLU | 55 × 55 × 96 |
| 4 | Cross Channel Normalization | 50 × 55 × 96 |
| 5 | Max pooling | 27 × 27 × 96 |
| 6 | Grouped Convolution | 27 × 27 × 256 |
| 7 | ReLU | 27 × 27 × 256 |
| 8 | Cross Channel Normalization | 27 × 27 × 256 |
| 9 | Max pooling | 13 × 13 × 256 |
| 10 | Convolution | 13 × 13 × 384 |
| 11 | ReLU | 13 × 13 × 384 |
| 12 | Grouped Convolution | 13 × 13 × 384 |
| 13 | ReLU | 13 × 13 × 384 |
| 14 | Grouped Convolution | 13 × 13 × 256 |
| 15 | ReLU | 13 × 13 × 256 |

**Table 4.** *Cont.*

| | Type | Activations |
|---|---|---|
| 16 | Max pooling | $6 \times 6 \times 256$ |
| 17 | Fully-Connected | $1 \times 1 \times 4096$ |
| 18 | ReLU | $1 \times 1 \times 4096$ |
| 19 | Dropout | $1 \times 1 \times 4096$ |
| 20 | Fully-Connected | $1 \times 1 \times 4096$ |
| 21 | ReLU | $1 \times 1 \times 4096$ |
| 22 | Dropout | $1 \times 1 \times 4096$ |
| 23 | Fully-Connected | $1 \times 1 \times 2$ |
| 24 | Softmax | $1 \times 1 \times 2$ |
| 25 | Classification Output | $1 \times 1 \times 2$ |

After deep learning, images from the validation set are randomly imported into the model. The model classifies the images according to the feature results obtained from the previous training and creates a confusion matrix by calculation to get the classification results and the accuracy of the model.

### 2.3.2. Adjust Hyperparameter

In the training phase, the setting of hyperparameters determines the success of the model. Each parameter represents a different meaning such as the number of layers of the neural network, the loss function, the size of the convolution kernel and the learning rate. This study describes the three modified parameters, including Initial Learning Rate, Max Epoch and Mini Batch Size. In addition, the detailed information of each parameter is listed in Table 5.

**Table 5.** Hyperparameters in CNN model.

| Hyperparameters | Value |
|---|---|
| Initial Learning Rate | 0.0001 |
| Max Epoch | 50 |
| Mini Batch Size | 64 |
| Validation Frequency | 10 |
| Learning Drop Period | 3 |
| Learning Rate Drop Factor | 0.02 |

A. *Initial Learning Rate*

In machine deep learning, the learning rate is a tuning parameter in the optimization algorithm. This means that the model needs an appropriate parameter which is the learning rate to get the best point of convergence. If the model has difficulty converging, it is most likely caused by the use of a too large learning rate. On the contrary, the convergence rate is too slow, which makes the model easy to overfit. Therefore, it is very important to choose an appropriate learning rate. After multiple tests on tooth images, the ideal learning rate is 0.0001.

B. *Max Epoch*

When an integrated database has passed through the CNN and has returned once, the whole process is referred to as an Epoch. However, if the Epoch is too large, it needs to be broken up into smaller pieces. With the increase of Epoch, the number of weight updates in the neural networks is also increased. The curve changed from under-fitting to over-fitting in the process of training. In general, if a CNN model has an appropriate increase in Epoch, it will lead to a better accuracy and in turn will also add training time. After repeated testing, choosing 50 as the Epoch value in each CNN model was determined by the control variable method.

C. *Mini Batch Size*

Mini Batch Size is a subset of the training set. Usually, the weights are updated and the gradient which is from the loss function is evaluated. In general, it affects the convergence of the optimization algorithm and how much memory is used in the calculation. Within a reasonable range, when the Batch Size is larger, the descending direction is more accurate and the oscillation is smaller. However, if it exceeds this range, the Batch Size is too large and local optimization or memory overflow may occur. Mini Batch Size introduces larger randomness making it difficult to achieve convergence. In this research, adjusting the approximate Mini Batch Size value to 64 can produce an ideal training result.

## 3. Results

This chapter presents the performance results of the proposed CNN model algorithm and compares it with the methods proposed in [20,23]. The proposed method for advanced symptom enhancement is also analyzed. The comparison of the image processing effect of the dataset with the results of the three CNN networks is presented for further discussion of the results.

One significant goal of this research is to enable the system to be employed in therapeutic settings. Figure 8 depicts the most common clinical workflow nowadays. Manual identification by doctors and the establishment of cumbersome medical records is a time-consuming process. The purpose of the system in this proposal is to obtain objective data for physicians prior to diagnosis and therapy after the patient takes the PA film, as shown in Figure 9.

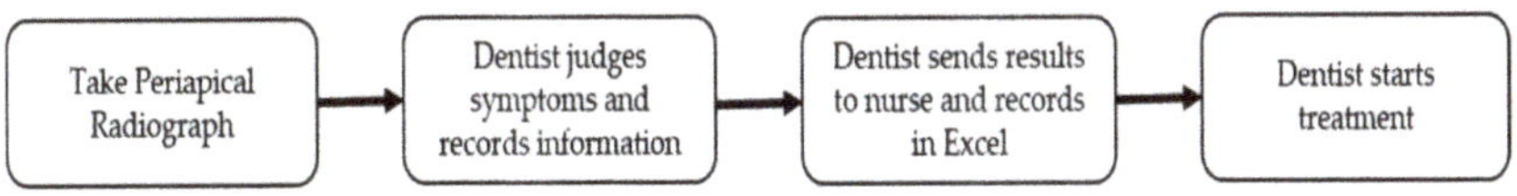


**Figure 8.** The original flow chart of clinical medicine.

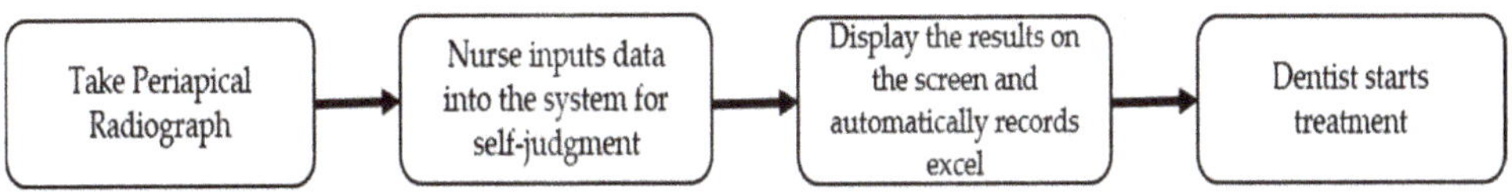


**Figure 9.** The flow chart of using this system.

In terms of model accuracy, this study uses the network input validation set for evaluation. The predictions obtained from the monitoring model are compared with the correct answers from the images to obtain the accuracy of the CNN. Table 6 presents the detailed training process of AlexNet and this is illustrated in Figures 10 and 11. The confusion matrix and truth table according to the network model are shown in Table 7.

**Table 6.** AlexNet training process.

| Epoch | Iteration | Time Elapsed | Mini-Batch Accuracy | Validation Accuracy | Mini-Batch Loss | Validation Loss |
|---|---|---|---|---|---|---|
| 1 | 1 | 00:00:02 | 48.44% | 53.03% | 1.4716 | 0.7940 |
| 5 | 40 | 00:00:15 | 70.31% | 81.82% | 0.5114 | 0.4379 |
| 10 | 80 | 00:00:27 | 90.62% | 85.61% | 0.2726 | 0.3277 |
| 15 | 120 | 00:00:39 | 90.62% | 88.64% | 0.2668 | 0.2648 |
| 20 | 160 | 00:00:42 | 89.06% | 90.91% | 0.2776 | 0.2422 |
| 25 | 200 | 00:01:03 | 87.50% | 91.67% | 0.3722 | 0.2230 |
| 30 | 240 | 00:01:16 | 90.62% | 93.94% | 0.1955 | 0.1787 |
| 35 | 280 | 00:01:28 | 95.31% | 95.31% | 0.1313 | 0.1883 |
| 40 | 320 | 00:01:41 | 90.62% | 95.45% | 0.2768 | 0.1585 |
| 45 | 360 | 00:01:53 | 96.88% | 95.45% | 0.0896 | 0.1424 |
| 50 | 400 | 00:02:05 | 93.72% | 96.21% | 0.1520 | 0.1201 |

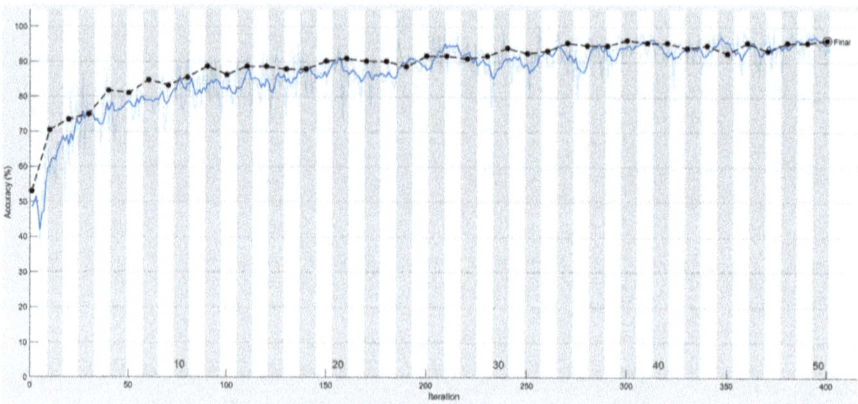

**Figure 10.** The accuracy of Alexnet model in test set which is black line and training set which is blue line during training process.

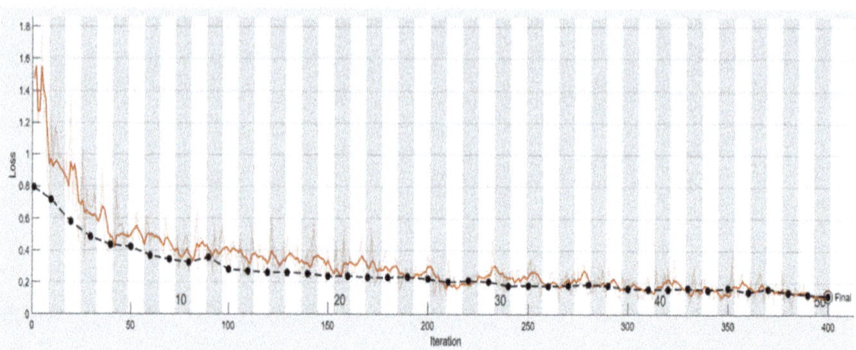

**Figure 11.** The accuracy of Alexnet model in test set (black line) and training set (orange line) during loss training process.

**Table 7.** The confusion matrix of AlexNet training result.

| | | Target Class | | |
|---|---|---|---|---|
| Category Name | | Lesion | Normal | Subtotal |
| Output Class | Lesion | 49.2% | 3.0% | 94.2% |
| | Normal | 0.8% | 47.0% | 98.4% |
| | subtotal | 98.5% | 93.9% | 96.2% |

Figure 12 shows the training process of this paper using the symptom enhancement technique at different stages. From the results, it can be seen that when the number of iterations increases, the three curves representing different preprocessing methods all show an upward trend in accuracy. The blue line is using the Gaussian high-pass filter and discoloration at the suspected lesion, the gray line is only discoloring the lesion without using the filter, and the orange line is the no enhancement technique. The experimental results show that although all three curves show an upward trend, the results of the enhanced two curves, the blue line and the gray line, are significantly higher than the unprocessed curves. This means that preprocessing has a significant impact on the verification accuracy. In addition, the model accuracy of the technique combining the Gaussian high-pass filter with discoloration at the lesion is about 1% and 5% higher than the other two methods. These results show that the method proposed in this paper can improve the final accuracy of the model.

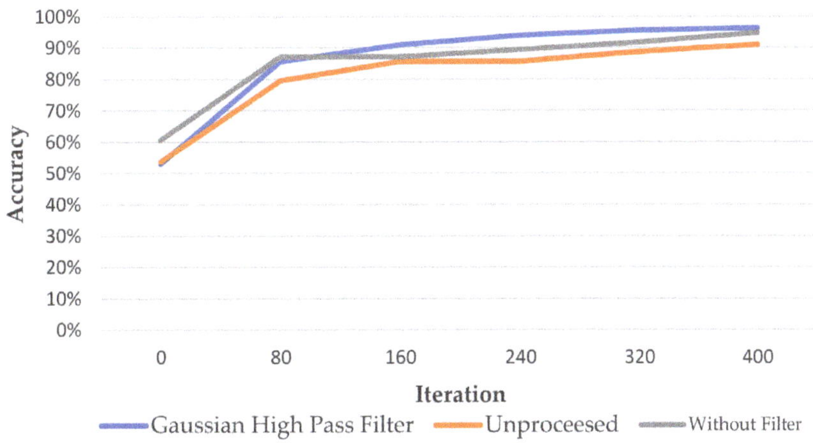


**Figure 12.** Comparison of the accuracy of AlexNet's training process for the unprocessed image, applied Gaussian high pass filter and without filter.

The technology proposed in this study is applied to clinical image judgment. Figure 13 shows the image used as the target image for clinical image judgment of symptoms. Figure 13 shows the two tooth X-rays in the red frame. The left side is the normal healthy tooth while the one on the right side is the apical diseased tooth. After implementing this technology, the classification accuracy results obtained according to the model are listed in Table 8. The accuracy of the image classification results after enhancement in this work is higher than that before disease enhancement. In clinical medicine, excellent medical quality requires high-precision judgment. The image recognition ability of CNN is exceptional. The results show that the recognition using the proposed method in this study are all above 90%.

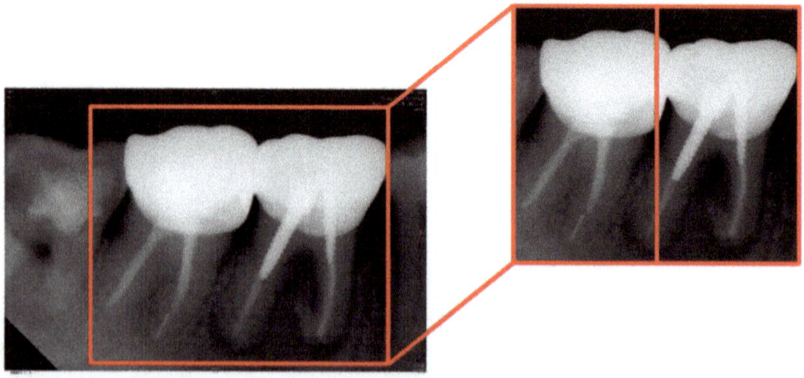

**Figure 13.** The example for validation with cropping image.

**Table 8.** Comparison of the clinical data and the resulting image.

| Tooth Position in Figure 13 | Left | Right |
|---|---|---|
| Clinical Data | Normal | Lesion |
| This Work Before Enhancement | 90.91% Normal | 94.70% Lesion |
| This Work After Enhancement | 93.93% Normal | 97.35% Lesion |

From the research results listed in Table 9, the diagnostic accuracy of AlexNet for apical lesions can reach 96.21% which is higher than the other three models in the literature. This presents a significant improvement of more than 3% compared with 92.91% in [23] which also uses the same AlexNet architecture. Furthermore, the results of the apical lesion detection technique proposed in this paper are in stark contrast to the 75.53% accuracy reported in the tooth identification study in [20]. The research results show that the method proposed in this work is very effective and successful for apical lesions. Furthermore, it can be shown that enhancing symptoms through image preprocessing improves classification accuracy.

**Table 9.** Image recognition accuracy obtained from a different CNN model.

| | Method in [20] | Method in [23] | This Work | | | |
|---|---|---|---|---|---|---|
| | | | AlexNet | ResNet101 | ResNet 50 | Google Net |
| Accuracy | 75.53% | 92.91% | 96.21% | 94.70% | 93.94% | 87.88% |

## 4. Discussion

In this proposal, the apical slices of multiple teeth are cut into pictures of a single tooth before training to improve the accuracy of these models. However, in the process of image cropping, this study discovered that the cutting accuracy obtained for the image by adaptive thresholding is higher than the one obtained by simple binary processing which reduces the possibility that many images contain non-target areas. The improvement of the cutting accuracy can make the effect of symptom enhancement more and thus improve the accuracy of the model. In addition, this paper uses a different method in the preprocessing of image symptoms to increase the dark area of the possible lesion area which actually helps the model accuracy to increase to more than 96%. Compared to other papers, the Gaussian high-pass filter is a tool for residual noise area to reduce noise in other projects. Changing the color of the lesion area is a different approach, and learning the features in the movie is instinctive and easy in the machine learning step. In addition, this paper proposes a hypothesis, that is the enhancement of apical lesions. The lesion area was preprocessed before importing the images into training. It can be found that the preprocessed images can further improve the recognition accuracy of CNN which is based on the premise of the quantity and quality of models and databases. The accuracy of the AlexNet model used in this research can reach up to 96.21%. Furthermore, the system's sensitivity and specificity on clinical apical radiographs were 98.5% and 93.9%, respectively.

## 5. Conclusions

The main purpose of this study is to achieve automatic and accurate diagnosis of apical teeth, and to help dentists improve treatment efficiency. The final experimental results show that the accuracy of AlexNet can reach 96.21% which provides confidence for this project to expand the research scope, improve the accuracy and realize the clinical medical application. In the future, the research team has formulated three objectives. Firstly, the project will continue to explore the possibility of identifying multiple symptoms and achieving the classification of different symptoms. Secondly, it will try to make the model more comprehensive and improve its accuracy. Thirdly, it will develop a GUI interface integrating the functions of picture cutting, disease strengthening and disease detection which can simplify the operation process and enhance the practicability of the plan at the same time.

**Author Contributions:** Conceptualization, Y.C. and W.-M.L.; Data curation, Y.C., W.-M.L.; Formal analysis, M.-L.C.; Funding acquisition, S.-L.C. and C.-A.C.; Methodology, T.-Y.C. and M.-L.C.; Resources, C.-A.C., S.-L.C. and P.A.R.A.; Software, S.-L.C., Y.-H.S., Y.-S.C., Y.-J.L. and Y.-R.L.; Supervision, C.-A.C. and S.-L.C.; Validation, Y.-H.S., Y.-S.C. and Y.-J.L.; Visualization, Y.-R.L. and P.A.R.A.; Writing—original draft, T.-Y.C.; Writing—review & editing, M.-L.C., C.-A.C. and P.A.R.A. All authors have read and agreed to the published version of the manuscript.

**Funding:** This work was supported by the Ministry of Science and Technology (MOST), Taiwan, under grant numbers of MOST-111-2221-E-033-041, MOST-111-2823-8-033-001, MOST-111-2622-E-131-001, MOST-110-2223-8-033-002, MOST-110-2221-E-027-044-MY3, MOST-110-2218-E-035-007, MOST-110-2622-E-131-002, MOST-109-2622-E-131-001-CC3, MOST-109-2221-E-131-025, MOST-109-2410-H-197-002-MY3, and the National Chip Implementation Center, Taiwan.

**Institutional Review Board Statement:** Institutional Review Board Statement: Chang Gung Medical Foundation Institutional Review Board; IRB number: 202002030B0; Date of Approval: 2020/12/01; Protocol Title: A Convolutional Neural Network Approach for Dental Bite-Wing, Panoramic and Periapical Radiographs Classification; Executing Institution: Chang-Geng Medical Foundation Taoyuan Chang-Geng Memorial Hospital of Taoyuan; Duration of Approval: From 2020/12/1 To 2021/11/30; The IRB reviewed and determined that it is expedited review according to case research or cases treated or diagnosed by clinical routines. However, this does not include HIV-positive cases.

**Informed Consent Statement:** The IRB approves the waiver of the participants' consent.

**Data Availability Statement:** Data sharing not applicable.

**Conflicts of Interest:** The authors declare no conflict of interest.

## References

1. Panetta, K.; Rajendran, R.; Ramesh, A.; Rao, S.P.; Agaian, S. Tufts Dental Database: A Multimodal Panoramic X-Ray Dataset for Benchmarking Diagnostic Systems. *IEEE J. Biomed. Health Inform.* **2022**, *26*, 1650–1659. [CrossRef] [PubMed]
2. Karamifar, K.; Tondari, A.; Saghiri, M.A. Endodontic Periapical Lesion: An Overview on the Etiology, Diagnosis and Current Treatment Modalities. *Eur. Endod. J.* **2020**, *5*, 54–67. [CrossRef] [PubMed]
3. Luongo, R.; Faustini, F.; Vantaggiato, A.; Bianco, G.; Traini, T.; Scarano, A.; Pedullà, E.; Bugea, C. Implant Periapical Lesion: Clinical and Histological Analysis of Two Case Reports Carried Out with Two Different Approaches. *Bioengineering* **2022**, *9*, 145. [CrossRef] [PubMed]
4. Jiménez-Sánchez, M.C.; Cabanillas-Balsera, D.; Areal-Quecuty, V.; Velasco-Ortega, E.; Martín-González, J.; Segura-Egea, J.J. Cardiovascular Diseases and Apical Periodontitis: Association Not Always Implies Causality. *Med. Oral Patol. Oral Cir. Bucal* **2020**, *25*, e652–e659. [CrossRef] [PubMed]
5. Kamberi, B.; Hoxha, V.; Stavileci, M.; Dragusha, E.; Kuçi, A.; Kqiku, L. Prevalence of Apical Periodontitis and Endodontic Treatment in a Kosovar Adult Population. *BMC Oral Health* **2011**, *11*, 32. [CrossRef] [PubMed]

6. Nair, P.N.R. On the Causes of Persistent Apical Periodontitis: A Review. *Int. Endod. J.* **2006**, *39*, 249–281. [CrossRef]
7. Ridao-Sacie, C.; Segura-Egea, J.J.; Fernández-Palacín, A.; Bullón-Fernández, P.; Ríos-Santos, J.V. Radiological Assessment of Periapical Sta-tus Using the Periapical Index: Comparison of Periapical Radiography and Digital Panoramic Radiography. *Int. Endod. J.* **2007**, *40*, 433–440. [CrossRef]
8. De Paula-Silva, F.W.G.; Wu, M.-K.; Leonardo, M.R.; da Silva, L.A.B.; Wesselink, P.R. Accuracy of Periapical Radiography and Cone-Beam Computed Tomography Scans in Diagnosing Apical Periodontitis Using Histopathological Findings as a Gold Standard. *J. Endod.* **2009**, *35*, 1009–1012. [CrossRef]
9. Wallace, J.A.; Nair, M.K.; Colaco, M.F.; Kapa, S.F. A Comparative Evaluation of the Diagnostic Efficacy of Film and Digital Sensors for Detection of Simulated Periapical Lesions. *Oral Surg. Oral Med. Oral Pathol. Oral Radiol. Endod.* **2001**, *92*, 93–97. [CrossRef]
10. Chazel, J.-C.; Mafart, B. Apical Lesions. *Br. Dent. J.* **2004**, *196*, 2. [CrossRef]
11. Chen, S.-L.; Chang, C.E.; Chen, C.A.; Abu, P.A.R.; Lin, T.L.; Lin, S.Y.; Chiang, W.Y.; Tu, W.C. Real-Time Image Contrast Enhancement VLSI Design for Intelligent Autonomous Vehicles. *J. Imaging Sci. Technol.* **2020**, *64*, 10504-1–10504-11. [CrossRef]
12. Chi, T.-K.; Chen, H.-C.; Chen, S.-L.; Abu, P.A.R. A High-Accuracy and Power-Efficient Self-Optimizing Wireless Water Level Monitoring IoT Device for Smart City. *Sensors* **2021**, *21*, 1936. [CrossRef]
13. Hsia, C.-H.; Lin, T.-Y.; Lin, J.-L.; Prasetyo, H.; Chen, S.-L.; Tseng, H.-W. System for Recommending Facial Skincare Products. *Sens. Mater.* **2020**, *32*, 3235. [CrossRef]
14. Lin, S.-Y.; Wu, C.-M.; Chen, S.-L.; Lin, T.-L.; Tseng, Y.-W. Continuous Facial Emotion Recognition Method Based on Deep Learning of Academic Emotions. *Sens. Mater.* **2020**, *32*, 3243. [CrossRef]
15. Adedigba, A.P.; Adeshina, S.A.; Aibinu, A.M. Performance Evaluation of Deep Learning Models on Mammogram Classification Using Small Dataset. *Bioengineering* **2022**, *9*, 161. [CrossRef]
16. Madan, P.; Singh, V.; Singh, D.P.; Diwakar, M.; Pant, B.; Kishor, A. A Hybrid Deep Learning Approach for ECG-Based Arrhythmia Classification. *Bioengineering* **2022**, *9*, 152. [CrossRef]
17. Zhou, R.; Wang, P.; Li, Y.; Mou, X.; Zhao, Z.; Chen, X.; Du, L.; Yang, T.; Zhan, Q.; Fang, Z. Prediction of Pulmonary Function Parameters Based on a Combination Algorithm. *Bioengineering* **2022**, *9*, 136. [CrossRef]
18. Chen, S.-L.; Zhou, H.-S.; Chen, T.-Y.; Lee, T.-H.; Chen, C.-A.; Lin, T.-L.; Lin, N.-H.; Wang, L.-H.; Lin, S.-Y.; Chiang, W.-Y.; et al. Dental Shade Matching Method Based on Hue, Saturation, Value Color Model with Machine Learning and Fuzzy Decision. *Sens. Mater.* **2020**, *32*, 3185–3207. [CrossRef]
19. Lakshmi, M.M.; Chitra, P. Tooth Decay Prediction and Classification from X-Ray Images Using Deep CNN. In Proceedings of the 2020 International Conference on Communication and Signal Processing (ICCSP), Chennai, India, 28–30 July 2020; pp. 1349–1355.
20. Lin, N.-H.; Lin, T.-L.; Wang, X.; Kao, W.-T.; Tseng, H.-W.; Chen, S.-L.; Chiou, Y.-S.; Lin, S.-Y.; Villaverde, J.; Kuo, Y.-F. Teeth Detection Algorithm and Teeth Condition Classification Based on Convolutional Neural Networks for Dental Panoramic Radiographs. *J. Med. Imaging Health Inform.* **2018**, *8*, 507–515. [CrossRef]
21. Yilmaz, E.; Kayikçioğlu, T.; Kayipmaz, S. Semi-Automatic Segmentation of Apical Lesions in Cone Beam Computed Tomography Images. In Proceedings of the 2017 25th Signal Processing and Communications Applications Conference (SIU), Antalya, Turkey, 15–18 May 2017; pp. 1–4.
22. Jayalakshmi, G.S.; Kumar, V.S. Performance Analysis of Convolutional Neural Network (CNN) Based Cancerous Skin Lesion Detection System. In Proceedings of the 2019 International Conference on Computational Intelligence in Data Science (ICCIDS), Chennai, India, 21–23 February 2019; pp. 1–6.
23. Li, C.-W.; Lin, S.-Y.; Chou, H.-S.; Chen, T.-Y.; Chen, Y.-A.; Liu, S.-Y.; Liu, Y.-L.; Chen, C.-A.; Huang, Y.-C.; Chen, S.-L.; et al. Detection of Dental Apical Lesions Using CNNs on Periapical Radiograph. *Sensors* **2021**, *21*, 7049. [CrossRef]
24. Mao, Y.-C.; Chen, T.-Y.; Chou, H.-S.; Lin, S.-Y.; Liu, S.-Y.; Chen, Y.-A.; Liu, Y.-L.; Chen, C.-A.; Huang, Y.-C.; Chen, S.-L.; et al. Caries and Restoration Detection Using Bitewing Film Based on Transfer Learning with CNNs. *Sensors* **2021**, *21*, 4613. [CrossRef] [PubMed]
25. Chen, S.-L.; Tsai, H.-J. A Novel Adaptive Local Dimming Backlight Control Chip Design Based on Gaussian Distribution for Liquid Crystal Displays. *J. Disp. Technol.* **2016**, *12*, 1494–1505. [CrossRef]

*bioengineering*

MDPI

*Article*

# Cervical Net: A Novel Cervical Cancer Classification Using Feature Fusion

**Hiam Alquran [1,2], Mohammed Alsalatie [3], Wan Azani Mustafa [4,5,*], Rabah Al Abdi [2] and Ahmad Rasdan Ismail [6,*]**

1 Department of Biomedical Systems and Informatics Engineering, Yarmouk University, Irbid 21163, Jordan
2 Department of Biomedical Engineering, Jordan University of Science and Technology, Irbid 21163, Jordan
3 The Institute of Biomedical Technology, King Hussein Medical Center, Royal Jordanian Medical Service, Amman 11855, Jordan
4 Faculty of Electrical Engineering & Technology, Campus Pauh Putra, Universiti Malaysia Perlis, Arau 02000, Perlis, Malaysia
5 Advanced Computing, Centre of Excellence (CoE), Universiti Malaysia Perlis (UniMAP), Arau 02000, Perlis, Malaysia
6 Mechanical Engineering Department, Faculty of Engineering, Universiti Teknologi PETRONAS, Seri Iskandar 32610, Perak, Malaysia
* Correspondence: wanazani@unimap.edu.my (W.A.M.); ahmadrasdan.ismail@utp.edu.my (A.R.I.)

**Abstract:** Cervical cancer, a common chronic disease, is one of the most prevalent and curable cancers among women. Pap smear images are a popular technique for screening cervical cancer. This study proposes a computer-aided diagnosis for cervical cancer utilizing the novel Cervical Net deep learning (DL) structures and feature fusion with Shuffle Net structural features. Image acquisition and enhancement, feature extraction and selection, as well as classification are the main steps in our cervical cancer screening system. Automated features are extracted using pre-trained convolutional neural networks (CNN) fused with a novel Cervical Net structure in which 544 resultant features are obtained. To minimize dimensionality and select the most important features, principal component analysis (PCA) is used as well as canonical correlation analysis (CCA) to obtain the best discriminant features for five classes of Pap smear images. Here, five different machine learning (ML) algorithms are fed into these features. The proposed strategy achieved the best accuracy ever obtained using a support vector machine (SVM), in which fused features between Cervical Net and Shuffle Net is 99.1% for all classes.

**Keywords:** pap smear; cervical net; shuffle net; canonical correlation analysis (CCA); support vector machine (SVM); random forest (RF); k-nearest neighbour (KNN); artificial neural network (ANN)

**Citation:** Alquran, H.; Alsalatie, M.; Mustafa, W.A.; Abdi, R.A.; Ismail, A.R. Cervical Net: A Novel Cervical Cancer Classification Using Feature Fusion. *Bioengineering* **2022**, *9*, 578. https://doi.org/10.3390/bioengineering9100578

Academic Editors: Paolo Zaffino and Maria Francesca Spadea

Received: 8 September 2022
Accepted: 30 September 2022
Published: 19 October 2022



## 1. Introduction

According to the World Health Organization (WHO), cervical cancer is the fourth most common cancer among women globally, with an estimated 604,000 new cases and 342,000 deaths in 2020. About 90% of the new cases and deaths in 2020 occurred in low- and middle-income countries worldwide [1,2]. Cervical cancer begins with no overt signs and has a long latent period, making early detection through regular checkups important. Cancer is a disease in which the body's cells grow rapidly, generally termed after the part where it originates, even if it spreads to other parts of the body [3–5]. Cervical cancer denotes cancer that begins in the cervix [6,7]. In the year 2018, an estimation of more than 500,000 women worldwide were diagnosed with cervical cancer, resulting in approximately more than 300,000 women dying due to cancer. Infection with high-risk human papillomaviruses (HPV), an immensely prevalent virus spread via sexual contact, is associated with almost all cervical cancer cases (99%). Therefore, cervical cancer may be prevented via screening tests and getting a vaccine that defends against HPV infection. In addition, cervical cancer is usually detected with a Pap smear test. It is a painless, fast screening test for precancer or cancer of the uterine cervix. Moreover, the regular Pap test system lowers the cervical cancer incidence rate [8–11].

Cervical cancer is a fatal condition of which individuals who possess a low level of awareness. Thus, although it is a life-threatening condition, early diagnosis and treatment may assist in its prevention [12]. Nevertheless, most nations lack efficient screening techniques to encounter this kind of cancer. Hence, in this study we provide a comparison of performance indicators. For example, in terms of accuracy, several machine learning (ML) and deep learning (DL) models for cancerous and normal cervical cells were categorised, including their subtypes. The following literature reviews are related to prior studies on classifying cervical cancer cells.

## 2. Review of Study

In 2015, Mbaga et al. [13] explained cervical cancer detection classification utilising a support vector machine (SVM) classifier gaining around 92.961% accuracy. Furthermore, Win et al. [14] suggested a technique for computer-assisted screening of Pap smear images utilising digital image processing. They utilised texture, shape, and colour features to classify Pap smear images with an accuracy of 94.09%. An investigation by Plissiti et al. [15] found a new method for cervical cancer detection using handcrafted cell features and deep learning (DL) features utilising multi-layer perceptron (MLP) and an SVM classifier, which resulted in the best accuracy obtained, 95.35%. On the other hand, Basak et al. [16] found that a fully automated framework that employs feature selection and DL utilising evolutionary optimisation for cytology image classification obtains an accuracy of 97.87%. With the same objective of recognising cervical cancer's indications utilising cervicography images, Park et al. [17] examined the performance of two distinct models, DL and ML. Applying the ResNet-50 DL, Random Forest (RF), XGboot (XGB), and SVM and ML models, 4119 cervicography images were identified as negative or positive for cervical cancer by employing square images by omitting the vaginal wall areas. Note that the ResNet-50 model outperformed the average (0.82) of the three ML techniques by 0.15 points ($p < 0.05$). Since this process necessitates segmentation and the acquisition of handcrafted characteristics, a mix of ML and DL techniques is the most efficient. Furthermore, the findings of Tripathi et al. [18] are congruent with the findings of this research. They demonstrated DL classification methods utilising the SIPaKMeD Pap smear image dataset to provide a foundation for new classification strategies. The ResNet-152 architecture achieved the greatest classification accuracy of 94.89% utilising this technique.

Alternatively, Al Mubarak et al. [19] used a hybrid, fusion-based, localised imaging and DL technique to categorise squamous epithelium into cervical intraepithelial neoplasia (CIN) grades, utilising a dataset of 83 digitised histology images. For each segment, 27 handmade image features and a rectangular patch comprising sliding window-based convolutional neural network (CNN) features were computed after partitioning the epithelium region into ten vertical segments. Meanwhile, the DL and imaging patch characteristics are merged and utilised as inputs to a secondary classifier for the individual segment and total epithelium classification. With an accuracy of 80.72% in terms of the whole epithelium CIN classification, the hybrid technique outperformed the imaging and DL techniques alone by 15.51% and 11.66%, respectively. On the other hand, Alyafeai and Ghouti [20] discovered variances, proposing that the suggested pipeline comprises two pre-trained DL models for cervix identification and cervical tumour categorisation. The first model discovers the cervix region 1000 times quicker compared to current data-driven algorithms, with a detection accuracy of 0.68 with respect to the intersection of the union (IoU) scale. The second model utilises self-extracted characteristics to categorise cervical cancers. Here, two lightweight models relying on CNN are employed to learn these characteristics. Moreover, the suggested DL classifier outshines prior models in terms of speed and classification accuracy. The area under the curve (AUC) score of our classifier is 0.82, classifying every cervical region 20 times more quickly. In the most recent published research, Alquran et al. [21] proposed an automated system to classify cervical cancer into seven classes on the Harvel dataset. Their approach exploited the benefits of DL with a model of a cascading SVM classifier to achieve the highest accuracy among all previous studies working on a similar dataset, namely, up to 92% for seven classes. Moreover, their method is fast because the image preprocessing step is skipped.

Missed diagnoses and misdiagnoses often occur due to the high similarity in pathological cervical images, the large number of readings, the long reading time, and the insufficient experience levels of pathologists. In addition, existing models have insufficient feature extraction and representation capabilities, and they suffer from insufficient pathological classification. In 2021, Park et al. [17] mentioned the significant differences between two different models, ML and DL, in identifying signs of cervical cancer using cervicography images. They concluded that the ResNet-50 DL algorithm could perform better than current ML models in identifying cervical cancer using cervicography images. This is supported by Dhawan et al.'s [22] study, which reveals improved techniques for cervical cancer predictive models based on DL and transfer learning techniques. They classify the cervix images into three classes (Type1/Type2/Type3) by creating a Con-vet structure from combinations between pretrained models (InceptionV3, ResNet-50, and VGG19) were used to create ConvNet that can classify the cervix images. The result of the experiment revealed that the InceptionV3 model performs better than VGG19 and ResNet-50, with an accuracy of 96.1% on the cervical cancer dataset.

In another study, Huang et al. [23] suggest extracting deep convolutional features by fine-tuning pre-trained deep network models, including ResNet-50V2, DenseNet-121, InceptionV3, VGG19 Net, and Inception ResNet, and then local binary patterns and a histogram of the oriented gradient are used to extract traditional image features. The serial fusion effect of the deep features extracted by ResNet-50V2 and DenseNet-121 (C5) is the best, with the average classification accuracy reaching 95.33%, which is 1.07% higher than ResNet-50V2 and 1.05% higher than DenseNet-121. Furthermore, the recognition ability is significantly improved to 90.89%, which is 2.88% higher than ResNet-50V2 and 2.1% higher than DenseNet-121. Thus, this method significantly improves the accuracy and generalisation ability of pathological cervical whole slice image (WSI) recognition by fusing deep features [23]. Mulmule and Kanphade [24] proposed method that employs adaptive fuzzy k-means clustering to separate cell from the unwanted background of the pathological Pap smear image. The 40 features are extracted from the segmented images based on the shape, size, intensity, orientation, colour, energy, and entropy of the nucleus and cytoplasm individually. Finally, the performance of the supervised classification approach utilising an MLP with three kernels and an SVM with five different kernels as the classifiers to predict the cancerous cells is on par with the existing techniques. The classifier is trained and tested on a benchmark database with 280 Pap smear images. Furthermore, the performance of these two classifiers are evaluated and it is found that the MLP classifier with hyperbolic tangent activation function outperforms the SVM classifier in all the performance criteria, with a classification accuracy of 97.14%, sensitivity of 98%, specificity of 95%, and positive predictive value (PPV) of 98% [24].

A particular image can be used by computer-aided diagnosis (CAD) systems that are trained using artificial intelligence (AI) algorithms to predict the possibility of cervical cancer, which has been highlighted in several cervical cancer studies. For example, Nikookar et al. [25] found that a cervical cancer predictor model, which incorporates the result of different classification algorithms and ensemble classifiers, is more effective for cervical cancer stages. They investigated different aggregation strategies to find the best formula for the aggregation function. They then evaluated our method using the quality assessment of the digital colposcopies dataset. Our approach, performing with 96% sensitivity and 94% specificity values, yields a significant improvement in the field. It can now be used in a supporting clinical decision-making strategy by providing more reliable information to the clinical decision makers. With the same objective, Yaman and Tuncer [26] performed a comprehensive review to classify cervical cells in Pap smear images based on two datasets, SIPaKMeD and Mendeley Liquid Based Cytology (LBC). The 1000 features selected by neighbourhood component analysis (NCA) were classified with the SVM algorithm. Both five-fold cross-validation and hold-out validation (80:20) have been utilised as validation techniques. The best accuracies for the SIPaKMeD and Mendeley LBC datasets have been computed as 98.26% and 99.47%, respectively. The obtained results illustrate that the proposed exemplar pyramid model successfully diagnoses cervical cancer using Pap smear images [26].

According to literature reviews, cancer detection in the early stages is crucial for the treatment process. Therefore, early diagnosis/detection is essential for the treatment of cervical cancer. Note that the gold standard for diagnosing cervical cancer is the Pap smear test. In recent years, there has been an increasing interest in artificial intelligence approaches in medical imaging, such as ML, DL, and CNN [27]. ML is a good solution to automatically diagnose cervical cancer, and many computer vision/DL-based models have been presented in the literature. However, the morphological changes and their entanglement in the structural sections of the cells is one of the constraints. DL and ML algorithms possess a substantial improvement in the healthcare industry. Furthermore, advances in deep learning have led to the development of neural network algorithms that today rival human performance in vision tasks, such as image classification or segmentation. The translation of these techniques into clinical science has also significantly advanced medical image analysis [28]. Research has shown that machine learning can improve the effectiveness of medical image analysis. Algorithms can be developed and trained to remove image noise, improve quality, and gather image data in greater quantities and at a faster rate than standard techniques [29]. Moreover, these algorithms enhance the consistency and accuracy of cancer diagnoses. They also aid medical practitioners in terms of work complexity, minimising labour time, and prognosis.

This study aimed to build a highly accurate computer-aided diagnosis model for cervical cancer. We obtained features from pre-trained CNN models utilising Shuffle Net, applying different classifiers to discriminate the Pap smear images. Subsequently, we created our DL model called Cervical Net with a simple and light structure, in which its features are passed to different ML classifiers. The key point of this paper is not only the novel DL model but the fusion features between the DL descriptors from various structures to obtain a high level of accuracy. The remainder of this article is structured as follows: Section 3 is devoted to the materials and methods, Section 4 focuses on the results and discussion, and the last section concludes.

## 3. Materials and Methods

The proposed method of cervical cytology is displayed in the system flow diagram in Figure 1.

### 3.1. Image Acquisition

For multi-cell classification, SIPaKMeD datasets were utilised for image acquisition [13]. There were 966 photos in the multi-cell dataset, while 4049 cells were cropped from these images. Note that cells were separated into three stages: normal, benign, and abnormal. Dyskeratotic cells, metaplastic cells, parabasal cells, superficial–intermediate cells, and koilocytotic cells were the five cell types. Table 1 has been created to describe the specifics of each dataset. Table 1 and Figure 2 represent a Pap smear image from the SIPaKMeD dataset.

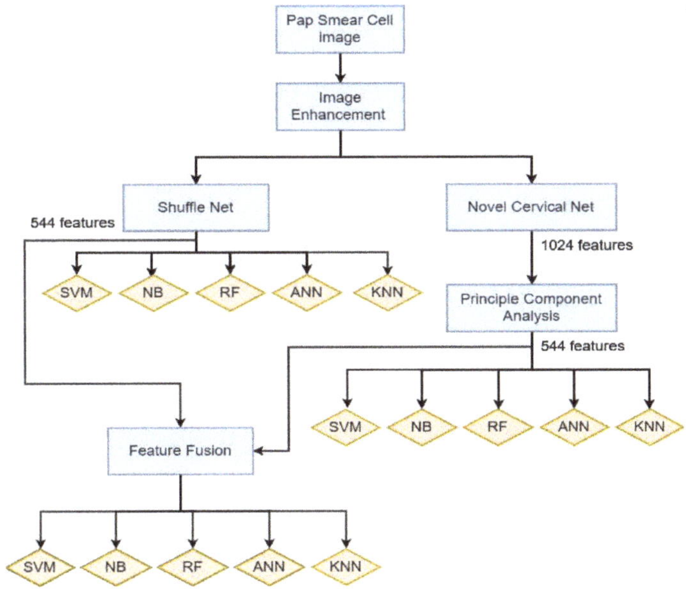


**Figure 1.** Design of the proposed method.

**Table 1.** Specification of five classes of cells obtained from the SIPaKMeD (multi-cell) dataset.

| Class | Number of Images | Number of Cells |
|---|---|---|
| **Normal Class** | | |
| 1. Superficial–Intermediate Cells | 126 | 831 |
| 2. Parabasal Cells | 108 | 787 |
| **Benign Cell** | | |
| 3. Metaplastic Cells | 271 | 793 |
| **Abnormal Cells** | | |
| 4. Dyskeratotic Cells | 223 | 813 |
| 5. Koilocytotic Cells | 238 | 825 |
| **Total** | **966** | **4049** |

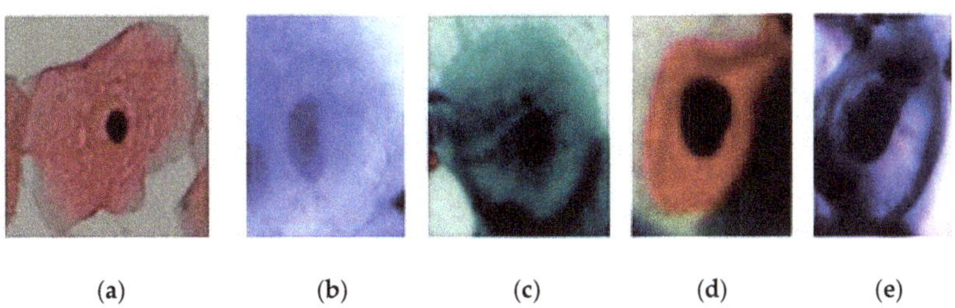

**Figure 2.** Example images from each class: (**a**) superficial, (**b**) parabasal, (**c**) metaplastic, (**d**) dyskeratotic, (**e**) koilocytotic.

### *3.2. Image Enhancement*

As shown in Figure 3a, most Pap smear images were low-contrast and noisy. As a result, image processing was required to reduce noise and raise contrast [30]. To eliminate the noise, a median filter was utilised. The median filter used here is more effective than convolution filters because it removes the noise while preserving the edges. The kernel size in this paper was 3 × 3. Figure 3b shows the image after applying a median filter.

Histogram equalisation and normalisation are some of the most common techniques used to enhance the contrast of images, which stretches the histogram of the intensity values into wider ranges. Increasing the contrast leads to extracting more representative features for each class. Figure 3c shows the image after median filtering and histogram equalisation.

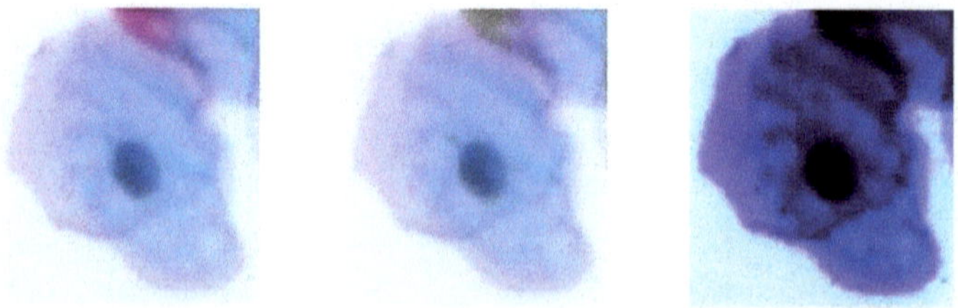

**(a)** **(b)** **(c)**

**Figure 3.** Image enhancement: (**a**) original image, (**b**) noise removal via the median filter, and (**c**) contrast enhancement via histogram equalisation.

### 3.3. Cervical Net

Cervical Net is a novel DL structure that was designed in this study. Figure 4 shows the layout of its layers with distinguished group convolutional layers. The structure starts with an input layer of an image size of 224 × 224 × 3. Consequently, the coloured image is passed to a convolutional layer with 64 filters, kernel size 7 × 7 and stride 2 × 2. The output is passed to the rectified linear unit (ReLU) layer, which maps the resultant output from the convolutional layer into 1 or -1. To downsample the image feature, it is passed to the average pooling layer with size 3 × 3 and stride 2 × 2. The output is passed to a two-dimensional (2D) grouped convolutional layer, which separates the input into groups and then is applied to slide convolutional filters. The convolution is performed vertically and horizontally, combining the layer of each group independently. In this layer, two groups are used and 94 filters with size 5 × 5 and padding size 2 × 2 × 2 × 2 for all groups. Note that the main goal behind grouping convolutional layers is to obtain higher accuracy than traditional ones. The grouped output is then passed to the ReLU layer and average pooling layer to downsample it with kernel size 3 × 3 and padding 2 × 2. The output is passed to the second convolutional neural network (CNN) for extracting more depth features using 128 filters, kernel size 3 × 3, and padding size 1 × 1 × 1 × 1. Subsequently, the output is passed to the ReLU layer to map it into 1 or −1. The grouped convolutional network is applied to the resultant output with two groups of convolutions using 196 filters, and the kernel size is 3 × 3. The combined output from the depth-wise separable channel is mapped to -1 and 1 using another ReLU function. For extracting depth features and obtaining a higher accuracy, another two groups of the convolutional layer are applied to the mapping output with 128 filters and kernel size 3 × 3. The output is passed to the ReLU layer. The downsampling is performed on the resultant mapping output using the global average pooling layer. The fully connected layer is added to the last output with five neurons compatible with the number of classes, and the softmax layer ends the fully connected layer. This can be defined by the corresponding equation [31–33].

$$f(x_i) = \frac{\boldsymbol{exp}(x_i)}{\sum_j \boldsymbol{exp}(x_j)},$$

where $x$ refers to the input vector of the layer with size $\boldsymbol{K}$, denoted by $j$, in the range of $1:K$. Further, $x_i$ indicates the $i_{th}$ individual input. The output of this layer is expressed as probabilities commonly used in multi-classification tasks. Here, the proposed network is terminated by the classification layer. The detailed information regarding the proposed Cervical Net is displayed in Table 2.

**Table 2.** Structure summaries of Cervical Net.

| Layer | Information |
|---|---|
| Input Layer | Size: 224 × 224 × 3 |
| conv1 | Number of Filters: 64<br>Kernel Size: 7 × 7<br>Stride: 2 × 2<br>Padding: 0 |
| Activation Layer | ReLU |
| Pooling Layer | Type: Average Pooling<br>Kernel size: 3 × 3<br>Stride: 2 × 2<br>Padding: 0 |
| Grouped Convolutional Layer | Number of Groups: 2<br>Number of Filters: 94<br>Kernel Size: 5 × 5<br>Padding: 2 × 2 × 2 × 2 |
| Activation Layer | ReLU |
| Pooling Layer | Type: Average Pooling<br>Kernel Size: 3 × 3<br>Stride: 2 × 2<br>Padding: 0 |
| Convolutional Layer | Number of Filters: 128<br>Kernel Size: 3 × 3<br>Padding: (1 × 1 ×1 × 1) |
| Activation Layer | ReLU |
| Grouped Convolutional Layer | Number of Groups: 2<br>Number of Filters: 192<br>Kernel Size: 3 × 3<br>Padding: (1 × 1 × 1 × 1) |
| Activation Layer | ReLU |
| Grouped Convolutional Layer | Number of Groups: 2<br>Number of Filters: 128<br>Kernel Size: 3 × 3<br>Padding: (1 × 1 × 1 × 1) |
| Activation Layer | ReLU |
| Pooling Layer | Type: Global Average Pooling |
| Fully connected Layer | 5 neurons |
| Softmax Layer | |
| Classification Layer | |

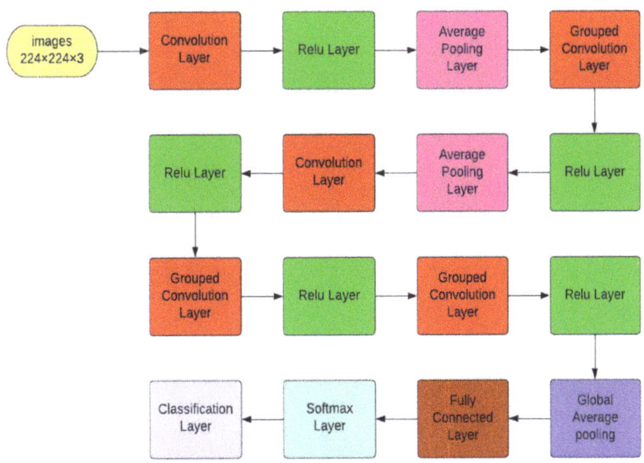


**Figure 4.** Cervical Net structure.

### 3.4. Pre-Trained Shuffle Net

Convolutional, pooling and fully linked layers are components of traditional CNN models. The use of large pooling layers and convolution kernels increases the computational complexity of the model. The model's size and depth increase to enhance the model's accuracy [34]. Because of the limited performance of some specific applications, the model demands a small size and high accuracy.

Shuffle Net V2 tackles the aforementioned issues without resorting to large pooling layers or convolution kernels. A depth-wise convolution and a $1 \times 1$ tiny convolution kernel replace the traditional convolutional layer. Since one convolution kernel is accountable for one input channel with a depth-wise convolution kernel size of $3 \times 3$, the number of convolution kernels is the same as the number of input channels. To combine characteristics of the depth-wise convolution output, a $1 \times 1$ convolution is utilised. This increases the network's expressiveness and nonlinearity without increasing the size of the output feature graph. Furthermore, Shuffle Net downsamples the feature via modifying the depth-wise convolution step instead of utilising the traditional pooling layer [34]. Figure 5 describes the structure of the Shuffle Net basic unit.

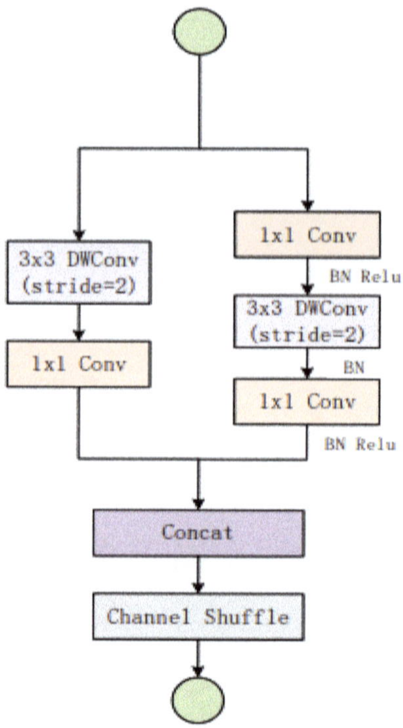


**Figure 5.** Shuffle Net basic unit [28].

After the convolutional layer, a new layer known as a pooling layer is added. Specifically, after a nonlinearity is employed for the feature map output via a convolutional layer, the pooling layer functions on each feature map independently to construct a new set of pooled feature maps with the same number of characteristics. Moreover, global pooling [35] is another type that occasionally utilises downsamples of the entire feature map to a single value rather than downsampling sections of the input feature map. In our study, we extract features from global pooling and employ them in the classification task.

### 3.5. Deep Features Extraction

Traditional machine learning (ML) algorithms for handcrafted or manual feature extraction have limitations in terms of the correlations and their feature number. With the introduction of artificial intelligence (AI) and deep learning (DL) in the domains of healthcare and the medical sciences, it has become rather common to rely on the findings projected via this decision support system to prevent issues of observer bias. Backpropagation is utilised in DL models to determine the key features, which removes the time-consuming procedure of employing handmade features [36,37].

We utilised both our own structure—Cervical Net—and the pre-trained model to alter the CNN by employing our data, allowing each image to propagate across the layers in a forwarding manner, finishing at the pre-final layer and extracting the output of this layer as the feature vector. Because biological data are inadequate and sparse for DL models to perform effectively if trained from the beginning, we employed pre-learned weights (transfer learning) in this research. For the present study, we have used Cervical Net and Shuffle Net for feature extraction from the model's global average pooling layer.

### 3.6. Feature Selection

The major goal of utilising a feature selection approach was to determine the crucial features while improving the classifier's accuracy. Note that the feature selection technique may help ML algorithms train faster by reducing the complexity of the classification model [14]. There are plenty of feature selection algorithms to choose from, and principal component analysis (PCA) is one of them. It is known as a linear dimensionality reduction technique that maximises the variance of the lower dimension into higher dimensional data [16]. PCA is used in this paper to reduce the extracted features of Cervical Net from 1024 to 544 most significant features.

The number of components in the down-selection stage is chosen based on the number of extracted features from the pre-trained Shuffle Net structure. This procedure is performed using PCA with 95% variance between the selected components.

### 3.7. Feature Fusion

Canonical correlation analysis (CCA) is a standard tool in statistical analysis that measures the linear relationship between two datasets. CCA is an unsupervised representation learning technique for correlating multi-view data by learning a set of projection matrices. The analysis and methods based on CCA are often used in traditional feature fusion methods. It only considers the correlated information of the paired data but ignores the correlated information between the samples in the same class. Furthermore, these methods generally have great deficiencies in exploring the influence of non-negative constraints, feature dimensions, sample size, and noise power. Being complementary to CCA, many discriminant methods have been proposed to extract discriminative features of multi-view data by introducing the supervised class information. However, the learned projection matrices in these methods are mathematically constrained to be of equal rank to the class number and thus cannot represent the original data comprehensively [38]. Canonical correlation analysis (CCA ) considers intraclass and interclass correlations and solves the problem of computation and information redundancy with simple series or parallel feature fusion [39]. Deep CCA based on the encoder–decoder network is designed to extract cross-modal correlations by maximising the relevance between multimodal data [40]. Moreover, CCA is an important method for multiple feature extraction and fusion in which the canonical projective vectors in the classical CCA method satisfy conjugated orthogonality constraints. Class information is useful for CCA, but there is little class information in the scenarios of real applications.

### 3.8. Machine Learning Classifiers

DL features extracted from Cervical Net are passed to various ML classifiers to obtain the best classifier's accuracy. The same experiment is performed using the pre-trained

Shuffle Net features. The combined features between the novel Cervical Net and Shuffle Net are fused using CCA. The resultant fused features are passed to various ML classifiers to obtain the highest level of accuracy. Subsequently, a comparison is performed between different classifiers for the same features and methods, such as Cervical Net features, Shuffle Net features, or using CCA techniques.

### 3.8.1. Support Vector Machine (SVM)

A support vector machine (SVM) refers to a supervised learning model that appropriately labels distinct classes in a set of training samples. The feature plane plot representation of the training data in the SVM model denotes a distinction between the prominent instances representing various classes. A curve that fits in the space between two classes and maintains maximum distances from each class point and SVM can be seen [41,42].

### 3.8.2. Artificial Neural Networks (ANN)

An artificial neural network (ANN) is a well-known ML technique based on the biological neural network found in the human brain. For example, feedforward neural networks are a typical form of ANN. Once the inputs from neurons are processed in the previous layer, it yields the weight values of each artificial neuron to the proceeding layer. Note that the backpropagation algorithm is the most extensively utilised multi-layer perceptron (MLP) training technique. To reduce inaccuracy, the weights between neurons are altered. Hence, when it comes to learning patterns, this model performs excellently. It can quickly adjust to new data values, but it might be sluggish to converge and runs the risk of a local optimum [43,44].

### 3.8.3. Naive Bayes

The Naive Bayes technique is a basic probability classifier that calculates probabilities by counting the number of different value and frequency combinations in a dataset. The technique focuses on Bayes' theorem and assumes that all variables are unaffected by the class variable's value. Since this conditional independence assumption is hardly valid in real-world applications, it is labelled Naive. Nevertheless, the algorithm learns swiftly in various controlled classification situations [45].

### 3.8.4. k-Nearest Neighbour (KNN)

Fix and Hodges invented the supervised k-nearest neighbour (KNN) classification technique in 1951 [46], which categorises a data point depending on the class of its neighbours. Moreover, the classification findings are provided depending on the nearest neighbour's k-value, which was set to 1. Here, the closest k-samples from the training set are chosen to categorise the new sample depending on its attribute vector. As a result, the new vector is directed at it via examining the classes into which the candidate's samples are categorised [47].

### 3.8.5. Random Forest (RF)

The random forest (RF) classifier comprises numerous decision trees [48], where every node in the tree contains a set of training cases and a predictor. At each attribute split, a random selection of features is chosen depending on the bagging approach. The trees continue to grow until they attain a certain depth, where a class voting system is established when a large number of trees have been generated [47].

## 4. Results and Discussion

The SIPaKMeD (multi-cell) dataset was utilised to test the efficiency of our suggested method. There was a total of 996 images, with 4049 cells cropped. These cells were categorised into five classes: class 1, superficial–intermediate cells; class 2, parabasal cells; class 3, metaplastic cells; class 4, dyskeratotic cells; and class 5, koilocytotic cells. After processing the images using the convolutional neural network (CNN) architectures, deep features were extracted from global pooling layers.

### 4.1. *Shuffle Net Features*

Utilising the extracted features from the global pooling layer, different classifiers were used to classify the images into five classes, including support vector machine (SVM), random forest (RF), k-nearest neighbour (KNN), Naive Bayes, and artificial neural network (ANN). At the same time, we utilised 70% of the data as training and 30% as testing. Figure 6 illustrates the confusion matrices result of classifiers, where the test accuracy reaches 98.9%, 96.5%, 97.3%, 89.7%, and 98.7%, respectively, and the training accuracy reaches 100% for all the different classifiers.

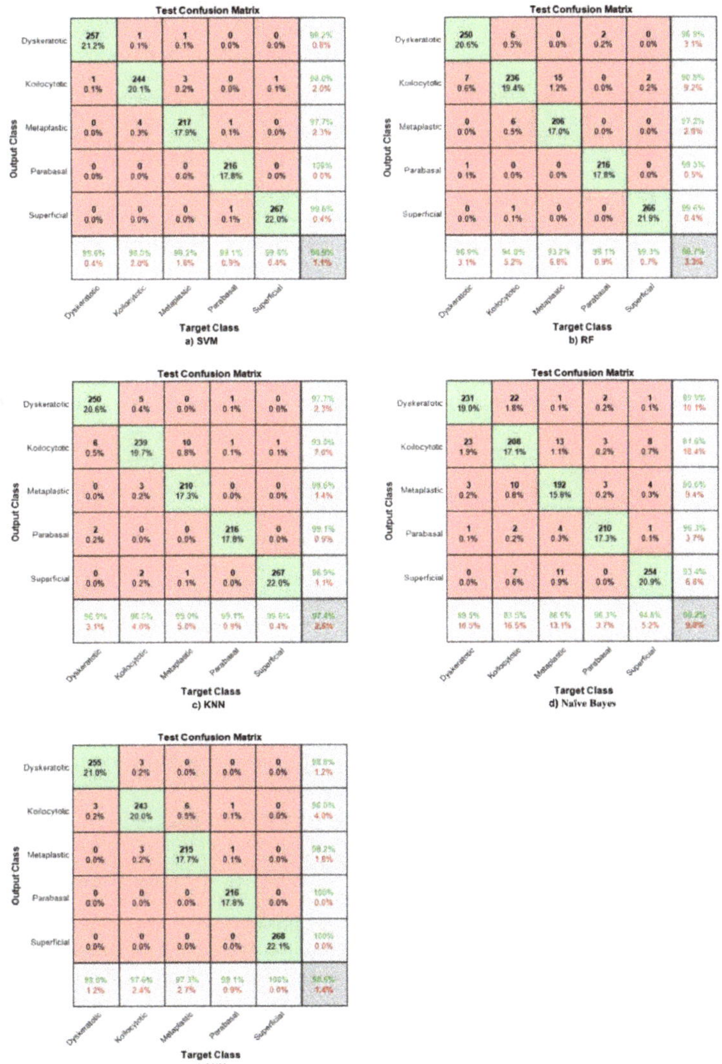


**Figure 6.** Confusion matrix with respect to Shuffle Net features for different ML classifiers. (**a**) SVM, (**b**) RF, (**c**) KNN, (**d**) Naïve Bays, (**e**) ANN.

In Figure 6a, the diagonal represents the correctly classified observations, whereas the off-diagonal cells indicate incorrectly classified observations. Note that the column on the far right of the plot shows the precision or positive predictive value (PPV). The row at the bottom of the plot refers to the recall or true positive rate (TPR) or sensitivity. Meanwhile, the cell in the bottom right of the plot shows the overall accuracy. The overall accuracy here is 98.9%, and metaplastic benign cells obtain the highest sensitivity and precision of 100%.

The same Shuffle Net features are fed to the RF classifier. The overall accuracy reaches 96.5%, and parabasal malignant cells reach the highest PPV, which is 100%. However, dyskeratotic normal cells obtain the highest sensitivity, which reaches 97.8%, as shown in Figure 6b.

The accuracy of the hybrid model between Shuffle Net features and KNN does not exceed 97.3%. Meanwhile, the highest precision is in parabasal malignant cells, and the highest recall is in superficial malignant cells, which is represented in Figure 6c.

Naive Bayes is exploited to classify five cells whose highest accuracy does not exceed 89.7%, and the best sensitivity is obtained by metaplastic benign cells, reaching 95.2%. The parabasal PPV is 99.5%. This is clearly shown in Figure 6d.

An ANN was used in this study and was fed with Shuffle Net features to obtain the second highest accuracy, reaching 98.7%. Dyskeratotic cells have the highest sensitivity, and parabasal cells the highest precision. This is shown in Figure 6e.

Previous confusion matrices have shown that the SVM has the highest accuracy for all five classes. Other than that, numerous cervical cell classification models have been developed in the literature using the same datasets. However, this study differs from previous ones in that it focuses on handcrafted features, such as shape, texture, and colour, to classify Pap smear images into five classes.

### 4.2. Novel Cervical Net Features

The proposed network was utilised to extract features from the global average pooling layer, in which the number of extracted features was 1024 graphical features. These features were fed to various machine learning (ML) classifier models to obtain the best model using the novel features. Furthermore, the time taken to extract the features for all test images did not exceed 60 s. The corresponding confusion matrices clarify the test phase for each classifier using novel Cervical Net features.

An SVM was fed with 1024 features to discriminate between various classes. The overall accuracy reached 96%, with higher sensitivity for parabasal normal cells and high precision for parabasal malignant cells. The lowest sensitivity is appeared in malignant cells, namely, dyskeratotic and koilocytotic malignant cells, and it is found in the cells that are very similar in shape and colour, as well. Moreover, the same features were used to design an RF classifier, and the results are clearly shown in Figure 7a. The overall accuracy for the whole system does not exceed 94.2%. The sensitivity of the malignant cells is the lowest in the case of the SVM and the highest in normal cells. Therefore, investigating other methods to enhance the classification process is necessary to discriminate between various classes, either normal or abnormal.

The KNN classifier was utilised in this study for testing its performance in distinguishing between five classes using the extracted features from Cervical Net. Figure 7b shows that the overall test accuracy is 93.7%. The highest sensitivity was obtained by superficial normal classes. On the other hand, the lowest sensitivity appears in the koilocytotic abnormal class. The highest precision appears in the parabasal normal class, and the lowest PPV is in the koilocytotic abnormal cell. As shown in the KNN confusion matrix, the koilocytotic cell has the lowest TPR and precision.

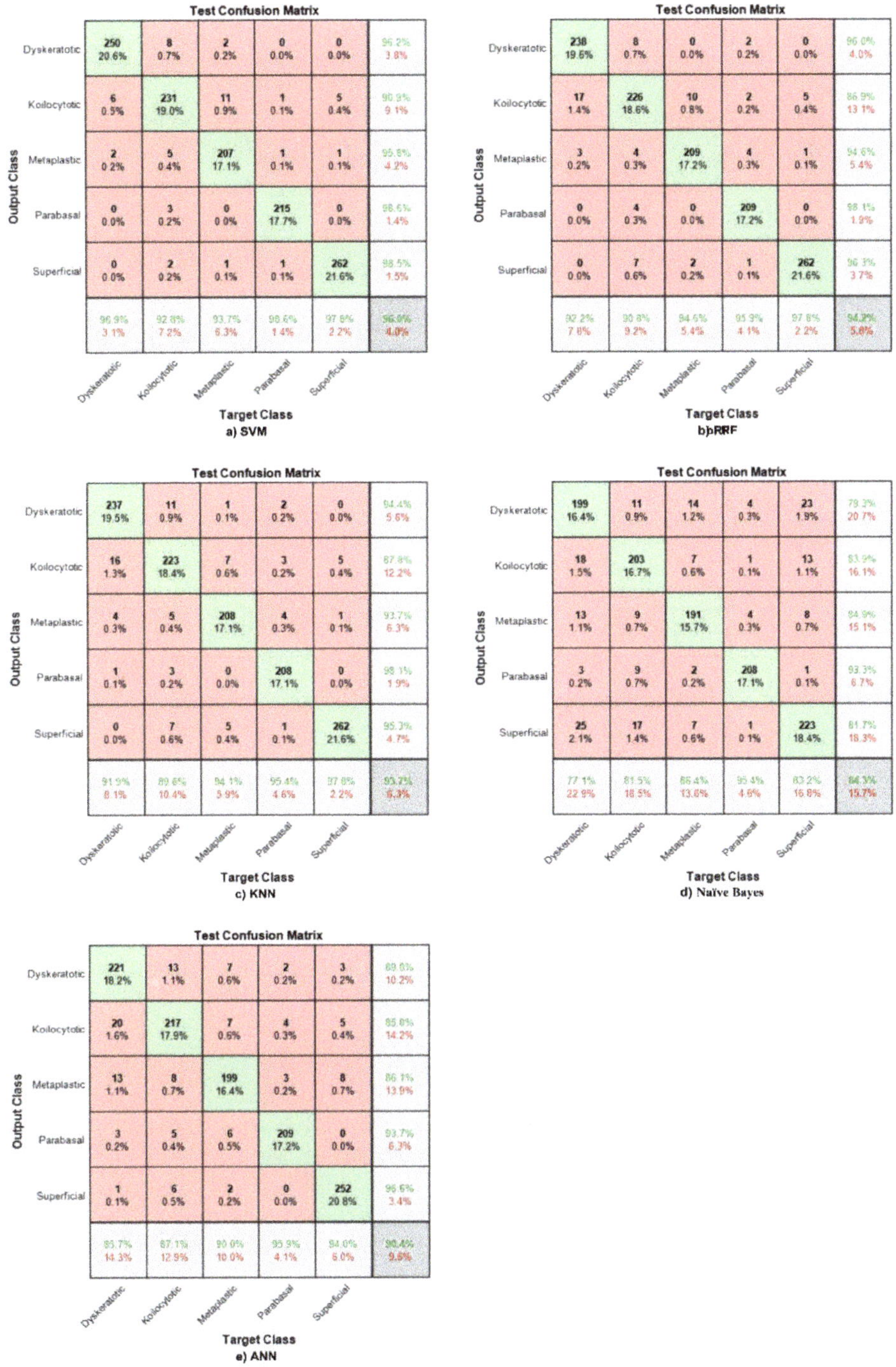


**Figure 7.** Confusion matrix with respect to Cervical Net features for different ML classifiers. (**a**) SVM, (**b**) RF, (**c**) KNN, (**d**) Naïve Bays, (**e**) ANN.

In Figure 7, the green color indicates to the correctly classified cells. Furthermore, the red color represents the misclassified cases.

The lowest accuracy and sensitivity, and even precision, are obtained using the Naive Bayes classifier, in which the accuracy does not exceed 85%. The sensitivity is poor in all classes, as well as the precision. Figure 7c clarifies the confusion matrix generated using test features with the lowest sensitivity appearing among malignant cells. On top of that, the precision of malignant cells is also low.

An ANN was used in this paper to evaluate the efficiency of the extracted features from the global average pooling from Cervical Net for classifying the five classes. The highest accuracy obtained here does not exceed 90.4% for all classes. Figure 7d describes the test confusion matrix, showing that the two abnormal classes have the lowest sensitivity and PPV.

Utilising the extracted features from Cervical Net shows that the SVM has the highest accuracy for all five classes and behaves the best among all classifiers.

### 4.3. Feature Fusion (CCA)

Feature fusion is a technique used for combining features from various structures, which strengthens the capability of the designed classifier to discriminate between different classes. The extracted features from the global average pooling layer are reduced to 544 features using the principal component analysis (PCA) algorithm to find the most significant features and then combine the resultant descriptors with the graphical features extracted from the global average pooling layer in Shuffle Net. Note that the total number of features after fusion is 544. These features are used to design different ML classifiers. Figure 8 illustrate the confusion matrices for the most known classifiers (SVM, RF, KNN, NB, ANN). Figure 8a shows the maximum accuracy obtained using the SVM's fusion features, 99.1%. The sensitivity for all classes is elevated to almost 100% for all classes, which helps clinicians in diagnosing even abnormal classes.

The same features are used to design the RF classifier, and the overall accuracy is 94.7%. The achieved accuracy is higher using the fusion features than the DL descriptors alone. Figure 8b shows that the sensitivity and precision for abnormal classes are enhanced.

The maximum accuracy obtained using fusion features and the KNN classifier is 91.1%, as shown in Figure 8c. Nevertheless, the sensitivity in the normal superficial class is the lowest among all cell types, and abnormal koilocytotic cells have the lowest precision among all classes.

The same procedure is applied to design the Naive Bayes classifier, and the overall accuracy for the test phase is 93.3%. Although the accuracy is not high, it is better than using DL features solely. Figure 8d shows that the highest recall appears in metaplastic benign cells, which reaches 95.5%, and the highest PPV value appears in koilocytotic abnormal cells. However, the TPR for all abnormal and normal classes exceeds 90%, and the precision is also higher for abnormal and normal classes.

An ANN classifier was designed with the proposed fusion features, and the accuracy was enhanced to 94.9%, with the best precision obtained is in parabasal normal cells almost 100%. The sensitivity is the best for normal classes discrimination. Note that the sensitivity for all classes exceeds 90%, as shown in Figure 8e.

Figure 9 illustrates the accuracies of all the ML classifiers using various scenarios (Shuffle Net features only, novel Cervical Net features only, and the feature fusion from Shuffle Net and Cervical Net). The highest accuracy is obtained by the SVM with feature fusion.

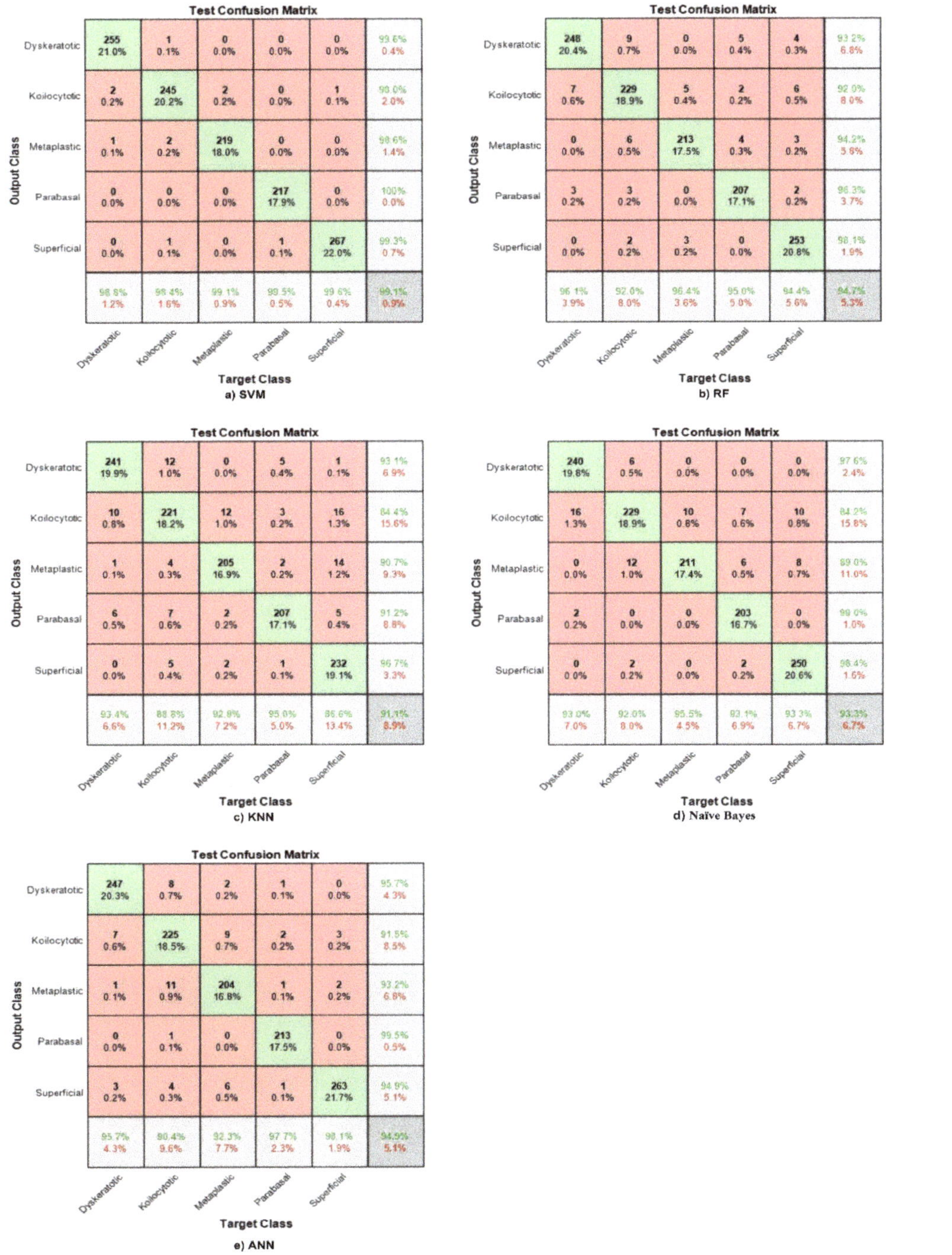


**Figure 8.** Confusion matrix with respect to CCA features for different ML classifiers. (**a**) SVM, (**b**) RF, (**c**) KNN, (**d**) Naïve Bays, (**e**) ANN.

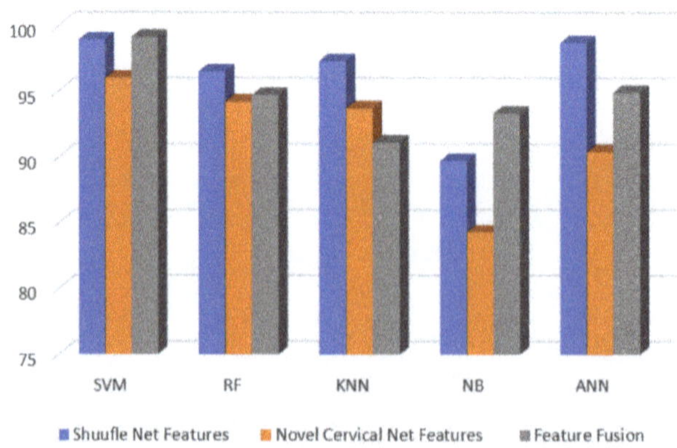


**Figure 9.** Comparison between various scenarios.

When the proposed method is compared with all previous studies, the obtained results are significant because 99.1% is the highest accuracy achieved using the same dataset. This accuracy was obtained using the CCA method with an SVM classifier. Even though the literature has focused on traditional methods, this study proposed a new structure and utilised the existing method to enhance the resultant accuracy, sensitivity, and precision for all classes. Moreover, the proposed method is fast and accurate. The time needed for testing one new image does not exceed milliseconds, which is acceptable in medical applications, and the proposed structure is simple, unique, and accurate. Table 3 and Figure 10 summarise the results for all the methods.

**Table 3.** The results obtained for all proposed methods.

| | Shuffle Net | Cervical Net | Feature Fusion (CCA) |
|---|---|---|---|
| **SVM** | 98.90% | 96.00% | <u>99.10%</u> |
| **RF** | 96.70% | 94.20% | 94.70% |
| **KNN** | 97.40% | 93.70% | 91.10% |
| **Naïve Bayes** | 90.20% | 84.30% | 93.30% |
| **ANN** | 98.60% | 90.40% | 94.90% |

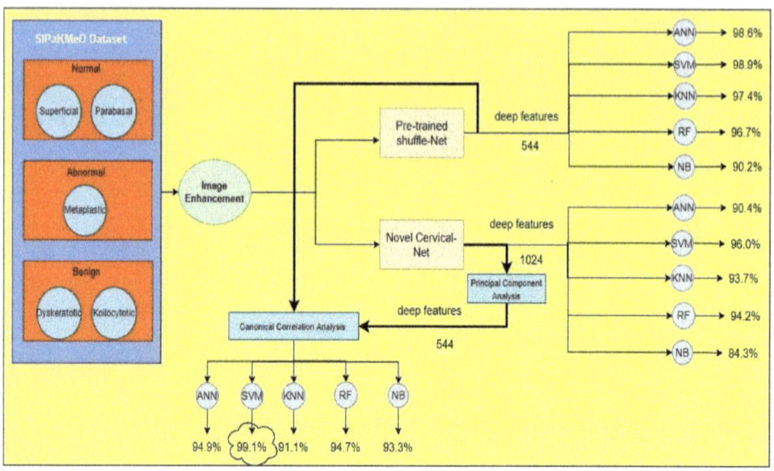


**Figure 10.** The proposed method with the highest accuracy that has been obtained.

This method is now compared with previous studies. The proposed method is distinguished from literature due to its simplicity beside to exploiting new features to obtain the highest accuracy for the SIPaKMeD dataset for five classes. Table 4 summarises the results of all previous studies.

**Table 4.** Comparison of the proposed method with previous studies.

| Study | Method | Dataset | Classes | Accuracy |
|---|---|---|---|---|
| Mbaga et al. [11] | SVM | Herlev dataset | 7 classes | 92.96% |
| Win et al. [12] | SVM, KNN, boosted trees, bagged trees, and major voting | SIPaKMeD dataset | 2 classes<br>5 classes | 98.27%<br>94.09% |
| Plissiti et al. [13] | MLP and SVM | SIPaKMeD dataset | 5 classes | 95.35% |
| Basak et al. [14] | feature selection and DL | SIPaKMeD dataset | 5 classes | 97.87% |
| Park et al. [15] | ResNet-50 and SVM | Cervicography images | 2 classes | 82.00% |
| Tripathi et al. [16] | ResNet-152 | SIPaKMeD dataset | 5 classes | 94.89% |
| Al Mubarak et al. [17] | Fusion based and CNN | | 4 classes | 80.72% |
| Alquran et al. [19] | DL and cascading SVM | Herlev dataset | 7 classes | Up to 92% |
| Dhawan et al. [20] | InceptionV3 | Kaggle dataset | 3 classes | 96.10% |
| Huang et al. [21] | ResNet-50V2 and DenseNet-121 | Tissue biopsy image dataset | 4 classes | 95.33% |
| Mulmule and Kanphade [22] | MLP with three kernels and SVM | Benchmark database | | 97.14% |
| Nikookar et al. [23] | Artificial intelligence | Digital colposcopy dataset | 2 classes | 96% for sensitivity and 94% for specificity |
| Yaman and 155 Tuncer [24] | SVM | SIPaKMeD<br>Mendeley | 2 classes | 98.26%<br>99.47% |
| **This study** | **Cervical Net and feature fusion with ML classifiers** | **SIPaKMeD** | **5 classes** | **99.1%** |

The highest accuracy obtained for the same dataset is given in [14]. Nevertheless, this study achieved the highest accuracy in the literature and proposed a novel DL structure that can extract a new feature. Feature engineering is employed here to find the most significant features and combine them with existing features from the pre-trained DL structure.

## 5. Conclusions

Cervical cancer is the second most frequent cancer among women globally, with a 60% mortality rate. Cervical cancer has no outward symptoms and a long latent period. Therefore, early identification via frequent examinations is critical to counter the high death rate and necessitates using automation in cervical cancer detection. This paper proposed an automated system for cervical cancer using a novel deep learning (DL) structure to extract the features and find the most significant ones. Subsequently, it fused these features with existing pre-trained structures' graphical descriptors. We suggested a system comprising six steps: image acquisition, image enhancement, feature extraction, feature selection, feature fusion, and classification. This system reached the highest accuracy for five classes at 99.1% in the support vector machine (SVM) classifier after selecting the 544 most significant features from the novel Cervical Net and combining them with 544 from Shuffle Net. The key benefit of our technique is its improved prediction performance in separating classes of Pap smear images and showing better classification accuracy. Furthermore, the obtained result is the best among all previous studies, with the largest dataset for single cells. To summarise, a novel DL structure with modifications to the extracted features can outperform existing machine learning (ML) models when detecting cervical cancer from cervicography images.

The presented study can be applied in medical fields because it is built based on a huge dataset, making the results more reliable and confidential. Furthermore, this method combines deep learning features and machine learning classifiers, making it easy, fast, and reliable.

**Author Contributions:** Conceptualization, H.A. and M.A.; methodology, H.A.; validation, W.A.M. and R.A.A.; formal analysis, W.A.M.; investigation, M.A.; resources, R.A.A.; writing—original draft preparation, H.A. and M.A.; writing—review and editing, H.A. and W.A.M.; visualisation, M.A.; supervision, H.A.; project administration, W.A.M. and H.A.; funding acquisition, W.A.M. and A.R.I. All authors have read and agreed to the published version of the manuscript.

**Funding:** This work was supported by the Ministry of Higher Education Malaysia under the Fundamental Research Grant Scheme (FRGS/1/2021/SKK0/UNIMAP/02/1) and Universiti Teknologi PETRONAS.

**Institutional Review Board Statement:** Not applicable.

**Informed Consent Statement:** Not applicable.

**Data Availability Statement:** The dataset analysed during the current study was derived from the SIPaKMeD database, which consists of 4049 manually isolated pap smear cell images. This dataset has been publicly available online since 2018. It is available on the corresponding website: https://www.cs.uoi.gr/~marina/sipakmed.html (accessed on 15 March 2022).

**Acknowledgments:** Thanks to the Ministry of Higher Education Malaysia under the Fundamental Research Grant Scheme (FRGS/1/2021/SKK0/UNIMAP/02/1). The authors thank the authors of the dataset for making it available online, and they would also like to thank the anonymous reviewers for their contribution to enhancing this paper.

**Conflicts of Interest:** The authors declare that they have no conflict of interest to report regarding the present study.

## References

1. World Health Organization. *WHO Cancer Regional Profile 2020*; International Agency for Research on Cancer: Lyon, France, 2020; pp. 1–2.
2. Sung, H.; Ferlay, J.; Siegel, R.L.; Laversanne, M.; Soerjomataram, I.; Jemal, A.; Bray, F. Global Cancer Statistics 2020: GLOBOCAN Estimates of Incidence and Mortality Worldwide for 36 Cancers in 185 Countries. *CA Cancer J. Clin.* **2021**, *71*, 209–249. [CrossRef] [PubMed]
3. Ferlay, J.; Colombet, M.; Soerjomataram, I.; Mathers, C.; Parkin, D.M.; Piñeros, M.; Znaor, A.; Bray, F. Estimating the global cancer incidence and mortality in 2018: GLOBOCAN sources and methods. *Int. J. Cancer* **2019**, *144*, 1941–1953. [CrossRef] [PubMed]
4. Siegel, R.L.; Miller, K.D.; Fuchs, H.E.; Jemal, A. Cancer Statistics, 2021. *CA Cancer J. Clin.* **2021**, *71*, 7–33. [CrossRef]
5. Siegel, R.L.; Miller, K.D.; Jemal, A. Cancer statistics, 2020. *CA Cancer J. Clin.* **2020**, *70*, 7–30. [CrossRef]
6. Mustafa, W.A.; Halim, A.; Nasrudin, M.W.; Rahman, K.S.A. Cervical cancer situation in Malaysia: A systematic literature review. *Biocell* **2022**, *46*, 367–381. [CrossRef]
7. Nahrawi, N.; Mustafa, W.A.; Kanafiah, S.N.A.M. Knowledge of Human Papillomavirus ( HPV ) and Cervical Cancer among Malaysia Residents: A Review. *Sains Malays.* **2020**, *49*, 1687–1695. [CrossRef]
8. William, W.; Ware, A.; Basaza-Ejiri, A.H.; Obungoloch, J. A pap-smear analysis tool (PAT) for detection of cervical cancer from pap-smear images. *Biomed. Eng. Online* **2019**, *18*, 16. [CrossRef]
9. Nkwabong, E.; Badjan, I.L.B.; Sando, Z. Pap smear accuracy for the diagnosis of cervical precancerous lesions. *Trop. Doct.* **2019**, *49*, 34–39. [CrossRef]
10. Mustafa, W.A.; Halim, A.; Jamlos, M.A.; Idrus, Z.S.S. A Review: Pap Smear Analysis Based on Image Processing Approach. *J. Phys. Conf. Ser.* **2020**, *1529*, 022080. [CrossRef]
11. Mustafa, W.A.; Halim, A.; Rahman, K.S.A. A Narrative Review: Classification of Pap Smear Cell Image for Cervical Cancer Diagnosis. *Oncologie* **2020**, *22*, 53–63. [CrossRef]
12. Varalakshmi, P.; Lakshmi, A.A.; Swetha, R.; Rahema, M.A. A Comparative Analysis of Machine and Deep Learning Models for Cervical Cancer Classification. In Proceedings of the 2021 International Conference on System, Computation, Automation and Networking (ICSCAN), Puducherry, India, 30–31 July 2021. [CrossRef]
13. Mbaga, A.H.; ZhiJun, P. Pap Smear Images Classification for Early Detection of Cervical Cancer. *Int. J. Comput. Appl.* **2015**, *118*, 10–16. [CrossRef]
14. Win, K.P.; Kitjaidure, Y.; Hamamoto, K.; Aung, T.M. Computer-assisted screening for cervical cancer using digital image processing of pap smear images. *Appl. Sci.* **2020**, *10*, 1800. [CrossRef]
15. Plissiti, M.E.; Dimitrakopoulos, P.; Sfikas, G.; Nikou, C.; Krikoni, O.; Charchanti, A. Sipakmed: A New Dataset for Feature and Image Based Classification of Normal and Pathological Cervical Cells in Pap Smear Images. In Proceedings of the International Conference on Image Processing, ICIP, Athens, Greece, 7–10 October 2018; pp. 3144–3148. [CrossRef]
16. Basak, H.; Kundu, R.; Chakraborty, S.; Das, N. Cervical Cytology Classification Using PCA and GWO Enhanced Deep Features Selection. *SN Comput. Sci.* **2021**, *2*, 369. [CrossRef]

17. Park, Y.R.; Kim, Y.J.; Ju, W.; Nam, K.; Kim, S.; Kim, K.G. Comparison of machine and deep learning for the classification of cervical cancer based on cervicography images. *Sci. Rep.* **2021**, *11*, 16143. [CrossRef] [PubMed]
18. Tripathi, A.; Arora, A.; Bhan, A. Classification of cervical cancer using Deep Learning Algorithm. In Proceedings of the 5th International Conference on Intelligent Computing and Control Systems, ICICCS 2021, Madurai, India, 6–8 May 2021; pp. 1210–1218. [CrossRef]
19. AlMubarak, H.A.; Stanley, J.; Guo, P.; Long, R.; Antani, S.; Thoma, G.; Zuna, R.; Frazier, S.; Stoecker, W. A hybrid deep learning and handcrafted feature approach for cervical cancer digital histology image classification. *Int. J. Healthc. Inf. Syst. Inform.* **2019**, *14*, 66–87. [CrossRef]
20. Alyafeai, Z.; Ghouti, L. A fully-automated deep learning pipeline for cervical cancer classification. *Expert Syst. Appl.* **2020**, *141*, 112951. [CrossRef]
21. Alquran, H.; Mustafa, W.A.; Qasmieh, I.A.; Yacob, Y.M.; Alsalatie, M.; Al-Issa, Y.; Alqudah, A.M. Cervical Cancer Classification Using Combined Machine Learning and Deep Learning Approach. *Comput. Mater. Contin.* **2022**, *72*, 5117–5134. [CrossRef]
22. Dhawan, S.; Singh, K.; Arora, M. Cervix image classification for prognosis of cervical cancer using deep neural network with transfer learning. *EAI Endorsed Trans. Pervasive Health Technol.* **2021**, *7*, e5. [CrossRef]
23. Huang, P.; Tan, X.; Chen, C.; Lv, X.; Li, Y. AF-SENet: Classification of cancer in cervical tissue pathological images based on fusing deep convolution features. *Sensors* **2021**, *21*, 122. [CrossRef]
24. Mulmule, P.V.; Kanphade, R.D. Supervised classification approach for cervical cancer detection using Pap smear images. *Int. J. Med. Eng. Inform.* **2021**, *1*, 1. [CrossRef]
25. Nikookar, E.; Naderi, E.; Rahnavard, A. Cervical cancer prediction by merging features of different colposcopic images and using ensemble classifier. *J. Med. Signals Sens.* **2021**, *11*, 67–78. [CrossRef] [PubMed]
26. Yaman, O.; Tuncer, T. Exemplar pyramid deep feature extraction based cervical cancer image classification model using pap-smear images. *Biomed. Signal Process. Control* **2022**, *73*, 103428. [CrossRef]
27. Coppola, F.; Faggioni, L.; Gabelloni, M.; De Vietro, F.; Mendola, V.; Cattabriga, A.; Cocozza, M.A.; Vara, G.; Piccinino, A.; Lo Monaco, S.; et al. Human, All Too Human? An All-Around Appraisal of the 'Artificial Intelligence Revolution' in Medical Imaging. *Front. Psychol.* **2021**, *12*, 710982. [CrossRef] [PubMed]
28. Wang, R.; Lei, T.; Cui, R.; Zhang, B.; Meng, H.; Nandi, A.K. Medical image segmentation using deep learning: A survey. *IET Image Process.* **2022**, *16*, 1243–1267. [CrossRef]
29. Erickson, B.J.; Korfiatis, P.; Akkus, Z.; Kline, T.L. Machine learning for medical imaging. *Radiographics* **2017**, *37*, 505–515. [CrossRef] [PubMed]
30. Mustafa, W.A.; Sam, S.; Jamlos, M.A.; Khairunizam, W. Effect of different filtering techniques on medical and document image. *Lect. Notes Electr. Eng.* **2021**, *666*, 727–736. [CrossRef]
31. Alqudah, A.; Alqudah, A.M.; Alquran, H.; Al-zoubi, H.R.; Al-qodah, M.; Al-khassaweneh, M.A. Recognition of handwritten arabic and hindi numerals using convolutional neural networks. *Appl. Sci.* **2021**, *11*, 1573. [CrossRef]
32. Alsharif, R.; Al-Issa, Y.; Alqudah, A.M.; Qasmieh, I.A.; Mustafa, W.A.; Alquran, H. Pneumonianet: Automated detection and classification of pediatric pneumonia using chest X-ray images and cnn approach. *Electronics* **2021**, *10*, 2949. [CrossRef]
33. Alawneh, K.; Alquran, H.; Alsalatie, M.; Mustafa, W.A.; Al-Issa, Y.; Alqudah, A.; Badarneh, A. LiverNet: Diagnosis of Liver Tumors in Human CT Images. *Appl. Sci.* **2022**, *12*, 5501. [CrossRef]
34. Liu, H.; Yao, D.; Yang, J.; Li, X. Lightweight convolutional neural network and its application in rolling bearing fault diagnosis under variable working conditions. *Sensors* **2019**, *19*, 4827. [CrossRef]
35. Brownlee, J. A Gentle Introduction to Pooling Layers for Convolutional Neural Networks. *Mach. Learn. Mastery* **2019**, *22*, 1–16.
36. Basak, H.; Kundu, R. Comparative Study of Maturation Profiles of Neural Cells in Different Species with the Help of Computer Vision and Deep Learning. *Commun. Comput. Inf. Sci.* **2021**, *1365*, 352–366. [CrossRef]
37. Basak, H.; Ghosal, S.; Sarkar, M.; Das, M.; Chattopadhyay, S. Monocular Depth Estimation Using Encoder-Decoder Architecture and Transfer Learning from Single RGB Image. In Proceedings of the IEEE 7th Uttar Pradesh Section International Conference on Electrical, Electronics and Computer Engineering (UPCON), Prayagraj, India, 27–29 November 2020. [CrossRef]
38. Wang, Z.; Wang, L.; Huang, H. Sparse additive discriminant canonical correlation analysis for multiple features fusion. *Neurocomputing* **2021**, *463*, 185–197. [CrossRef]
39. Shi, J.; Chen, C.; Liu, H.; Wang, Y.; Shu, M.; Zhu, Q. Automated Atrial Fibrillation Detection Based on Feature Fusion Using Discriminant Canonical Correlation Analysis. *Comput. Math. Methods Med.* **2021**, *2021*, 6691177. [CrossRef] [PubMed]
40. Zhang, K.; Li, Y.; Wang, J.; Wang, Z.; Li, X. Feature fusion for multimodal emotion recognition based on deep canonical correlation analysis. *IEEE Signal Process. Lett.* **2021**, *28*, 1898–1902. [CrossRef]
41. Pisner, D.A.; Schnyer, D.M. Support vector machine. In *Machine Learning: Methods and Applications to Brain Disorders*; Academic Press: Cambridge, MA, USA, 2019; pp. 101–121. [CrossRef]
42. Alquran, H.; Qasmieh, I.A.; Alqudah, A.M.; Alhammouri, S.; Alawneh, E.; Abughazaleh, A.; Hasayen, F. The melanoma skin cancer detection and classification using support vector machine. In Proceedings of the 2017 IEEE Jordan Conference on Applied Electrical Engineering and Computing Technologies, AEECT 2017, Aqaba, Jordan, 11–13 October 2017; pp. 1–5. [CrossRef]
43. Haykin, S. *Neural Networks and Learning Machines*, 3rd ed.; Prentice Hall: Hoboken, NJ, USA, 2009.
44. Rumelhart, D.E.; Hinton, G.E.; Williams, R.J. Learning representations by back-propagating errors. *Nature* **1986**, *323*, 533–536. [CrossRef]

45. Chen, S.; Webb, G.I.; Liu, L.; Ma, X. A novel selective naïve Bayes algorithm. *Knowl.-Based Syst.* **2020**, *192*, 105361. [CrossRef]
46. Fix, E.; Hodges, J.L. Discriminatory Analysis. Nonparametric Discrimination: Consistency Properties. *Int. Stat. Rev. Rev. Int. Stat.* **1989**, *57*, 238. [CrossRef]
47. Alquran, H.; Alsleti, M.; Alsharif, R.; Qasmieh, I.A.; Alqudah, A.M.; Harun, N.H.B. Employing texture features of chest x-ray images and machine learning in covid-19 detection and classification. *Mendel* **2021**, *27*, 9–17. [CrossRef]
48. Sun, G.; Li, S.; Cao, Y.; Lang, F. Cervical cancer diagnosis based on random forest. *Int. J. Perform. Eng.* **2017**, *13*, 446–457. [CrossRef]

bioengineering

MDPI

Article

# NDG-CAM: Nuclei Detection in Histopathology Images with Semantic Segmentation Networks and Grad-CAM

**Nicola Altini [1,*,†], Antonio Brunetti [1,2,†], Emilia Puro [1], Maria Giovanna Taccogna [1], Concetta Saponaro [3], Francesco Alfredo Zito [4], Simona De Summa [5] and Vitoantonio Bevilacqua [1,2]**

1. Department of Electrical and Information Engineering (DEI), Polytechnic University of Bari, Via Edoardo Orabona n.4, 70126 Bari, BA, Italy
2. Apulian Bioengineering s.r.l., Via delle Violette n.14, 70026 Modugno, BA, Italy
3. Laboratory of Preclinical and Translational Research, Centro di Riferimento Oncologico della Basilicata (IRCCS-CROB), Via Padre Pio n.1, 85028 Rionero in Vulture, PZ, Italy
4. Pathology Department, IRCCS Istituto Tumori "Giovanni Paolo II", Via O. Flacco n.65, 70124 Bari, BA, Italy
5. Molecular Diagnostics and Pharmacogenetics Unit, IRCCS Istituto Tumori "Giovanni Paolo II", Via O. Flacco n.65, 70124 Bari, BA, Italy

* Correspondence: nicola.altini@poliba.it

† These authors contributed equally to this work.

**Abstract:** Nuclei identification is a fundamental task in many areas of biomedical image analysis related to computational pathology applications. Nowadays, deep learning is the primary approach by which to segment the nuclei, but accuracy is closely linked to the amount of histological ground truth data for training. In addition, it is known that most of the hematoxylin and eosin (H&E)-stained microscopy nuclei images contain complex and irregular visual characteristics. Moreover, conventional semantic segmentation architectures grounded on convolutional neural networks (CNNs) are unable to recognize distinct overlapping and clustered nuclei. To overcome these problems, we present an innovative method based on gradient-weighted class activation mapping (Grad-CAM) saliency maps for image segmentation. The proposed solution is comprised of two steps. The first is the semantic segmentation obtained by the use of a CNN; then, the detection step is based on the calculation of local maxima of the Grad-CAM analysis evaluated on the nucleus class, allowing us to determine the positions of the nuclei centroids. This approach, which we denote as NDG-CAM, has performance in line with state-of-the-art methods, especially in isolating the different nuclei instances, and can be generalized for different organs and tissues. Experimental results demonstrated a precision of 0.833, recall of 0.815 and a Dice coefficient of 0.824 on the publicly available validation set. When used in combined mode with instance segmentation architectures such as Mask R-CNN, the method manages to surpass state-of-the-art approaches, with precision of 0.838, recall of 0.934 and a Dice coefficient of 0.884. Furthermore, performance on the external, locally collected validation set, with a Dice coefficient of 0.914 for the combined model, shows the generalization capability of the implemented pipeline, which has the ability to detect nuclei not only related to tumor or normal epithelium but also to other cytotypes.

**Keywords:** nuclei segmentation; histopathology; deep learning; Grad-CAM; semantic segmentation; instance segmentation; nuclei detection

**Citation:** Altini, N.; Brunetti, A.; Puro, E.; Taccogna, M.G.; Saponaro, C.; Zito, F.A.; De Summa, S.; Bevilacqua, V. NDG-CAM: Nuclei Detection in Histopathology Images with Semantic Segmentation Networks and Grad-CAM. *Bioengineering* **2022**, *9*, 475. https://doi.org/10.3390/bioengineering9090475

Academic Editors: Paolo Zaffino and Maria Francesca Spadea

Received: 11 August 2022
Accepted: 13 September 2022
Published: 15 September 2022



## 1. Introduction

In the healthcare scenario, artificial intelligence is exploited in medical imaging as a powerful tool with which to characterize objects of interest and lesions in anatomical regions under consideration. Traditionally, pathologists manually analyze numerous biopsies or tissue samples to diagnose complex pathologies, such as cancer. Even though it is tedious and time-consuming, this approach remains the gold standard [1,2].

Computational pathology attempts to overcome the main challenges arising from manual histological image evaluation, such as inter- and intraobserver variability or the inability to evaluate the smallest visual features and the time required to examine whole slide images (WSIs) [1,3,4].

The nuclei of cells provide a great deal of information for the analysis of histopathological tissue. For instance, immunohistochemistry-marked nuclei can be exploited for the estimation of cellular proliferation in cancer (e.g., Ki-67). Hence, nuclei segmentation is a fundamental first step toward the automated analysis of WSIs [5]. However, the difficulties associated with variable coloring arising from hematoxylin and eosin (H&E)-stained images, overlapped nuclei, the presence of artifacts, and differences in cell morphology and texture, represent obstacles for computer-based segmentation algorithms [2,3]. Moreover, WSIs have very high resolutions and contain an enormous number of nuclei, adding peculiarity to the task [6]. A critical aspect in several computational pathology pipelines is to achieve accurate segmentation of nuclei both for subsequent extraction and classification of nucleus features, but also for analyzing cellular distribution, useful for classifying tissue subtypes and identifying abnormalities [3].

Several studies focused on nuclei detection because of its importance in the pathologic diagnostic pipeline, in particular in the field of oncology. As an example, nuclei detection could be helpful to distinguish nuclei undergoing changes, indicating a progression of squamous epithelium cervical intraepithelial neoplasia [7]. Moreover, the estimation of tumor cellularity is very important, particularly in the era of precision medicine. Indeed, bioinformatic pipelines for copy number variation analysis require tumor cellularity as input and for a correct evaluation of variant allelic frequency [8].

Machine learning-based nuclear segmentation methods are typically the most efficient, as they can learn to identify variations in the shape and coloration of nuclei. In the semantic segmentation [9,10] approach, all image pixels are labeled as nuclear or background through a deep learning model. Nevertheless, these methods often fail to distinguish the different instances of objects of interest, i.e., nuclei, which then need to be addressed with ad hoc post-processing techniques, such as clustering [11].

The detection task can be approached by exploiting morphological features. CRImage [12] profits from thresholding as the first step for nuclei detection. Centroids of segmented nuclei are used as the point of detection. Then, a list of statistics for each segmented nucleus is utilized as a feature vector, and classification involves a support vector machine with radial basis kernel. Finally, spatial density smoothing is used to correct false detections.

LIPSyM [13] introduces the local isotropic phase symmetry measurement, designed to give high values to cell centers and nearby pixels; on the other hand, it cannot precisely detect spindle-like and other irregularly shaped nuclei such as fibroblasts and malignant epithelial nuclei.

In the last several years, convolutional neural networks (CNN) are emerging as the most effective way to tackle the nuclei detection task. In particular, the spatially constrained convolutional neural network (SC-CNN) [14] uses spatial regression for localizing the nuclei centers; the regression in SC-CNN is model-based, which explicitly constrains the output form of the network.

Xu et al. [6] used a stacked sparse autoencoder (SSAE) to learn a high-level representation of nuclear and non-nuclear objects by means of a softmax classifier.

Finally, the R2U-Net-based regression model named "UD-Net" [4] is proposed for end-to-end nuclei detection from pathological images. The recurrent convolutional operations help the model learn and represent features better than the feed-forward convolutional operations, and the robustness of the R2U-Net model has been demonstrated previously in several studies [15].

Methodologies prior to the advent of deep learning demonstrate worse performance on the nuclei detection task. Moreover, handcrafted feature extraction is a tedious and complex process, which can lead to different results depending on the experience of the feature engineers and domain experts. It is worth noting that CNN-based approaches require datasets with a distinct label for every nucleus, based on observations made in the last several years. Simple existing semantic segmentation methods, trained without the knowledge of different instances, cannot be reliably adopted for nuclei detection.

Many cell nuclei detection methods share a basic approach that includes generating an intermediate map through a CNN that indicates the presence of a nucleus, called the probability or proximity map (P-Map) [3,16], or have specialized architectures that are trained to individuate the centers of the nuclei, such as SC-CNN [14]. Indeed, the P-Map represents proximities as a monochromatic image: the intensities have high values near the centroid of the nucleus, and gradually lower going toward the boundaries.

By following the idea of determining a structure similar to a P-Map, we propose a novel method for nuclei detection, without the need for specialized architectures or handcrafted feature extraction; rather, only semantic segmentation networks and explainable artificial intelligence (XAI) techniques are used. The proposed method is quick to train, and is extensible because it can be plugged on top of existing semantic segmentation networks.

The presence of clustered or overlapped nuclei with semantic segmentation models can be spotted on visual inspection of the images. In order to overcome this issue, we exploited the potentialities of the gradient-weighted class activation mapping (Grad-CAM) for segmentation, which made it possible to highlight the activation of the nucleus class (compared to the background class), thus obtaining a saliency map with properties similar to the classic P-Map. The locations of the nuclei are subsequently determined by looking for local maxima in the activation map. Starting from the identified centroids, it is possible to associate all the pixels belonging to the considered nucleus, with a proximity criterion. This model alone, which we denote as nuclei detection with Grad-CAM (NDG-CAM), was capable of achieving performance in line with state-of-the-art methods. Because the Mask R-CNN [17] instance segmentation architecture is widely employed and constitutes a standard baseline for these tasks, we also realized a combined model for further enhancing the results, surpassing the state of the art.

To summarize, our contributions can be considered as follows: (i) we introduce a novel detection method for nuclei—NDG-CAM—which exploits Grad-CAM for semantic segmentation; (ii) we collected and annotated a local dataset of patients diagnosed with colorectal cancer to show the applicability of the proposed method in a local hospital; (iii) we examined and compared different state-of-the-art techniques to show the effectiveness of the proposed approach; (iv) we trained and evaluated an instance segmentation architecture as the baseline; and (v) we proposed a combined model which, exploiting both NDG-CAM and Mask R-CNN, can surpass the current literature performance concerning nuclei detection.

The remainder of the manuscript is organized as follows. Section 2 first describes the datasets adopted for the analysis. Then, semantic segmentation configurations and architectures are presented. The NDG-CAM is proposed, and its workflow is delineated. An instance segmentation is also considered as the baseline. Lastly, implementation details, the combined model, and the evaluation metrics employed for the analysis are presented. Results are portrayed in Section 3 and discussed in Section 4. A comparison with other state-of-the-art approaches is considered here. Lastly, final remarks, conclusions, and ideas for future works are drawn in Section 5.

## 2. Materials and Methods

### 2.1. Datasets

For the tasks of nuclei segmentation and detection, different datasets were considered in order to find the best-performing model. In particular, we considered the latest and largest publicly available datasets for nuclei detection and segmentation. Moreover, a local dataset has been collected, to prove the feasibility of the proposed system on new data from a local hospital.

- **MoNuSeg** [1,18,19]. The cell nucleus segmentation dataset used in this work is publicly accessible from the 2018 Data Science Bowl challenge [20]. The dataset contains a large number of segmented nuclei images and includes different cell types; there are 30 training H&E images containing 21,623 hand-annotated nuclear boundaries from the breast, kidney, prostate, liver, colon, bladder, and stomach. Moreover, there are also 14 H&E test images containing 7000 nuclear boundary annotations from the breast, kidney, prostate, colon, bladder, lung, and brain. All images, each of size 1000 × 1000 , were captured at 40× magnification. The nuclear contour annotations are provided through XML files.
- **CRCHistoPhenotypes**: Labeled Cell Nuclei Data [14,21]. This publicly available dataset contains 100 H&E-stained histology images of colon cell nuclei obtained from WSI of 10 patients with a magnification factor of 20×. Tiles have a size of 500 × 500. Nuclear annotations are provided through the coordinates of the centroids in `.mat` format, resulting in a total of 29,756 annotated nuclei for detection purposes.
- **NuCLS** [22]. The dataset contains over 220,000 labeled nuclei from breast cancer images from TCGA, obtained from 125 patients with breast cancer (1 slide per patient) and captured with a magnification factor of 40×. These nuclei were annotated through the collaborative effort of pathologists, pathology residents, and medical students. Data from both single-rater and multi-rater studies are provided. For single-rater data, there are both pathologist-reviewed and uncorrected annotations. For multi-rater datasets, there are annotations generated with and without suggestions from weak segmentation and classification algorithms. We used only the single-rater dataset, which is already split into train and test sets. The annotations for the single-rater dataset include 59,485 nuclei and 19,680 boundaries, extracted from 1744 H&E image tiles of variable dimensions between 200 and 400 pixels.
- **Local dataset** from Pathology Department of IRCCS Istituto Tumori Giovanni Paolo II [23]. This consists of 19 H&E image tiles which overall contain more than 6378 nuclei from patients with colorectal cancer. Images have a size of 512 × 512 and were captured at 40× magnification. Annotations have been provided by a biologist with experience in analyzing histopathological data.

Hereafter, we will denote with T1 and V1 the training and test sets of MoNuSeg (D1), and with D2 the overall dataset of CRCHistoPhenotypes. The Mask R-CNN model has been trained on the NuCLS (D3) dataset, being the largest publicly available dataset with annotations formatted for instance segmentation. Because D1 already includes a validation set, we have used that one for the first validation stage. As an independent external validation set, we collected other image tiles from the Pathology Department of IRCCS Istituto Tumori Giovanni Paolo II [23], which will be denoted as V4, in order to assess the generalization capability of the best semantic segmentation network configuration individuated with the D1 and D2 datasets, and the Mask R-CNN model trained on the D3 dataset. Figure 1 summarizes the pipeline implemented for training and validating the models.

A summary of the details for the employed datasets is reported in Table 1, whereas sample images are depicted in Figure 2.

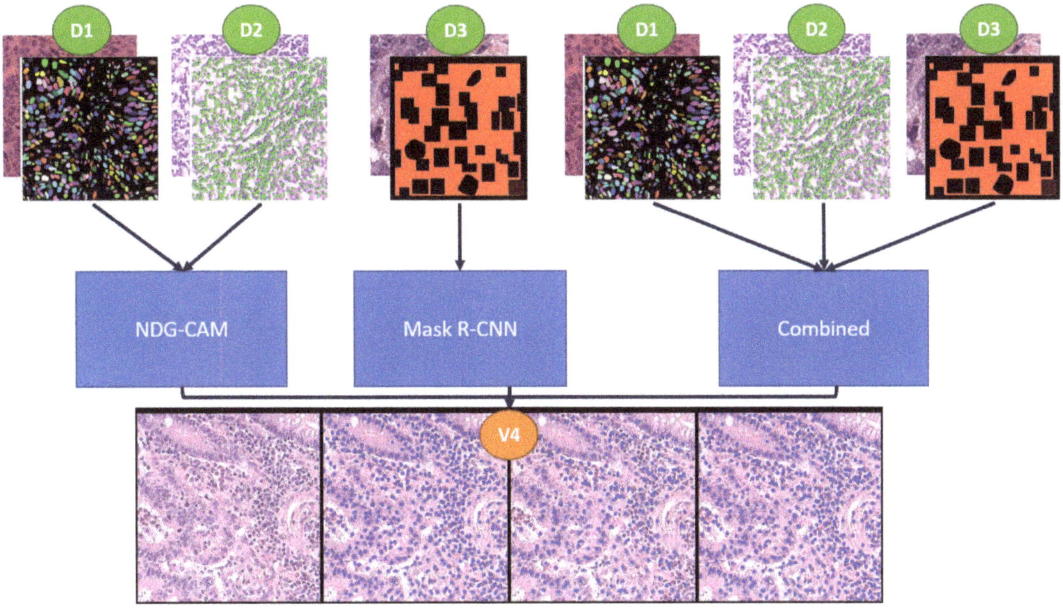


**Figure 1.** Pipeline adopted for training and validation. D1 and D2 datasets have been used to train and select the best semantic segmentation network. D3 dataset has been exploited to train the Mask R-CNN instance segmentation architecture. Finally, external validation has been conducted on the local validation dataset V4.

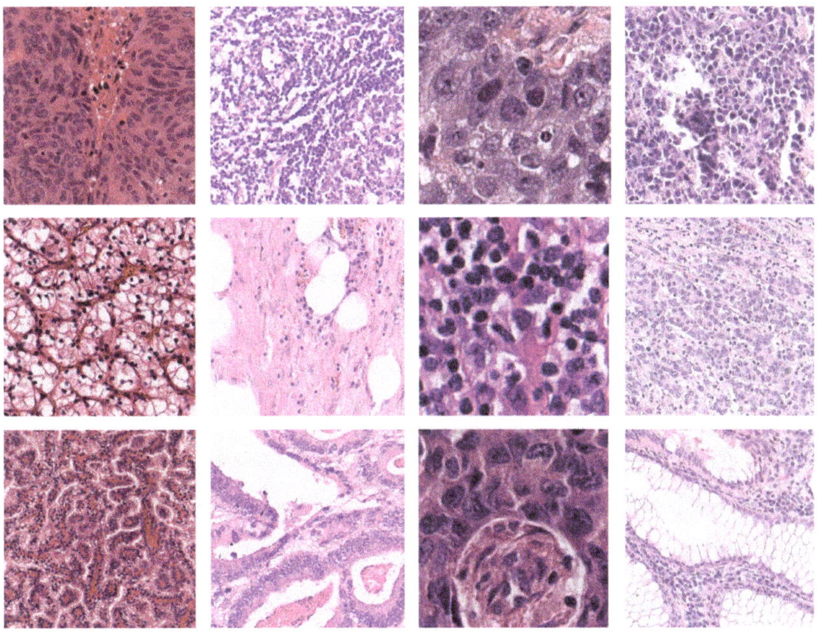

**Figure 2.** Sample images of datasets for nuclei detection. (First column) D1—MoNuSeg [18]; (second column) D2—CRCHistoPhenotypes [21]; (third column) D3—NuCLS [22]; (fourth column) V4—local dataset [23].

**Table 1.** Summary of datasets for nuclei.

| Dataset | Publication Year | Organs | Resolution | Number of H&E images | Number of Nuclei | Size (pixels) | Annotations Format |
|---|---|---|---|---|---|---|---|
| MoNuSeg—Train (T1) [1] | 2017 | breast, kidney, prostate, liver, colon, bladder, stomach | 40× | 30 | 21,623 | 1000 × 1000 | Nuclei Contours |
| MoNuSeg—Test (V1) [1] | 2017 | breast, kidney, prostate, colon, bladder, lung, brain | 40× | 14 | 7000 | 1000 × 1000 | Nuclei Contours |
| CRCHistoPhenotypes (D2) [14] | 2016 | colon | 20× | 100 | 29,756 | 500 × 500 | Nuclei Centroids |
| NuCLS (D3) [22] | 2019 | breast | 40× | 1744 | 59,485 | 200–400 per side | Nuclei Contours or Bounding Boxes |
| Local (V4) | 2022 | colon | 40× | 19 | 6378 | 512 × 512 | Nuclei Centroids |

## 2.2. NDG-CAM

In this section, we introduce the methodology adopted for NDG-CAM. Several steps have been carried out. As the first step, a semantic segmentation architecture trained for nuclear segmentation is required. Different experimental configurations of the datasets and network architectures have been compared in order to find the most suitable model, with details reported in Sections 2.2.1 and 2.2.2. Then, the Grad-CAM technique for semantic segmentation, which is still underexplored if compared to Grad-CAM for classification, has been employed to obtain saliency maps of the nuclei, with higher values of intensity corresponding to positions nearest to the centroids. Subsequently, a search for local maxima, combined with post-processing and clustering, allowed for the detection and eventually instance segmentation of the nuclei. This process is presented in Section 2.2.3. Compared to specialized architectures, such as those used for instance segmentation, semantic segmentation networks are simpler and faster to train. In addition, our system can be trained if labels do not distinguish between different nuclear instances, which would not be possible for instance segmentation models.

### 2.2.1. Semantic Segmentation Workflow

Starting from the datasets described in the previous sections, the following experiments were carried out, all with images at a size of 512 × 512:

a Train on D2 and validation on V1 at 20× resolution.
b Train on T1 and validation on V1 at 20× resolution.
c Train on T1 and validation on V1 at 40× resolution.

In the first two experiments, images were padded from 500 × 500 to 512 × 512 exploiting the mirror padding. Instead, in the last experiment, the images were padded from 1000 × 1000 to 1024 × 1024 with mirror padding and subsequently divided into 4 tiles of 512 × 512. For each experiment, different deep network architectures were trained and compared: U-Net [24], SegNet [25], and DeepLab v3+ [26] in three different backbone configurations, namely ResNet18, ResNet50 [27], and MobileNet-v2 [28]. The aforementioned experiments were carried out in MATLAB R2021a.

### 2.2.2. Network Architectures

The segmentation phase is a milestone for the detection phase; this step aims to discriminate between cell nuclei and the background. semantic segmentation architectures play a role of pivotal importance in deep learning-based medical image analysis [9,29–31]. It is a process that associates a label or a category to each pixel of an input image, thus allowing the pixelwise spatial localization of each object category appearing in the scene.

In the specific case under analysis, the goal was to segment the cell nuclei in a robust way, so as to provide satisfactory results even when the algorithm would have been applied to different images of the same type. For this reason, it was decided to carry out the same experiments with several convolutional architectures.

The considered architectures include:

- U-Net [24]. It is a fully convolutional network to perform the semantic segmentation task. The U-Net architecture consists of a series of encoding layers and contractions that are used to extract the context of the image, followed by a sequence of symmetrical decoding layers and expansions to recover the spatial information. In our MATLAB setting, the network is characterized by 58 convolutional layers; the first layer deals with a z-score normalization of the inputs, whereas the last one presents the Dice function as a loss function.
- SegNet [25]. This is another encoder–decoder architecture. In this case, the decoding blocks exploit max pooling indices received from the corresponding contraction block to perform the oversampling, instead of using trainable upsampling layers as transposed convolutions. In our MATLAB setting, this CNN consists of 31 layers with a cross-entropy loss function.
- DeepLab v3+ [26]. This architecture features atrous spatial pyramid pooling (ASPP) and the encoder–decoder paradigm. The first aspect concerns a particular way of combining layers of atrous and depthwise convolution, with which the model captures and concatenates features at different scales. For this network, the backbone is customizable. Three different basic CNN encoders were used: ResNet18, ResNet50, and MobileNet-v2. The DeepLab v3+ has 100 layers, of which the last is a softmax layer that is used to obtain the probabilities that each pixel belongs to the nucleus or background class; in this case, the chosen loss function is the Dice loss.

An example of semantic segmentation prediction from DeepLab v3+ with backbone ResNet18 is shown in Figure 3.

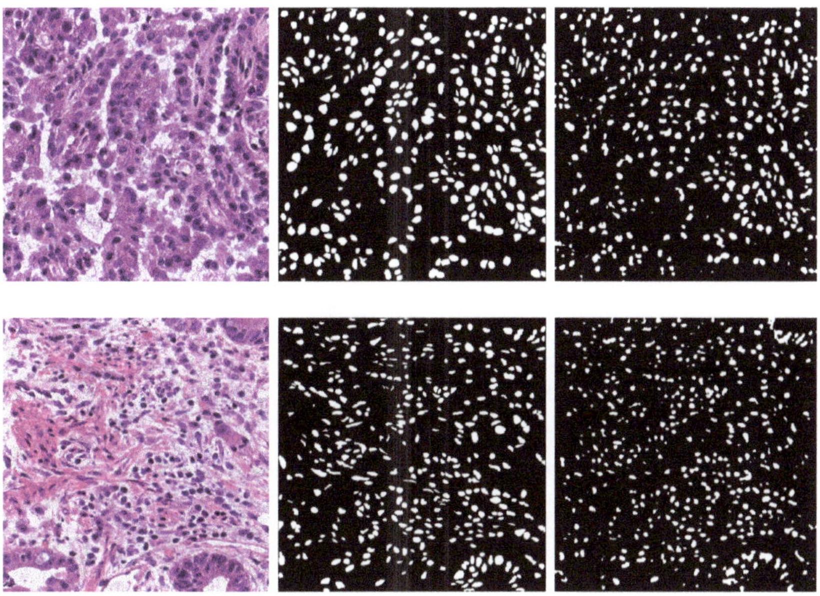

**Figure 3.** Semantic segmentation output for nuclei images. (**Left**) Original image. (**Middle**) Ground truth. (**Right**) Prediction of experiment (b) with DeepLab v3+ and backbone ResNet18.

### 2.2.3. Nuclei Detection with Grad-CAM

After the best performing network has been identified, the output returned by the semantic segmentation was a mask in which the pixels of the input image were classified into pixels belonging to the foreground, i.e., nucleus, or background class. As mentioned previously, this did not allow us to distinguish multiple instances of the same object and therefore to distinguish multiple nuclei adjacent to each other.

In this scenario, the detection phase begins. In fact, after the semantic segmentation, post-processing was carried out in order to solve this problem. The first step was to calculate the Grad-CAM of the input image according to the chosen network. A CNN is often seen as a black box, or rather, as a model with parameters $W$ that, given an image of input $X$, through a function $f(X, W)$, is able to map to the related output $y$. XAI techniques have been designed in order to unveil the underlying mechanisms involved in the processing stages of deep neural networks, and are recently gaining a lot of attention in medical imaging and clinical decision support systems [32–35].

During the training phase, even if we are capable of achieving high performance according to the considered metrics, we do not know which image features are more determinant for the network to make its choices. One of the ways to visually solve this problem is Grad-CAM [35].

Grad-CAM is typically used in image-classification scenarios [36], but it can also be extended to semantic segmentation problems [37]. In general, the heatmap $L^c$ for class $c$ is generated by using $a_c^k$ (as defined in Equation (1)) to sum the feature maps $A^k$, as in Equation (2).

$$a_c^k = \frac{1}{N} \sum_{u,v} \frac{\partial y^c}{\partial A_{uv}^k} \quad (1)$$

$$L^c = \mathrm{ReLU}\left( \sum_k a_c^k A^k \right) \quad (2)$$

$N$ is the number of pixels and $(u, v)$ are the indices. ReLU is applied pixelwise to clip negative values at zero, to only highlight areas that positively contribute to the decision for class $c$. The difference with the classification task is that for semantic segmentation $y^c$, the scalar class score, is obtained by reducing the pixelwise class scores for the class of interest to a scalar [37], as in Equation (3).

$$y^c = \sum_{(u,v) \in P} Y_{(u,v)}^c \quad (3)$$

$P$ is a set of pixel indices of interest in the output layer: in our case, the softmax layer before the pixel classification layer. Higher values of $L^c$ map indicate which areas of the image are important for the decision to classify pixels.

In the proposed approach, the activation of the network for the nucleus class was analyzed, obtaining a probability map with values that we denote as CAM-Map. Therefore, activations greater in correspondence with the centroids of the nuclei (even when adjacent to each other) are visible from Figure 4C.

From CAM-Map, we applied a morphological grayscale dilation operator with a spherical shape factor of radius 7. The result is depicted in Figure 4D. This step allowed the enlargement of the activation areas so that no false nuclei were identified in the nearby regions where activations were not high enough compared to the maximum point.

Then, as portrayed in Figure 4E, we proceeded with the calculation of the local maximum of the regions and the localization of all the connected components, with the related geometric centroids, which correspond to the identified nuclei.

Once the centroids were found, K-means clustering, with K equal to the number of connected components, has been exploited to associate the adjacent pixels to each nucleus, so as to have the overall predicted mask of the original starting image. The final mask is reported in Figure 4F.

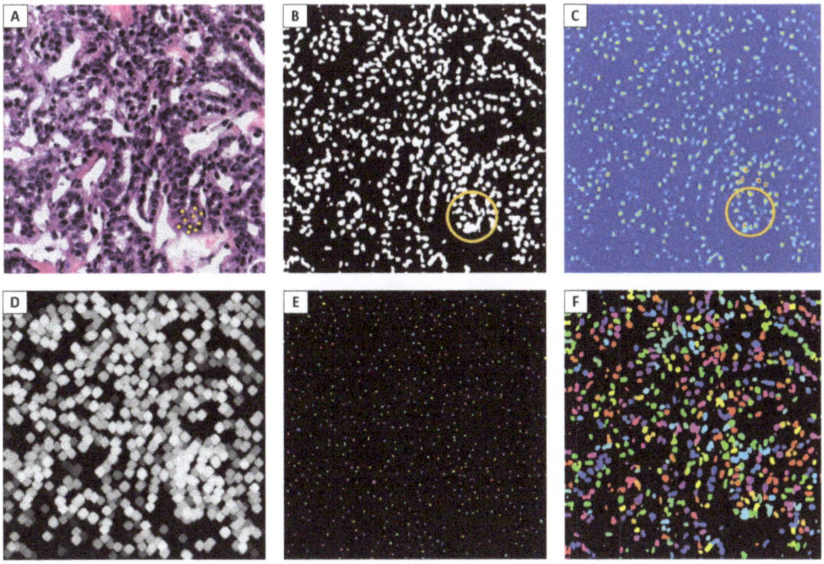


**Figure 4.** NDG-CAM Detection workflow. (**A**) Zone with multiple neighboring instances of nuclei. (**B**) Failure to recognize adjacent nuclei. (**C**) Grad-CAM for semantic segmentation. (**D**) Dilated image. (**E**) Connected components. (**F**) Detection prediction.

### 2.3. Instance Segmentation

Object detection involves the detection, with a bounding box, of all the different objects of interest present in a scene. Instance segmentation further extends this task, by also considering the problem of delineating a precise mask around each object. Architectures for object detection are usually divided into one-stage and two-stage models, with the first being faster and the former being more accurate. Inside the realm of methods for two-stage object detectors, a pivotal role has been played by architectures from the R-CNN family [38].

Mask R-CNN evolves the R-CNN family by adding a semantic segmentation branch, making the model capable of performing instance segmentation [17]. The overall Mask R-CNN architecture is composed of two parts: the backbone architecture, which performs feature extraction, and the head architecture, which performs classification, bounding box regression, and mask prediction.

We employed the Detectron2 [39], a platform powered by the Pytorch framework, that provides state-of-the-art detection and segmentation algorithms. It includes high-quality implementations of the most popular object detection algorithms, comprising different variants of the pioneering Mask R-CNN model. Detectron2 has an extensible design so that it can be easily employed to implement cutting-edge research projects.

The NuCLS dataset [22] was chosen to train the network, the instance segmentation model `mask_rcnn_R_50_DC5_1x`. Annotations were converted into the COCO annotation format for adoption in the Detectron2 framework.

### 2.4. Implementation Details

All the semantic segmentation networks have been trained on a laptop with a GeForce GTX960M. For carrying out the training, the chosen optimizer was SGDM, with a starting learning rate of 0.05. The learning rate schedule was piecewise with a drop factor of 0.94 and a drop period of 2. $L_2$ regularization parameter was set to 0.0005. With a batch size of 2, 15 epochs lasted roughly 105 min for the best performing architecture, DeepLab v3+ with ResNet18 as the backbone.

The Mask R-CNN model, being heavier, has been trained on a Google Colab Pro environment. With a Tesla P100, 20,000 iterations were carried out in roughly 110 min. The chosen optimizer was SGDM, as set by default in the Detectron2 environment, with a starting learning rate of 0.00025.

### 2.5. Combined Model

In order to obtain the advantages of both approaches, a combined model has been developed.

It exploits a criterion for obtaining merged outputs from NDG-CAM detection and Mask R-CNN. In detail, a distance criterion was used to check if a nucleus was found by only one of the approaches. In that case, the nucleus was simply retained. Instead, if more nuclei centroids are found in proximity, only the ones found by Mask R-CNN are retained. The combined methodology has the idea to increase the recall, which is very important because nuclei detection is the first stage for further analyses.

### 2.6. Evaluation Metrics

Each semantic segmentation architecture described in Section 2.2.1 was tested in all three experimental configurations mentioned. In order to assess the goodness of pixelwise classification performed by semantic segmentation networks, the pixelwise precision, recall, and Dice coefficient were considered as performance indices. Given pixelwise true positives (TP), false positives (FP) and false negatives (FN), then precision, recall, and Dice coefficient can be defined as in Equations (4)–(6), respectively:

$$Precision = \frac{TP}{TP + FP} \tag{4}$$

$$Recall = \frac{TP}{TP + FN} \tag{5}$$

$$Dice = \frac{2 \cdot TP}{2 \cdot TP + FP + FN}. \tag{6}$$

For all these metrics, a higher value denotes a better segmentation result; that is, predicted masks are more similar to ground truth ones.

Instead, for assessing the detection procedure, we considered two kinds of metrics. The first is based on the simple calculation of the number of detected nuclei with respect to the ground truth. The error ($e_a$), defined in Equation (7), is given by the difference in absolute value between the number of nuclei found and the real number, divided by the latter. An example of the prediction vs. ground truth result, which is the basis for enumerating nuclei, is depicted in Figure 5A. Because we were also interested in understanding if our algorithm was more prone toward overdetection or underdetection, a signed error ($e_s$), defined in Equation (8), was also evaluated:

$$e_a = \frac{|d - g|}{g} \tag{7}$$

$$e_s = \frac{d - g}{g}. \tag{8}$$

In these two equations, $d$ denotes the number of detected nuclei, whereas $g$ is the number of ground truth nuclei.

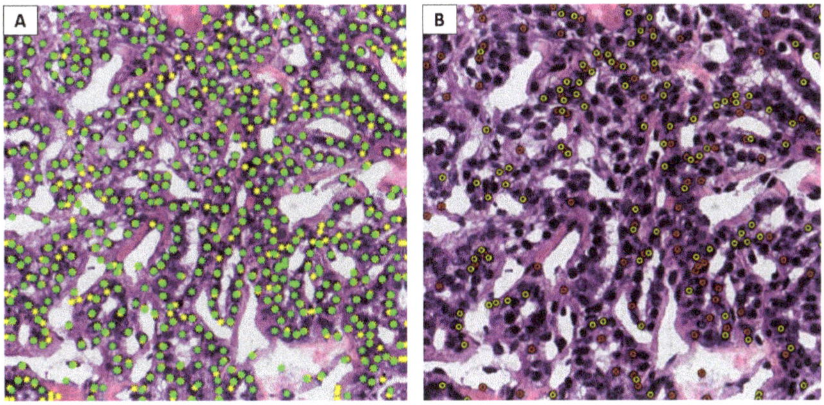

**Figure 5.** Example of calculation of evaluation metrics for object detection. (**A**) Prediction vs ground truth. Yellow, ground truth; green, prediction; (**B**) Differences between prediction and ground truth. Yellow, detection FN; red, detection FP.

The second category of metrics includes Dice coefficient, precision, and recall for object detection, which can provide more information about the quality of the detection results. In this case, we are not simply rewarding our prediction of as many nuclei as are present in the ground truth, but we also want to ensure that detected nuclei are in the right place. In order to achieve this result, we need to discover object detection FP and FN, as can be seen in Figure 5B. In order to determine these quantities, as the first step, we computed the distance matrix between the centroids of the detected nuclei and the real ones. In order to decide whether a detection actually corresponds to a nucleus centroid, a distance threshold $\xi$ was considered, equal to the mean radius of the nuclei of each image [16]. If the distance between a prediction and a ground truth annotation is less than or equal to $\xi$, the prediction is counted as a TP. If more than one detection verifies this condition, the one closest to the ground truth position is counted as TP and the others as FP. The detections further than $\xi$ from any ground truth location are counted as FP, and all ground truth annotations without close detections are marked as FN. Lastly, the following control condition was added. If the distance between an FP and an FN is less than an $\epsilon$ threshold, set to 6 (a value close to the nuclear radius), the count of FP and FN will each be decreased by one, whereas TP will be increased by one. The pseudocode for determining TP, FP, and FN is reported in Algorithm 1.

In order to assess the statistical significance of the obtained results calculated per case, we determined the $p$-value with the two-tailed Wilcoxon signed-rank test. The threshold for significance has been set to 0.05.

**Algorithm 1:** Object Detection TP, FP, FN calculation.

```
input  : gt, the ground truth nuclei centroids, an array of g coordinate pairs
         pred, the predicted nuclei centroids, an array of d coordinate pairs
         ξ, the mean radius of the ground truth nuclei
         ε, the distance threshold                              // set to 6
output : TP, the true positives
         FP, the false positives
         FN, the false negatives
g = size(gt)
TP = 0
FP = 0
FN = 0
idx_FP = list()                          // a list of false positive indexes
idx_FN = list()                          // a list of false negative indexes
δ = distance(gt, pred)                               // the distance matrix
i = 0
while i < g do
    v = δ[:, i]
    idx = where(v < ξ)                 // a (possibly empty) array of indexes
    if size(idx) == 1 then
        TP = TP + 1
    else if size(idx) > 1 then
        TP = TP + 1
        FP = FP + (size(idx) − 1)
        idx_FP.extend(idx)
    else if size(idx) == 0 then
        FN = FN + 1
        idx_FN.append(i)
    end
    i = i + 1
end
arr_FN = filter(gt, idx_FN)                    // extract the false negatives
p = 0
while p < size(idx_FP) do
    a = 0
    while a < size(arr_FN) do
        Δ = distance(pred[p], arr_FN[a])
        if (Δ ≤ ε) then
            FP = FP − 1
            FN = FN − 1
            TP = TP + 1
        end
        a = a + 1
    end
    p = p + 1
end
```

## 3. Results

The automatic segmentation of cell nuclei attracted significant interest from the scientific community, as their identification is an important starting point for many medical analyses based on histopathological images. In this work, for the semantic segmentation phase, different architectures were elaborated and tested on different datasets, for a total of 15 experiments. For each of them, performance indices were calculated to identify the best model with which to proceed for the subsequent phases. From this comparison, it emerged that the best performance can be obtained by referring to the experimental configuration (b) defined in Section 2.2.1.

Table 2 reports the results obtained for each network architecture in the semantic segmentation task. For DeepLab v3+, the backbone architecture is included within square brackets.

**Table 2.** Performance comparison between considered network architectures for semantic segmentation.

| Network | Metric | Experiment (a) | Experiment (b) | Experiment (c) |
|---|---|---|---|---|
| U-Net | DICE | 66.74 ± 3.44 | 65.71 ± 8.57 | 60.74 ± 11.65 |
| | PRECISION | 57.13 ± 8.15 | 52.69 ± 11.96 | 45.43 ± 11.77 |
| | RECALL | 83.56 ± 10.61 | 91.65 ± 6.57 | 96.46 ± 2.44 |
| SegNet | DICE | 56.44 ± 9.31 | 65.05 ± 6.32 | 62.02 ± 12.28 |
| | PRECISION | 67.09 ± 8.01 | 58.93 ± 14.23 | 51.67 ± 14.96 |
| | RECALL | 52.60 ± 16.20 | 81.35 ± 17.69 | 85.05 ± 13.24 |
| DeepLab v3+ [ResNet18] | DICE | 52.21 ± 11.99 | **74.23 ± 4.85** | 72.17 ± 8.03 |
| | PRECISION | 76.78 ± 6.60 | 76.42 ± 8.69 | 62.76 ± 11.78 |
| | RECALL | 41.76 ± 13.55 | 74.25 ± 11.23 | 87.17 ± 5.64 |
| DeepLab v3+ [ResNet50] | DICE | 57.87 ± 6.88 | 61.68 ± 8.75 | 65.98 ± 7.84 |
| | PRECISION | 59.70 ± 6.35 | 63.69 ± 7.51 | 54.14 ± 13.81 |
| | RECALL | 57.10 ± 10.43 | 60.71 ± 11.94 | 90.95 ± 10.02 |
| DeepLab v3+ [mobilenetv2] | DICE | 56.64 ± 6.60 | 73.01 ± 7.56 | 66.31 ± 13.80 |
| | PRECISION | 66.49 ± 5.56 | 73.50 ± 11.76 | 57.52 ± 16.31 |
| | RECALL | 50.66 ± 10.50 | 75.07 ± 10.38 | 85.35 ± 9.43 |

It therefore emerges that the best solution coincides with experiment (b) conducted with DeepLab v3+ using the ResNet18 network as the backbone. It allowed us to obtain a pixelwise Dice coefficient of 74.23 ± 4.85%, a precision of 76.42 ± 8.69%, and a recall of 74.25 ± 11.23%.

DeepLab v3+ was hence chosen as the base model to be exploited in the detection phase. By exploiting the Grad-CAM for semantic segmentation, it was possible to retrieve nuclei centroids via local maxima of the obtained saliency maps.

On the V1 dataset, the experimental results demonstrated an $e_a$ of the identified nuclei equal to 2.11%, 2.43%, and 11.50% for the NDG-CAM, Mask R-CNN, and combined method, respectively. When calculated per case, the values for $e_s$ were 1.84 ± 13.05%, 3.46 ± 6.15%, and 14.45 ± 11.22%, indicating that the models generally tend to overdetect on this dataset.

In the V4 dataset, the $e_a$ had a value of 15.26%, 59.22%, and 14.10% for the NDG-CAM, Mask R-CNN, and combined method, respectively. When calculated per case, the values for $e_s$ were −16.86 ± 13.79%, −60.13 ± 13.88%, and −14.88 ± 12.86%, showing that the models have a tendency to underdetect on this dataset. In particular, it was noticed that very small nuclei, such as those of lymphocytes, and elongated ones, such as those of fibrocytes, were underdetected.

For the detection task, the results are reported in Table 3. In the V1 dataset, NDG-CAM, Mask R-CNN, and the combined method were capable of achieving a Dice coefficient of 0.824, 0.878, and 0.884, respectively. Thus, the combined method obtained slightly better results than the other methods. As for the recall, the combined method decisively surpasses the other approaches, with a value of 0.934.

In the V4 dataset, the combined method proves to be the best, achieving a recall of 0.850 and a Dice coefficient of 0.914. Mask R-CNN performs poorly in this case, with a recall of 0.403 and a Dice coefficient of 0.573.

**Table 3.** Comparison of detection methods, extending the one proposed by Alom et al. [4] and Sirinukunwattana et al. [14].

| Method | Precision | Recall | Dice |
|---|---|---|---|
| CRImage [12] | 0.657 | 0.461 | 0.542 |
| CNN [12] | 0.783 | 0.804 | 0.793 |
| SSAE [6] | 0.617 | 0.644 | 0.630 |
| LIPSyM [13] | 0.725 | 0.517 | 0.604 |
| SC-CNN (M = 1) [14] | 0.758 | 0.827 | 0.791 |
| SC-CNN (M = 2) [14] | 0.781 | 0.823 | 0.802 |
| UD-Net [4] | 0.822 | 0.842 | 0.828 |
| NDG-CAM (V1) | 0.833 | 0.815 | 0.824 |
| NDG-CAM (V4) | **0.992** | 0.841 | 0.910 |
| Mask R-CNN (V1) | **0.867** | 0.888 | 0.878 |
| Mask R-CNN (V4) | 0.989 | 0.403 | 0.573 |
| Combined (V1) | 0.838 | **0.934** | **0.884** |
| Combined (V4) | 0.986 | **0.850** | **0.914** |

The violin plots calculated per tile are reported in Figure 6 for the V1 and V4 datasets, comparing the NDG-CAM detection method, Mask R-CNN, and the combined approach. It is worth noting that the Mask R-CNN model works very well on the V1 dataset but performs poorly on the V4 one. On the other hand, the NDG-CAM and the combined methods maintain high levels of performance in all scenarios.

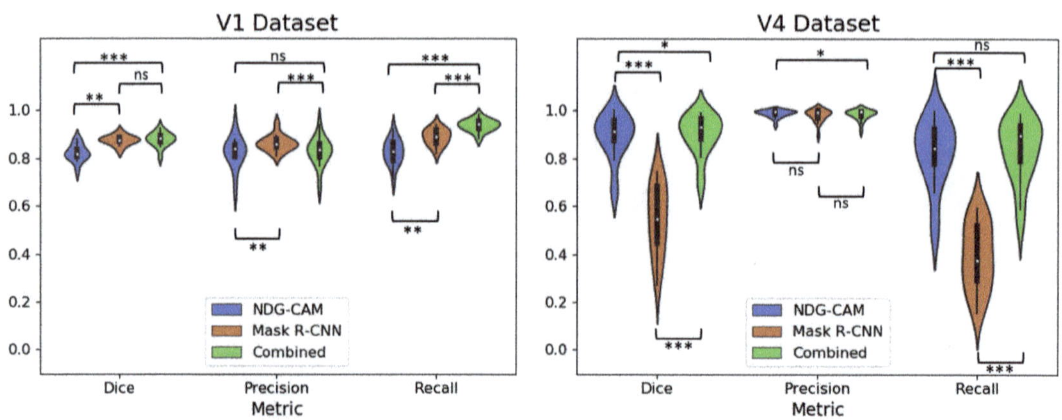


**Figure 6.** Violin plots for the detection metrics calculated per case. (**Left**) V1 dataset. (**Right**) V4 dataset. In the figure, ns stands for nonsignificant; * denotes $p$-value < 0.05; ** indicates $p$-value < 0.01; and *** means $p$-value < 0.001.

In the V1 dataset, the combined model does not show a Dice coefficient that is higher in a statistically significant way than the Mask R-CNN approach, with a $p$-value of 0.07. On the other hand, the recall was much higher for the combined method, resulting in a $p$-value < 0.001 for both NDG-CAM and Mask R-CNN. In the V4 dataset, both the NDG-

CAM and the combined method showed much stronger results than Mask R-CNN, with a $p$-value less than 0.001 in both cases for Dice coefficient and recall. Moreover, the combined approach shows a statistically significant advantage over NDG-CAM ($p$-value = 0.048) for the Dice coefficient.

## 4. Discussion

In order to show the effectiveness of the proposed method, we compared it with existing state-of-the-art approaches. It has to be noted that our method allows exploiting semantic segmentation architectures to realize nuclei detection, whereas other approaches usually involve networks specialized for this task. Several approaches proposed in the literature try to localize centers of the nuclei or proximity maps to those centers [3,14,16]. These approaches require instance-level annotations, although the results are promising. On the other hand, the proposed method exploits an XAI technique, Grad-CAM for semantic segmentation, to reconstruct post hoc saliency maps that are related to the centers of the nuclei, showing that semantic segmentation networks can perform detection tasks without specialized modifications.

The most widespread metrics employed for assessing algorithms for object detection involve precision, recall, and Dice coefficient. Namely, they are the metrics that are also related to the position of the detected nuclei, and not only on the counts.

A quantitative comparison between considered approaches and existing ones from the literature is presented in Table 3.

From this comparative analysis, it emerges that the proposed method is perfectly aligned with the state of the art, without the need to implement specific kinds of specialized loss functions [24] or architectures for detection [17,40].

Indeed, the NDG-CAM method alone was capable of achieving a Dice coefficient for object detection of 0.824, whereas the UD-Net [4] method, the top-performing method among the selected from the literature, had a Dice coefficient of 0.828. When the proposed NDG-CAM detection method is used in combined usage with Mask R-CNN, the recall increases to 0.934, and the Dice coefficient to 0.884, surpassing the current state-of-the-art methods for nuclei detection. On the collected external validation set, metrics are even higher, with a Dice coefficient of 0.914, showing the generalization capabilities of the proposed workflow.

Qualitative results for the the object detection pipeline involving semantic segmentation and Grad-CAM on the images of the independent external validation set V4 are depicted in Figure 7. Instead, Figure 8 shows the final detection results on the validation datasets V1 and V4 with the NDG-CAM method, the Mask R-CNN architecture, and the combined adoption of both methods.

It can be seen from the images of Figure 7, taken from the V4 dataset, that precision is very high. Indeed, virtually all detected nuclei are real. Some small or elongated nuclei, such as lymphocytic or fibrocytic nuclei, are underdetected. This may be due to a lack of proper training datasets with a large variety of nuclear shapes.

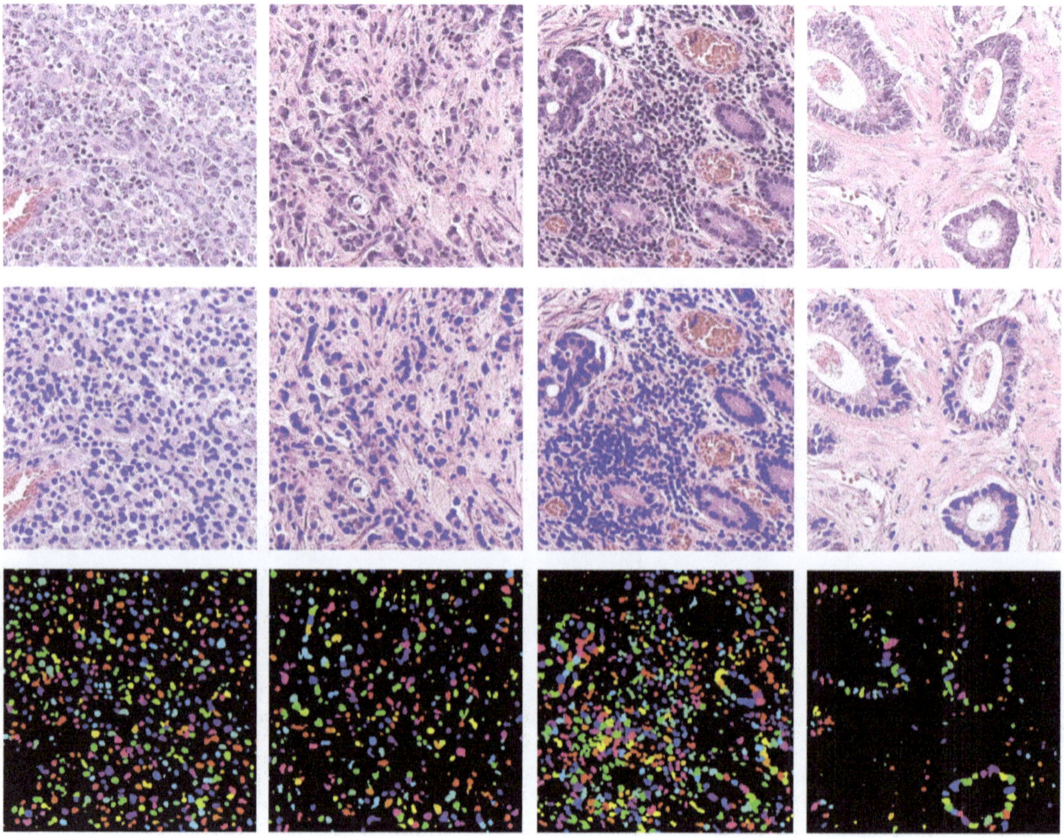

**Figure 7.** Examples of the NDG-CAM method on the data from the Pathology Department of IRCCS Istituto Tumori Giovanni Paolo II. Results are shown for the best architecture (DeepLab v3+ with ResNet18 backbone). (**First row**) Original images. (**Second row**) Semantic segmentation. (**Third row**) Instance segmentation after detection of centroids of the nuclei, with each color denoting a different nuclear instance.

The two methods show similar performance on the V1 dataset, as can be observed from Figure 8. Mask R-CNN achieves slightly better performance on this dataset, and considering that it has been trained on a larger training set, the combined method proved to be superior. From the same figure, it is possible to observe that, in the V4 dataset, Mask R-CNN does not properly generalize, resulting in the missing of many nuclei (low recall).

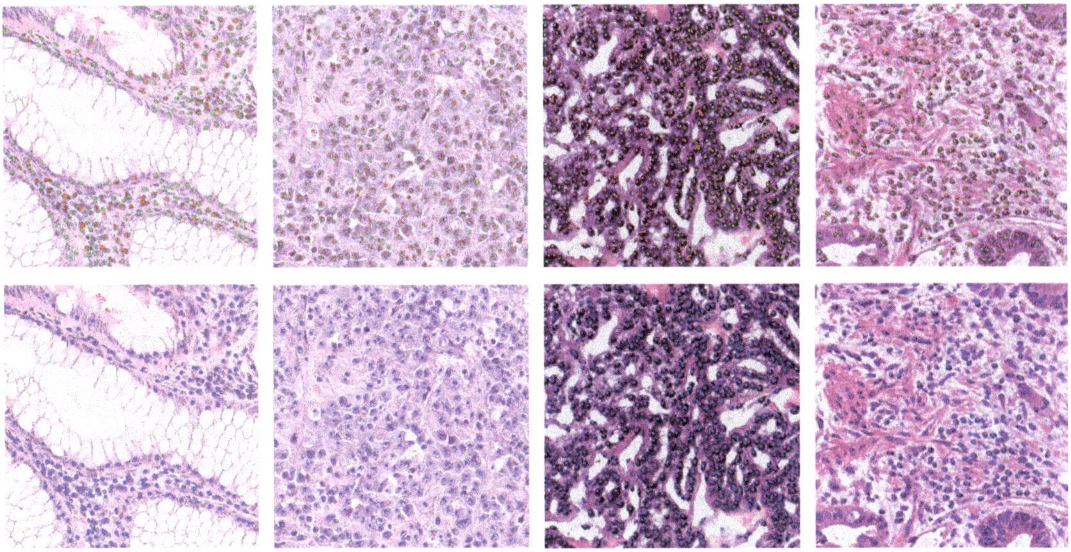

**Figure 8.** Examples of centroid detection on the validation sets V1 and V4. (**Top row**) Green, NDG-CAM method detections; red, Mask R-CNN detections. (**Bottom row**) Blue, combined method detections. First and second columns show data from V4, whereas the third and fourth columns depict data from V1.

## 5. Conclusions and Future Works

In this work, a novel method was presented with the aim of nuclei identification from histological H&E images. In our multi-stage pipeline, the first phase involved semantic segmentation. After various experiments, DeepLab v3+ (ResNet18 backbone) emerged as the best-performing architecture. Subsequently, because this analysis did not allow the distinction of multiple instances of the same object, we proposed a novel detection algorithm, NDG-CAM, which exploited Grad-CAM to solve the problem of separating the instances. Even without the need to use specialized loss functions or architectures, it allowed us to achieve satisfactory results in the detection task, comparable to or even better than more sophisticated training setups [3,6,12,16]. When the method is combined with the Mask R-CNN instance segmentation architecture, results exceed the state-of-the-art methods for nuclei detection.

Even though the local validation set includes only colorectal cancer H&E slides, it has to be considered that in each slide there are several tissue types present (e.g., stroma, immune infiltration) and the proposed method has the ability to detect nuclei not only related to the tumor or normal epithelium of colon but also to other cytotypes.

Indeed, we noticed underdetection of lymphocytic or fibrocytic nuclei, and this could be explained by a lack of datasets enriched in these nuclei subtypes. For such a reason, a direction for future works includes the collection of a dataset with multiple and balanced nuclei annotations.

On the clinical side, the proposed workflow could be a valid tool to support pathologists in the detection and reporting of histological samples, thus allowing a considerable saving of time and resources, besides providing an objective tool that is more reliable than manual assessment. Future works will concern the classification of the detected nuclei, in order to estimate how many are malignant or subjected to specific lesions, so that important clinical parameters, such as neoplastic cellularity, can be determined quantitatively.

**Author Contributions:** Conceptualization, N.A., A.B. and V.B.; methodology, N.A., E.P., M.G.T., A.B. and V.B.; software, N.A., E.P. and M.G.T.; validation, C.S., F.A.Z. and S.D.S.; data curation, C.S., F.A.Z. and S.D.S.; writing—original draft preparation, N.A., E.P. and M.G.T.; writing—review and editing, all authors; visualization, N.A., E.P. and M.G.T.; supervision, N.A., A.B. and V.B. All authors have read and agreed to the published version of the manuscript.

**Funding:** The study has been partially funded by the projects "Tecnopolo per la Medicina di Precisione, CUP: B84I18000540002" and "CustOm-made aNTibacterical/bioactive/bioCoated prostheses (CONTACT), CUP: B99C20000300005".

**Institutional Review Board Statement:** The institutional Ethic Committee of the IRCCS Istituto Tumori Giovanni Paolo II approved the study (Prot n. 780/CE).

**Informed Consent Statement:** Patient consent was waived due to the fact that this was a retrospective observational study with anonymized data, already acquired for medical diagnostic purposes.

**Data Availability Statement:** The MoNuSeg [18], CRCHistoPhenotypes [21], and NuCLS [22] datasets are publicly available. The local dataset from IRCCS Istituto Tumori Giovanni Paolo II presented in this study is available upon request from the corresponding author.

**Conflicts of Interest:** The authors declare no conflict of interest.

## References

1. Kumar, N.; Verma, R.; Sharma, S.; Bhargava, S.; Vahadane, A.; Sethi, A. A dataset and a technique for generalized nuclear segmentation for computational pathology. *IEEE Trans. Med. Imaging* **2017**, *36*, 1550–1560.
2. Mahmood, F.; Borders, D.; Chen, R.J.; McKay, G.N.; Salimian, K.J.; Baras, A.; Durr, N.J. Deep adversarial training for multi-organ nuclei segmentation in histopathology images. *IEEE Trans. Med. Imaging* **2019**, *39*, 3257–3267.
3. Höfener, H.; Homeyer, A.; Weiss, N.; Molin, J.; Lundström, C.F.; Hahn, H.K. Deep learning nuclei detection: A simple approach can deliver state-of-the-art results. *Comput. Med. Imaging Graph.* **2018**, *70*, 43–52.
4. Alom, Z.; Asari, V.K.; Parwani, A.; Taha, T.M. Microscopic nuclei classification, segmentation, and detection with improved deep convolutional neural networks (DCNN). *Diagn. Pathol.* **2022**, *17*, 38.
5. Shu, J.; Fu, H.; Qiu, G.; Kaye, P.; Ilyas, M. Segmenting overlapping cell nuclei in digital histopathology images. In Proceedings of the 2013 35th Annual International Conference of the IEEE Engineering in Medicine and Biology Society (EMBC), Osaka, Japan, 3–7 July 2013; pp. 5445–5448.
6. Xu, J.; Xiang, L.; Liu, Q.; Gilmore, H.; Wu, J.; Tang, J.; Madabhushi, A. Stacked sparse autoencoder (SSAE) for nuclei detection on breast cancer histopathology images. *IEEE Trans. Med. Imaging* **2015**, *35*, 119–130.
7. Sornapudi, S.; Stanley, R.J.; Stoecker, W.V.; Almubarak, H.; Long, R.; Antani, S.; Thoma, G.; Zuna, R.; Frazier, S.R. Deep learning nuclei detection in digitized histology images by superpixels. *J. Pathol. Inform.* **2018**, *9*, 5.
8. Larson, N.B.; Fridley, B.L. PurBayes: Estimating tumor cellularity and subclonality in next-generation sequencing data. *Bioinformatics* **2013**, *29*, 1888–1889.
9. Prencipe, B.; Altini, N.; Cascarano, G.D.; Brunetti, A.; Guerriero, A.; Bevilacqua, V. Focal Dice Loss-Based V-Net for Liver Segments Classification. *Appl. Sci.* **2022**, *12*, 3247.
10. Altini, N.; Brunetti, A.; Napoletano, V.P.; Girardi, F.; Allegretti, E.; Hussain, S.M.; Brunetti, G.; Triggiani, V.; Bevilacqua, V.; Buongiorno, D. A Fusion Biopsy Framework for Prostate Cancer Based on Deformable Superellipses and nnU-Net. *Bioengineering* **2022**, *9*, 343.
11. Altini, N.; Cascarano, G.D.; Brunetti, A.; Marino, F.; Rocchetti, M.T.; Matino, S.; Venere, U.; Rossini, M.; Pesce, F.; Gesualdo, L.; et al. semantic segmentation framework for glomeruli detection and classification in kidney histological sections. *Electronics* **2020**, *9*, 503.
12. Yuan, Y.; Failmezger, H.; Rueda, O.M.; Ali, H.R.; Gräf, S.; Chin, S.F.; Schwarz, R.F.; Curtis, C.; Dunning, M.J.; Bardwell, H.; et al. Quantitative image analysis of cellular heterogeneity in breast tumors complements genomic profiling. *Sci. Transl. Med.* **2012**, *4*, 157ra143.
13. Kuse, M.; Wang, Y.F.; Kalasannavar, V.; Khan, M.; Rajpoot, N. Local isotropic phase symmetry measure for detection of beta cells and lymphocytes. *J. Pathol. Inform.* **2011**, *2*, 2.
14. Sirinukunwattana, K.; Raza, S.E.A.; Tsang, Y.W.; Snead, D.R.; Cree, I.A.; Rajpoot, N.M. Locality sensitive deep learning for detection and classification of nuclei in routine colon cancer histology images. *IEEE Trans. Med. Imaging* **2016**, *35*, 1196–1206.
15. Alom, M.Z.; Yakopcic, C.; Hasan, M.; Taha, T.M.; Asari, V.K. Recurrent residual U-Net for medical image segmentation. *J. Med. Imaging* **2019**, *6*, 014006.
16. Kainz, P.; Urschler, M.; Schulter, S.; Wohlhart, P.; Lepetit, V. You should use regression to detect cells. In Proceedings of the International Conference on Medical Image Computing and Computer-Assisted Intervention, Munich, Germany, 5–9 October 2015; pp. 276–283.

17. He, K.; Gkioxari, G.; Dollár, P.; Girshick, R. Mask r-cnn. In Proceedings of the IEEE International Conference on Computer Vision, Venice, Italy, 22–29 October 2017; pp. 2961–2969.
18. MoNuSeg—Grand Challenge. Available online: https://monuseg.grand-challenge.org/Data/ (accessed on 7 April 2022).
19. Kumar, N.; Verma, R.; Anand, D.; Zhou, Y.; Onder, O.F.; Tsougenis, E.; Chen, H.; Heng, P.A.; Li, J.; Hu, Z.; et al. A multi-organ nucleus segmentation challenge. *IEEE Trans. Med. Imaging* **2019**, *39*, 1380–1391.
20. Caicedo, J.C.; Goodman, A.; Karhohs, K.W.; Cimini, B.A.; Ackerman, J.; Haghighi, M.; Heng, C.; Becker, T.; Doan, M.; McQuin, C.; et al. Nucleus segmentation across imaging experiments: The 2018 Data Science Bowl. *Nat. Methods* **2019**, *16*, 1247–1253.
21. CRCHistoPhenotypes—Labeled Cell Nuclei Data, Tissue Image Analytics (TIA) Centre, Warwick. Available online: https://warwick.ac.uk/fac/cross_fac/tia/data/crchistolabelednucleihe (accessed on 7 April 2022).
22. Amgad, M.; Elfandy, H.; Hussein, H.; Atteya, L.A.; Elsebaie, M.A.; Abo Elnasr, L.S.; Sakr, R.A.; Salem, H.S.; Ismail, A.F.; Saad, A.M.; et al. Structured crowdsourcing enables convolutional segmentation of histology images. *Bioinformatics* **2019**, *35*, 3461–3467.
23. Altini, N.; Marvulli, T.M.; Caputo, M.; Mattioli, E.; Prencipe, B.; Cascarano, G.D.; Brunetti, A.; Tommasi, S.; Bevilacqua, V.; Summa, S.D.; et al. Multi-class Tissue Classification in Colorectal Cancer with Handcrafted and Deep Features. In Proceedings of the International Conference on Intelligent Computing, Shenzhen, China, 12–15 August 2021; pp. 512–525.
24. Ronneberger, O.; Fischer, P.; Brox, T. U-net: Convolutional networks for biomedical image segmentation. In Proceedings of the International Conference on Medical Image Computing and Computer-Assisted Intervention, Munich, Germany, 5–9 October 2015; pp. 234–241.
25. Badrinarayanan, V.; Kendall, A.; Cipolla, R. Segnet: A deep convolutional encoder-decoder architecture for image segmentation. *IEEE Trans. Pattern Anal. Mach. Intell.* **2017**, *39*, 2481–2495.
26. Chen, L.C.; Zhu, Y.; Papandreou, G.; Schroff, F.; Adam, H. Encoder-decoder with atrous separable convolution for semantic image segmentation. In Proceedings of the European conference on computer vision (ECCV), Munich, Germany, 8–14 September 2018; pp. 801–818.
27. He, K.; Zhang, X.; Ren, S.; Sun, J. Deep residual learning for image recognition. In Proceedings of the IEEE Conference on Computer Vision and Pattern Recognition, Las Vegas, NV, USA, 27–30 June 2016; pp. 770–778.
28. Sandler, M.; Howard, A.; Zhu, M.; Zhmoginov, A.; Chen, L.C. Mobilenetv2: Inverted residuals and linear bottlenecks. In Proceedings of the IEEE Conference on Computer Vision and Pattern Recognition, Salt Lake City, UT, USA, 18–23 June 2018; pp. 4510–4520.
29. Altini, N.; Prencipe, B.; Cascarano, G.D.; Brunetti, A.; Brunetti, G.; Triggiani, V.; Carnimeo, L.; Marino, F.; Guerriero, A.; Villani, L.; et al. Liver, kidney and spleen segmentation from CT scans and MRI with deep learning: A survey. *Neurocomputing* **2022**, *490*, 30–53.
30. Altini, N.; Prencipe, B.; Brunetti, A.; Brunetti, G.; Triggiani, V.; Carnimeo, L.; Marino, F.; Guerriero, A.; Villani, L.; Scardapane, A.; et al. A Tversky loss-based convolutional neural network for liver vessels segmentation. In Proceedings of the International Conference on Intelligent Computing, Bari, Italy, 2–5 October 2020; pp. 342–354.
31. Bevilacqua, V.; Altini, N.; Prencipe, B.; Brunetti, A.; Villani, L.; Sacco, A.; Morelli, C.; Ciaccia, M.; Scardapane, A. Lung Segmentation and Characterization in COVID-19 Patients for Assessing Pulmonary Thromboembolism: An Approach Based on Deep Learning and Radiomics. *Electronics* **2021**, *10*, 2475.
32. Tjoa, E.; Guan, C. A survey on explainable artificial intelligence (xai): Toward medical xai. *IEEE Trans. Neural Netw. Learn. Syst.* **2020**, *32*, 4793–4813.
33. Gunning, D.; Stefik, M.; Choi, J.; Miller, T.; Stumpf, S.; Yang, G.Z. XAI—Explainable artificial intelligence. *Sci. Robot.* **2019**, *4*, eaay7120.
34. Antoniadi, A.M.; Du, Y.; Guendouz, Y.; Wei, L.; Mazo, C.; Becker, B.A.; Mooney, C. Current challenges and future opportunities for XAI in machine learning-based clinical decision support systems: A systematic review. *Appl. Sci.* **2021**, *11*, 5088.
35. Hussain, S.M.; Buongiorno, D.; Altini, N.; Berloco, F.; Prencipe, B.; Moschetta, M.; Bevilacqua, V.; Brunetti, A. Shape-Based Breast Lesion Classification Using Digital Tomosynthesis Images: The Role of Explainable Artificial Intelligence. *Appl. Sci.* **2022**, *12*, 6230.
36. Selvaraju, R.R.; Cogswell, M.; Das, A.; Vedantam, R.; Parikh, D.; Batra, D. Grad-cam: Visual explanations from deep networks via gradient-based localization. In Proceedings of the IEEE International Conference on Computer Vision, Venice, Italy, 22–29 October 2017; pp. 618–626.
37. Vinogradova, K.; Dibrov, A.; Myers, G. Towards interpretable semantic segmentation via gradient-weighted class activation mapping (student abstract). In Proceedings of the AAAI Conference on Artificial Intelligence, New York, NY, USA, 7–12 February 2020; Volume 34, pp. 13943–13944.
38. Du, L.; Zhang, R.; Wang, X. Overview of two-stage object detection algorithms. *J. Phys. Conf. Ser.* **2020**, *1544*, 012033.
39. Wu, Y.; Kirillov, A.; Massa, F.; Lo, W.Y.; Girshick, R. Detectron2. 2019. Available online: https://github.com/facebookresearch/detectron2 (accessed on September 7, 2022).
40. Altini, N.; Cascarano, G.D.; Brunetti, A.; De Feudis, I.; Buongiorno, D.; Rossini, M.; Pesce, F.; Gesualdo, L.; Bevilacqua, V. A deep learning instance segmentation approach for global glomerulosclerosis assessment in donor kidney biopsies. *Electronics* **2020**, *9*, 1768.

bioengineering

MDPI

*Article*

# A Fully Unsupervised Deep Learning Framework for Non-Rigid Fundus Image Registration

Giovana A. Benvenuto [1], Marilaine Colnago [2], Maurício A. Dias [1], Rogério G. Negri [3], Erivaldo A. Silva [1] and Wallace Casaca [4,*]

1 Faculty of Science and Technology (FCT), São Paulo State University (UNESP), Presidente Prudente 19060-900, Brazil
2 Institute of Mathematics and Computer Science (ICMC), São Paulo University (USP), São Carlos 13566-590, Brazil
3 Science and Technology Institute (ICT), São Paulo State University (UNESP), São José dos Campos 12224-300, Brazil
4 Institute of Biosciences, Letters and Exact Sciences (IBILCE), São Paulo State University (UNESP), São José do Rio Preto 15054-000, Brazil
* Correspondence: wallace.casaca@unesp.br

**Abstract:** In ophthalmology, the registration problem consists of finding a geometric transformation that aligns a pair of images, supporting eye-care specialists who need to record and compare images of the same patient. Considering the registration methods for handling eye fundus images, the literature offers only a limited number of proposals based on deep learning (DL), whose implementations use the supervised learning paradigm to train a model. Additionally, ensuring high-quality registrations while still being flexible enough to tackle a broad range of fundus images is another drawback faced by most existing methods in the literature. Therefore, in this paper, we address the above-mentioned issues by introducing a new DL-based framework for eye fundus registration. Our methodology combines a U-shaped fully convolutional neural network with a spatial transformation learning scheme, where a reference-free similarity metric allows the registration without assuming any pre-annotated or artificially created data. Once trained, the model is able to accurately align pairs of images captured under several conditions, which include the presence of anatomical differences and low-quality photographs. Compared to other registration methods, our approach achieves better registration outcomes by just passing as input the desired pair of fundus images.

**Keywords:** fundus image; image registration; deep learning; computer vision applications

**Citation:** Benvenuto, G.A.; Colnago, M.; Dias, M.A.; Negri, R.G.; Silva, E.A.; Casaca, W. A Fully Unsupervised Deep Learning Framework for Non-Rigid Fundus Image Registration. *Bioengineering* **2022**, *9*, 369. https://doi.org/10.3390/bioengineering9080369

Academic Editors: Paolo Zaffino and Maria Francesca Spadea

Received: 29 June 2022
Accepted: 3 August 2022
Published: 5 August 2022



## 1. Introduction

In ophthalmology, computing technologies such as computer-assisted systems and content-based image analysis are indispensable tools to obtain more accurate diagnoses and detect signals of diseases. As a potential application, we can cite the progressive monitoring of eye disorders, such as glaucoma [1] and diabetic retinopathy [2], which can be conveniently performed by inspecting retina fundus images [3]. In fact, in follow-up examinations conducted by eye specialists, a particularly relevant task is image registration [4,5], where the goal is to assess the level of agreement between two or more fundus photographs captured at different instants or even by distinct acquisition instruments. In this kind of application, issues related to eye fundus scanning, such as variations in lighting, scale, angulation, and positioning, are properly handled and fixed when registering the images.

In more technical terms, given a pair of fundus images, $I_{Mov}$ and $I_{Ref}$, the registration problem comprises determining a geometric transformation that best aligns these images and maximizing their overlap areas while facilitating the visual comparison between them. As manually verifying with the naked eye possible changes between two or more fundus photographs is arduous and error-prone, there is a necessity to automate such a procedure [6,7]. Moreover, the difficulty in comparing large fundus datasets by a human expert and the time spent by ophthalmologists to accomplish manual inspections are commonly encountered challenges in the medical environment.

In recent years, machine and deep learning (DL) have paved their way into image registration and other related applications, such as computer-aided diagnosis [8,9], achieving very accurate and stable solutions. However, despite the existence of several proposals in the image registration literature, Litjens et al. [10], and Haskins et al. [11] recently indicated that there is a lack of consensus on a categorical technique that benefits from the robustness of deep learning towards providing high-accuracy registrations regardless of the condition of the acquired image pair. In addition, among methods specifically developed to cope with eye fundus registration, there is only a limited number of proposals that apply DL strategies, and most of them are focused on the supervised learning paradigm, i.e., the methods usually assume ground-truth reference data to train an alignment model. As reference data can be automatically generated by specific techniques or acquired through manual notes by an eye professional, both cases may suffer from the following drawbacks: (a) synthetically generating benchmark data can affect the accuracy of the trained models [12], and (b) manually annotating data are prone to failure due to the high number of samples to be labeled by a human agent, which includes the complication of creating full databases, large and representative enough in terms of ground-truth samples to be used to train a DL model effectively [11,13]. Lastly, dealing with ethical issues is another difficulty imposed when one tries to collect a large database of labeled medical images.

Aiming to address most of the issues and drawbacks raised above, in this paper, we propose a new methodology that combines two DL-based architectures into a fully unsupervised approach for retina fundus registration. More specifically, a U-shaped fully convolutional neural network (CNN) [14] and a spatial-transformer-type network [15] are integrated, so that the former produces a set of matching points from the fundus images, while the latter utilizes the mapped points to obtain a correspondence field used to drive geometric bilinear interpolation. Our learning scheme takes advantage of a benchmark-free similarity metric that gauges the difference between fixed and moving images, allowing for the registration without taking any prelabeled data to train a model or a specific technique to synthetically create training data. Once the integrated methodology is fully trained, it can achieve one-shot registrations by just passing the desired pair of fundus images.

A preliminary study of our learning scheme appears in our recently published ICASSP paper [16]. Going beyond our previous investigation, several enhancements are put forward. First, we extend our integrated DL framework to achieve more accurate outcomes, leading to a more assertive and stable registration model. We also provide a comprehensive literature review classifying popular and recent DL-based registration methods according to their network types, geometric transformations, and the general category of medical images (see Section 2). An extensive battery of new experiments and assessments are now given, in particular, the analysis of two additional fundus databases, the inclusion of new registration methods in the comparisons, and an ablation study covering the refinement step of our registration framework (see Section 3). Lastly, we also show that our learning registration pipeline can succeed with multiple classes of eye fundus images (see Section 4), a trait hard to be found in other fundus image registration methods.

In summary, the main contributions introduced by our approach are:

- A fully automatic learning strategy that unifies a context-aware CNN, a spatial transformation network and a label-free similarity metric to perform fundus image registration in one-shot without the need for any ground-truth data.
- Once trained, the registration model is capable of aligning fundus images of several classes and databases (e.g., super-resolution, retinal mosaics, and photographs containing anatomical differences).
- The combination of multiple DL networks with image analysis techniques, such as isotropic undecimated wavelet transform and connected component analysis, allow-

ing for the registration of fundus photographs even with low-quality segments and abrupt changes.

## 2. Related Work

The literature covers a large number of DL-driven applications for clinical diagnosis in ophthalmology. Recently, several studies have been conducted on deep learning for the early detection of diseases and eye disorders, which include diabetic retinopathy detection [17,18], glaucoma diagnosis [19,20], and the automated identification of myopia using eye fundus images [21]. All these DL-based applications have high clinical relevance and may prove effective in supporting the design of suitable protocols in ophthalmology. Going deeper into DL-based applications, the image translation problem has also appeared in different ophthalmology image domains, such as image super resolution [22], denoising of retinal optical coherence tomography (OCT) [23], and OCT segmentation [24]. For instance, Mahapatra et al. [22] introduced a generative adversarial network (GAN) to increase the resolution of fundus images in order to enable more precise image analysis. Aiming at solving the issue of image denoising in high- and low-noise domains for OCT images, Manakov et al. [23] developed a model on the basis of the cycleGAN network to learn a mapping between these domains. Still on image translation, Sanchez et al. [24] combined two CNNs, the Pix2Pix and a modified deep retinal understanding network, to achieve the segmentation of intraretinal and subretinal fluids, and hyper-reflective foci in OCT images. For a comprehensive survey of image translation applications, see [25].

We now focus on discussing particular approaches for solving the image registration task. We split the registration methods into two groups: those that do not use DL (traditional methods), and those that do. Since our work seeks to advance the DL literature, we focus our discussion on this particular branch.

Considering the general application of image registration in the medical field, the literature has recently explored DL as a key resolution paradigm, including new approaches to obtain highly accurate results for various medical image categories, as discussed by Litjens et al. [10], Haskings et al. [11], and Fu et al. [26]. Most of these approaches rely on supervised learning, requiring annotated data to train a model. For example, Yang et al. [27] introduced an encoder–decoder architecture to carry out the supervised registration of magnetic resonance images (MRI) of the brain. Cao et al. [28] covered the same class of images, but they employed a guided learning strategy instead. Eppenhof and Pluim [29] also applied a supervised approach, but for registering chest computed tomography (CT) images through a U-shaped encoder-decoder network [30]. Still concerning supervised learning, several works attempted to compensate for the lack of labeled data by integrating new metrics into an imaging network. Fan et al. [31] induced the generation of ground-truth information used to perform the registration of brain images. Hering et al. [32] utilized a weakly supervised approach to align cardiac MRI images, and Hu et al. [33] took two networks: the former applied an affine transformation, while the latter gave the final registration of patients with prostate cancer.

More recently, new registration methods were proposed to circumvent the necessity of annotated data when training neural networks [15,34–38]. Jun et al. [34] presented a registration method that relied on a spatial transformer network (STN) network and a resampler for inspiration or expiration images of abdominal MRI. Zhang [35] covered the specific case of brain imaging, implementing two fully convolutional networks (FCNs), one to predict the parameters of a deformable transformation to align the fixed image to the moving image, and the other to proceed with the opposite alignment from moving image to a fixed one. Kori et al. [36] proposed a method that focused on exploring specific features of multimodal images by using a pretrained CNN followed by a keypoint detector, while the framework designed by Wang et al. [37] learn a modality-independent representation from an architecture composed of five subnets: an encoder, two decoders, and two transformation networks. Still on the registration of nonretinal cases, the method developed by Vos et al. [15] aligned cardiac images by comparing similar pixels to optimize the parameters of a CNN applied during the learning process. The method presented by Balakrishnan et al. [38] is another example of nonretinal registration, where the authors took a spatial transformation and U-Shaped learning scheme to explore brain MR data.

Concerning the DL-based methods specifically designed to handle retinal fundus images, Mahapatra et al. [39] presented a generative adversarial network (GAN) to align fundus photographs formed by two networks, a generator and a discriminator. While the former maps data from one domain to the other, the latter is tasked with discerning between true data and the synthetic distribution created by the generator [11]. Wang et al. [40] introduced a framework composed of two pretrained networks that perform the segmentation, detection, and description of retina features. Recently, Rivas-Villar et al. [41] have proposed a feature-based supervised registration method for fundus images where a network is trained using reference points transformed into heat maps to learn how to predict these maps in the inference step. The predicted maps are converted back into point locations and then used by a RANSAC-based matching algorithm to create the transformation models. Despite their capability in specifically solving the fundus registration problem, the methods described above employ reference data to compose the loss function.

In summary, most registration methods rely on supervised learning or take synthetically generated data in order to be effective. While generating new labels can overcome the scarcity of reference data, it also introduces an additional complication in modeling the problem, raising the issue of the reliability of artificially induced data in the medical image domain [42]. Another common trait shared by most DL registration methods is that they only produce high-accuracy outputs for a certain class of medical images or even subcategories of fundus photographs, such as super-resolution and retinal mosaics.

Table 1 summarizes the main DL registration methods discussed above.

**Table 1.** Survey of DL studies. Blue lines refer to works that specifically cover fundus registration.

| Papers | Ref. | Images Type | Network | Architecture | Transformation |
|---|---|---|---|---|---|
| Yang et al. | [27] | Brain MRI (3D) | Supervised | Encoder + Decoder | Affine + Nonrigid (LDDMM) |
| Cao et al. | [28] | Brain MRI (3D) | Supervised | Network preparation + network learning | Affine + Nonrigid (TPS) |
| Eppenhof and Pluim | [29] | Chest CT (3D) | Supervised | Adapted U-Net | Nonrigid (B-Spline) |
| Fan et al. | [31] | Brain MRI (3D) | Weakly supervised | BIRNet | Nonrigid |
| Hering et al. | [32] | Cardiac MRI (3D) | Weakly supervised | Adapted U-Net | Nonrigid (B-Spline) |
| Hu et al. | [33] | TRUS and prostate MRI (3D) | Weakly supervised | Global Net + Local Net | Affine + Non-rigid |
| Mahapatra et al. | [39] | Retinal FA images + cardiac MRI (2D) | Weakly supervised | GAN | Nonrigid |
| Wang et al. | [40] | Multimodal retinal image | Weakly supervised | Segmentation network + feature detection and description network + outlier rejection network | Affine |
| Rivas-Villar et al. | [41] | Color fundus images | Weakly supervised | U-Net + RANSAC | Similarity transformation |
| Jun et al. | [34] | Abdominal MRI (2D and 3D) | Unsupervised | CNN + STN | Nonrigid (B-Spline) |
| Zhang | [35] | Brain MRI (3D) | Unsupervised | Adapted U-Net + 2 FCN | Nonrigid (B-Spline) |
| Vos et al. | [15] | Cardiac MRI and chest CT (3D) | Unsupervised | CNN Affine + CNN nonrigid | Affine + Nonrigid (B-Spline) |
| Wang et al. | [37] | Brain MRI (2D and 3D) | Unsupervised | Encoder + decoders + transformation networks | Affine + Nonrigid |
| Kori et al. | [36] | Brain MRI (3D) | Unsupervised | VGG-19 + transformation estimator | Affine |
| Balakrishnan et al. | [38] | Brain MRI (3D) | Unsupervised | Adapted U-Net + STN (+ information optional auxiliary) | Nonrigid (linear) |

## 3. Materials and Methods

### 3.1. *Overview of the Proposed Approach*

The proposed framework seeks to align a pair of fundus images, $I_{Mov}$ and $I_{Ref}$, without the need for any labeled data. First, we extract the blood veins, bifurcations, and other relevant compositions of the eye, producing images $B_{Mov}$ and $B_{Ref}$ that are passed through a U-shaped fully convolutional neural network that outputs a correspondence grid between the images. In the next learning step, a matching grid is taken as input by a spatial transformation layer that computes the transformation model used to align the moving image. In our integrated architecture, the learning occurs through an objective function that measures the similarity between the reference and transformed images. As a result, the unified networks learn the registration task without the need for ground-truth annotations and reference data. Lastly, as a refinement step, we apply a mathematical morphology-based technique to remove noisy pixels that may appear during the learning process. Figure 1 shows the proposed registration approach.

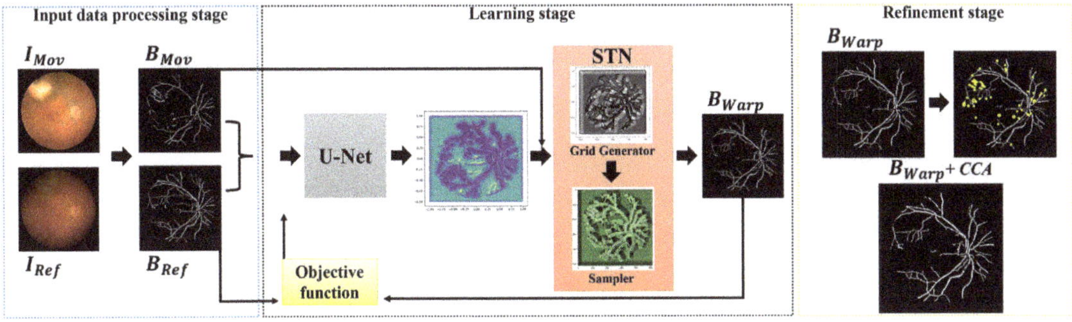


**Figure 1.** Overview of the proposed registration workflow.

### 3.2. Network Input Preparation

This step aims to handle the image pairs, $I_{Ref}$ and $I_{Mov}$, to improve the performance of the networks. In our approach, the images were resized to 512 × 512 to reduce the total number of network parameters related to the image sizes, thus leveraging the process of training the registration model. Next, a segmentation step was performed to obtain the eye's structures that may be more relevant to the resolution of the registration problem. These include the blood vessels and the optic disc, as we can see from images $B_{Ref}$ and $B_{Mov}$ in the leftmost frame in Figure 1. To maximize the segmentation accuracy, we applied the isotropic undecimated wavelet transform (IUWT) [43] technique, which was developed specifically for the detection and measurement of retinal blood vessels.

### 3.3. Learning a Deep Correspondence Grid

As mentioned before, the first implemented learning mechanism assumes a U-Net-type structure whose goal is to compute a correspondence grid for the reference and moving images. The network input is formed by the pair $B_{ref}$ and $B_{Mov}$, which is passed through the first block of convolutional layers. This network comprises two downsample blocks: a max pooling layer and two convolution layers, as illustrated in Figure 2. In each block, the size of the input is decremented in half according to the resolution of the images, while the total number of analyzed features doubles.

In the second stage, two blocks are added as part of the network upsampling process. These are composed of a deconvolution layer, which accounts for increasing the input size while decreasing the number of features processed by the network, and two convolutional layers. The resultant data from the deconvolution are then concatenated with the data obtained by the output of the convolution block at the same level from the previous step (see the dashed arrows in Figure 2). In our implementation, the ReLU activation function and a batch normalization layer were used in each convolutional layer except for the last one. The last convolutional layer enables to return a correspondence field compatible with the dimension of the input data.

The network outputs a grid of points (i.e., the correspondence grid), which is used to drive the movement of each pixel when aligning the pair of images. The rightmost quiver plot in Figure 2 displays the correspondence grid, where the arrows moved from the coordinates of the regular grid to the positions produced by the network, while the purple and yellow maps show the points of highest and lowest mobility, respectively.

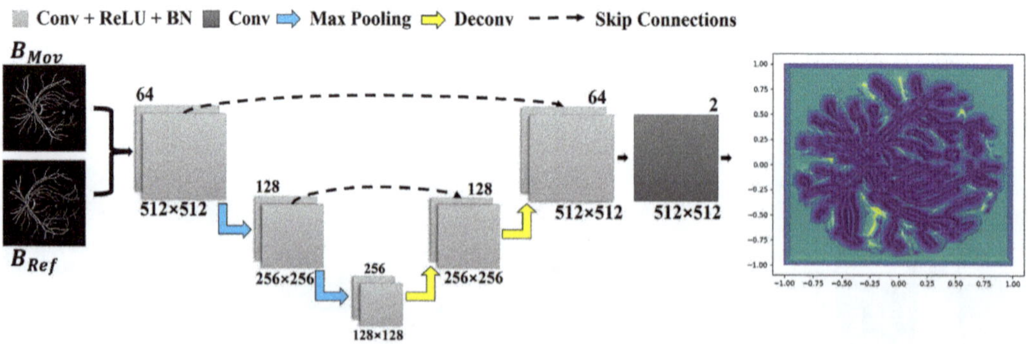


**Figure 2.** The implemented network architecture, used to obtain a correspondence grid. Each layer is represented by a block with a distinct color. Below each block, the data resolution is described, while in the upper-right corner, the number of kernels per layer is shown. The correspondence grid is the network's output, as displayed in the rightmost corner.

### 3.4. *Learning a Spatial Transformation*

In this step, we took an adaptation of the spatial transformer network architecture [44] to obtain a transformation model for mapping $B_{Mov}$. Particularly, the STN structure allows for the network to dynamically apply scaling, rotation, slicing, and nonrigid transformations on the moving image or feature map without the requirement for any additional training supervision or lateral optimization process.

The STN network incorporated as part of our integrated learning scheme consists of two core modules: grid generator and sampler. The goal of the grid generator is to iterate over the matching points previously determined by the U-shaped network to align the correspondence positions in target image $B_{Mov}$. Once the matches are properly found, the sampler module extracts the pixel values at each position through a bilinear interpolation, thus generating the definitive transformed image $B_{Warp}$. Figure 1 (middle frame) illustrates the implemented modules of STN.

### 3.5. *Objective Function*

Since registration is performed without using any set of labeled data, the objective function used to train our approach consists of an independent metric that gauged the similarity degree between the images. In more mathematical terms, we took the normalized cross-correlation (NCC) as a measure of similarity for the objective function:

$$NCC(x,y) = \frac{\sum_{i=0}^{m}\sum_{j=0}^{n} T_{i,j}R_{i,j}}{\sqrt{\left(\sum_{i=0}^{m}\sum_{j=0}^{n} T_{i,j}^2\right)\left(\sum_{i=0}^{m}\sum_{j=0}^{n} R_{i,j}^2\right)}}. \tag{1}$$

In Equation (1), $T_{i,j} = t(x+i, y+j) - \bar{t}_{x,y}$, $R_{i,j} = r(i,j) - \bar{r}$, and $t(i,j)$ and $r(i,j)$ are the pixel values at $(i,j)$ regarding the warped and reference images, $B_{Warp}$ and $B_{Ref}$, respectively, while $\bar{r}$ and $\bar{t}$ give the average pixel values w.r.t. $B_{Ref}$ and $B_{Warp}$ [45]. In Equation (1) the objective (fitness) function is maximized, as the higher the NCC is, the more similar (correlated) the two images are.

The NCC metric can also be defined in terms of a dot product where the output is equivalent to the cosine of the angle between the two normalized pixel intensity vectors. This correlation allows for standard statistical analysis to ascertain the agreement between two datasets, which is frequently chosen as a similarity measure due to its robustness [46], high-accuracy and adaptability [47].

### 3.6. Refinement Process

Since our approach allows for nonrigid registrations, transformed image $B_{Warp}$ may hold some noisy pixels, especially for cases where the images to be aligned are very different from each other. In order to overcome this, we applied a mathematical morphology technique called connected component analysis (CCA) [48].

CCA consists of creating collections of objects formed by groups of adjacent pixels of similar intensities. As a result, eye fundus structures are represented in terms of their morphologically continuous structures, such as connected blood vessels. We, therefore, can identify and filter out small clusters of noisy pixels (see the yellow points in the rightmost frame in Figure 1) from a computed set of connected morphological components.

### 3.7. Datasets and Assessment Metrics

In order to assess the performance of the registration methodology, we took three retina fundus databases. The specification of each data collection is described below.

- **FIRE**—A full database containing several classes of high-resolution fundus images, as detailed in [49]. This data collection comprises 134 pairs of images, grouped into three categories: A, S, and P. Categories A and S covers 14 and 71 pairs of images, respectively, whose fundus photographs present an estimated overlap of more than 75%. Category A also includes images with anatomical differences. Category P, on the other hand, is formed by image pairs with less than 75% of estimated overlap.
- **Image Quality Assessment Dataset (Dataset 1)**—this public dataset [50] is composed of 18 pairs of images captured from 18 individuals, where each pair is formed by a poor-quality image (blurred and/or with dark lighting with occlusions), and a high-quality image of the same eye. There are also pairs containing small displacements caused by eye movements during the acquisition process.
- **Preventive Eye Exams Dataset: (Dataset 2)**—a full database containing 85 pairs of retinal images provided by an ophthalmologist [7]. This data collection gathers real cases of acquisitions such as monitoring diseases, the presence of artifacts, noise, and excessive rotations, i.e., several particular situations typically faced by ophthalmologists and other eye specialists in their routine examinations with real patients.

Aiming at quantitatively assessing the registration results, four validation metrics were adopted: mean squared error (MSE) [36,39], structural similarity index measure (SSIM) [36], Dice coefficient (Dice) [15,28,31,37,40,51] and gain coefficient (GC) [7,52].

The MSE is a popular risk metric that computes the squared error between expected and real values, as shown in Equation (2):

$$MSE(B_{Ref}, B_{Warp}) = \frac{1}{H \times W} \sum_{x=0}^{W} \sum_{y=0}^{H} (B_{Ref_{(x,y)}} - B_{Warp_{(x,y)}})^2, \quad (2)$$

where $H$ and $W$ represent the dimensions of the images $B_{Ref}$ and $B_{Warp}$. The values of the MSE range from 0 to infinite. The closer MSE is to zero, the better.

The SSIM metric takes the spatial positions of the image pixels to calculate the so-called similarity score, as determined by Equation (3):

$$SSIM(B_{Ref}, B_{Warp}) = \frac{(2\mu_{B_{Ref}}\mu_{B_{Warp}} + c_1)(2\sigma_{B_{Ref}B_{Warp}} + c_2)}{(\mu^2_{B_{Ref}} + \mu^2_{B_{Warp}} + c_1)(\sigma^2_{B_{Ref}} + \sigma^2_{B_{Warp}} + c_2)}. \tag{3}$$

In Equation (3), $\mu$ represents the mean value of the image pixels, $\sigma$ is the variance, $\sigma^2$ gives the covariance of $B_{Ref}$ and $B_{Warp}$, and $c_1$ and $c_2$ are variables used to stabilize the denominators. The results are concentrated into a normalized range of 0 and 1, with 0 being the lowest score for the metric, and 1 the highest.

The Dice coefficient is another metric extensively used in the context of image registration, which varies between 0 and 1, where 1 indicates an overlap of 100% . Equation (4) rules the mathematical calculations of this metric:

$$Dice(B_{Ref}, B_{Warp}) = \frac{2 \times B_{Ref} \cap B_{Warp}}{B_{Ref} \cup B_{Warp}}. \tag{4}$$

The GC metric, as described by Equation (5), compares the overlap between the images $B_{Ref}$ and $B_{Warp}$, and the pair of images $B_{Ref}$ and $B_{Mov}$ [52]. Thus, if the number of pixels aligned after the transformation is equal to the number of pixels before the image is transformed, the result is equal to 1. The more pixels are aligned compared to the original overlap, the greater the overlapping value.

$$GC(B_{Ref}, B_{Mov}, B_{Warp}) = \frac{|B_{Ref} \cap B_{Warp}|}{|B_{Ref} \cap B_{Mov}|}. \tag{5}$$

### 3.8. Implementation Details and Training

Our computational prototype was implemented using Python language with the support of libraries for image processing and artificial intelligence routines such as OpenCV [53], Scikit-learn [54] and Tensorflow [55].

The module of integrated networks was trained with batches of eight pairs of images for 5000 epochs. The plot in Figure 3 shows the learning curve of the integrated networks. The curve exponentially increased with a few small oscillations, converging in the first 2000 epochs and remaining stable towards the end of this phase. The learning process was optimized with the ADAM algorithm [56], a mathematical method based on the popular stochastic descending gradient algorithm. The training was performed on a cluster with 32GB of RAM and two Intel(R) Xeon(R) E5-2690 processors.

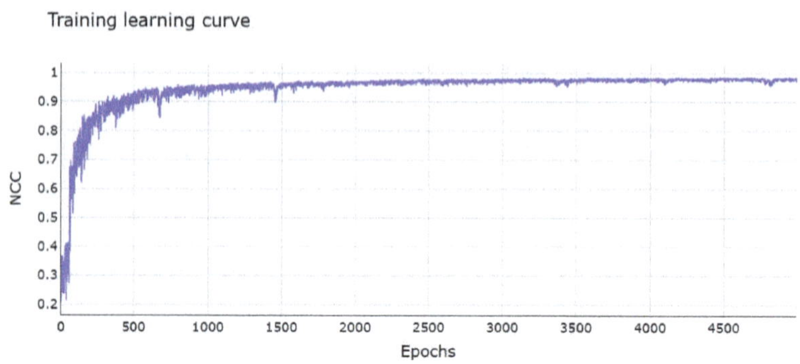


**Figure 3.** Network learning curve after 5000 epochs. The vertical axis represents the fitness value, which is maximized during training, for each epoch on horizontal axis.

The images used in the training step were taken from the category S testing set of the FIRE database, which gathers fundus images of 512 × 512 pixels. This particular category was chosen for training because it comprised the largest and most comprehensive collection of images in the FIRE database, covering pairs of retina images that are more similar to each other (see Figure 4 for an illustrative example). An exhaustive battery of tests showed that this full dataset is effective for training, as the conducted tests revealed that the presence of images with low overlapping levels avoids oscillations in the learning curve of the network, leading to a smaller number of epochs for convergence.

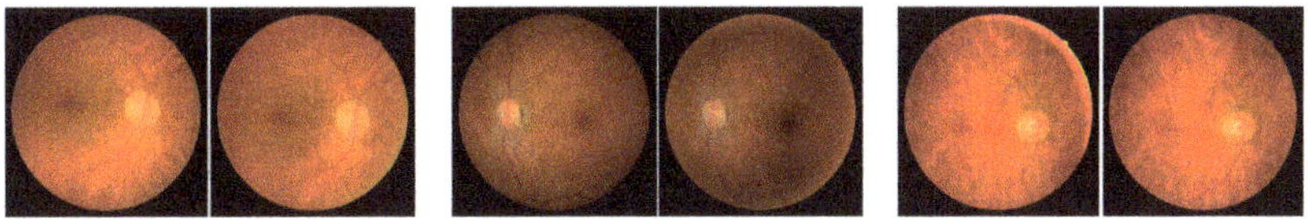

**Figure 4.** Fundus image pairs typically used for training.

Another observable aspect when using our approach is that the registration model was trained by taking a moderately sized dataset of fundus images—a trait that can also be found in other fundus photography related applications, such as landmark detection [41] and even for general applications of DL-type networks [57].

## 4. Results and Discussion

In this section, we present an ablation study concerning the refinement stage of our methodology, which includes the analysis of different settings to increase the quality of the registration results. We also provide and discuss a comprehensive experimental evaluation of the performance of our approach by comparing it with recent image registration methods from both quantitative as well as qualitative aspects.

### *4.1. Ablation Study*

We start by investigating whether the CCA technique can be applied to improve the registration results. We thus incorporated CCA as part of our framework, verifying its impact quantitatively and visually. We compared the application of such a technique by taking three distinct threshold values used to discard clusters with noisy pixels. We also compared the submodels derived from CCA + registration networks against two popular digital image processing techniques: opening and closing morphological filters.

Table 2 lists the average of the evaluation metrics for each submodel and database. The standard deviation is also tabulated in parentheses. By verifying the scores achieved by the morphological transformations (network + opening and network + closing), one can conclude that they did not lead to an improvement in quality for the registered image pairs, even for those containing noise. Moreover, the application of these morphology-based filters may alter the contour of the structures present in the images, as shown in Figure 5a,c.

On the other hand, by comparing the results output by submodels network + CCA, we noticed that they clearly contributed to a substantial gain in registration quality in all examined datasets, as one can see from the scores highlighted in bold in Table 2.

In Figure 5, the image registered by the integrated networks without any refinement process appears in green (Figure 5a), while the others are comparisons between these and the images after applying each denoising technique, and they assume a magenta color so that when added to the green image lead to white pixels. In this way, the noise data in green indicate the pixels that were treated in these images. Visually speaking, when comparing the results in Figure 5e,f, the noise was substantially reduced after applying the CCA technique.

From the conducted ablation analysis, we included as part of our full registration framework the application of CCA algorithm with a threshold value of 20 pixels.

**Table 2.** Comparison of registration submodels created as variations of our framework. Values in bold indicate the best scores, and values in italics the second best.

| Metrics | Methods | FIRE A | FIRE S | FIRE P | Dataset 1 | Dataset 2 |
|---|---|---|---|---|---|---|
| MSE (↓) | Network | 0.0080 (0.0017) | 0.0074 (0.0019) | 0.0143 (0.0026) | 0.0095 (0.0034) | 0.0093 (0.0039) |
| | Network + Opening | 0.0287 (0.0030) | 0.0319 (0.0023) | 0.0343 (0.0031) | 0.0324 (0.0037) | 0.0268 (0.0035) |
| | Network + Closing | 0.0284 (0.0029) | 0.0316 (0.0023) | 0.0337 (0.0030) | 0.0321 (0.0035) | 0.0265 (0.0034) |
| | Network + CCA 10 | *0.0068 (0.0015)* | **0.0062 (0.0017)** | *0.0121 (0.0027)* | **0.0079 (0.0034)** | **0.0071 (0.0038)** |
| | Network + CCA 20 | **0.0068 (0.0014)** | **0.0062 (0.0017)** | **0.0120 (0.0027)** | *0.0079 (0.0035)* | **0.0071 (0.0038)** |
| | Network + CCA 30 | 0.0069 (0.0015) | 0.0063 (0.0017) | *0.0121 (0.0027)* | 0.0080 (0.0035) | **0.0071 (0.0038)** |
| SSIM (↑) | Network | 0.9586 (0.0086) | 0.9638 (0.0104) | 0.9290 (0.0080) | 0.9539 (0.0130) | 0.9572 (0.0162) |
| | Network + Opening | 0.8928 (0.0110) | 0.8807 (0.0094) | 0.8773 (0.0107) | 0.8797 (0.0130) | 0.9001 (0.0118) |
| | Network + Closing | 0.8923 (0.0103) | 0.8818 (0.0092) | 0.8752 (0.0104) | 0.8800 (0.0124) | 0.8998 (0.0119) |
| | Network + CCA 10 | *0.9731 (0.0055)* | **0.9749 (0.0068)** | 0.9575 (0.0076) | **0.9682 (0.0128)** | *0.9733 (0.0106)* |
| | Network + CCA 20 | **0.9732 (0.0053)** | *0.9748 (0.0068)* | **0.9585 (0.0075)** | *0.9681 (0.0133)* | **0.9734 (0.0103)** |
| | Network + CCA 30 | 0.9727 (0.0054) | 0.9744 (0.0068) | *0.9580 (0.0073)* | 0.9678 (0.0133) | *0.9733 (0.0102)* |
| Dice (↑) | Network | 0.9399 (0.0121) | 0.9484 (0.0143) | 0.8915 (0.0237) | 0.9363 (0.0268) | 0.9295 (0.0425) |
| | Network + Opening | 0.7814 (0.0101) | 0.7743 (0.0121) | 0.7367 (0.0173) | 0.7807 (0.0359) | 0.8046 (0.0382) |
| | Network + Closing | 0.7874 (0.0090) | 0.7798 (0.0117) | 0.7465 (0.0171) | 0.7860 (0.0331) | 0.8086 (0.0369) |
| | Network + CCA 10 | *0.9502 (0.0100)* | *0.9579 (0.0120)* | *0.9103 (0.0238)* | *0.9476 (0.0265)* | *0.9466 (0.0404)* |
| | Network + CCA 20 | **0.9505 (0.0097)** | **0.9580 (0.0122)** | **0.9109 (0.0238)** | **0.9477 (0.0270)** | **0.9467 (0.0404)** |
| | Network + CCA 30 | 0.9496 (0.0100) | 0.9573 (0.0123) | 0.9097 (0.0236) | 0.9471 (0.0270) | 0.9463 (0.0404) |
| GC (↑) | Network | 3.4237 (0.9921) | 3.2125 (1.3424) | 6.7499 (0.8029) | 3.4786 (0.9630) | 3.0494 (1.6853) |
| | Network + Opening | 2.8025 (0.8065) | 2.5910 (1.0920) | 5.4621 (0.6265) | 2.8544 (0.7680) | 2.6075 (1.4265) |
| | Network + Closing | 2.8733 (0.8394) | 2.6515 (1.1326) | 5.6395 (0.6508) | 2.9203 (0.7960) | 2.6565 (1.4714) |
| | Network + CCA 10 | *3.5511 (1.0343 )* | **3.3379 (1.3973)** | **7.0506 (0.8443)** | **3.5963 (0.9943)** | **3.1755 (1.7625)** |
| | Network + CCA 20 | **3.5520 (1.0361)** | *3.3378 (1.3965)* | *7.0410 (0.8410)* | *3.5956 (0.9940)* | *3.1716 (1.7571)* |
| | Network + CCA 30 | 3.5443 (1.0345) | 3.3321 (1.3920) | 7.0160 (0.8373) | 3.5892 (0.9888) | 3.1672 (1.7517) |

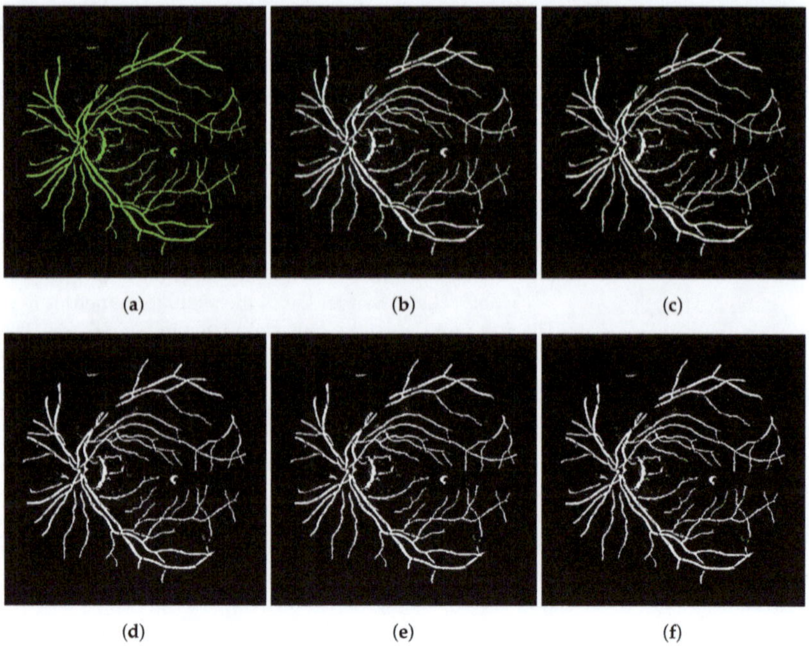

**Figure 5.** Visual comparison for several denoising strategies applied on transformed images generated by the integrated networks. (**a**) Network – SSIM: 0.9338; (**b**) Opening – SSIM: 0.8640; (**c**) Closing – SSIM: 0.8625; (**d**) CCA 10 – SSIM: 0.9613; (**e**) CCA 20 – SSIM: 0.9611; (**f**) CCA 30 – SSIM: 0.9598.

### 4.2. Comparison with Image Registration Methods

We compare the outputs obtained by our approach against the ones produced by four modern image registration methods. Within the scope of keypoint-based techniques, the algorithms proposed by Wang et al. [58] and Motta et al. [7], called GFEMR and VOTUS, were considered in our analysis. For comparisons covering DL-based methods, we ran the techniques proposed by Vos et al. [59], DIRNet, and the weakly supervised strategy introduced by Hu et al. [33]. These DL-driven algorithms were tuned following the same experimental process performed by our approach, i.e., they were fully trained with the same group of training samples, taking into account the same amount of epochs.

Figure 6a–d show box plots for each validation metric and registration dataset. The generated plots show that the proposed framework outperformed both conventional and DL-based techniques in all instances, demonstrating consistency and stability for different categories of fundus images. The MSE, SSIM and Dice metrics exhibited similar behavior while still holding the smallest variation in the box plots, thus attesting to the capability of our approach in achieving high-accuracy registrations regardless of the pair of fundus images. Lastly, concerning the GC metric (Figure 6d), since such a measure gauges the overlap segments before and after the registration, the datasets holding more discrepant images were the ones that produced higher scores, as one can check for Category P of FIRE database. DIRNet and VOTUS remain competitive for Category S of FIRE, but they were still outperformed by the proposed methodology. A similar outcome was found when DIRNet was compared to our approach for Dataset 2.

A two-sided Wilcoxon test at 5% significance level was applied to verify the statistical validity of the registrations produced by our approach against the ones delivered by other methods. From the $p$-values in Table 3, the results from our approach were statistically more accurate than others in all datasets for at least three of the four evaluation metrics (MSE, SSIM and DICE). Moreover, we can check that our approach was statically superior ($p < 0.05$) in 96 of the 100 tests conducted, thus attesting to the statistical validation of the obtained results.

**Table 3.** $p$-values from two-sided Wilcoxon test at 5% significance level applied to compare the proposed approach against other registration methods.

| Metric | Method | Fire A | FIRE S | FIRE P | Dataset 1 | Dataset 2 |
|---|---|---|---|---|---|---|
| MSE | Before | $<10^{-7}$ | 0.0 | 0.0 | $<10^{-9}$ | 0.0 |
| | GFEMR | $<10^{-7}$ | 0.0 | 0.0 | $<10^{-9}$ | 0.0 |
| | VOTUS | $<10^{-7}$ | 0.0 | 0.0 | $<10^{-9}$ | 0.0 |
| | DIRNet | $<10^{-7}$ | 0.0 | 0.0 | $<10^{-9}$ | 0.0 |
| | HU et al. | $<10^{-7}$ | 0.0 | 0.0 | $<10^{-9}$ | 0.0 |
| SSIM | Before | $<10^{-7}$ | 0.0 | 0.0 | $<10^{-9}$ | 0.0 |
| | GFEMR | $<10^{-7}$ | 0.0 | 0.0 | $<10^{-9}$ | 0.0 |
| | VOTUS | $<10^{-7}$ | 0.0 | 0.0 | $<10^{-9}$ | 0.0 |
| | DIRNet | $<10^{-7}$ | 0.0 | 0.0 | $<10^{-9}$ | 0.0 |
| | HU et al. | $<10^{-7}$ | 0.0 | 0.0 | $<10^{-7}$ | 0.0 |
| DICE | Before | $<10^{-7}$ | 0.0 | 0.0 | $<10^{-9}$ | 0.0 |
| | GFEMR | $<10^{-7}$ | 0.0 | 0.0 | $<10^{-9}$ | 0.0 |
| | VOTUS | $<10^{-7}$ | 0.0 | 0.0 | $<10^{-9}$ | 0.0 |
| | DIRNet | $<10^{-7}$ | 0.0 | 0.0 | $<10^{-9}$ | 0.0 |
| | HU et al. | $<10^{-7}$ | 0.0 | 0.0 | $<10^{-9}$ | 0.0 |
| GC | Before | $<10^{-7}$ | 0.0 | 0.0 | $<10^{-9}$ | 0.0 |
| | GFEMR | 0.0017 | 0.0028 | 0.0 | 0.0001 | 0.0253 |
| | VOTUS | 0.0058 | 0.1206 | 0.0 | 0.0224 | 0.0 |
| | DIRNet | 0.0 | 0.0 | 0.0 | 0.0 | 0.0 |
| | HU et al. | 0.1139 | 0.1994 | 0.0 | 0.0037 | 0.1594 |

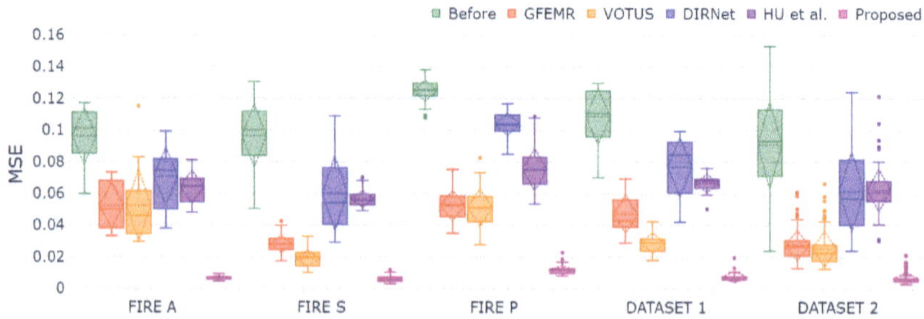


(**a**)

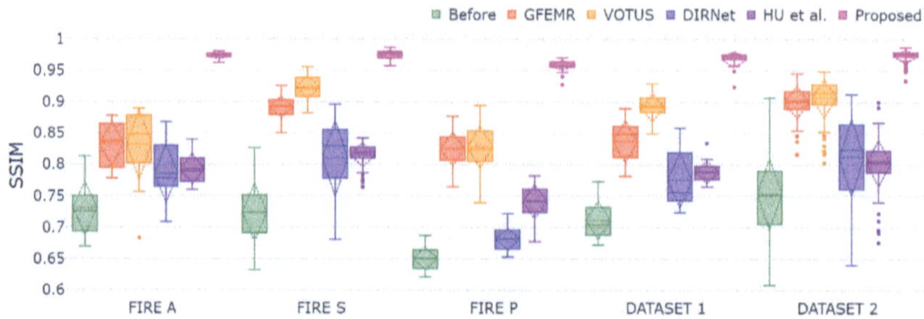


(**b**)

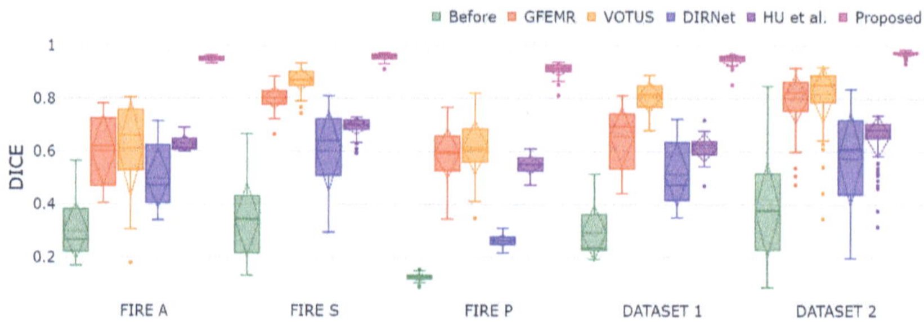


(**c**)

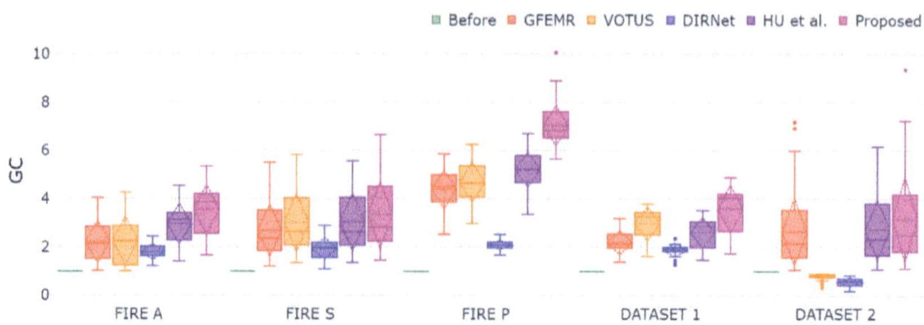


(**d**)

**Figure 6.** Box-plot charts for each evaluation metric and dataset. Symbols (↓) and (↑) indicate that "lower is better" and "higher is better", respectively. (**a**) Box-plot distribution for MSE metric (↓); (**b**) box-plot distribution for SSIM metric (↑); (**c**) box-plot distribution for Dice metric (↑); (**d**) box-plot distribution for GC metric (↑).

In addition to the four registration methods already assessed in our validation study, we provide new assessments involving two new methods: the recent registration through eye modelling and pose estimation (REMPE) technique [60], and the well-established scale-invariant feature transform (SIFT) algorithm [61]. Figure 7 shows the box-plot distribution for each validation metric applied to categories A, S and P from FIRE database. The plotted box plot shows that our framework outperformed the REMPE and SIFT methods, achieving the smallest variations between outputs, which are visually represented by the tightest clusters in each plot.

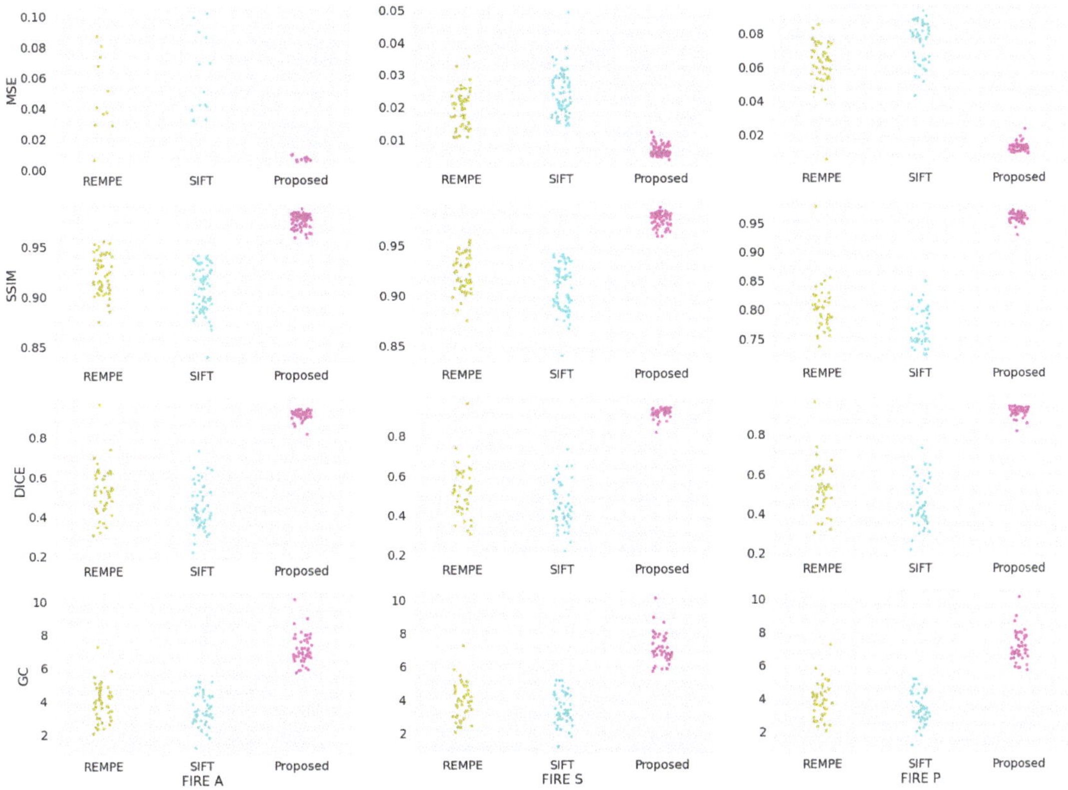


**Figure 7.** Sample distribution analysis for REMPE, SIFT, and our framework for the FIRE datasets.

A visual qualitative analysis of the registrations produced by the competing methods is presented in Figure 8. Here, we followed [7,16,52] to represent the aligned images in terms of color compositions to increase the visual readability and interpretation of the results. More specifically, images $B_{Ref}$ and $B_{Warp}$ were rendered in green and magenta, while the overlap of both images is in white, giving the level of agreement between them.

Keypoint-based approaches GEEMR and VOTUS produced acceptable results for most image pairs, but they are not yet able to satisfactorily deal with the blood veins located farther away from the eye globe. DL-based methods DIRNET and Hu et al. performed nonrigid registrations, causing deformations in the output images (e.g., see the misalignment and distortions in the first, third, and fourth images from Figure 8). Our framework also performs nonrigid registration; however, the implemented networks ensure that the transformation applied to moving image $B_{Mov}$ uniformly distorts the image structures, rendering $B_{Mov}$ closer to the reference image $B_{Ref}$. Lastly, one can verify that our registration model and that of Hu et al. were the ones that were capable of aligning the very hard images from Category P of the FIRE database.

Another relevant observation when inspecting Figure 8 is the role of vessels in our framework. Indeed, such a procedure allows for the method to carry out the registration under the most diverse conditions. For instance, the fundus images from Dataset 1 are composed of dark lighting, blur, and smoky occlusions. By handling the eye's vessels, it is possible to highlight the vascular structure of these images, accurately performing the registration while avoiding the need for new exams to replace poorly captured photographs.

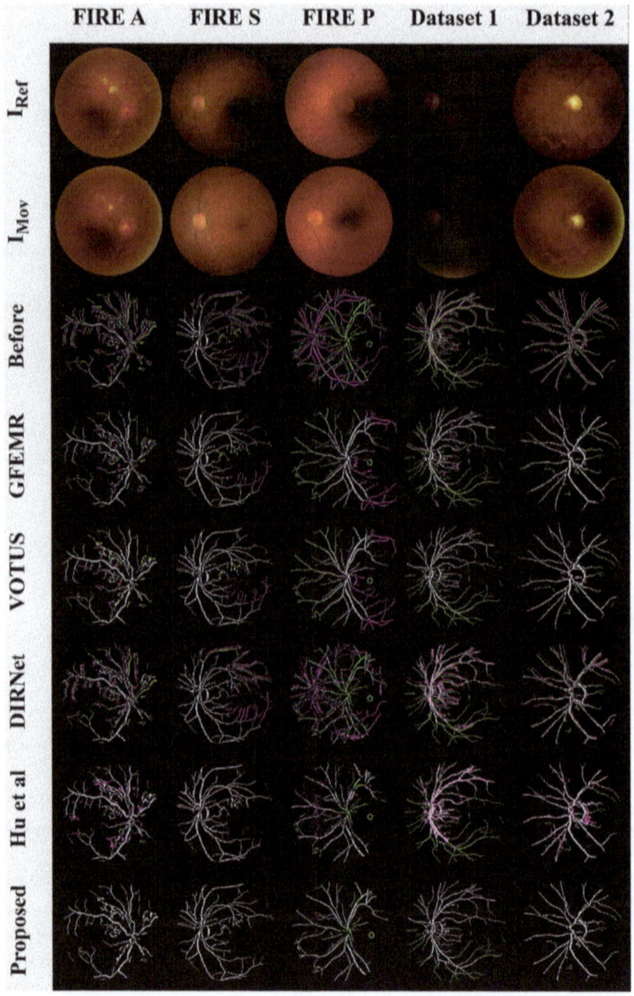


**Figure 8.** Visual analysis of the results. Lines 1 and 2: original images from each examined database, Line 3: the images before the registration process, Lines 4-9: the overlapping areas between $B_{Ref}$ (in green) and $B_{Warp}$ (in magenta) produced by each registration method.

## 5. Conclusions

This paper introduced an end-to-end methodology for fundus image registration using unsupervised deep learning networks and morphological filtering. As shown by the conducted experiments, our approach was able to operate in a fully unsupervised fashion, requiring no prelabeled data or side computational strategy to induce the creation of synthetic data for training. After being trained, the current model produced one-shot registrations by just inputting a pair of fundus images.

From the battery of conducted experiments, it was verified that the proposed methodology produced very stable and accurate registrations for five representative datasets of fundus images, most of them covering several challenging cases, such as images with anatomical differences and very low-quality acquisitions. Furthermore, the methodology performed better than several modern existing registration methods in terms of the accuracy, stability, and capability of generalization for several datasets of fundus photographs. Visual representations of the registration results also revealed a better adherence achieved by the introduced framework in comparison with keypoint-based and DL-based methods.

As future work, we plan to: (i) analyze the effects of applying other fitness functions beyond NCC; (ii) investigate the use of other DL neural networks, for example, SegNet, X-Net and adversarial networks; and (iii) extend our framework to cope with specific clinical problems, including its adaptation for domain transformation, from fundus images to ultra-wide-field fundus photography [25], and 3D stereoscopic reconstruction of retinal images, which is another application related to the context of diagnostic assistance.

**Author Contributions:** Conceptualization, G.A.B., M.C., M.A.D., R.G.N., E.A.S. and W.C.; funding acquisition, R.G.N., E.A.S. and W.C.; investigation, G.A.B. and W.C.; methodology, G.A.B. and W.C.; resources, M.C. and W.C.; validation, G.A.B., M.A.D. and W.C.; writing—original draft, G.A.B., R.G.N., E.A.S. and W.C. All authors have read and agreed to the published version of the manuscript.

**Funding:** This research was funded by the São Paulo Research Foundation (FAPESP), grants #2013/07375-0, #2019/26288-7, #2021/03328-3, and #2021/01305-6; and the National Council for Scientific and Technological Development (CNPq), grants #304402/2019-2, #164326/2020-0 and #316228/2021-4. The APC was funded by São Paulo State University (UNESP-PROPG).

**Data Availability Statement:** The computational framework was implemented in Python language using libraries provided by OpenCV: https://opencv.org (accessed on 11 August 2021), Scikit-learn: https://scikit-learn.org/stable/ (accessed on 14 September 2021) and TensorFlow: https://www.tensorflow.org/ (accessed on 22 September 2021). The public databases cited in the Section 3.7 are freely available at: https://projects.ics.forth.gr/cvrl/fire/ (accessed on 15 July 2021) and https://www5.cs.fau.de/research/data/fundus-images/index.html (accessed on 15 July 2021).

**Conflicts of Interest:** the authors declare no conflict of interest.

## References

1. Weinreb, R.N.; Aung, T.; Medeiros, F.A. The pathophysiology and treatment of glaucoma: A review. *J. Am. Med. Assoc. (JAMA)* **2014**, *311*, 1901–1911. [CrossRef] [PubMed]
2. Kim, K.M.; Heo, T.Y.; Kim, A.; Kim, J.; Han, K.J.; Yun, J.; Min, J.K. Development of a Fundus Image-Based Deep Learning Diagnostic Tool for Various Retinal Diseases. *J. Pers. Med.* **2021**, *11*, 321. [CrossRef] [PubMed]
3. Shabbir, A.; Rasheed, A.; Shehraz, H.; Saleem, A.; Zafar, B.; Sajid, M.; Ali, N.; Dar, S.H.; Shehryar, T. Detection of glaucoma using retinal fundus images: A comprehensive review. *Math. Biosci. Eng.* **2021**, *18*, 2033–2076. [CrossRef] [PubMed]
4. Saha, S.K.; Xiao, D.; Bhuiyan, A.; Wong, T.Y.; Kanagasingam, Y. Color fundus image registration techniques and applications for automated analysis of diabetic retinopathy progression: A review. *Biomed. Signal Process. Control* **2019**, *47*, 288–302. [CrossRef]
5. Ramli, R.; Hasikin, K.; Idris, M.Y.I.; Karim, N.K.A.; Wahab, A.W.A. Fundus Image Registration Technique Based on Local Feature of Retinal Vessels. *Appl. Sci.* **2021**, *11*, 11201. [CrossRef]
6. Karali, E.; Asvestas, P.; Nikita, K.S.; Matsopoulos, G.K. Comparison of Different Global and Local Automatic Registration Schemes: An Application to Retinal Images. In Proceedings of the International Conference on Medical Image Computing and Computer-Assisted Intervention (MICCAI), Saint-Malo, France, 26–29 September 2004; pp. 813–820.
7. Motta, D.; Casaca, W.; Paiva, A. Vessel Optimal Transport for Automated Alignment of Retinal Fundus Images. *IEEE Trans. Image Process.* **2019**, *28*, 6154–6168. [CrossRef]
8. Dasariraju, S.; Huo, M.; McCalla, S. Detection and classification of immature leukocytes for diagnosis of acute myeloid leukemia using random forest algorithm. *Bioengineering* **2020**, *7*, 120. [CrossRef]
9. Bechelli, S.; Delhommelle, J. Machine Learning and Deep Learning Algorithms for Skin Cancer Classification from Dermoscopic Images. *Bioengineering* **2022**, *9*, 97. [CrossRef]
10. Litjens, G.; Kooi, T.; Bejnordi, B.E.; Setio, A.A.A.; Ciompi, F.; Ghafoorian, M.; van der Laak, J.A.; van Ginneken, B.; Sánchez, C.I. A survey on deep learning in medical image analysis. *Med. Image Anal.* **2017**, *42*, 60–88. [CrossRef]
11. Haskins, G.; Kruger, U.; Yan, P. Deep learning in medical image registration: A survey. *Mach. Vis. Appl.* **2020**, *31*, 1–18. [CrossRef]

12. Chen, X.; Diaz-Pinto, A.; Ravikumar, N.; Frangi, A. Deep learning in medical image registration. *Prog. Biomed. Eng.* **2020**, *3*, 012003. [CrossRef]
13. Pluim, J.P.; Muenzing, S.E.; Eppenhof, K.A.; Murphy, K. The truth is hard to make: Validation of medical image registration. In Proceedings of the 2016 23rd International Conference on Pattern Recognition (ICPR), Cancun, Mexico, 4–8 December 2016; pp. 2294–2300. [CrossRef]
14. Punn, N.S.; Agarwal, S. Modality specifc U-Net variants for biomedical image segmentation: A survey. *Artif. Intell. Rev.* **2022**, *3*, 1–45.
15. de Vos, B.D.; Berendsen, F.F.; Viergever, M.A.; Sokooti, H.; Staring, M.; Išgum, I. A deep learning framework for unsupervised affine and deformable image registration. *Med. Image Anal.* **2019**, *52*, 128–143. [CrossRef]
16. Benvenuto, G.A.; Colnago, M.; Casaca, W. Unsupervised Deep Learning Network for Deformable Fundus Image Registration. In Proceedings of the ICASSP 2022—IEEE International Conference on Acoustics, Speech and Signal Processing, Singapore, 22–27 May 2022; pp. 1281–1285.
17. Oh, K.; Kang, H.M.; Leem, D.; Lee, H.; Seo, K.Y.; Yoon, S. Early detection of diabetic retinopathy based on deep learning and ultra-wide-field fundus images. *Sci. Rep.* **2021**, *11*, 1–9. [CrossRef] [PubMed]
18. Mushtaq, G.; Siddiqui, F. Detection of diabetic retinopathy using deep learning methodology. In Proceedings of the IOP Conference Series: Materials Science and Engineering, Tamil Nadu, India, 4–5 December 2020; Volume 1070, p. 012049.
19. Ajitha, S.; Akkara, J.D.; Judy, M. Identification of glaucoma from fundus images using deep learning techniques. *Indian J. Ophthalmol.* **2021**, *69*, 2702. [PubMed]
20. Deperlioglu, O.; Kose, U.; Gupta, D.; Khanna, A.; Giampaolo, F.; Fortino, G. Explainable framework for Glaucoma diagnosis by image processing and convolutional neural network synergy: Analysis with doctor evaluation. *Future Gener. Comput. Syst.* **2022**, *129*, 152–169. [CrossRef]
21. Du, R.; Xie, S.; Fang, Y.; Igarashi-Yokoi, T.; Moriyama, M.; Ogata, S.; Tsunoda, T.; Kamatani, T.; Yamamoto, S.; Cheng, C.Y.; et al. Deep learning approach for automated detection of myopic maculopathy and pathologic myopia in fundus images. *Ophthalmol. Retin.* **2021**, *5*, 1235–1244. [CrossRef] [PubMed]
22. Mahapatra, D.; Bozorgtabar, B.; Hewavitharanage, S.; Garnavi, R. Image Super Resolution Using Generative Adversarial Networks and Local Saliency Maps for Retinal Image Analysis. In Proceedings of the Medical Image Computing and Computer Assisted Intervention—MICCAI, Quebec City, QC, Canada, 11–13 September 2017; Descoteaux, M., Maier-Hein, L., Franz, A., Jannin, P., Collins, D.L., Duchesne, S., Eds.; Springer International Publishing: Cham, Switzerland, 2017; pp. 382–390.
23. Manakov, I.; Rohm, M.; Kern, C.; Schworm, B.; Kortuem, K.; Tresp, V. Noise as Domain Shift: Denoising Medical Images by Unpaired Image Translation. In Proceedings of the Domain Adaptation and Representation Transfer and Medical Image Learning with Less Labels and Imperfect Data, Shenzhen, China, 13–17 October 2019; Wang, Q., Milletari, F., Nguyen, H.V., Albarqouni, S., Cardoso, M.J., Rieke, N., Xu, Z., Kamnitsas, K., Patel, V., Roysam, B., et al., Eds.; Springer International Publishing: Cham, Switzerland, 2019; pp. 3–10.
24. Sanchez, Y.D.; Nieto, B.; Padilla, F.D.; Perdomo, O.; Osorio, F.A.G. Segmentation of retinal fluids and hyperreflective foci using deep learning approach in optical coherence tomography scans. *Proc. SPIE* **2020**, *11583*, 136–143. [CrossRef]
25. You, A.; Kim, J.K.; Ryu, I.H.; Yoo, T.K. Application of Generative Adversarial Networks (GAN) for Ophthalmology Image Domains: A Survey. *Eye Vis.* **2022**, *9*, 1–19. [CrossRef]
26. Fu, Y.; Lei, Y.; Wang, T.; Curran, W.J.; Liu, T.; Yang, X. Deep learning in medical image registration: A review. *Phys. Med. Biol.* **2020**, *65*, 20TR01. [CrossRef]
27. Yang, X.; Kwitt, R.; Styner, M.; Niethammer, M. Fast predictive multimodal image registration. In Proceedings of the 2017 IEEE 14th International Symposium on Biomedical Imaging (ISBI 2017), Melbourne, Australia, 18–21 April 2017; pp. 858–862. [CrossRef]
28. Cao, X.; Yang, J.; Zhang, J.; Nie, D.; Kim, M.; Wang, Q.; Shen, D. Deformable Image Registration Based on Similarity-Steered CNN Regression. In Proceedings of the Medical Image Computing and Computer Assisted Intervention—MICCAI 2017, Quebec City, QC, Canada, 11–13 September 2017; Descoteaux, M., Maier-Hein, L., Franz, A., Jannin, P., Collins, D.L., Duchesne, S., Eds.; Springer International Publishing: Cham, Switzerland, 2017; pp. 300–308.
29. Eppenhof, K.A.J.; Pluim, J.P.W. Pulmonary CT Registration Through Supervised Learning With Convolutional Neural Networks. *IEEE Trans. Med. Imaging* **2019**, *38*, 1097–1105. [CrossRef] [PubMed]
30. Ronneberger, O.; Fischer, P.; Brox, T. U-Net: Convolutional Networks for Biomedical Image Segmentation. *arXiv* **2015**, arXiv:1505.04597.
31. Fan, J.; Cao, X.; Yap, P.T.; Shen, D. BIRNet: Brain image registration using dual-supervised fully convolutional networks. *Med. Image Anal.* **2019**, *54*, 193–206. [CrossRef] [PubMed]
32. Hering, A.; Kuckertz, S.; Heldmann, S.; Heinrich, M.P. Enhancing Label-Driven Deep Deformable Image Registration with Local Distance Metrics for State-of-the-Art Cardiac Motion Tracking. In *Bildverarbeitung für die Medizin 2019*; Handels, H., Deserno, T.M., Maier, A., Maier-Hein, K.H., Palm, C., Tolxdorff, T., Eds.; Springer: Wiesbaden, Germany, 2019; pp. 309–314.
33. Hu, Y.; Modat, M.; Gibson, E.; Li, W.; Ghavami, N.; Bonmati, E.; Wang, G.; Bandula, S.; Moore, C.M.; Emberton, M.; et al. Weakly-supervised convolutional neural networks for multimodal image registration. *Med. Image Anal.* **2018**, *49*, 1–13. [CrossRef]
34. Lv, J.; Yang, M.; Zhang, J.; Wang, X. Respiratory motion correction for free-breathing 3D abdominal MRI using CNN-based image registration: A feasibility study. *Br. J. Radiol.* **2018**, *91*, 20170788. [CrossRef]

35. Zhang, J. Inverse-Consistent Deep Networks for Unsupervised Deformable Image Registration. *arXiv* **2018**, arXiv:1809.03443.
36. Kori, A.; Krishnamurthi, G. Zero Shot Learning for Multi-Modal Real Time Image Registration. *arXiv* **2019**, arXiv:1908.06213.
37. Wang, C.; Yang, G.; Papanastasiou, G. FIRE: Unsupervised bi-directional inter- and intra-modality registration using deep networks. In Proceedings of the 2021 IEEE 34th International Symposium on Computer-Based Medical Systems (CBMS), Virtual, 7–9 June 2021; pp. 510–514. [CrossRef]
38. Balakrishnan, G.; Zhao, A.; Sabuncu, M.; Guttag, J.; Dalca, A.V. VoxelMorph: A Learning Framework for Deformable Medical Image Registration. *IEEE TMI Trans. Med. Imaging* **2019**, *38*, 1788–1800. [CrossRef]
39. Mahapatra, D.; Antony, B.; Sedai, S.; Garnavi, R. Deformable Medical Image Registration using Generative Adversarial Networks. In Proceedings of the IEEE 15th International Symposium on Biomedical Imaging (ISBI 2018), Washington, DC, USA, 4–7 April 2018; pp. 1449–1453. [CrossRef]
40. Wang, Y.; Zhang, J.; An, C.; Cavichini, M.; Jhingan, M.; Amador-Patarroyo, M.J.; Long, C.P.; Bartsch, D.U.G.; Freeman, W.R.; Nguyen, T.Q. A Segmentation Based Robust Deep Learning Framework for Multimodal Retinal Image Registration. In Proceedings of the ICASSP 2020—2020 IEEE International Conference on Acoustics, Speech and Signal Processing (ICASSP), Barcelona, Spain, 4–8 May 2020; pp. 1369–1373. [CrossRef]
41. Rivas-Villar, D.; Hervella, Á.S.; Rouco, J.; Novo, J. Color fundus image registration using a learning-based domain-specific landmark detection methodology. *Comput. Biol. Med.* **2022**, *140*, 105101. [CrossRef]
42. Rohé, M.M.; Datar, M.; Heimann, T.; Sermesant, M.; Pennec, X. SVF-Net: Learning Deformable Image Registration Using Shape Matching. In Proceedings of the MICCAI 2017—The 20th International Conference on Medical Image Computing and Computer Assisted Intervention, Quebec City, QC, Canada, 11–13 September 2017; pp. 266–274. [CrossRef]
43. Bankhead, P.; Scholfield, C.; McGeown, J.; Curtis, T. Fast retinal vessel detection and measurement using wavelets and edge location refinement. *PLoS ONE* **2012**, *7*, e32435. [CrossRef]
44. Jaderberg, M.; Simonyan, K.; Zisserman, A.; Kavukcuoglu, K. Spatial transformer networks. In Proceedings of the NIPS, Montreal, QC, Canada, 7–12 December 2015.
45. Kaso, A. Computation of the normalized cross-correlation by fast Fourier transform. *PLoS ONE* **2018**, *13*, 1–16. [CrossRef] [PubMed]
46. Hisham, M.; Yaakob, S.N.; Raof, R.; Nazren, A.A.; Wafi, N. Template Matching using Sum of Squared Difference and Normalized Cross Correlation. In Proceedings of the 2015 IEEE Student Conference on Research and Development (SCOReD), Kuala Lumpur, Malaysia, 13–14 December 2015; pp. 100–104. [CrossRef]
47. Cui, Z.; Qi, W.; Liu, Y. A Fast Image Template Matching Algorithm Based on Normalized Cross Correlation. *J. Phys. Conf. Ser.* **2020**, *1693*, 012163. [CrossRef]
48. He, L.; Ren, X.; Gao, Q.; Zhao, X.; Yao, B.; Chao, Y. The connected-component labeling problem: A review of state-of-the-art algorithms. *Pattern Recognit.* **2017**, *70*, 25–43. [CrossRef]
49. Hernandez-Matas, C.; Zabulis, X.; Triantafyllou, A.; Anyfanti, P.; Douma, S.; Argyros, A. FIRE: Fundus Image Registration Dataset. *J. Model. Ophthalmol.* **2017**, *1*, 16–28. [CrossRef]
50. Köhler, T.; Budai, A.; Kraus, M.F.; Odstrčilik, J.; Michelson, G.; Hornegger, J. Automatic no-reference quality assessment for retinal fundus images using vessel segmentation. In Proceedings of the 26th IEEE International Symposium on Computer-Based Medical Systems, Porto, Portugal, 20–22 June 2013; pp. 95–100. [CrossRef]
51. Che, T.; Zheng, Y.; Cong, J.; Jiang, Y.; Niu, Y.; Jiao, W.; Zhao, B.; Ding, Y. Deep Group-Wise Registration for Multi-Spectral Images From Fundus Images. *IEEE Access* **2019**, *7*, 27650–27661. [CrossRef]
52. Motta, D.; Casaca, W.; Paiva, A. Fundus Image Transformation Revisited: Towards Determining More Accurate Registrations. In Proceedings of the IEEE International Symposium on Computer-Based Medical Systems (CBMS), Karlstad, Sweden, 18–21 June 2018; pp. 227–232.
53. Bradski, G. The OpenCV Library. *Dr. Dobb's J. Softw. Tools* **2000**, *120*, 122–125.
54. Pedregosa, F.; Varoquaux, G.; Gramfort, A.; Michel, V.; Thirion, B.; Grisel, O.; Blondel, M.; Prettenhofer, P.; Weiss, R.; Dubourg, V.; et al. Scikit-learn: Machine Learning in Python. *J. Mach. Learn. Res.* **2011**, *12*, 2825–2830.
55. Abadi, M.; Agarwal, A.; Barham, P.; Brevdo, E.; Chen, Z.; Citro, C.; Corrado, G.S.; Davis, A.; Dean, J.; Devin, M.; et al. TensorFlow: Large-Scale Machine Learning on Heterogeneous Systems. *arXiv* **2015**, arXiv:1603.04467.
56. Kingma, D.; Ba, J. Adam: A Method for Stochastic Optimization. *arXiv* **2014**, arXiv:1412.6980.
57. Brigato, L.; Iocchi, L. A Close Look at Deep Learning with Small Data. In Proceedings of the IEEE International Conference on Pattern Recognition (ICPR), Milan, Italy, 10–15 January 2021; pp. 2490–2497.
58. Wang, J.; Chen, J.; Xu, H.; Zhang, S.; Mei, X.; Huang, J.; Ma, J. Gaussian field estimator with manifold regularization for retinal image registration. *Signal Process.* **2019**, *157*, 225–235. [CrossRef]
59. de Vos, B.; Berendsen, F.; Viergever, M.; Staring, M.; Išgum, I. End-to-end unsupervised deformable image registration with a convolutional neural network. *arXiv* **2017**, arXiv:1704.06065.
60. Hernandez-Matas, C.; Zabulis, X.; Argyros, A. REMPE: Registration of Retinal Images Through Eye Modelling and Pose Estimation. *IEEE J. Biomed. Health Informat.* **2020**, *24*, 3362–3373. [CrossRef] [PubMed]
61. Lowe, D. Distinctive Image Features from Scale-Invariant Keypoints. *Int. J. Comput. Vis.* **2004**, *60*, 91–110.

MDPI
St. Alban-Anlage 66
4052 Basel
Switzerland
www.mdpi.com

*Bioengineering* Editorial Office
E-mail: bioengineering@mdpi.com
www.mdpi.com/journal/bioengineering

www.ingramcontent.com/pod-product-compliance
Lightning Source LLC
Chambersburg PA
CBHW061601220926
51850CB00005B/395
* 9 7 8 3 0 3 6 5 8 5 8 7 1 *